AF781815

EMERGING ISSUES IN CLIMATE SMART LIVESTOCK PRODUCTION

EMERGING ISSUES IN CLIMATE SMART LIVESTOCK PRODUCTION

Biological Tools and Techniques

Edited by

SUKANTA MONDAL

Principal Scientist, ICAR-National Institute of Animal Nutrition and Physiology, Adugodi, Bangalore, India

RAM LAKHAN SINGH

Vice Chancellor, Nilamber-Pitamber University, Palamu, Jharkhand, India

ELSEVIER

ACADEMIC PRESS

An imprint of Elsevier

Academic Press is an imprint of Elsevier
125 London Wall, London EC2Y 5AS, United Kingdom
525 B Street, Suite 1650, San Diego, CA 92101, United States
50 Hampshire Street, 5th Floor, Cambridge, MA 02139, United States
The Boulevard, Langford Lane, Kidlington, Oxford OX5 1GB, United Kingdom

Library of Congress Cataloging-in-Publication Data
A catalog record for this book is available from the Library of Congress

British Library Cataloguing-in-Publication Data
A catalogue record for this book is available from the British Library

ISBN: 978-0-12-822265-2

For information on all Academic Press publications
visit our website at https://www.elsevier.com/books-and-journals

Publisher: Nikki P. Levy
Acquisitions Editor: Nancy Maragioglio
Editorial Project Manager: Jai Marie Jose
Production Project Manager: Bharatwaj Varatharajan
Cover Designer: Victoria Pearson

Typeset by STRAIVE, India

Dedication

Dedicated to our beloved daughters

Anoushka (daughter of Mrs. Shrabanti Mondal and Dr. Sukanta Mondal), for making everything worthwhile and for cheerfully tolerating many hours of absence involved in writing this book

and

Suneha Bisen (daughter of Dr. (Mrs.) Uma Lakhan Singh and Professor Ram Lakhan Singh), who always functions as our lifeline and motivational force and remains concerned about every single aspect related to us.

Contents

Contributors

Satish Adusumilli
Freimann Life Science Center, University of Notre Dame, Notre Dame, IN, United States

P.K. Agarwal
NASF, ICAR, New Delhi, India

Akash
Division of Surgery, ICAR-Indian Veterinary Research Institute, Izatnagar, Uttar Pradesh, India

R.L. Babu
Department of Bioinformatics and Biotechnology, Karnataka State Akkamahadevi Women's University, Jnanashakthi Campus, Vijayapura, Karnataka, India

S. Lokeswara Balakrishna
Department of Biotechnology, Indira Gandhi National Tribal University, Amarkantak, MP, India

Arnab Banerjee
Department of Biological Sciences, KK Birla, BITS Pilani, Zuarinagar, Goa, India

Mamoni Banerjee
Rajendra Mishra School of Engineering Entrepreneurship, IIT, Kharagpur, West Bengal, India

Ramanuj Banerjee
Department of Scientific and Industrial Research, Ministry of Science and Technology, Government of India, New Delhi, India

Indrashis Bhattacharya
Department of Zoology, HNB Garhwal University, Uttarakhand, India

Dipanshu Bisht
Division of Surgery, ICAR Indian Veterinary Research Institute, Izatnagar, Uttar Pradesh, India

B.K. Biswas
West Bengal University of Animal and Fishery Sciences, Kolkata, West Bengal, India

Kavya Chandra
Department of Biological Sciences, KK Birla, BITS Pilani, Zuarinagar, Goa, India

D.N. Das
Genetics Lab, ICAR-National Dairy Research Institute, Southern Regional Station, Bengaluru, Karnataka, India

Moitreyi Das
Department of Zoology, Goa University, Goa, India

M. Devasena
Department of Civil Engineering, PSG Institute of Technology and Applied Sciences, Coimbatore, India

Neha Dobriyal
Department of Biological Sciences, Indian Institute of Science Education and Research Bhopal, Bhopal, MP, India

Joseph R.D. Fernandes
Department of Biological Sciences, KK Birla, BITS Pilani, Zuarinagar, Goa, India

Minal Garg
Department of Biochemistry, University of Lucknow, Lucknow, India

Bharti Gupta
Department of Biotechnology, Indira Gandhi National Tribal University, Amarkantak, MP, India

P.S.P. Gupta
Animal Physiology Division, ICAR-National Institute of Animal Nutrition and Physiology, Adugodi, Bengaluru, Karnataka, India

Mozammel Hoque
Division of Surgery, ICAR-Indian Veterinary Research Institute, Izatnagar, Uttar Pradesh, India

K.M. Kiran Kumar
Yen Onco Centre, Yenepoya Medical College, Yenepoya Deemed to be University, Mangaluru, Karnataka, India

Anyam VVNGSV Kishore
Department of Surgery and Radiology, West Bengal University of Animal and Fishery Sciences, Kolkata, India

Munish Kumar
Department of Biochemistry, University of Allahabad, Prayagraj, UP, India

Sukanta Mondal
Animal Physiology Division, ICAR-National Institute of Animal Nutrition and Physiology, Adugodi, Bengaluru, Karnataka, India

Rashmi Nagesh
Department of Biotechnology and Genetics, M.S. Ramaiah College of Arts, Science and Commerce, Bengaluru, India

S. Nandi
Animal Physiology Division, ICAR-National Institute of Animal Nutrition and Physiology, Adugodi, Bengaluru, Karnataka, India

M. Naveen Kumar
Department of Biotechnology and Genetics, M.S. Ramaiah College of Arts, Science and Commerce, Bengaluru, India

Darshika Nigam
School of Health Sciences, Quantum University, Roorkee, India

Hari Om Pandey
ICAR-Indian Veterinary Research Institute, Bareilly, Uttar Pradesh, India

D. Paul
Genetics Lab, ICAR-National Dairy Research Institute, Southern Regional Station, Bengaluru, Karnataka, India

Kingshuk Poddar
New Materials Business, Tata Steel Limited, Kolkata, India

S.J. Prashanth
Department of Food Processing and Nutrition, Karnataka State Akkamahadevi Women's University, Vijayapura, India

I.J. Reddy
Animal Physiology Division, ICAR-National Institute of Animal Nutrition and Physiology, Adugodi, Bengaluru, Karnataka, India

Ritambhara
Department of Biochemistry, University of Allahabad, Prayagraj, UP, India

S.I. Rizvi
Department of Biochemistry, University of Allahabad, Prayagraj, UP, India

V. Sangeetha
Department of Civil Engineering, Coimbatore Institute of Technology, Coimbatore, India

Bhoopendra Singh
Department of Veterinary Gynaecology & Obstetrics, College of Veterinary Science & Animal Husbandry, Acharya Narendra Deva University of Agriculture & Technology, Ayodhya, Uttar Pradesh, India

Kaushlendra Pratap Singh
Government Veterinary Hospital, Deornia, Bareilly, Uttar Pradesh, India

Kshitij R.B. Singh
Department of Chemistry, Govt. V. Y. T. PG. Autonomous College, Durg, Chhattisgarh, India

Pankaj Singh
Department of Biotechnology, Dr. Rammanohar Lohia Avadh University, Ayodhya, Uttar Pradesh, India

Pradeep Kumar Singh
Department of Biochemistry, Dr. Rammanohar Lohia Avadh University, Ayodhya, Uttar Pradesh, India

Ram Lakhan Singh
Nilamber-Pitamber University, Palamu, Jharkhand, India

Ravindra Pratap Singh
Department of Biotechnology, Indira Gandhi National Tribal University, Amarkantak, MP, India

N.P. Soumya
University of Kerala, Thiruvananthapuram, Kerala, India

Parikipandla Sridevi
Department of Biotechnology, Indira Gandhi National Tribal University, Amarkantak, MP, India

Deepak Upadhyay
ICAR-Indian Grassland and Fodder Research Institute, Jhansi, Uttar Pradesh, India

Kumari Bhoomika
School of Health Sciences, Quantum University, Roorkee, India

Preface

Livestock production constitutes the livelihood of the rural poor belonging to the lowest socioeconomic strata and generates gainful employment in the rural sector, particularly among landless laborers, small and trivial farmers, and women in developing countries. The growing demand for livestock products as a result of increasing population and urbanization is one of the major factors responsible for the balanced growth of the livestock sector. Biotechnology is a significant tool that can be used to meet the demand for the production of more food products from animals and fisheries to supply the food requirements of the increasing human population. Biotechnology has the potential to improve the productivity of animals via increased growth, carcass quality and reproduction, improved nutrition and feed utilization, improved quality and safety of food, improved health and welfare of animals, and reduced waste through more efficient utilization of resources. Several biotechnological approaches can be evaluated in the field of quality assurance programs, which should be beneficial for producing livestock products of assured quality and safety with respect to public health. Recent advances in reproductive biotechnologies have provided a powerful tool that can be used to improve animal production and address challenges of livestock production in the future. Application of these technologies will help cope with the limited resources available and simultaneously increase the need for food production. Producers can make use of rapid genetic changes to produce the next generation of superior animals. In addition, introducing new genes into a population via transgenic methodologies can lead to benefits to the environment, producers, and consumers. Although several of these technologies have been available for several decades, the use of these animals has been restricted to medical research and farm animal application of this technology is far behind biomedical areas. Implementation of these technologies for animal agriculture will continue to be slow due to their innate complexity, expense, and inefficiency. Production of cloned and transgenic livestock will continue, but realizing their full promise and impact on animal welfare, environmental issues, and food security will require consumer acceptance.

Advances in genomics offer the potential to speed up productivity in the livestock sector. With the combination of traditional and high-throughput sequencing platforms, there has been a tremendous increase in genomic resources available, including expressed sequence tags (ESTs), BAC end sequences, genetic sequence polymorphisms, gene expression profiling, whole-genome (re)sequencing, and genome-wide association studies. Advances in livestock and fish genomics have boosted the ability to enhance specific traits for production of consistently high-quality foods that are safe. Animal agriculture has entered an area of exciting possibilities for rapidly propagating and tailoring animals to meet product and environmental demands. This is made possible by the development of newer biotechnologies such as gene transfer, cloning, DNA vaccine, gene knockout, and gene therapy. The strength of these technologies lies in their complementariness and their ability to be readily scaled up for commercial use. These newer technologies along with older techniques such as estrous synchronization, artificial insemination, cryopreservation of gametes, embryo transfer, in vitro production of embryos, and sexing of embryo and semen have opened the realm of well-organized systems for wide-scale production, sale, and transfer of genetically superior animals. These biotechnological approaches and current bioinformatics tools would be of tremendous importance in understanding the complex mechanism of low productive and reproductive efficiency in various livestock species in order to evolve safer and potent gene-based strategies for enhancement of production. The application of molecular genetic technologies to animal agriculture will definitely bring about exciting changes in livestock production and the tailoring of animals to produce products needed by humans.

Emerging Issues in Climate-Smart Livestock Production is designed as an advanced book integrating both basic and advanced concepts of animal biotechnology along with future developments for sustainable livestock production. It will also serve as a reference book for scientists, researchers, educators, and professionals in this area. This book will enable them to solve problems concerning sustainable development with the help of current innovative biotechnologies such as recombinant DNA technology and genetic engineering technologies (genome editing, gene silencing, transgenesis), which have tremendous potential for impacting global food security, environmental health, human and animal health, and overall livelihood of mankind. The contributors to this book are internationally recognized experts in their respective fields representing reputed institutions across the globe.

Organization of this book

This book consists of 18 chapters that focus on current approaches and strategies for sustainable livestock production.

1. *Introduction* presents a brief introduction, scope, and applications of biotechnology for sustainable agriculture.
2. *Biotechnological tools to enhance sustainable livestock production* deals with approaches and traits to improve livestock production, marker-assisted selection (MAS), transgenic approaches, propagation, and use of growth enhancers.
3. *Livestock production systems* covers classification of livestock production systems, distribution of various livestock species under different production systems, other perspectives on its classification, and drivers.
4. *Sustainable livestock production and food security* deals with the specialized knowledge, understanding, and skills necessary to contribute effectively and ethically to strategic decision-making, opinion forming, and operational management for the sustainable development of agricultural and food supply systems.
5. *Livestock biodiversity and sustainability* covers sustainability and its dimensions, challenges in global meat production and consumption, global distribution of grassland resources, conversion of grassland ecosystem into agriculture, challenges and opportunities for livestock production systems, managing biomass production, biodiversity loss, management of greenhouse gases and climate change.
6. *Climate change impact on livestock production* deals with climate change and heat stress, impact of heat stress on animal health, production performances, reproduction, adaptation and mitigation strategies.
7. *Nanotechnology in animal production* illustrates its application in disease management and preventing pathogenesis, animal breeding and growth, treating reproductive problems, sterilization of animals, reducing transport and storage of animal protein (food packaging), food processing nanoapplications, biological waste management through nanotechnology for sustainable animal husbandry and livestock production, and toxicological and environmental safety concerns.
8. *Transgenic animals* describes the state of the art in animal transgenesis, its contribution to animal welfare, pitfalls and risks as well as ethical concerns. Genetically modified (transgenic) livestock have important roles in producing both better quality and quantity of food derived from livestock, improving strain of livestock, enhancing prolificacy and reproductive performance, increasing feed utilization and growth rate, improving milk production and/or composition, modifying hair or fiber, and increasing disease resistance and ethics in transgenic animal production.

9. *Potential implications of gonadotropin-releasing hormone (GnRH) analogues in assisted reproductive technology* discusses GnRH agonists–antagonists in female and male livestock, rodents and GnRH analogues and future directions.
10. *Oxidative stress in modulation of immune function in livestock* highlights the role of free radicals in evoking immune responses, implication of oxidative stress on immune responses in dairy cattle, immunological reflex in poultry birds on redox burden, impact of stress on the goat immune system and sequential effects of oxido-reductive stress in piglets.
11. *Cow urine: potential resource for sustainable agriculture* illustrates the composition of cow urine, diet effects on composition of cow urine, and direct and indirect applications of cow urine.
12. *Effect of climate change on vector-borne disease* (VBD) covers VBDs, types of VBD transmission, climate change and vector, as well as livestock management and insect control.
13. *Role of biotechnology in animal breeding and genetic improvement* focuses on animal breeding and biotechnologies, factors governing genetic gain, artificial insemination, embryo sexing, estrus synchronization, animal cloning, transgenesis, multiple ovulation and embryo transfer (MOET), (MAS and gene mapping, genome editing biotechnologies transforming the genome at the DNA level, and constraints on applying the biotechnologies in developing countries.
14. *Biotechnology in animal nutrition and feed utilization* describes the composition and nutrient requirements of livestock, feed of ruminants, feed availability, applications of biotechnology in livestock nutrition, feed improvement, forage breeding, and biotechnology products as feed additives. It also discusses the role of biotechnology in feed improvement, feed utilization and rationing, biological treatments to improve feed utilization, biotechnological interventions to improve feed utilization and bionanotechnology for animal health and feed utilization.
15. *Biotechnology in livestock reproduction* deals with artificial insemination, MAS, embryo technologies, in vitro maturation, fertilization and embryo production, embryo freezing, sexing of embryo and sperm, cloning, transgenesis and stem cell technology.
16. *Targeted genome editing by CRISPR/Cas9 for livestock improvement* deals with genome editors, journey of the CRISPR/Cas9 system from its discovery to development as a genome editing tool, CRISPR/Cas9 system components, mechanism of action, application of CRISPR/Cas9 in

livestock improvement, disease-resistant livestock, animal welfare, production of allergen-reduced/free and enriched animal-derived products and biomedical research.

17. *Stem cell technology: application in animal health and livestock production* describes the characteristics of mesenchymal stem cells, embryonic stem cells (ESCs), induced pluripotent stem cells (iPSCs), adult mammary stem cells, germ cell transfer technique to livestock production, applications of germ cell transplantation in livestock, transgenesis, common techniques used for the generation of transgenic animals, stem cell therapy application in veterinary medicine and future aspects and translation to human health.

18. *Way forward for sustainable livestock sector* outlines the importance of livestock production, adverse effects of climatic changes on livestock production, role of the livestock sector in climate change and mitigation approaches in animal husbandry practices.

Sukanta Mondal
Ram Lakhan Singh

Acknowledgments

We are deeply indebted to all the contributors who assisted materially in the preparation of this book. We are grateful to our families who cheerfully tolerated and supported many hours of our absence while finishing this book project. Thanks are also due to Heijkoop Liz, Maragioglio Nancy, Osborn Patricia, Jai Marie Jose, Indhumathi Mani, Bharatwaj Varatharajan, and the entire publishing team of Elsevier for their extra care and patience in helping the book take shape.

Sukanta Mondal
Ram Lakhan Singh

CHAPTER ONE

Introduction

N.P. Soumya[a], Sukanta Mondal[b], and Ram Lakhan Singh[c]
[a]University of Kerala, Thiruvananthapuram, Kerala, India
[b]Animal Physiology Division, ICAR-National Institute of Animal Nutrition and Physiology, Adugodi, Bengaluru, Karnataka, India
[c]Nilamber-Pitamber University, Palamu, Jharkhand, India

Introduction

Biotechnology research focuses on the production of good quality food animals with improved quality of their products. This could be achieved through various biotechniques, including cloning and transgenesis. Similarly, the alteration of essential genes aids in improving the quality of livestock products. Recombinant DNA technology (rDNA technology) provides large-scale production of hormones (e.g., somatostatin) and enzymes (e.g., chymosin), which are used for improved growth and carcass composition (Hui et al., 2001). The development and integration of antimicrobial agents yield auspicious results in extending food product shelf life. Several powerful tools in biotechnology have made great strides, including rDNA technology, genetic engineering, PCR (polymerase chain reaction), RAPD (random amplified polymorphic DNA), RFLP (restriction fragment length polymorphism), AFLP (amplified fragment length polymorphism), and microarray. The extensive development of nucleic acid-based technologies offers excellent help in the meat industry to authenticate species-specific meat. New frontiers can be explored using biotechnology to develop nutritious and healthy food products with longer shelf lives that are more acceptable to customers. The application of biotechnology tools in agriculture and livestock sector is summarized in Fig. 1.

Biotechnological tools to enhance sustainable livestock production

Biotechnologies are those processes that manipulate the cellular and biomolecular mechanisms of living organisms to advance human health. Modern biotechnological tools together with conventional technologies

Emerging Issues in Climate Smart Livestock Production
https://doi.org/10.1016/B978-0-12-822265-2.00014-4

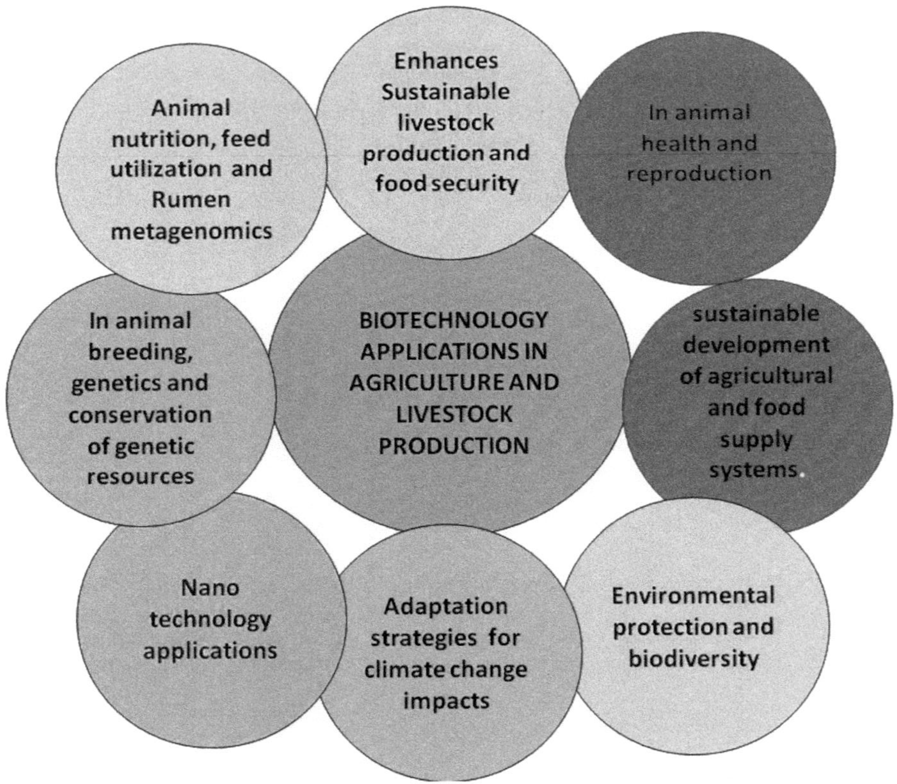

Fig. 1 Applications of biotechnology tools for sustainable agriculture and livestock production.

have granted increased productivity by influencing animal health, nutrition, genetics and breeding, livestock reproduction, and the quality and quantity of animal products.The cost of adoption and improvement in productivity has facilitated the wide acceptability of this branch of science and helped alleviate poverty and hunger, particularly in developed countries.

All the main sectors of animal sciences are extensively using biotechnology's comprehensive application. Recent advances in molecular markers combined with bioinformatics have opened the possibility of recognizing significant genes for genetic variation and genomic alterations. Polymerase chain reaction (PCR) is the backbone of most of the molecular marker systems. Specific enzymes and hormones have been produced by recombinant bacteria, improving nutrient utilization and increasing productivity. Similarly, fermentation technologies are used to improve the digestibility of animal feeds. Probiotics and microbial cultures are other techniques used to improve the quality and digestibility of silage. Fiber-degrading enzymes can be used to increase the productivity of the animal (FAO Stat, 2010).

Furthermore, biotechnologies improve medical diagnostics and disease prevention while enhancing treatment accuracy in the animal welfare aspect. Vaccination is an essential tool for preserving animal health. Recombinant DNA vaccines provide possible benefits in accuracy, durability, and efficacy compared with standard vaccines. The use of monoclonal antibodies and DNA/RNA probes creates substantial incentives for improved diagnostic tests. Enzyme-linked immunosorbent assays (ELISA) and radioimmunoassays (RIA) are some of the immunological tests performed using monoclonal antibodies. Consequently, biotechnology's ultimate goal applied to animal welfare is protecting both humans and animals with environmental conservation.

Livestock production systems

Worldwide, the livestock sector is highly dynamic and is responsible for producing animal protein from plants. These systems are remarkably diverse, extending from low-input, pastoral production systems in arid and semiarid regions to highly intensive production systems. After a period of marginal conceptual transformation in the late 1980s, lots of new scientific developments significantly changed the practice of land grazing.

In highly variable environments, producers employ conservative stockings to minimize risk over the years. Efficient technical developments and systemic changes in the livestock sector increase the competitiveness and sustainability of livestock production systems (Herrero et al., 2013). Options for sustainable intensification of livestock production systems include the overcrowding of native grass pastures with legumes, the development of "mosaic" forage irrigation "hot spots" on the landscape where limited areas of suitable soil on large properties are developed with irrigation to produce high-value forage that enables animals to be finished on the market, and potential low-cost on-farm production.

Livestock fertility is undoubtedly the most prominent element that could improve animal protein sustainability by global livestock producers. Greenhouse mitigation strategies and technologies that increase fertility rates offer significant incentives for reducing resource utilization, greenhouse gas emissions, and feed costs (White and Capper, 2013). The consequences of many less-measurable losses of production in livestock systems, such as male reproduction, clinical and subclinical morbidity, and replacement animal production, have yet to be quantified. The production of new vaccines provides personalized benefits and protects from animal-related casualties.

Recent selection efforts in livestock production have led to improved growth, production and carcass quality. Genomic selection is the current breeding model suggested (Meuwissen et al., 2001). A combination of new genomic information with conventional pedigree and quality data is vital to livestock producers, and hence it needs to be cost-effective and highly precise. This rapidly emerging scenario of genetic tools for DNA and RNA is expected to simultaneously allow the development of individual and multitrait genomic selection systems.

Integrated livestock–crop production systems are an emergent management strategy that can reduce enterprise risk, restore degraded land, increase productivity, and diversify production. By combining livestock with both crops and forests, animal manure can be used as a fertilizer to increase soil nutrients and organic. The carbon-constrained economy will increasingly affect production in the future (Thornton et al., 2009). Advancements in breeding, nutrition, and animal health will continue to contribute to increased potential production and further gains in genetic efficiency.

Sustainable livestock production and food security

Livestock contributes significantly to total food production, and its contribution also involves multipurpose uses such as hides, fibers, fertilizers, and power supporting economic growth. Sustainable agriculture research and strategy can be described in terms of resource availability and sustainability. Food safety has become a particular issue for the last few years. When conventional agri-food logistics practices no longer satisfy consumer requirements, developing a traceability framework for the agri-food supply chain becomes extremely urgent. It can be accomplished through livestock farmers' awareness of good management practices (GMPs) and good husbandry practices (GHPs), which will eventually result in secure, stable, and balanced food production for both producers and consumers.

Policy innovations will play a crucial role in all countries. Policy decisions should ensure that livestock practices improve human nutrition, address climate change, and ensure sustainability for the environment (Shenggen, 2018). The policy changes would also be necessary for the global livestock sector to contribute in a sustainable way toward food safety and nutrition. Developed nations would continue to focus on mitigating GHG pollution. Moreover, significant changes may be made to the sector, especially in animal and herd productivity for ruminants through better breeding and grazing practices as well as managing manure to ensure nutrient recovery and recycling. Therefore, ensuring more targeted and productive social security policies for smallholder livelihoods is a crucial prerequisite.

Livestock biodiversity and sustainability

Agricultural biodiversity encompasses all biodiversity components relevant to food and agriculture and all biological diversity components that constitute and sustain essential agroecosystem functions. While recognizing livestock sustainability and management at the lowest level, farmers are usually involved in farm products for their own nutritional needs and survival. Agricultural biodiversity has several aspects, such as the variety and variability of plants, animals, micro-organisms at DNA, the organism level, habitats, biotic conditions, and factors of socioeconomic and cultural factors. The financial, supporting, and regulating services (Millenium Ecosystem Assessment

(MEA), 2005) rendered by the livestock are similarly relevant but economically undervalued. Besides performing significant roles in nutrition and feeding, livestock products have high socioeconomic and cultural responsibilities in many societies. They assist in nutrient cycling, waste product recycling, seed dispersal, transport, and draft power and serve as a means of wealth accumulation (FAO, 2009).

Genetic variation is the foundation of global food production, and livestock genetic diversity is represented in breeds. These group populations represent distinct combinations of genes for production and reproduction characteristics. The sustainable use of genetic resources for livestock is based on the persistent utilization of genetic diversity within and between breeds. In traditional agricultural cultures, livestock exchange is also a crucial factor in maintaining social assets.

Globally, genetic diversity in livestock is steadily decreasing due to specialization in plant and animal breeding and the harmonizing effect of globalization. Perhaps the reason for this is the trade-off between farm and herd genetic management. Cattle production's significant adverse environmental externalities are not evident in livestock costs and products. There are trade-offs between goals at each level, affected by property regimes and scaling. Global meat production increased between 1980 and 2009 to 3.7% per annum, 1.7% for milk, and 5.0% for eggs (FAO Stat, 2010). Overall, this increase in global meat production is projected to be 16% higher in 2025 than from 2013 to 15 (OECD/FAO, 2016). World meat production increased by 1.0% to 327 Mt in 2018 (OECD/FAO, 2020). Similarly, world milk production is projected to grow at 1.6% p.a. over the projection period to 997 Mt by 2029, faster than most other primary agricultural commodities (OECD/FAO, 2020).

Livestock kept in large-scale projects, using advanced techniques, depends on quality feed, and animal genetics plays a crucial role in global food distribution. Recent work into agrifood in developing countries discusses genomic collection, animal health, consistency of food security, weather, climate change, and protein richness. Studies through sustainable use of animal genetics diversity are complex, as accurate details are sparse for breed status and the drivers that influence them, the multilevel parties involved, and the trade-offs between the various international public interests responsible for culture and the environment (Hoffmann, 2011).

Impact of climate change on livestock production

The livestock sector is essential to global food security and health by contributing to nutritious diets, economic growth, and livelihoods. However, climate change causes severe threats to the livestock sector, ultimately reducing the profitability of farmers. Competition for natural resources will adversely affect livestock production and food security (Thornton, 2010). Reducing feed intake is an adaptive mechanism to decrease heat production in warm environments. Accordingly, animals will be in negative energy balance (NEB) due to depressed appetites, resulting in reduced body condition score and weight gain (Lacetera et al., 1996). Climate change's direct impact on health includes temperature-related illness and death.

The prevalence of infectious and metabolic diseases may be greater in hot weather, as cell-mediated immune response decreases during heat stress. Similarly, high temperature and reduced feed intake will adversely affect the performance of broilers and layers. Increased air temperature along with a high temperature-humidity index (THI) causes reduced dry matter intake (DMI), causing decreased milk yield, reduced body weight, poor reproductive performance, and altered immunity (West et al., 1991). High environment temperatures may compromise the reproductive efficiency of farm animals in both sexes. Many reproductive processes may be affected, such as spermatogenesis, ovarian follicular growth, oocyte competence, formation of the corpus luteum, embryonic growth, gonadotropin secretion, steroidogenesis, uterine endometrial responses, etc. Climate change will also affect the quality and quantity of meat produced by livestock.

The risk of emerging complex diseases can be best tackled by advanced veterinary care with improved animal husbandry practices. Mitigation strategies are human interventions to ameliorate the adverse effects of climate change on living things. There can be a great variety of adaptation options available, including technological, behavioral, managerial, and policy options such as planning regulations and infrastructural development. It is necessary to adopt climate-smart livestock production strategies. Adaptation to climate change can be done by substituting climate-resilient livestock species and modifying animal husbandry practices such as improved nutrition as well as assisted reproductive strategies and proper animal health.

Nanotechnology in animal production

Nanotechnology is described as the science of materials and structures where the size is in nanometers. The nanoparticle's biocompatibility is highly advantageous for the medical and food industries, with significantly fewer or no side effects on patients or consumers. With few alterations, natural compounds are more likely to be biocompatible, biodegradable, and distributable in the body. When materials are scaled down to nanosizes, the material changes physical properties due to alterations in surface chemistries (Albanese et al., 2012). Nanoparticles can be metals, polymers, natural compounds, or nanostructured materials (Hill and Li, 2017). With applications in various fields such as physics, chemistry, and biology, the study of materials at the nanoscale is referred to as nanotechnology. Nanoparticles have been used in the medical field for diagnostic and therapeutic purposes for quite a long time. However, their application in veterinary medicine is still relatively new. Possible applications of nanotechnology in animal production include:

1. Administration of medication, nutrients, probiotics, supplements, and other substances.
2. Diagnosis and treatment of diseases, detection, and elimination of the cause of the disease without the need for surgery.
3. Identity registry that allows a follow-up on the history of an animal and its products (meat, milk, eggs, mainly).
4. Management of reproduction in livestock with nanoparticle-incorporated hormonal immunosensors.

Continuous use of antibiotics as growth promoters is not recommended nowadays due to the fear of antibiotic resistance. Many reports suggest nanoparticles as the right alternative for animal growth promotion and antimicrobials. However the meat quality of nanoparticle-incorporated animals used for human consumption must be determined due to the possibility of residues. Hence, the health and well-being of both animals and humans are to be considered.

The future applications of nanotechnology in veterinary science can be numerous, such as fortification of livestock feeds with nanosupplements; replacement of antibiotics, antiseptics, and other antiseptic wound dressings; and cancer research (Ramírez-Mella and Hernández-Mendo, 2010). Nanoparticles are now commercially available, and their properties, with rapid improvement, are more optimized for a greater variety of applications.

The use of nanotechnology in livestock production is in its infancy, yet it is providing inspiring results. However, it also requires a considerable amount of research to promote nanotechnology's feasibility and safety as well as preventing any risk to the atmosphere or human beings.

Transgenic animals

Transgenic animals are proficient in enhancing nutrients in animal products regarding their quantity, quality, and nutritional composition. Enhanced omega-3 fatty acids have been produced by transgenic pigs. Transgenic production provides methods for the rapid introduction of new or modified genes and DNA sequences into livestock without cross-breeding or hybridization (Wheeler, 2013). Livestock genetic modification can increase animal welfare by raising healthy animals, as animal health is a crucial concern for livestock production. It is possible to manipulate growth factors, growth factor receptors, and growth modulators through transgenic technology. For example, Merino sheep have been identified for the presence of the Boroola fecundity (FecB) gene, which allows for an increased ovulation rate. Similarly, milk's quantity and composition could be altered by introducing novel proteins in milk through advances in transgenic technology. In the same way, an increase in porcine growth hormone (GH) leads to enhanced growth and feed efficiency in pigs. Transgenic livestock can reduce the environmental impact of animals on agriculture. Improving hair and fiber quality as well as yield are other areas of focus for transgenic manipulation in livestock.

Rumen metagenomics

The rumen microflora has a vital role in normal ruminant physiology and nutrition. In the rumen, microbial alterations have significant physiological and pathological implications. Numerous studies showed their relationship with feed efficiency in beef and dairy cattle, reporting the differences in relative abundance between efficient and inefficient individuals of several rumen microbial phylotypes (Myer et al., 2015). The majority of these studies used microbiota sequencing to 16S rRNA. Metagenomics and metatranscriptomics have now become powerful tools for estimating the rumen microbiome's functional potentials (DNA-based) and functional activities (RNA-based). Metagenomics provides culture-free analyses of microbial genetic structure and can be categorized into functional and computational metagenomics sequencing. The introduction of next-generation sequencing technology and the rapid growth of computational tools and reference databases offer potent tools in the rumen microbiome.

Furthermore, while mRNA enrichment increased the sequencing depth of mRNA and the precision of metatranscriptomics on functional analysis, a comparison of total-RNA-based and mRNA-enriched metatranscriptomes showed possible biases in estimating certain gene expressions. Systematic analyses of the rumen microbiota have identified previously unrecognized species in the rumen, including its viral (virome) and plasmid (plasmidome) fractions. The rumen microbial community's resistance and resilience to fluctuation are also addressed. Besides, the need for mechanistic models and the implementation of general environmental theories and concepts in rumen should be emphasized.

Biotechnology in animal health

Prompt, accurate, and instantaneous disease diagnosis is essential for the maintenance of good animal health. In this context, molecular diagnostic approaches offer significant potential for improving the speed and accuracy of diagnosis and aid in effectively controlling and treating infectious diseases. The maintenance of good animal health focuses on early, reliable, and timely identification of pathogens and infectious agents. Recently, substantial progress has been made in biotechnology applications, both in diagnosing and preventing infectious and parasitic diseases. The use of monoclonal antibodies and DNA/RNA probes creates substantial incentives for improved diagnostic tests. It is helpful in the rapid and simultaneous identification of multiple pathogens in a given sample (Zhao et al., 2014). Immunoassays such as ELISA can be used for the widespread detection and identification of a variety of bacteria, viruses, and parasites. Quantitative real-time PCR (qPCR) is the latest development in the conventional PCR technique that significantly reduces cross-contamination between samples.

Developments in rDNA technology have benefited greatly from expanding advanced vaccines and diagnostic tools. Modern advancements in immunology, immunochemistry, cellular biology, and biochemistry have resulted in new vaccines (subunit vaccines, recombinant vaccines, hybrid vaccines, and antiidiotype vaccines). The primary basis for PCR applications in veterinary diagnosis is by the detection of infectious agents and pathogenic discrimination strains through specific genes. Nested PCR is a modification of traditional PCR, intended to reduce nonspecific binding in products due to the amplification of undesirable binding primary sites. Multiplex PCR is another commonly used method that comprises multiple primers, which allows the amplification of multiple targets within a single PCR reaction. DNA sequencing and chip-based technologies such as microarrays allow large-scale diagnostic and genetic screening of genomes. They are currently used to detect a wide range of pathogens of veterinary importance. Next-generation sequencing (NGS) diagnosis has emerged recently due to the demand for rapid, cheap, and precise DNA sequencing data. Similarly, NGS platforms have helped successfully sequence many large animal genomes and genomic variations (Bai et al., 2012). The growth of novel antiinfective, antiparasitic, immunomodulatory, therapeutic, and medicinal drugs as well as new diagnostic techniques and vaccinations is the most vital biotechnology application that continues to pursue new uses in the animal health sector.

Climate change and infectious disease

Global warming is a subcomponent of globally evolving climate change. Vectors are highly influenced by climatic factors that cause disease transmission into other susceptible hosts. The indirect effects of climate change on disease epidemics are due to changes in the temporal and spatial distribution of microbial communities (pathogens or parasites) in addition to the change in host resistance (Nardone et al., 2010). Poor feeding and watering can make livestock more vulnerable to infectious diseases. There can be an altered distribution and transmission of animal pathogens and heat stress can improve pathogens or their vectors. The change in host distributions due to climate change and increased exposure of hosts to harmful ultraviolet radiation due to ozone depletion have been observed to cause immune suppression and thus increased predisposition to infectious diseases. Many bacterial disease outbreaks such as anthrax are often associated with alternating heavy rainfall and drought and high temperatures (Parker et al., 2002) as high humidity and rainfall are suitable for the successful germination of the spores.

Heavy rainfall, particularly after drought, can cause insect population booms by raising larval habitats. Likewise, floods can increase waterborne diseases such as cholera and leptospirosis; storms can increase wastewater transportation to groundwater. The rodent population in temperate regions is known to increase after mild/wet winters, thereby increasing rodent-borne diseases that can include Lyme disease/borreliosis, tick-borne encephalitis, and hantavirus pulmonary syndrome. The forced migration of animals in search of water due to altered climate plays an essential role in disease transmission. Similarly, increased high-speed winds/cyclones/storms can spread several bacterial and viral diseases over long distances. Virus-infected culicoide vectors transmit Bluetongue and African horse sickness over hundreds of kilometers through winds. Biting rates or the length of the transmission period of vectors are also affected due to temperature changes. Thus, the global temperature increase has not only affected environmental factors such as biodiversity, rain, heat cycles, air pollution, and sea level, but has also produced a proliferation of current and emerging infectious diseases.

Biotechnology in animal breeding, genetics, and conservation of genetic resources

The impact of biotechnology in animal breeding and genetics can be assessed by genetic progress. The essential biotechnological tools that affect male and female reproduction proficiency are artificial insemination, multiple ovulation, in-vitro fertilization, ova pick-up, embryo transfer, sexing of semen and embryo cloning, and twinning. These technologies chiefly improved the distribution of superior germplasms as well as the selection intensity and accuracy. Multiple ovulation embryo transfers (MOET) and open nucleus breeding system (ONBS) schemes are alternatives to the traditional breeding system for genetic improvement. MOET is a combined technology comprised of superovulation, estrus synchronization among recipients, artificial insemination of the donor, in vitro culture of embryos, embryo recovery from the donor, and embryo transfer into the recipient. Based on the gene flow, there can be a closed nucleus breeding scheme (CNBS) and an open nucleus breeding scheme (ONBS).

In traditional selection, genetic progress is made by increasing the frequencies of favorable alleles by selecting the phenotype and mutation. Marker-assisted selection is an indirect way of selection, where the selection will be made based on a marker associated with a trait of interest rather than on the trait itself. Identification of QTL will allow for implementing novel marker-assisted selection schemes. Much MAS research in the current period uses markers dependent on DNA. This includes microsatellites,

RFLP, random amplification of polymorphic DNA (RAPD), AFLP, and single nucleotide polymorphisms (SNPs).

Over the past decade, various genotyping strategies have been established to identify SNPs and other common polymorphic variants. Microarray genotyping uses hybridization on solid support with fluorescence detection. Pyrosequencing is another type of DNA sequencing focused on identifying pyrophosphate production, a DNA polymerization byproduct. Knowledge of marker inheritance patterns combined with chip-based genotyping assays as well as information on linkage disequilibrium (LD) has encouraged genome-wide association studies (GWAS). Current achievements of GWAS in identifying specific genes that affect common diseases have led to scientific breakthroughs with improved technology. All measurable trait loci are identified with at least one marker in the disequilibrium. The early selection of animals enables new breeding approaches to improve genetic development, thus reducing costs (Ibtisham et al., 2017). Genomic selection is the potential of livestock breeding programs by reducing the genetic interval and genetic gain. Further research is warranted to manage long-term genetic gains and improve genomic selection accuracy in animal breeding.

Biotechnology in animal nutrition and feed utilization

Biotechnology plays a crucial role in enhancing livestock nutrition, mainly through the value addition of forage, the development of feed additives, and the modification of rumen microbiota. Nutrients such as protein, amino acids, and fats may be produced or preserved according to the need of various physiological statuses of animals. Moreover, enzymes may increase feed nutrient supply and reduce feed and fodder wastage. Also, prebiotics and probiotics or other immune supplements may be beneficial in inhibiting pathogenic enteric bacteria. Probiotics are live microbial feed supplements that support the host by improving gut microbial balance (Ekinci et al., 2002). The inclusion of vaccines and antibodies in feeds helps to guard livestock against disease.

Nowadays, the insertion of different genes into gut microorganisms has been thoroughly explored. The genetically engineered bacteria can ingest fibrous materials in forage, destroy toxins, synthesize essential amino acids, reduce methane synthesis, or resist acids. Likewise, genetically modified crops were found to minimize the excretion of urea into the environment. Thus, biotechnology has a full application in enhancing livestock feed efficiency through increasing the digestibility of low-quality forage, improving the nutritive value of cereals, removing antinutritive factors from available feeds, improving the nutritive value of conserved feed, improving rumen function, and modifying the metabolism.

Biotechnology in livestock reproduction

Recent years have seen the advent of reproductive biotechnologies, which start replacing traditional techniques. Among these, artificial insemination (AI) is the most widely applied animal biotechnology, and in combination with cryopreservation, endures considerable genetic improvement. Furthermore, advanced technologies of estrus synchronization, simultaneous/multiple ovulation, embryo transfer (ET), production of in vitro embryos, sperm, and embryo sexing accompanied those advances. Multiple ovulation and embryo transfer (MOET) programs and reproductive hormone assays also assisted in the faster propagation of superior females and farm animals' genetic improvement. Semen sexing is another technology used to produce offspring of the desired sex from a particular mating. Similarly, sexing of embryos is also possible so that all offspring would be of the desired sex, which has excellent potential for the livestock industry. Embryo cloning is a technique of making a physical copy of a fertilized egg that has begun development. Somatic cell nuclear transfer is the most common technique used in embryo cloning. Embryonic cloning can improve food availability by increased the production of beneficial plants and animals. Also, embryonic cloning helps prevent the extinction of rare or endangered species.

Introducing a foreign gene into animal cells that leads the input transgene to the succeeding generation is called transgenesis. Simultaneously, a gene knockout is a technique used to make an organism's gene nonfunctional. Knockout organisms can be used to learn gene function by investigating the effect of the lost gene. Knockouts are mainly used to identify a specific gene or DNA region's role by comparing a wild type with the knockout organism. They are also used as screening tools in the development of drugs.

Targeted genome editing by CRISPR/Cas9 for livestock improvement

The CRISPR-Cas9 mechanism (clustered regularly interspaced short palindromic repeats) is an adaptive, microbial immune response present in many bacteria and the CRISPR-associated proteins or CAS proteins. Bacteria and archaea produce a cellular memory of past invaders by incorporating DNA sequences similar to past invaders into their genomes. An RNA-driven nuclease from a bacterial immune system named Cas 9 is an easily exploited enzyme that can create double-strand breaks when inserted into eukaryotic cells or organisms (Jinek et al., 2013). It could be designed to act as a fast, low-cost, effective, and scalable method for manipulating genomic sequences in eukaryotic cells. Researchers conclude that by constructing a targeted double-strand break at a site of interest, errors can occur during the repair process, resulting in a desirable site mutation. In comparison, endogenous double-strand break repair mechanisms could also stimulate the introduction of exogenous DNA, producing different edits as designed (Thurtle-Schmidt and Lo, 2018). Combining this with recent genome-editing methods and genomic sequencing technologies has increased the possibilities in genetic manipulation in multiple species.

Genome editing is mainly suitable for addressing qualitative traits controlled by a single gene rather than quantitative traits. This technology for livestock improvements depends on the regulatory system implemented, which would differ according to country. Indeed, genome editing is a well-suited tool for altering single-gene qualitative traits and conventional selection methods to tackle concerns such as disease resistance as well as improve welfare traits. Editing genomes could be incorporated into the genomic selection to alter genetic variation and generation interval to speed up genetic gain. So far, work on genome editing has concentrated mainly on disease resistance such as TB in cattle,the Meat industry for Knocking out myostatin,allergens removal by knockout of beta- and in animal welfare traits. Genome editing may be used to insert beneficial alleles such as heat tolerance and disease resistance into native cattle breeds developed locally, thereby adding to their efficiency.

Conclusion

Although animal production is being changed significantly by revolutionary breakthroughs in biotechnology around the world, benefits are reaching developing countries in only a few areas such as conservation, animal improvement, disease diagnostics, and the augmentation of feed resources. Understanding the genes influencing animal growth; the efficiency of growth; environmental adaptation; meat, milk, or egg composition; or disease resistance are of very essential. The application of molecular genetic technologies to animals agriculture should drastically bring about exciting changes in animal production and the tailoring of animals to produce products needed by man. Genetic engineering has the potential to provide compelling benefits to transform public health, including improved foods, advances for human health, enhanced animal welfare, and reduced environmental impact.

References

Albanese, A., Tang, P.S., Chan, W.C., 2012. The effect of nanoparticle size, shape, and surface chemistry on biological systems. Annu. Rev. Biomed. Eng. 14, 1–16.

Bai, Y., Sartor, M., Cavalcoli, J., 2012. Current status and future perspectives for sequencing livestock genomes. J. Anim. Sci. Biotechnol. 3 (1), 1–6.

Ekinci, M.S., Martin, J.C., Flint, H.J., 2002. Expression of a cellulase gene, celA, from the rumen fungus Neocallimastix patriciarum in Streptococcus bovis by means of promoter fusions. Biotechnol. Lett. 24 (9), 735–741.

FAO, 2009. Livestock Keepers—Guardians of Biodiversity. Animal Production and Health Paper 167, Rome.

FAO, 2020. Environment Statistics, FAOSTAT Livestock Primary Data: 1961–2019. FAO, Rome.

FAO Stat, 2010. FAO Statistical Database 2010.

Herrero, M., Havlík, P., Valin, H., Notenbaert, A., Rufino, M.C., Thornton, P.K., Blümmel, M., Weiss, F., Grace, D., Obersteiner, M., 2013. Biomass use, production, feed efficiencies, and greenhouse gas emissions from global livestock systems. Proc. Natl. Acad. Sci. 110 (52), 20888–20893.

Hill, E.K., Li, J., 2017. Current and future prospects for nanotechnology in animal production. J. Anim. Sci. Biotechnol. 8 (1), 1–13.

Hoffmann, I., 2011. Livestock biodiversity and sustainability. Livest. Sci. 139 (1–2), 69–79.

Hui, Y.H., Nip, W.K., Rogers, R. (Eds.), 2001. Meat Science and Applications. CRC Press.

Ibtisham, F., Zhang, L., Xiao, M., An, L., Ramzan, M.B., Nawab, A., Zhao, Y., Li, G., Xu, Y., 2017. Genomic selection and its application in animal breeding. Thai J. Vet. Med. 47 (3), 301.

Jinek, M., East, A., Cheng, A., Lin, S., Ma, E., Doudna, J., 2013. RNA-programmed genome editing in human cells. Elife 2, e00471.

Lacetera, N., Bernabucci, U., Ronchi, B., Nardone, A., 1996. Body condition score, metabolic status and milk production of early lactating dairy cows exposed to warm environment. Riv. Agric. Subtrop. Trop. 90 (1), 43–55.

Meuwissen, T.H., Hayes, B.J., Goddard, M.E., 2001 Apr 1. Prediction of total genetic value using genome-wide dense marker maps. Genetics 157 (4), 1819–1829.
Millenium Ecosystem Assessment (MEA), 2005. Ecosystems and Human Well-Being. Biodiversity Synthesis. World Resources Institute, Washington, DC.
Myer, P.R., Smith, T.P., Wells, J.E., Kuehn, L.A., Freetly, H.C., 2015. Rumen microbiome from steers differing in feed efficiency. PLoS One 10 (6), e0129174.
Nardone, A., Ronchi, B., Lacetera, N., Ranieri, M.S., Bernabucci, U., 2010. Effects of climate changes on animal production and sustainability of livestock systems. Livest. Sci. 130 (1–3), 57–69.
OECD/FAO, 2016. OECD-FAO Agricultural Outlook 2016–2025. OECD Publishing, Paris.
OECD/FAO, 2020. OECD-FAO Agricultural Outlook. OECD Publishing, Paris.
Parker, R., Mathis, C., Looper, M., Sawyer, J., 2002. Anthrax and Livestock. Guide B-120. Cooperative Extension Service, College of Agriculture and Home Economics, University of New Mexico, Las Cruces, Nuevo Mexico.
Ramírez-Mella, M., Hernández-Mendo, O., 2010. Nanotechnology on animal production. Trop. Subtrop. Agroecosyst. 12 (3), 423–429.
Shenggen, Fan, 2018. Livestock Production for Food Security and Nutrition 10th Berlin Agriculture Ministers' Conference, Berlin, Germany.
Thornton, P.K., 2010. Livestock production: recent trends, future prospects. Philos. Trans. R. Soc. B Biol. Sci. 365 (1554), 2853–2867.
Thornton, P.K., van de Steeg, J., Notenbaert, A., Herrero, M., 2009. The impacts of climate change on livestock and livestock systems in developing countries: a review of what we know and what we need to know. Agr. Syst. 101 (3), 113–127.
Thurtle-Schmidt, D.M., Lo, T.W., 2018. Molecular biology at the cutting edge: a review on CRISPR/CAS9 gene editing for undergraduates. Biochem. Mol. Biol. Educ. 46 (2), 195–205.
West, J.W., Mullinix, B.G., Sandifer, T.G., 1991. Changing dietary electrolyte balance for dairy cows in cool and hot environments. J. Dairy Sci. 74 (5), 1662–1674.
Wheeler, M.B., 2013. Transgenic animals in agriculture. Nat. Educ. Knowledge 4 (11), 1–9.
White, R.R., Capper, J.L., 2013. An environmental, economic, and social assessment of improving cattle finishing weight or average daily gain within US beef production. J. Anim. Sci. 91 (12), 5801–5812.
Zhao, X., Lin, C.W., Wang, J., Oh, D.H., 2014. Advances in rapid detection methods for foodborne pathogens. J. Microbiol. Biotechnol. 24 (3), 297–312.

CHAPTER TWO

Biotechnological tools to enhance sustainable livestock production

Minal Garg
Department of Biochemistry, University of Lucknow, Lucknow, India

Introduction

The associations between humans and animals of agriculture importance have been well established for ages. Humans are heavily dependent in daily life on domesticated animals of diverse species, including goats, sheep, horses, cattle, swine, and chickens. These animals are not only human companions but are also reliable sources of food. Owing to the increased demand for animals, conscious efforts are being made toward improvements in animal production efficiency, especially breeding, growth, disease resistance, and animal product composition.

Livestock is commonly defined as domestic animals that are raised in agricultural settings that produce labor and commodities such as fur, meat (pork, beef, veal, and lamb), milk, eggs, wool, and leather. This term refers to farmed ruminants such as cattle and goats or those bred mainly for consumption. Traditionally, pigs, cattle, horses, sheep, and goats are considered livestock while poultry and fish are not included as livestock. Livestock farming practices are intensively carried out and hence are also known as factory farming. In most countries, livestock on a large scale are raised this way.

Traits to improve livestock production

Some of the major challenges in improving the leaner and more efficient meat-animal livestock production are reduced carcass fat, enhanced growth rate, and increased efficiency of conversion of ingested nutrients to lean tissue. Increased growth rate, large mature body sizes, and reduced body fat content at acceptable slaughter weights are the few major constitutional changes introduced in swine, cattle, and sheep (Robelin et al., 1978; Jones et al., 1981; Berg and Walters, 1983). Improvements in housing and

Emerging Issues in Climate Smart Livestock Production
https://doi.org/10.1016/B978-0-12-822265-2.00005-3

management practices, disease prevention and control, reproduction, nutrition, and milking practices play significant roles in increasing milk yield per cow with a rate of 4% per year.

Genetic modification of livestock allows the secured supply of nutritious, wholesome, and cost-effective animal products. This includes the increased quantity and quality of animal products by changing the product composition. For example, transgenic pigs with elevated levels of omega 3 fatty acids produce pork of improved nutritional quality (Lai et al., 2006). The production of transgenics with modified milk composition has a number of advantages, including greater quantity of milk production, milk of improved nutrient content, milk containing beneficial nutraceutical protein, and improved livestock growth and survivability. Elevating the levels of protein, fat, and/or lactose in milk has a greater impact on the growth and health of the offspring of meat-producing animals (sheep, goats, and cattle). Heat-tolerant livestock breeds such as *Bos indicus* cattle are essential for agriculture expansion in tropical areas. This cattle breed produces enough milk. Improvement in the milk yield by 2–4 L every day is realized by the increase in weaning weights of Nelore or Guzerat cattle breeds in Brazil. Similarly, Texel sheep and Boer goats with improved wearing weights are consumed for good-quality meat. In order to produce transgenic livestock with modified carcass composition and lowered fat and cholesterol content in eggs, meat, and cheese, the low-density lipoprotein receptor gene and hormones such as leptin can be potentially targeted. Sheep and goat breeds are manipulated/modified to obtain good quality, color, yield, strength, length, fineness, and crimp of the wool and hair fibers as well as the easy harvest of wool, hair, and fiber (Powell et al., 1994).

Animal productivity and reproductive efficiency share a positive relationship with each other. The reproductive efficiency of livestock is defined as the number of offspring per breeding animal in unit time (Dziuk and Bellows, 1983). Reproductive traits for higher fertility, superior growth, and increased product yield are identified, selected, and manipulated. Traits that influence livestock production are breeding per conception, scrotal circumference, teat placement, fecundity (increased ovulation rate), age at puberty, birth weight, litter size, litter weaning weight, multiple births, hatchability, egg production, and their weight. Transgenic technology has been instrumental in enhancing the growth rate of livestock by manipulating growth hormones, growth hormone receptors, and growth modulators. The production of growth hormone (GH) transgenic pigs leads to improvement in the growth rates as well as feed efficiency in pigs.

Much progress has been made in producing genetically superior or healthier animals by introducing traits for disease resistance. When introduced into transgenic dairy cows, an antimicrobial peptide called lysostaphin is secreted into the milk and confers protection/resistance against a major mastitis-causing pathogen, *Staphylococcus aureus*. Prions, the causative agents for bovine spongiform encephalopathy (BSE) or mad cow disease in cattle, are being targeted to produce prion-free livestock (Richt et al., 2007).

Approaches to improve livestock production

Time, cost, ethics, benefits, and risks render the projects based on genetic modification of livestock a difficult task. Transgenic techniques/procedures are currently being employed for the production of superior breeds. Nevertheless, animal welfare, animal heath, animal care, social benefits, consumer health, safety, and ethics are major concerns that cannot be ignored. Testing of offspring for the selection of desirable traits incorporated in modified animals (individual animal/group of animals) not only differentiates them from their traditional counterparts but is also an essential procedure from a safety point of view. Once tested safe, these animals can be potentially used as breeding animals during propagation for the production of superior offspring. Trait selection can be natural if it is influenced by natural forces or artificial if it is influenced by human efforts. Thus, trait selection may result in increased frequency of desirable genes in individuals. Offspring obtained from these breeding animals with superior features are subjected to evaluation/testing.

Marker-assisted selection

Quantitative traits such as carcass composition, growth rate, and milk and egg production are controlled by multiple pairs of genes. These many heterozygous pairs of genes, therefore, may lead to the extrapolation of billions of possible genetic combinations. This makes it difficult to screen and select individuals as breeding stock. However, the use of molecular markers in combination with quantitative traits can overcome the limitation of the successful selection of superior species/individuals. Molecular markers are used as tags for chromosome segments and therefore, they are conveniently used to identify and locate genes of interest (Kennedy et al., 1990). Marker-assisted selection enables the rapid screening of populations and improved selection of breeding animals for desirable quantitative/production traits (Gelderman, 1975). Potential gene loci markers are exploited to select the breed and genetically engineer livestock for high reproductive efficiency, quantitative traits, and disease resistance. These molecular markers include growth hormone and growth factors, including: (i) prolactin; casein (α, β, and κ); dystrophin; interferon receptor; interferon (α, β, and γ); cytochrome P450; immunoglobulin genes; parathyroid hormone; insulin; follicle-stimulating hormone beta polypeptide; bovine leukocyte antigen [BoLA: major histocompatibility complex (MHC) system in cattle, mapped to bovine autosome 23]; and swine leukocyte antigen [SLA, the MHC system in swine, mapped to pig chromosome 7] (Wheeler and Campion, 1993). Variations in the traits among livestock species can be explained on the basis of association between polymorphisms in the above-mentioned genes (allelic differences) and the type of quantitative trait.

Transgenic approaches

Tools of genetics, nutrition, and animal breeding play significant roles in the selection, propagation, and management of economically important characteristics in livestock, including improvements in product composition, growth, reproduction, and resistance to disease. Genetically modified or transgenic livestock can be produced by allowing the stable integration of transgene [gene(s) of interest] into their germ line for successful transfer into the offspring down through multiple generations. Many methods are available that allow the rapid transfer of modified gene(s) into livestock without carrying out cross-breeding or hybridization. These include: (i) the use of recombinant retroviruses; (ii) pronuclear injection; (iii) embryonic stem cells; (iv) sperm-mediated gene transfer (SMGT); (v) male germ cell transplantation; and (vi) nuclear transfer/somatic cell cloning (Wheeler and Walters, 2001; Wall, 2002; Wheeler et al., 2003, 2010).

Retroviral vectors are produced by replacing the retroviral sequences to create replication-defective vectors. Large amount of foreign DNA can be accommodated in replication-defective vectors and retroviral proteins can be produced. These vectors can infect the host germ cell and transfer the gene of interest for its stable integration into the host genome. Germline chimeras are inbred for 10–20 generations to obtain homozygous transgenic animals. Transgenic embryos can be cryopreserved for subsequent implantation.

The pronuclear injection method is a pioneering method that has proved to be effective for mammals. The pronucleus of fertilized ovum is directly microinjected with the gene construct derived from either the same species or different species. The manipulated fertilized ovum is then placed into the oviduct of the recipient female/foster mother for the embryo to develop (Gordon and Ruddle, 1981).

In embryonic stem cell-mediated gene transfer, the desired gene sequence is inserted into an in vitro culture of embryonic stem cells by homologous recombination. These cells are incorporated into an embryo at the blastocyst stage of development. This method is generally used to study the genetic control of developmental processes.

Sperm-mediated gene transfer is based on the efficiency of sperm cells to bind and internalize the exogenous/foreign DNA of interest and transport it into an oocyte during fertilization to produce modified animals. Owing to the poor ability of spermatozoa to uptake exogenous DNA, the chances of fertilization of an oocyte with transfected spermatozoa are reduced (Lavitrano et al., 2013).

Male germ cell transplantation uses testis stem cells that are obtained from a donor animal. Upon injection into seminiferous tubules, these donor stem cells migrate from the lumen and relocate to the basement membrane. These cells retain the capability to produce donor sperm in their new host. This technique has been reported to be successful in goats but not yet in cattle and pigs. Culturing the testis stem cells on a large scale may further enhance the efficiency of transgenic production (Hill and Dobrinski, 2006).

The technique of somatic cell cloning is based on the transfer of the DNA/nucleus of the somatic cell into an enucleated metaphase-II oocyte. The resultant embryo is allowed to develop to a blastocyst in vitro and then transferred into a recipient for the cloned animal to be born after the completion of gestation (Fig. 1).

The new individual thus produced is genetically identical to the somatic cell donor. The cloning of an entire sheep, such as the famous Dolly, from a differentiated adult mammary epithelial cell is possible by nuclear reprogramming, where differentiated cells revert back to a totipotent status. This method of cloning makes it possible to generate multiple copies of genetically elite farm animals (Tian et al., 2003; Menchaca et al., 2016a).

Propagation

Following the screening and selection of individuals with desirable traits, individuals may be propagated by natural mating or artificial mating to produce next-generation offspring. Owing to the limited number of offspring produced by cows (one per year), sheep (two per year), swine (litters

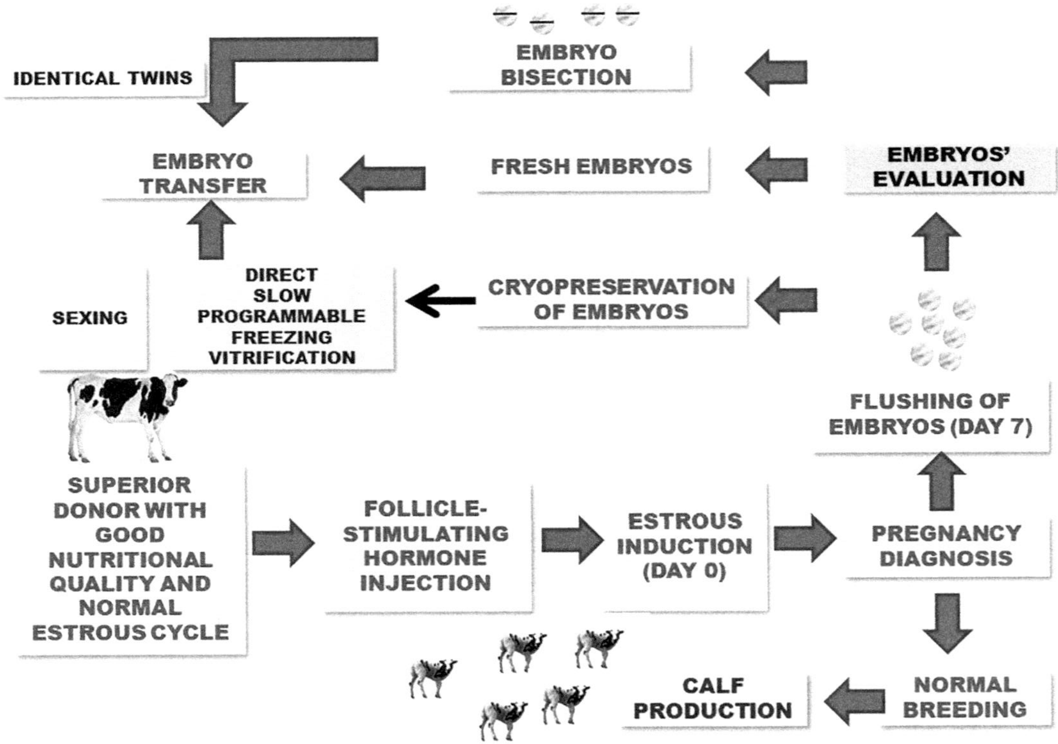

Fig. 1 Somatic cell cloning.

of 8–12 pigs twice per year), mares (one foal per year), and natural mating (number of females may be bred to a single male), this may not be suitable to produce many offspring. However, 500–100 females may be bred to a single diluted semen sample collected from a fertile bull by artificial mating or artificial insemination. This has been practiced in the cattle industry for many years to increase the frequency of desirable traits in beef and dairy populations (Wheeler and Campion, 1993). Globally, the number of embryo transfers reported in small ruminants (sheep and goats) is less compared to that in cattle. Under conventional breeding conditions of sheep and goats, the female produces 1–2 offspring per year or 6–8 offspring during the female's reproductive lifetime. However, embryo transfer procedures drastically enhance the reproductive potential of genetically superior females via increased oocyte reserve in the ovary. The first report on embryo transfer in sheep and goats dates back 86 years (Warwick et al., 1934). With continual genetic improvements in different breeds, embryo transfer methods ensure an alternate mechanism for milk, wool, meat, and hair production. Currently, embryo collection and transfer in sheep and goats exclusively depend upon surgical and/or laparoscopic methods. Embryo collection and transfer in pigs has not been intensively taken up on a commercial basis. Through 2012, Canada, Ireland and France were the only countries that reported the commercial collection and transfer of swine embryos. The introduction of superior genes into the sow herds with a minimal risk of disease by means of artificial insemination is practiced in countries with intensive pig production (Maes et al., 2008).

In the equine industry, the primary purpose of embryo transfer is to obtain offspring from those mares who have limited reproductive potential. The mares of limited reproductive potential either are of old age (enough old to continue to compete), or are diagnosed with uterine pathologies. Sometimes, these mares are identified with undiagnosed subfertility. Owing to the recent acceptance of foals born through embryo transfer procedures by many breed associations/registries, horse breeders obtain several foals per donor mare in one breeding season.

The subsequent section discusses in detail the strategies/techniques for the commercial production of farm animals.

Superovulation treatment, estrous synchronization, and embryo collection

Drugs/hormones are administrated in most of the mammals to superovulate the females to increase the probability of obtaining and collecting multiple embryos per procedure. The administration of hormones to superovulate donor cows is done according to standardized protocols/methods. The intramuscular injection of equine chorionic gonadotropin (eCG) hormone (2000–2500 IU) on the 10th day of the estrous cycle (zero day: the day when estrous is observed in cows). After 2–3 days, two injections of prostaglandin $F_{2\alpha}$ (dinoprost or cloprostenol) with a gap of 12–24 h are administrated. Another common method of superovulation is based on the administration of follicle-stimulating hormone (FSH). The superovulatory response induced by eCG treatment is greater than that from FSH treatment. Nevertheless, FSH treatment results in the production of good-quality transferable embryos; therefore, FSH treatment has become the method of choice at the commercial level. The low level of contamination of commercial FSH preparations made from porcine pituitary extracts with luteinizing hormone (LH) may be needed for superovulation. Differences in follicular counts and the presence of a dominant follicle in the donors could be responsible for the differential response of individual donors to superovulation treatment (Bó et al., 2010). Many advancements/variations to this method are known to increase superovulation. A new follicle wave can be induced by injecting estradiol or gonadotropin releasing hormone (GnRH) or transvaginal ablation of dominant follicle (follicle ablation), also known as dominant follicle removal (DFR) are some of the strategies to induce a new follicle wave and thereby to increase the ovary response to superovulatory treatments. These treatments promote the recruitment and growth of more uniform follicles and result in more viable embryos per procedure (Fig. 2).

Perfectly synchronized recipients are preferred for embryo transfer. Recipient cows ovulating the same day as donor cows exhibit high pregnancy rates. However, an asynchrony of ± 1 day in the recipient cows compared to the donor cows has also shown high pregnancy rates. Synchronization in recipient cows can be done by either applying synchronization protocols or by observing the natural estrous cycle. In both cases, the recipients are injected with luteolytic doses of $PGF_{2\alpha}$ 12 h prior to the first treatment of donor cows with $PGF_{2\alpha}$. Following this treatment, 70%–90% of the flushed cows are expected to produce embryos of transferable quality. Cows that respond to FSH treatment may produce none to more than 20 good-quality embryos per flush. An average production of 5–7 good-quality embryos per embryo collection is considered a commercially successful outcome. Uterine stage embryos (morula and blastocysts) are expected to recover on the 7th day of the cycle. At this point, the cow uterus is flushed and good-quality embryos are collected. The quality of the embryos depends upon the morphologic

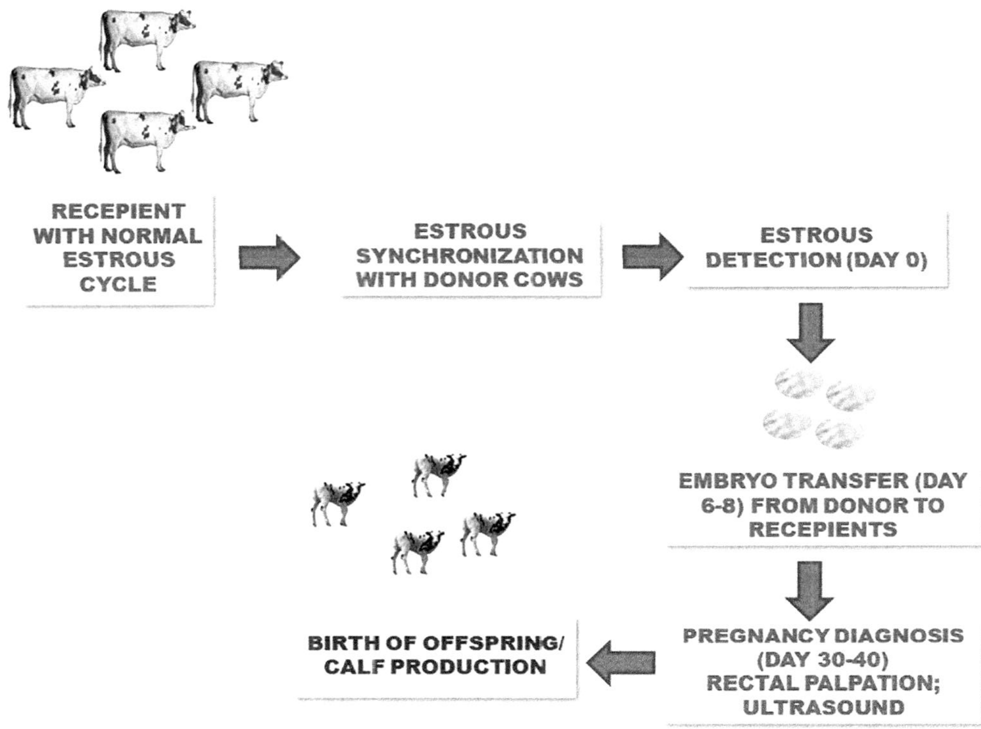

Fig. 2 Superovulation treatment.

assessment of the physical integrity of the embryos. The stage of embryonic development, the compaction status, the size and sphericity of the embryo, the number of extruded blastomeres, the size of the perivitelline space, the color of the cytoplasm, and the areas of cellular degeneration are some of the morphological characteristics used to assess the quality score of the embryos (Fig. 3).

Synchronization of donors and recipients with small ruminants (sheep and goats) is done using progestagens by injecting prostaglandin F_{2alpha} ($PGF_{2\alpha}$) on the day the intravaginal progestagen device is inserted. FSH is commonly used in small ruminants (sheep and goats) for superovulation (Jabbour et al., 1991). Like cattle, it is administered two times for 3 days with decreasing doses (1st day: 5, 5 mg; 2nd day: 3, 3 mg; and 3rd day: 2, 2 mg). Removal of the progestagen source and luteolytic injection of $PGF_{2\alpha}$ is given on the last day of FSH treatment (Cognié et al., 2003). Estrous is detected by vasectomized bucks/rams. Embryos are collected via surgical methods. Nonsurgical transcervical catheterization and laparoscopic procedures are constantly evolving. Although these methods yield good results, embryo collection is poorer when compared to surgical methods. Collection of embryos is done on the 7th–8th day after the estrous cycle.

Superovulation treatment in pigs may not be useful in embryo transfer procedures due to the high ovulatory rate in sows (female pigs). A genetically valuable sow is slaughtered to collect the reproductive tract for embryo retrieval. Ventral laparotomy of the caudal abdomen can be performed to surgically collect the embryos. Embryos are typically collected 6–7 days postovulation.

Owing to difficulties with manufacturing suitable gonadotropin preparations and the differential responses of mares to FSH stimulation, the commercial application of superovulation treatment is restricted to produce equine embryos. An inability of the ovarian fossa to accommodate multiple ovulations could be responsible for the low recovery rate of embryos in mares ovulating more than four follicles per ovary. However, according to another school of thought, the low recovery rate of embryos is due to dysmatured oocytes because of the excessive response to FSH stimulation in horses. In vitro equine embryos have recently been produced by using intracytoplasmic sperm injection (ICSI) in commercial settings. Only a portion of a frozen semen straw from a deceased stallion can be used to provide the sperm needed to produce in vitro equine embryos by ICSI. There has been a sharp rise in the number of ICSI-derived embryos produced over the past few years. Breeding soundness evaluations are necessary for both donors

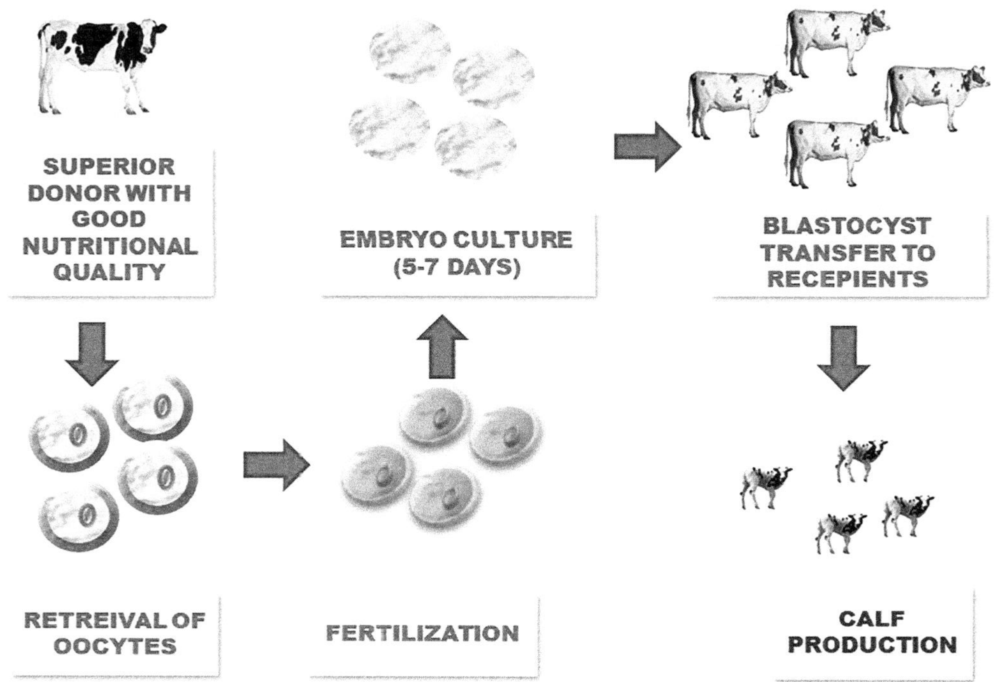

Fig. 3 Estrous synchronization.

and recipients. One of the two embryos is generally recovered from the donor mare. Accurate information about the time of ovulation is essential to determine the day of ovulation and the day for embryo collection. This information also helps in the selection of qualified recipient candidates for equine embryo transfer. Ultrasonography of the uterus and ovaries of the donor and potential recipient mares on a daily basis can provide critical information about the timing of ovulation and embryo collection. Before embryo collection, palpations per rectum and ultrasonography are necessary to observe the presence of corpus luteum, the absence of any fluid in the uterus, and the degree of cervical tone. Nonsurgical methods are applied to collect and transfer the equine embryos. Embryo collection is generally done on day 7 or day 8 postovulation. With good standard techniques, the embryo recovery rate is generally 75%; however, it could be as high as 90% with young maidens or fertile mares or as low as 10%–20% in subfertile mares.

While working with uncooperative mares, if sedation is required, then acepromazine should be avoided. However, an α_2-agonist (xylazine, detomidine) is preferred. Acepromazine results in the relaxation of the uterus, potentially impairing fluid recovery, and therefore, manipulation of the reproductive tract during flushing of the uterus. Embryo collection is done by the transcervical uterine lavage procedure, which uses a sterile silicone catheter of variable size with an inflatable cuff (the size depends upon the size of the reproductive tract of the mare). The sterile silicone catheter is then introduced through the cervix into the uterus. It is then inflated like a balloon containing 60–80 mL of air or collection medium. The catheter is gently pulled backward to allow the cuff to seal the internal cervical os (cervical opening into the vagina). The uterus is infused with 1–2 L of flush medium, and then that medium is allowed to flow backward to recover it. To facilitate fluid recovery, the uterus is gently manipulated per the rectum. Sometimes, an injection of oxytocin (10 IU) is also given to minimize the retention of a large volume of flush media in the uterus. The complete embryo collection procedure (flush) requires the infusion of the uterus with a total of 4–8 L of flush medium, followed by its 100% recovery from the uterus. Transrectal ultrasonography may be performed to check for the presence of any intraluminal fluid.

Cryopreservation

The international trading of frozen embryos results in the worldwide diffusion of the germplasm without any health risks. Good-quality embryos can be stored in a holding medium at room temperature until they are transferred to recipients or can be stored at low temperature for future use by applying various cryopreservation methods. Refrigeration of embryos is possible for 24 h without an appreciable loss of viability. In Asia, about 70% of the transferred embryos are cryopreserved/frozen compared to the other regions of the world. The direct transfer method as a freezing method uses ethylene glycol as the cryoprotectant (Dochi et al., 1995). This method is considered the standard cryopreservation method for the storage of bovine embryos worldwide. Survival rates of cryopreserved embryos after freezing and thawing are lower compared to in vivo derived embryos when these embryos are taken for in vitro fertilization (IVF), the determination of gender by embryo sexing, and somatic cell nuclear transfer. Therefore, a low pregnancy rate is expected for in vitro manipulated embryos. Another method of cryopreservation can overcome this problem. The vitrification method is more efficient and requires the use of liquid nitrogen. High concentrations of cryoprotectants are required to vitrify the embryos without the formation of ice crystals by rapid cooling (Kuwayama et al., 2005; Sripunya et al., 2010). Conception rates of 40% for in vitro produced (IVP) embryos and 50% for sexed embryos can be efficiently obtained (Inaba et al., 2011; Sano et al., 2010). Artificial surgical splitting/bisection of embryos at the cleavage of the blastocyst stage results in the formation of identical twins. This results in double the number of embryos with only a minor reduction in pregnancy rates (Martinez et al., 2006).

The relatively large diameter of equine embryos and the presence of an embryonic capsule limit the interaction of embryos with cryoprotectant agents. Hence, the cryopreservation of equine embryos is difficult. Embryos of the morula or early blastocyst stage obtained 6.5 days postovulation are preferred for cryopreservation. Older embryos with poor morphological characteristics show low pregnancy rates. The cryopreservation of equine embryos using commercially available vitrification media has shown to be an efficient procedure for their storage. Approximately 40%–50% pregnancy rates are reported with cryopreserved embryos. Embryos with the collapsed blastoceles during embryo biopsy procedures are also reported to be successfully cryopreserved.

In vitro fertilization

Many animal breeders are using in vitro fertilization to maximize pregnancy rates. This technique is advantageous if the uterine is damaged or the fallopian tubes are blocked. It is the process of harvesting oocytes from superior donor animals followed by the fertilization of oocytes with semen in a petri dish to create embryos. These embryos are either taken for cryopreservation until further use or are placed in the uterus of recipient/surrogate mothers.

In cattle, eggs are removed from the donor cows before the oocytes get naturally released from the ovary down the fallopian tube. The process of harvesting unfertilized oocytes directly from the ovaries of the donor cow is known as aspiration or ovum pick up (OPU). The dominant follicle in the ovary inhibits the rest of the follicle during normal pregnancy. All the follicles are left for 6 days for their uninhibited growth within the ovary by a process called dominant follicle regression (DFR). The dominant follicle is removed and then the donor cows are taken for superovulation stimulation. Oocytes are harvested by the transvaginal recovery process. Ten eggs are collected on average per collection and then the eggs are matured for 20 h. Eggs are placed into insulated chambers at 37°C, which mimics the cow body temperature. Embryos are produced by carrying out in vitro fertilization with good-quality semen. Embryos are matured for a week under laboratory conditions and passed through eight different media, which mimic the gas levels and changes in pH in the uterus. Embryos can be transferred directly or frozen indefinitely. Seven days after the recipient's standing heat or estrous, the embryos are transferred into the recipient cows. Because the collection of eggs from donor cows is possible from the age of 10 months to 20 years or more, in vitro fertilization extends the breeding life (Fig. 4).

The success of the in vitro production of embryos in sheep and goats mainly depends on the availability of good-quality oocytes. Follicular aspiration by laparoscopy (laparoscopic ovum pick-up; LOPU) is performed to retrieve oocytes from live animals. This technique is better over the traditional laparotomy and was first described by Tervit et al. and Baldassarre et al. in 1990s (Tervit et al., 1992; Baldassarre et al., 1994; Menchaca et al., 2016b). With the help of forceps, the ovarian surface is exposed by pulling from the fimbria in different directions, and follicles are punctured one by one by the aspiration pipette. It takes 1–20 min to complete the procedure with a recovery rate of the cumulus-oocyte complex (COC) of 50%–90%. LOPU can be performed in prepubertal females of 1–3 months and can be repeatedly done in the same female with a time gap of 4 days (Gibbons et al., 2007; Paramio, 2010). An in vitro maturation medium (IVM) is used for the incubation of good-quality oocytes in groups of 25–30 per 100 μL under mineral oil for 24 h at 39°C in a humidified atmosphere containing 5%

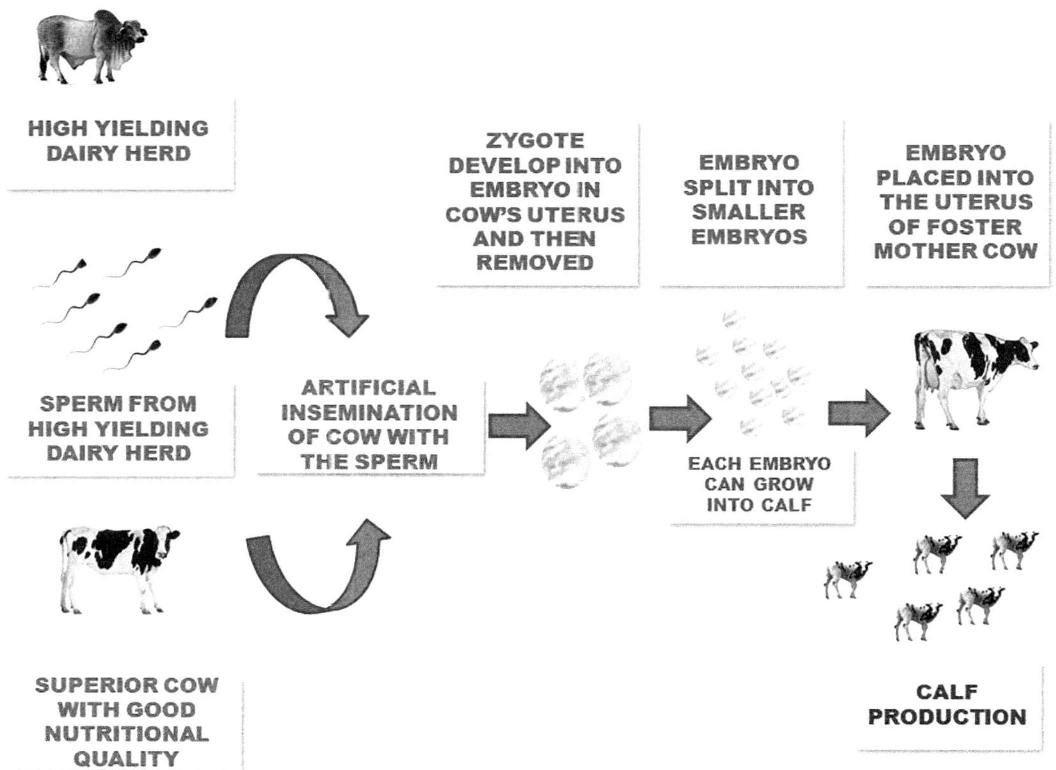

Fig. 4 In vitro fertilization (IVF).

CO_2. Strategies to prepare ovine/caprine semen for in vitro fertilization are similar to those used in cattle. The maintenance of cumulus cells during in vitro fertilization has been found to be associated with greater blastocyst yield and high developmental rates (Menchaca et al., 2012; Souza et al., 2013). The insemination dose consists of 10^6 total sperm per drop of 100 µL synthetic oviduct fluid under mineral oil and 20–30 oocytes. The coincubation of oocytes and sperm is done for 18–22 h at 39°C in a humidified atmosphere containing 5% CO_2. Embryos are cultured under standard conditions and are transferred into the recipient animal blastocyst stage instead of the morulae stage. The recipient female should produce sufficient progesterone and provide a suitable uterine environment for embryo development. The number of embryo transferred per recipient depends upon the species, breed, and reproductive system (Menchaca et al., 2016b).

The first report on the in vitro production of piglets came in 1986 (Cheng et al., 1986). During in vitro production of pig embryos, the good-quality eggs or oocytes as well as nurse cells that surround them are extracted from the female pigs, either by surgical procedures or from slaughterhouses. Follicles between 3 and 6 mm should be aspirated. They are placed in a chemical environment for maturation under proper conditions. Meiosis, the reduction of chromosomes to half, the extrusion of the first polar body, and the reorganization of all organelles are some of the changes that take place in immature cumulus oocyte complexes, which prepare them to finally receive the spermatozoa (Mehlmann, 2005). A highly motile sperm population is isolated, which reduces the occurrence of polyspermy in pig oocytes during in vitro fertilization and results in the improved quality of porcine embryos (Park et al., 2009; Holt et al., 2010). At the end of pig oocyte maturation, oocyte denudation is performed to separate the cumulus oocyte complexes from cumulus cells. This is followed by the incubation of oocytes with sperms of concentration 1 to 4.5×10^5 cells/mL in a defined media composition at 38.8°C under a humidified atmosphere of 5% CO_2, for 18–24 h. Zygotes undergo cleavage, then develop into morulae and, finally, into a blastocyst (Cheng et al., 1986; Mattioli et al., 1989; Nagai et al., 2006). The ideal culture time for the porcine embryos is 5–6 days. High-quality blastocysts/embryos (with numerous nuclei and a lower apoptotic index), if transferred, increase the success rates of pregnancy. In vitro production procedures provide viable porcine embryos more successfully and efficiently in less time on a large scale compared to the surgical collection of in vivo derived embryos from sows (Romar et al., 2019).

Assisted reproductive techniques in horses allow the production of foals from subfertile stallions and mares who are otherwise not capable of producing offspring with conventional breeding. Unfertilized eggs or oocytes are harvested either directly from the ovaries of mares postmortem or by performing ultrasound-guided transvaginal oocyte aspiration (TVA) in live mares. TVA can be done when appropriate follicle sizes and numbers are present for aspiration or every 14 days (Hinrichs, 2012). Mature oocytes from large preovulatory follicles (oocyte recovery rate is 75% or more) and immature oocytes from small to medium follicles (oocyte recovery rate is 50%–60%) can be aspirated (Carnevale et al., 2005; Colleoni et al., 2007; Jacobson et al., 2010). The tight adherence of the immature oocyte to the cumulus cells within the follicle could be the possible reason for the difference in oocyte recovery rates. Because in vitro fertilization is difficult to repeat in horses and is unpredictable, intracytoplasmic sperm injection (ICSI) is performed for the in vitro production of embryos. After injecting a sperm into a mature oocyte with the help of a micromanipulator, the oocyte is placed into a culture medium and incubated for the embryo to develop. The embryo then can either be transferred into a recipient mare or cryopreserved. The invasive nature of the procedures, TVA and ICSI, and the cost involved are limitations. Intraabdominal hemorrhages, rectal tears, peritonitis, ovarian abscesses, and colic are some of the complications associated with TVA (Carnevale, 2008).

Artificial insemination

The high success rates of pregnancy in animals depend upon the insemination of females at the correct time. The fertilization of female donors can be done naturally or by artificial insemination with fresh or frozen semen. The fertilization efficiency in multiovulated females varies. It depends on the fertilization technique used, the individual ovulatory response to hormonal treatment, and the time of artificial insemination. A ow fertility rate is possible in donors with a high ovulatory response rate (more than 10–12 ovulations). This could be due to a decrease in the sperm transport in the reproductive tract.

In cattle, the correct positioning of the cow is very important. Artificial insemination is successful when cows are relaxed and provided with adequate facilities, especially food and water. Extra care should be taken with the sexed semen. Thawing of the straw using lukewarm water is followed by placing it in the gun. The gun is inserted as soon as it is primed to preserve the quality of the semen. It is inserted almost vertically past the vestibule and into the vagina, then finished horizontally. The gun is lined up with the cervix to manipulate the cervix back over the tip of the gun. Care should be taken while slowly placing or depositing semen in the short chamber of the uterine horns, which are located on the other side of the cervix. Poor semen storage, incorrect timing of service, incorrect gun placement, serving cows that are already pregnant, and poor record keeping minimize the potential benefits of artificial insemination in cattle (Fig. 5).

Artificial insemination can be done in sheep or goats 12–24 h after estrus detection or 40–50 h after progestogen removal. To execute intracervical or transcervical artificial insemination, advanced training and practice are required. Semen deposition in the cranial aspect of the uterine horns close to the fertilization site during laparoscopic artificial insemination increases fertilization rates, reduces the insemination doses required, and increases pregnancy rates (Vallet et al., 1987; Wolff et al., 1994).

The outcome of artificial insemination in swine depends on the semen quality and the insemination procedure. Contaminated/diseased semen, if collected from boars, could result in the transmission of pathogens and disease outbreaks in different sow herds. The fertilizing capacity of produced semen doses is linked to the quality of spermatozoa as well as the semen processing procedure (Waberski et al., 2008; Tsakmakidis et al., 2010). Semen collection in artificial insemination centers is performed twice a week. The collected sperm-rich fraction (40–100 mL) is chalky in appearance, contains 80%–90% of the sperm cells in the normal ejaculate, and is sufficient to inseminate 15–25 sows. Observed variations in the number of spermatozoa among different breeds of pig, including Landrace, Duroc, and Yorkshire, influence semen dose production (Kommisrud et al., 2002). Coulter counters (determine the number of particles within known volume), computer-assisted semen analysis (CASA, determines the sperm concentration based on an image analysis of the sperm within counting chambers), and nucleocounters (DNA is fluorescently labeled and counted by image analysis) are used to accurately determine the sperm concentration. Freshly diluted semen (liquid semen) on the day of collection or in the following days is preferred over frozen storage boar semen for artificial insemination. At the time of insemination, a high level of boar stimuli increases the frequency of uterine contractions and supports passive sperm transport through the long uterine horns. Ultrasound testing can be used to determine the time of ovulation in relation to estrous behavior, insemination, and fertilization. Compared to standard transcervical artificial insemination,

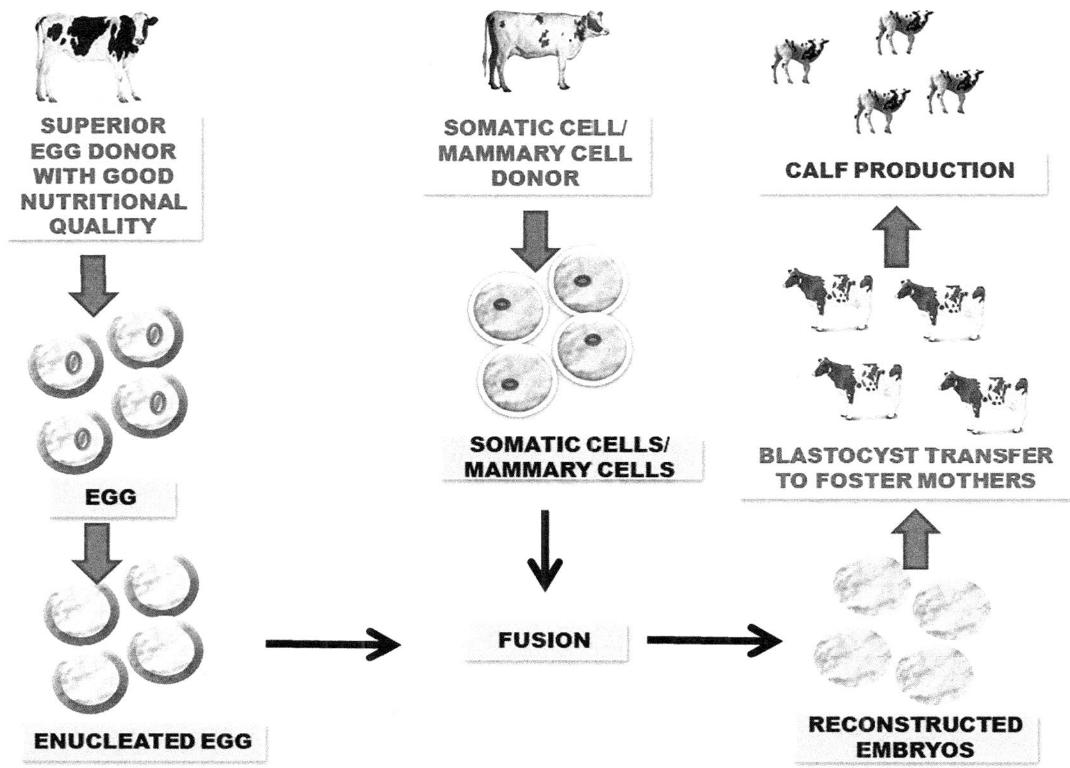

Fig. 5 Artificial insemination.

postcervical and deep intrauterine artificial insemination allow a 3-fold and a 5–20-fold reduction in the number of spermatozoa to be inseminated, respectively. Both methods of artificial insemination are developed to traverse the cervix and deposit the sperm in the uterine body or posterior horn of multiparous sows (Vazquez et al., 2008).

Higher conception and foaling rates are observed when normal mares are bred with artificial insemination compared to the breeding of normal mares by natural service. Before breeding, irrespective of age, mares should be examined for reproductive soundness, which includes an examination of the reproductive organs, cervical dysfunction, routine culturing of the uterus to assess uterine cytology, and an endometrial biopsy to check for any alterations that can affect fertility. A donor stallion with strong semen quality is chosen. The number (250–500 million motile sperm recommended) and volume (10–25 mL) of sperm; the number and timing of inseminations per cycle inseminated (viable sperm and oocyte must be in the reproductive tract of the mare at the same time); and the management (palpation and ultrasonography to manage the mare) and fertility of the mare and stallion all affect the conception rate. The highest fertility rate is recorded if insemination is done within 1 day before ovulation. The equine rectum is not as durable as in bovines, hence unlike artificial insemination in cattle, equine artificial insemination is completely carried out vaginally. Semen is aspirated into the breeding catheter and passed through the cervix into the uterus. Semen is deposited slowly by depressing the plunger of the syringe. The catheter is removed from the vagina and an ultrasound is performed in 12–14 days to check for pregnancy.

Embryo transfer procedures

Embryo transfer procedures via the use of ovulation-stimulating hormones and artificial insemination enable genetically screened superior females to produce hundreds of thousands of offspring per year (Seidel Jr., 1981). It helps to disseminate desirable genes from superior female animals of various species. Breeding programs are developed in cattle, particularly the dairy industry, to promote desirable traits from elite females to offspring through multiple ovulation embryo transfer (MOET) programs. Besides their utility in farm animals, embryo transfer protocols are also extensively used in biology and medicine to study fetal maternal interactions, develop models of humans and animals diseases, produce transgenic animals, and use animals as bioreactors for producing therapeutic proteins.

Embryo transfer procedures employ the transfer of recovered preimplanted embryos in the reproductive tract of donor animals to the reproductive tract of recipient animals to produce progenies of donor animals. The synchronization of estrous, superovulation treatment, embryo flushing, genetic testing for the evaluation of embryos, and the pretransfer testing of ovaries and uteruses are some of the necessary measures required during embryo transfer. Cryopreservation, sexing, and embryo bisection further improve the utility of this technique. Embryo bisection or splitting results in the production of identical twins from a single embryo. Nuclear transfer or cloning techniques are an extension to embryo transfer procedures. Cloning is based on the removal of DNA from the oocyte (unfertilized egg) followed by the transfer of the nucleus from mammary cells into the empty oocyte. These newly constructed cloned cells divide normally into embryos and have the potential to produce thousands or tens of thousands of offspring when implanted in the uteruses of female mammals (Prather et al., 1987).

Nonsurgical techniques are the current most preferred methods of cattle embryo transfer into synchronized recipients. Good-quality embryos are loaded into small plastic straws of small volume (0.25 mL) fitted into a specialized insemination gun/embryo transfer (ET) gun. They are then taken for transfer by placing the ET gun into the deep uterine horn by extending Teflon or silicon tubes from the tip of the guns on the ipsilateral side of the ovary with the corpus luteum through the cervix without damaging the endometrium. To prevent contamination with potential perineal and vaginal contaminants, the tip of the gun is covered with a disposable sterile sheath. Of the recipients, 60%–70% are expected to become pregnant upon direct transfer of single fresh embryos. However, pregnancy rates increase to 90% when two embryos are transferred. Despite precautions as well as good management by well-trained staff, the pregnancy rates derived from frozen-thawed embryos are expected to be 50%, which is 10% lower compared to those derived from fresh bovine embryos. Increased pregnancy rates could be possible with the transfer of multiple embryos. However, high pregnancy rates are associated with a risk of dystocia, the retention of fetal membranes, and the production of freemartin calves.

Surgical procedures yield an average efficiency of 60%–70% of the embryos in sheep and goats. A drawback to this procedure is the formation of postsurgical adhesions. This significantly reduces the recovery of embryos to 66%, 41%, and 35% during the first, second, and third surgical procedures, respectively. The nonsurgical or laparoscopic technique was developed by McKelvey et al. in sheep and by Vallet et al. in goats (McKelvey et al., 1986; Vallet et al., 1987). This procedure is expensive and requires skilled personnel. The yield of recovered embryos is around 60%. It reduces adhesion formation. Therefore, the percentage of embryo recovery does not decrease during subsequent operations. The average time for embryo recovery by a surgical procedure is 15–20 and 20–30 min by laparoscopy. The embryo survival rate is high when blastocysts are collected on day 5 compared with retarded morulae collected on day 6. Within 2 h of embryo collection in the preserving medium, they should be taken for transfer. The embryo transfer is performed in the uterine horn ipsilateral to the corpus luteum. Surgical or nonsurgical (laparoscopy) procedures can be used for embryo transfer (González et al., 1991). Embryos conditioned in phosphate buffer saline are in the uterine lumen by means of a micropipette. In surgical or nonsurgical procedures, a puncture is performed on the dorsal side of the uterine horn in its upper third portion. However, in the semisurgical method of embryo transfer, a small incision of 1 cm is made in the abdominal mid-line and the uterine horn is exposed using a clamp. In dairy goats, it is possible to obtain a 100% survival rate when two embryos are transferred per recipient (Tervit et al., 1983). However, in sheep, the global efficiency (lambs produced/transferred embryos) is high when one embryo is transferred per female recipient (Cseh and Seregi, 1993).

In most cases, a surgical technique including abdominal ventral laparotomy under general anesthesia is performed for the transfer of pig embryos. Nonsurgical techniques have recently been developed successfully to collect and transfer porcine embryos. Nevertheless, these methods require highly trained and skilled personnel as well as specialized equipment. Pregnancy rates are higher when embryos are transferred by surgical methods compared to nonsurgical methods. To achieve optimal pregnancy rates, it is highly recommended to transfer 16–22 embryos to each recipient.

Pregnancy rates in horses are comparable with either cryopreserved embryos (stored for a short time; embryos are shipped to the embryo transfer facility the day they are collected) or fresh embryos. Nonsurgical methods are preferred over surgical methods of equine embryo transfer. The nonsurgical procedure is performed using transcervical catheterization. Embryos are loaded onto plastic straws of 0.25 or 0.5 mL volume. Plastic straws holding the embryos are then fitted with the embryo transfer gun. During the procedure, embryo transfer sheaths with side delivery can minimize their damage.

Use of growth enhancers

Over the past many years, growth-promoting antibiotics, antibacterial drugs, antiparasitic drugs, xenobiotic agents, anabolic agents, and ionophores have been extensively used in livestock to increase total meat and milk production as well as to improve/modify the nutrient composition of animal products. The incorporation of tetracycline compounds including oxytetracycline and chlortetracycline into the diet of cattle/farm animals has resulted in an increase in g the growth rate and feed conversion rate by 5%. Besides the beneficial effects, antibiotics also protect animals from bacterial and parasitic infections as well as reduce the incidence of liver abscesses, thereby increasing the production efficiency (Wieser et al., 1966). Differences in sex contribute to the differences in growth rate, feed efficiency, and product composition in both ruminant and nonruminant animals. A number of experimental studies are being carried out to prove the effects of steroid hormones on livestock production as well as product composition. The implantation of diethylstilbestrol, a synthetic estrogen, has been shown to stimulate the growth of heifers. Its oral administration stimulates growth in cattle and lambs. The use of commercially available forms of androgens and estrogens has been shown to cause a daily gain by 8%–15% and improvement in feed efficiency by 5%–10% in beef cattle and lambs (Roche and Quirke, 1986). The use of anabolic steroids has been shown to increase the carcass yield and total carcass protein in ruminants. Nevertheless, no improvement in growth performance has been reported in swine upon administration of these compounds. The growth hormone somatotropin (a peptide) mediates the increase in growth rate and improvement in carcass composition. The naturally occurring catecholamines epinephrine and norepinephrine, also known as β-agonists, share structural and pharmacological properties with substituted phenylethanolamine compounds (Ricks et al., 1984). A profound effect of substituted phenylethanolamine compounds on the growth of skeletal muscle and fat deposition in livestock has been examined. The effect of β-agonists in repartitioning tissue growth is considered an important breakthrough in increasing the efficiency of livestock production. The supplementation of animal feed with certain additives called feed additives has been examined to increase animal productivity. Vitamins and trace minerals are considered essential feed additives as they are required for the maintenance of the biological function of animals. Another category of feed additives is nonessential feed additives, as they are not required to maintain the biological function of animals but do have positive effects on healthy animals. These include: (i) absorption enhancers; (ii) antimicrobial agents; (iii) ionophores; (iv) organic acids; (v) metabolic modifiers; (vi) growth promoters; (vii) probiotics; (viii) crystalline nonessential amino acids; (ix) antioxidants; (x) preservatives; (xi) dust preventives; (xii) prophylactics; (xiii) urea; and (xiv) propyleneglycol (Wheeler and Campion, 1993).

High-producing animals/ruminants require more energy. They derive energy from cereals/feed but cannot extract much energy from the fibrous or roughage portion. To reduce competition between animals and humans for high quality and quantity of cereals, it is important to improve their digestive ability for plant cell wall carbohydrates. Pretreatment of animal feed with sodium hydroxide or alkaline hydrogen peroxide improves the fiber digestion property of animals. Advancements in recombinant DNA technology open new avenues to enhance the digestive ability by manipulating the rumen microbial environment for enhanced cellulolytic activity and reduced proteolytic activity or by altering the fermentation characteristics of microorganisms for reduced methane production.

Conclusions

Success in the selection of desirable traits and the propagation of economically important domestic animals of agricultural significance relies on the use of biotechnological tools. Tools of animal breeding, genetics, and nutrition result in the production of superior animal breeds. This allows for the secured supply of nutritious, wholesome, and cost-effective animal products. Besides, the techniques of artificial mating and embryo transfer in the propagation of domestic animals are of great utility for not only conserving endangered species but also for producing genetically superior and healthy breeds on a commercial scale.

References

Baldassarre, H., de Matos, D.G., Furnus, C.C., Castro, T.E., Cabrera Fischer, E.I., 1994. Technique for efficient recovery of sheep oocytes by laparoscopic folliculocentesis. Anim. Reprod. Sci. 35, 145–150.

Berg, R.T., Walters, L.E., 1983. The meat animal: changes and challenges. J. Anim. Sci. 57 (suppl 2), 133–146.

Bó, G.A., Guerrero, D.C., Tríbulo, A., Tríbulo, H., Tríbulo, R., Rogan, D., Mapletoft, R. J., 2010. New approaches to superovulation in the cow. Reprod. Fertil. Dev. 22, 106–112.

Carnevale, E.M., 2008. Clinical considerations regarding assisted reproductive procedures in horses. J Equine Vet Sci. 28 (11), 686–690.

Carnevale, E.M., Coutinho da Silva, M.A., Panzani, D., Stokes, J.E., Squires, E.L., 2005. Factors affecting the success of oocyte transfer in a clinical program for subfertile mares. Theriogenology 64 (3), 519–527.
Cheng, W.T.K., Polge, C., Moor, R.M., 1986. The first instances of IVF using in vivo matured oocytes used freshly ejaculated. Theriogenology 25, 146.
Cognié, Y., Baril, G., Poulin, N., Mermillod, P., 2003. Current status of embryo technologies in sheep and goat. Theriogenology 59, 171–188.
Colleoni, S., Barbacini, S., Necchi, D., Duchi, R., Lazzari, G., Galli, C., 2007. Application of ovum pick-up, intracytoplasmic sperm injection and embryo culture in equine practice. Proc. Am. Assoc. Equine. Prac. 53, 554–559.
Cseh, S., Seregi, J., 1993. Practical experiences with sheep embryo transfer. Theriogenology 39, 207.
Dochi, O., Imai, K., Takakura, H., 1995. Birth of calves after direct transfer of thawed bovine embryos stored frozen in ethylene glycol. Anim. Reprod. Sci. 38, 179–185.
Dziuk, P.J., Bellows, R.A., 1983. Management of reproduction of beef cattle, sheep and pigs. J. Anim. Sci. 57 (suppl 2), 355–379.
Gelderman, H., 1975. Investigations on inheritance of quantitative characters in animals by gene markers. I. Methods. Theor. Appl. Genet. 46, 319–330.
Gibbons, A., Pereyra Bonnet, F., Cueto, M.I., Catala, M., Salamone, D.F., Gonzalez-Bulnes, A., 2007. Procedure for maximizing oocyte harvest for in vitro embryo production in small ruminants. Reprod. Domest. Anim. 42, 423–426.
González, R., García Vinent, J.C., Gibbons, A., Cueto, M.I., 1991. Laparoscopic embryo transfer in Merino sheep in Patagonia (Argentina). In: XXIV World Veterinary Congress, Río de Janeiro, Brazil.
Gordon, J.W., Ruddle, F.H., 1981. Integration and stable germ line transformation of genes injected into mouse pronuclei. Science 214, 1244–1246.
Hill, J.R., Dobrinski, I., 2006. Male germ cell transplantation in livestock. Reprod. Fertil. Dev. 18 (1–2), 13–18.
Hinrichs, K., 2012. Assisted reproduction techniques in the horse. Reprod. Fertil. Dev. 25 (1), 80–93.
Holt, W.V., Hernandez, M., Warrell, L., Satake, N., 2010. The long and the short of sperm selection in vitro and in vivo: swim-up techniques select for the longer and faster swimming mammalian sperm. J. Evol. Biol. 23, 598–608.
Inaba, Y., Aikawa, Y., Hirai, T., Hashiyada, Y., Yamanouchi, T., Misumi, K., Ohtake, M., Somfai, T., Kobayashi, S., Saito, N., Matoba, S., Konishi, K., Imai, K., 2011. In-straw cryoprotectant dilution for bovine embryos vitrified using Cryotop. J. Reprod. Dev. 57, 437–443.
Jabbour, H.N., Ryan, J.P., Evans, G., Maxwell, W.M.C., 1991. Effects of season, GnRH administration and lupin supplementation on the ovarian and endocrine responses of merino ewes treated with PMSG and FSH-P to induce superovulation. Reprod. Fertil. Dev. 3, 699–707.
Jacobson, C.C., Choi, Y.H., Hayden, S.S., Hinrichs, K., 2010. Recovery of mare oocytes on a fixed biweekly schedule, and resulting blastocyst formation after intracytoplasmic sperm injection. Theriogenology 73 (8), 1116–1126.
Jones, S.D.M., Price, M.A., Berg, R.T., 1981. The influence of breed on the efficiency of growth and muscle deposition in bulls and heifers. Can. J. Anim. Sci. 61, 237–242.
Kennedy, B.W., Vemnder Gibbins, A.M., Gibson, J.P., Smith, C., 1990. Coalesence of molecular and quantitative genetics for livestock improvement. J. Dairy Sci. 73, 2619–2627.
Kommisrud, E., Paulenz, H., Sehested, E., Grevle, I.S., 2002. Influence of boar and semen parameters on motility and acrosome integrity in liquid boar semen stored for five days. Acta Vet. Scand. 43 (1), 49–55.

Kuwayama, M., Vajta, G., Kato, O., Leibo, S.P., 2005. Highly efficient vitrification method for cryopreservation of human oocytes. Reprod. BioMed. Online 11, 300–308.

Lai, L., Kang, J.X., Li, R., Wang, J., Witt, W.T., Yong, H.Y., Hao, Y., Wax, D.M., Murphy, C.M., Rieke, A., Samuel, M., Linville, M.L., Korte, S.W., Evans, R.W., Starzl, T.E., Prather, R.S., Dai, Y., 2006. Generation of cloned transgenic pigs rich in omega-3 fatty acids. Nat. Biotechnol. 24, 435–436.

Lavitrano, M., Giovannoni, R., Cerrito, M.G., 2013. Methods for sperm-mediated gene transfer. Methods Mol. Biol. 92, 519–529.

Maes, D., Nauwynck, H., Rijsselaere, T., Mateusen, B., Vyt, P., de Kruif, A., Van Soom, A., 2008. Diseases in swine transmitted by artificial insemination: an overview. Theriogenology 70 (8), 1337–1345.

Martinez, A.G., Valcarcel, A., Furnus, C.C., De Matos, D.G., Iorio, G., De Las Heras, M.A., 2006. Cryopreservation of in vitro-produced ovine embryos. Small Rumin. Res. 63, 288–296.

Mattioli, M., Bacci, M.L., Galeati, G., Seren, E., 1989. Developmental competence of pig oocytes matured and fertilized in vitro. Theriogenology 31, 1201–1207.

McKelvey, W.A.C., Robinson, J.J., Aitken, R.P., Robertson, I.S., 1986. Repeated recoveries of embryos from ewes by laparoscopy. Theriogenology 25, 855–865.

Mehlmann, L.M., 2005. Stops and starts in mammalian oocytes: recent advances in understanding the regulation of meiotic arrest and oocyte maturation. Reproduction 130, 791–799.

Menchaca, A., Vilariño, M., dos Santos Neto, P.C., Wijma, R., Crispo, M., 2012. Cumulus cells are involved in oocyte maturation and fertilization in vitro produced ovine embryos. Reprod. Fertil. Dev. 47 (4).

Menchaca, A., Anegon, I., Whitelaw, C.B., Baldassarre, H., Crispo, M., 2016a. New insights and current tools for genetically engineered (GE) sheep and goats. Theriogenology 86, 160–169.

Menchaca, A., Barrera, N., dos Santos Neto, P.C., Cuadro, F., Crispo, M., 2016b. Advances and limitations of in vitro embryo production in sheep and goats. Anim. Reprod. 13, 273–278.

Nagai, T., Funahashi, H., Yoshioka, K., Kikuchi, K., 2006. Up date of in vitro production of porcine embryos. Front. Biosci. 11, 2565–2573.

Paramio, M.T., 2010. In vivo and in vitro embryo production in goats. Small Rumin. Res. 89, 144–148.

Park, C.H., Lee, S.G., Choi, D.H., Lee, C.K., 2009. A modified swim-up method reduces polyspermy during in vitro fertilization of porcine oocytes. Anim. Reprod. Sci. 115, 169–181.

Powell, B.C., Walker, S.K., Bawden, C.S., Sivaprasad, A.V., Rogers, G.E., 1994. Transgenic sheep and wool growth: possibilities and current status. Reprod. Fertil. Dev. 6, 615–623.

Prather, R.S., Barnes, F.L., Sims, M.M., Robl, J.M., Eyestone, W.H., First, N.L., 1987. Nuclear transplantation in the bovine embryo: assessment of donor nuclei and recipient oocyte. Biol. Reprod. 37, 859–866.

Richt, J.A., Kasinathan, P., Hamir, A.N., Castilla, J., Sathiyaseelan, T., Vargas, F., Sathiyaseelan, J., Wu, H., Matsushita, H., Koster, J., Kato, S., Ishida, I., Soto, C., Robl, J.M., Kuroiwa, Y., 2007. Production of cattle lacking prion protein. Nat. Biotechnol. 25, 132–138.

Ricks, C.A., Dalrymple, R.H., Baker, P.K., Ingle, D.L., 1984. Use of a LI-agonist to alter fat and muscle deposition in steers. J. Anim. Sci. 59, 1247–1255.

Robelin, J., Geay, Y., Bonati, B., 1978. Genetic variations in growth and body composition of male cattle. Curr. Top. Vet. Med. 2, 443–460.

Roche, J.F., Quirke, J.F., 1986. The effects of steroid hormones and xenobiotics on growth of farm animals. In: Buttery, P.J., Haynes, N.B., Lindsay, D.B. (Eds.), Control and Manipulation of Animal Growth. Butterworths & Company, London, pp. 39–51.
Romar, R., Cánovas, S., Matás, C., Gadea, J., Coy, P., 2019. Pig in vitro fertilization: where are we and where do we go? Theriogenology 137, 113–121.
Sano, F., Kitayama, T., Masuyama, R., Shiraishi, T., Kawano, Y., Inetani, K., Oda, Y., Hayashi, M., Moriyasu, S., Inaba, Y., Imai, K., 2010. Vitrification with ultra-rapid cooling and minimum volume technique for bovine biopsied embryos. Jap. J. Embryo Transfer 32, 113–119.
Seidel Jr., G.E., 1981. Superovulation and embryo transfer in cattle. Science 211, 351–358.
Souza, J.M., Duffard, N., Bertoldo, M.J., Locatelli, Y., Corbin, E., Fatet, A., Freitas, V.J., Mermillod, P., 2013. Influence of heparin or the presence of cumulus cells during fertilization on the in vitro production of goat embryos. Anim. Reprod. Sci. 138, 82–89.
Sripunya, N., Somfai, T., Inaba, Y., Nagai, T., Imai, K., Parnpai, R., 2010. A comparison of cryotop and solid surface vitrification methods for the cryopreservation of in vitro matured bovine oocytes. J. Reprod. Dev. 56, 176–181.
Tervit, H.R., Goold, P.G., McKenzie, R.D., Clarkson, D.T., 1983. Techniques and success of embryo transfer in angora goats. N. Z. Vet. J. 31, 67–70.
Tervit, H.R., Smith, J.F., McGowan, L.T., Wells, R.W., Parr, J., 1992. Laparoscopic recovery of oocytes from sheep. Proc. Aust. Soc. Reprod. Biol. 24, 26.
Tian, X.C., Kubota, C., Enright, B., Yang, X., 2003. Cloning animals by somatic cell nuclear transfer—biological factors. Reprod. Biol. Endocrinol. 1, 98.
Tsakmakidis, I.A., Lymberopoulos, A.G., Khalifa, T.A., 2010. Relationship between sperm quality traits and field-fertility of porcine semen. J. Vet. Sci. 11 (2), 151–154.
Vallet, J.C., Baril, G., Rougier, F., Chupin, D., Procureur, R., Corteel, J.M., 1987. Feasibility and repeatability of embryo recoveries from dairy goats under laparoscopy. In: Third Scientific Meeting of the European Embryo Transfer Association, Lyon (France), September 4 and 5 1987. Foundation Marcel Mérieux, Lyon, France. Page 60 (abstr.).
Vazquez, J.M., Roca, J., Gil, M.A., Cuello, C., Parrilla, I., Vazquez, J.L., Martínez, E.A., 2008. New developments in low-dose insemination technology. Theriogenology 70 (8), 1216–1224.
Waberski, D., Petrunkina, A.M., Töpfer-Petersen, E., 2008. Can external quality control improve pig AI efficiency? Theriogenology 70 (8), 1346–1351.
Wall, R.J., 2002. New gene transfer methods. Theriogenology 57, 189–201.
Warwick, B.L., Berry, R.O., Horlachwer, W.R., 1934. Results of mating rams to angora female goats. Proc. Am. Soc. Anim. Prod. 34, 225 (abstract).
Wheeler, M.B., Campion, D.R., 1993. Animal production-a longstanding biotechnological success. Am. J. Clin. Nutr. 58 (suppl), 276S–281S.
Wheeler, M.B., Walters, E.M., 2001. Transgenic technology and applications in swine. Theriogenology 56, 1345–1369.
Wheeler, M.B., Walters, E.M., Clark, S.G., 2003. Transgenic animals in biomedicine and agriculture: outlook for the future. Anim. Reprod. Sci. 79, 265–289.
Wheeler, M.B., Monaco, E., Bionaz, M., Tanaka, T., 2010. The role of existing and emerging biotechnologies for livestock production: toward holism. Acta Sci. Vet. 38, s463–s484.
Wieser, M.F., Prestom, T.R., MacDearmid, A., Powland, A.G., 1966. Intensive beef production. 8: the effect of chlortetracycline on growth, feed utilization and incidence of liver abscesses in barley fed beef cattle. Anim. Prod. 8, 411–423.
Wolff, M., Gibbons, A., Cueto, M., Willems, P., Arrigo, J., 1994. Results of artificial insemination with frozen semen in Australian Merino ewes multiovulated with FSHp. In: IV World Merino Conference, Montevideo, Uruguay. Page 269 (abstr.).

CHAPTER THREE

Global livestock production systems: Classification, status, and future trends

Hari Om Pandey[a] and Deepak Upadhyay[b]
[a]ICAR-Indian Veterinary Research Institute, Bareilly, Uttar Pradesh, India
[b]ICAR-Indian Grassland and Fodder Research Institute, Jhansi, Uttar Pradesh, India

Introduction

Throughout human history, animal husbandry has been a central component of agriculture; and animals have been central to agri-food systems. In fact, the association of human beings with domesticated animals goes back to around 10,000 BCE; eventually, several animal species became domesticated. Presently, there are around 8800 livestock breeds which belong to 38 different animal species in the world, providing a diversity of products and services to humankind (FAO, 2020a). Indeed, livestock are domesticated terrestrial animals that are raised to provide a broad array of goods and services including food (meat, milk, and eggs), energy (traction and transport), fibers, fertilizers, hides, etc. Livestock are also an important risk reduction strategy for vulnerable communities and smallholder systems through diversified production and year-round employment; additionally, livestock serve many social roles in different societies. Thus, they contribute to diverse agri-food systems globally, playing many roles for different groups of people (FAO, 2018a). Livestock systems used to occupy a significantly considerable area of our planet, almost 30% (Steinfeld et al., 2006), and are considered a significant global asset with a value of at least $1.4 trillion (Thornton, 2010). Livestock sector employs at least 1.3 billion people globally and directly supports the livelihoods of 600 million poor smallholder farmers in the developing world (Thornton et al., 2006). For instance, India is the largest producer of milk (187.75 million tonnes) in the world with 193.46 million dairy cattle and 109.85 million buffaloes in the country (BAHS, 2019) to cover the caloric needs of some 115 million people and the protein requirements of about 230 million people (FAO, 2018a). The rapidly growing demand for livestock products worldwide has been brought

Emerging Issues in Climate Smart Livestock Production
https://doi.org/10.1016/B978-0-12-822265-2.00017-X

about by the expanding human population pressure, growing income, and urbanization (Delgado, 2005). The livestock sector plays a vital role in the global economy, including the developing and developed worlds. In developing countries, it is evolving in response to the rapidly increasing demand for livestock products, whereas in developed countries, the demand for livestock products is stagnating with increased production efficiency. Livestock systems in developing countries produced about 50% of the beef, 41% of the milk, 72% of the lamb, 59% of the pork, and 53% of the poultry, globally (Herrero et al., 2013). These shares are likely to increase, as most future growth in livestock production is projected to occur in the developing world (Bruinsma, 2003; Herrero et al., 2013; Rosegrant et al., 2009). In fact, due to an anticipated human population of around 9.8 billion in 2050 and 11.2 billion in 2100 (UNPD, 2017), it is expected that between 2000 and 2050, the global cattle population may increase from 1.5 billion to 2.6 billion, and the global goat and sheep population from 1.7 billion to 2.7 billion (Thornton, 2010). Further, differences in the consumption of animal products are also observed that will change between regions. For example, the food demand for livestock products will nearly double in sub-Saharan Africa and South Asia, from some 200 kcal per person per day in 2000 to around 400 kcal per person per day in 2050 (Van Vuuren et al., 2009).

It is believed that the global livestock sector may undergo radical change in the future, due to continuously increasing demand for livestock products for the increasing human population with ever-increasing urbanization and enhanced purchasing power (Thornton, 2010; Van Vuuren et al., 2009). In the future, livestock production will increasingly be affected by mounting pressures on natural resources, particularly land and water, which will give rise to considerable competition between food and feed production. Rising population pressure may modify both the production and the consumption of livestock products, eventually leading to competition for land and water, climate change, the role of sociocultural drivers, and ethical concerns. Thus, it is important to understand land and livestock use concerning human population pressures that will impact the existing livestock production system. The livestock production systems embrace all aspects of the supply and use of livestock commodities, including the distribution and abundance of livestock, the different production systems in which they are raised, estimates of consumption and production now and in the future, the people engaged in livestock production, and the benefits and impacts of keeping livestock (FAO, 2020b). The scale, purpose, and nature of the farming enterprise are called the production system. Several livestock production systems in different corners of the globe grew and prevailed as a subset of farming systems. Seré and Steinfeld (1996) described a global livestock production system that has been widely used (Seré and Steinfeld, 1996). The livestock production response has been characterized by its systems and regional differences. The importance of these systems and livestock species, varies remarkably across these geographical regions. For example, Africa has vast livestock resources in semiarid and arid regions, and small ruminants play an important role. Asia possesses more than 90% of the world's buffalo inventory, whereas in eastern parts of Asia, pigs are a very important source of red meat. Similarly, wide differences in the intensity of production could also be observed. Intensive confined livestock production systems in developed countries are the source of much of the world's poultry and pig meat. However, such systems are being established in developing countries, particularly in Asia, to meet increasing demand. Bruinsma (2003) estimates that at least 75% of total production growth to 2030 will be in confined systems. Ruminant livestock, such as cattle, sheep, and goats, tend to be dependent directly on the land; their production systems, therefore, are largely determined by agro-ecology and land use. Further, production systems for monogastric species such as pigs and chickens depend more on consumer demand and the level of capital investment (FAO, 2020c).

The various driving forces will play out in different regions of the world in the coming decades pertaining to the dynamics of livestock production systems. Globally, augmented livestock production and efficiency in the recent past have been driven mostly by scientific and technological developments in livestock breeding, nutrition, and health. In future, many developed countries will see a continuing trend in which new tools of molecular genetics may have far-reaching impacts on livestock and livestock production. In the coming decades, livestock breeding may focus on other attributes in addition to production and productivity. Examples are DNA-based tests for genes or markers affecting traits that are difficult to measure currently, such as meat quality and disease resistance (Leakey et al., 2009). Another important domain is transgenic livestock for food production, which will be particularly useful and is also technically feasible. In combination with new dissemination methods such as cloning, such techniques could dramatically change livestock production (Lewin, 2009). There will always be a continuous quest, whether the tools used are competent enough in preserving animal genetic resources; because the preservation of farm animal genetic resources will be critical in helping livestock adaption to climate change and the changes that may occur in these systems.

Pursuit to develop livestock varieties/breeds with useful productive traits continually narrowing animal genetic resource base in several intensive livestock production systems, demonstrate a need to maintain as broad range of genetic resources as possible, which could provide genetic insurance against future challenges and shocks (Hoffmann et al., 2011). Thus, livestock production dynamics under the current scenario could pose considerable uncertainty as to how these driving factors will play out in different regions of the world in the coming decades and how these challenges could be resolved in a sustainable manner.

This chapter describes the classification, status, trends, and driving factors of the livestock production systems of the world. Further, brief description of the main features of the various livestock production systems, global driver of livestock production system such as economic growth, dietary preference and purchasing power of consumer, demographic and land-use changes, and technological integration in the farming system. We have covered cattle, buffalo, goat, sheep, pig, and poultry. The chapter concludes with a discussion of the dynamics of livestock production systems across the globe and future trends with the inclusion of recent studies.

Classification of livestock production systems

Livestock production systems (LPS) are essentially subsets within the agriculture sector. LPS can be classified in different ways based on various criteria. They may be the integration of livestock with crop production, the availability of land, the intensity of production, the type of product, the agro-ecological zone, commercialization, etc. Several attempts have been made to classify livestock production systems in the past using a combination of different criteria (Andreae, 1964a, 1972; Devendra, 1995; Dixon et al., 2001; Ruthenberg, 1971, 1980; Seré and Steinfeld, 1996; Thornton et al., 2002). Most of the attempts to define and map livestock production systems mention the classic classification system developed by Seré and Steinfeld (1996). Recently, Teufel et al. (2010) attempted to classify the livestock production systems of South Asian countries. It qualitatively and quantitatively describes each livestock system covering cattle, buffaloes, sheep, goats, pigs, and chickens. In their classification system, the following five criteria were used in the sequence given: integration with crops, land demand, irrigation, commercialization, and agro-ecological zone. According to Teufel et al. (2010), livestock production systems are mainly categorized into two types: solely livestock production systems and mixed farming systems. Both landless and mixed farming systems are further categorized based on commercialization according to the percentage of sales in relation to the gross returns (i.e., commercial or subsistence) and based on agro-ecological zones determined by the length of the growing period (LGP). The LGP is defined as the period in days during the year when rain-fed soil moisture is greater than half the potential evaporation (PET) (i.e., arid, semiarid, subhumid, humid, and tropical highlands). With this LPS could further be classified in 28 probable types (Fig. 1).

Solely livestock systems

In this system, less than 15% of the land is used for cropping; more than 90% of the dry matter fed to animals is derived from rangelands, pastures, annual forages and purchased feeds; and nonlivestock farming activities contribute less than 10% of the total value of production. The solely livestock systems were further split up into landless and grassland systems, according to their demand in land.

Landless livestock systems

They are found mostly in urban and periurban areas. In this system, less than 10% of the dry matter fed to animals is farm produced and the annual stocking rates are above 10 livestock units (LU) per hectare of agricultural land. They can be found in all agro-ecological zones where livestock rearing depends on external feed resources where on-farm feed production is nil. They can be further distinguished as landless small subsistence holdings and commercial enterprises. In general, smallholders keep their livestock for multiple purposes in backyard settings while commercial enterprises are demand-driven and specialize in meat, egg, or milk production. The best suitable livestock species for these systems are pigs and poultry. However, periurban dairy farms and sheep and goat farms are also growing these days due to rising demand in these areas. In terms of nutrient flow, they are open systems while waste disposal is a major concern, particularly in intensive farms near populated cities.

Grassland livestock systems

In this system, more than 10% of the dry matter fed to animals is produced on the farm itself and the annual average stocking rates are less than 10 LU per hectare of agricultural land. Grazing is the primary source of feed in these systems with varying nutritional quality of the fodder. They can also be found in different agro-ecological regions. In tropical highlands (mean

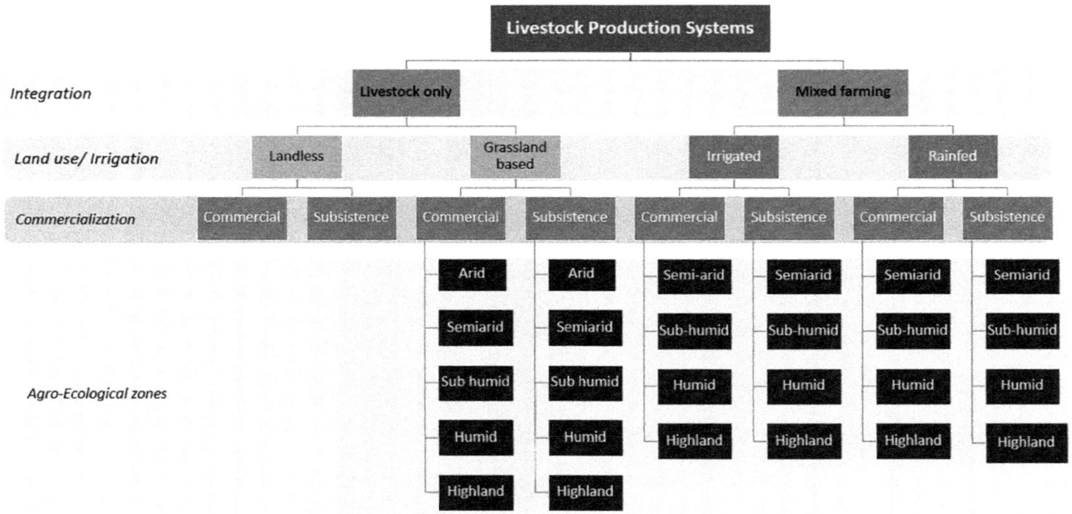

Fig. 1 Livestock production system classification. *(Adapted from Teufel, N., Markemann, A., Kaufmann, B., Zárate, A.V., Otte, J., 2010. Livestock Production Systems in South Asia and the Greater Mekong Sub-Region: A Quantitative Description of Livestock Production in Bangladesh, Cambodia, India, Lao PDR, Nepal, Pakistan, Sri Lanka, Thailand, and Vietnam (Working paper No. 48), Pro-Poor Livestock Policy Initiative (PPLPI). FAO.)*

temperature range 5–20°C and growing period <110 days), the transhumant system is found in which movement toward a higher altitude for access to summer pastures and movement toward valleys during winter seasons are practiced. Mainly sheep are reared in this system. In arid/semiarid zones, ranges are the main feed resource in grassland livestock systems. The availability of feed resources in these ranges varies per rainfall and seasonal variation. To overcome environmental constraints, herd mobility is practiced, hence the populations relying on these systems are recognized as pastoral groups. They include highly mobile nomadic groups as well as comparatively sedentary ranchers or agro-pastoralists. Livestock species mainly reared under these systems include camels, sheep, cattle, and a few goats. In subhumid/humid regions, mainly cattle and buffalo are reared for milk production due to the better availability of feed and forage.

Mixed farming systems

These systems may also be referred to as crop-animal systems. These systems are the backbone of agriculture in some developing countries such as India. They encompass systems where at least 15% of the land use is for crop cultivation. Further, more than 10% of the dry matter fed to the livestock is derived from crop byproducts or stubble and/or more than 10% of the total value of production comes from nonlivestock farming activities. It is estimated that more than 90% of the world's milk supply is produced in mixed farming systems. Further, on the basis of rainfall mixed farming system is again classified as follows:

Mixed rainfed system

This is a mixed farming system where less than 10% of the farm production comes from irrigated land. In tropical highland regions, livestock plays several vital roles in income generation, risk aversion, soil fertility, and traction. In semiarid regions, livestock plays a primary role due to low rainfall. The major species kept are indigenous breeds of cattle, buffalo, sheep, goat, and poultry. Further major crops cultivated in these systems are millets, sorghum, rice, groundnuts, pigeon peas, soybeans, cotton, and date palms. In these regions, grassland systems are evolving as mixed systems due to increasing population and scarcity of land. In humid and subhumid tropical regions, there are more heterogeneous systems under this classification as a range of socioeconomic conditions, soil types, and climate is involved. However, multiple functions of livestock still prevail there. Mainly annual cereals, soybeans and vegetable crops are cultivated in these regions.

Mixed irrigated system

In mixed farming, more than 10% of the arable land is irrigated. Because of the irrigation facility, the availability of feed for ruminants is better compared to a rainfed system. As rainfall is the major constraint for crop cultivation in semiarid and arid tropical regions, year-round crop production is possible in mixed irrigated systems. Irrigation further facilitates increased fodder production, which reduces the feed deficit and makes the intensification and commercialization of livestock production feasible, especially in urban and periurban areas (Dixon et al., 2001; Thornton et al., 2002). Here, mainly cattle and buffalo farming is practiced to get milk as a balanced food. Also, male animals are used for draft purposes, a practice that is continually diminishing in several countries due to increasing mechanization. In areas where marginal rangelands are available near an irrigated area, sheep and goats are also important. Further, because the land use of these regions is predominantly covered by irrigated crops, wheat and paddy straw serve as important feed resources despite their low digestibility.

Distribution of various livestock species under different livestock production systems

Livestock production systems, production intensities, and the uses of products and services vary greatly within and across regions/countries. Generally, these systems are shaped by the prevailing agro-ecological and sociocultural environments. A different production environment results in different levels of resource use intensity, size of operations, product orientation, marketing, etc. In different parts of the world, the same livestock species may serve different purposes. For example, cows are the major dairy animal in the world, but in South Asian countries, buffalo play that role. Therefore, it is important to understand the global spatial distribution and abundance of major livestock species such as cattle, buffaloes, sheep, goats, pigs, and poultry. This has been described by FAO in its map called the Gridded Livestock of the World (GLW), which represents the global distribution of livestock.

The type of ruminant population and animal products under different livestock production systems of the world is presented in Table 1 (Steinfeld et al., 2006).

The vast majority of the large and small ruminant populations under different systems are found in developing regions. Some 70% of small ruminants in grazing systems and more than 80% of large ruminants in grazing systems are in the developing world. These shares are respectively about

Table 1 Global livestock population and production in different production systems (averages 2001–2003).

Type of animal/animal product	**Livestock population (in millions) and production (million tonnes)**			
	Grazing	**Mixed rainfed**	**Mixed irrigated**	**Industrial (landless)**
Animal				
Cattle and buffalo	406	641	450	29
Sheep and goats	590	632	546	9
Animal products				
Total beef	14.6	29.3	12.9	3.9
Total mutton	3.8	4.0	4.0	0.1
Total pork	0.8	12.5	29.1	52.8
Total poultry meat	1.2	8.0	11.7	52.8
Total milk	71.5	319.2	203.7	–
Total eggs	0.5	5.6	17.1	35.7

Adapted from Steinfeld, H., Wassenaar, T., Jutzi, S., 2006. Livestock production systems in developing countries: status, drivers, trends. Rev. Sci. Tech. 25, 505–516.

80% and 70% in rainfed mixed systems and 87% and 92% in irrigated mixed systems (Steinfeld et al., 2006). Despite the low productivity of these systems, developing regions host the vast majority of the mixed rainfed ruminant population, which accounts for less than half of the system's production worldwide. In fact, beef productivity in these regions averages 26 kg/head compared to 46 kg/head at the world level, and their milk production represents only 22% of the world total (Steinfeld et al., 2006).

Globally, cattle are reared primarily for dairy or beef production; the system in which they are maintained varies with the purpose and geographical location. Dairy cattle are kept mostly under confinement, semiconfinement, pasture-intensive or -extensive conditions, and mixed systems. In South America, New Zealand, Australia, and India, the most common production systems for dairy cattle are extensive, pasture-intensive, or semiconfinement systems. On the other hand, in Europe and North America, confinement systems (free stall, tie-stall, and bedded pack) are common. The cattle farm size varies across the globe from a few to thousands of animals (Endres and Schwartzkopf-Genswein, 2018). Beef cattle production can be divided into three systems: the cow/calf (suckler calf), stocker, and feedlot systems, found in North and South America, Australia, New Zealand, and Europe. Of these, only the feedlot system is an intensive production system with a goal of fattening cattle until they reach slaughter weight.

A buffalo is also known as an Asian animal as it originated and is distributed mainly in Asian countries. About 97% of the world's buffaloes are found in the Asian region (FAO, 2018). India is home to more than half of the buffalo population. Buffalo milk is also preferred over cow milk in many regions; therefore, buffaloes are considered the main dairy animal and the backbone of Indian dairies. Further, buffaloes also produce meat and draft power, particularly in rice fields, transportation, and manure for crop farming (FAO, 2020d). India is home to the best buffalo milch breeds of the world such as the Murrah, Nili Ravi, Surti, and Jaffrabadi, and they have high potential for milk and fat production. Buffalo farming in the majority of Asian countries is mostly characterized by traditional small family farms, with 2–4 buffaloes that are fed on grasses available in communal paddocks and agricultural byproducts such as straw. However, in irrigated areas and areas closer to larger cities, semiintensive to intensive buffalo farms with several to hundreds of buffaloes can be seen (Wanapat and Kang, 2013). For example, in Haryana, Punjab, Uttar Pradesh, Rajasthan, and Gujarat as well as periurban areas of metro cities such as Delhi and Mumbai, there are several commercial buffalo farms for milk production. Buffaloes are good utilizers of inferior feed and fodder and therefore, they thrive well when fed with cultivated fodder, crop byproducts, and purchased concentrate feed. Land-based systems such as grazing or mixed farming accommodate most of the cattle and buffalo across the world. These systems support about 1.5 billion cattle and buffalo. A very small proportion of the world's cattle and buffalo population is maintained in industrial feedlots. Mainly in North American countries, cattle are reared in feedlot systems for beef production.

Sheep are found in all the parts of the globe; however, 60% of the sheep population dwells in temperate climates and pasture conditions (i.e., Europe and Asia within latitudes 35–55 degrees N and in South America, Australia, and New Zealand between 30 and 45 degrees S) (Zygoyiannis, 2006). The rest of the sheep are reared in the semiarid tropics of Africa and Asia between 5 and 35 degrees N, including India, the Middle East, and the highlands of East Africa. Wool and sheep meat are the major products obtained from sheep, although their share in the global red meat and textile markets is negligible. Most of the exported sheep meat worldwide is from New Zealand and Australia (Noordhuizen and Gunnarsson, 2006). Moreover, sheep can thrive and produce on land unfavorable to other forms of agriculture. Many sheep breeds across the globe are well adapted for survival on extensive unimproved, seminatural pastures in difficult climate conditions. The systems used for farming sheep vary between and within different countries.

In temperate regions, meat is the major product as in many countries, wool production is declining. For example, in Australia the importance of meat production is also increasing and the Merino ewe is increasingly regarded as a maternal breed (Morris, 2017). In the more arid regions of Asia, Africa, and the Middle East, sheep are used for wool, meat, and milk as well as for sustaining the livelihoods of sheep farmers. Kilgour et al. (2008) described the three major management systems for sheep production that exist in the world, namely extensive production for wool and meat, intensive dairy production, and traditional pastoralism. In all major sheep-producing countries (Australia, New Zealand, and the United Kingdom (UK)), the extensive management system is most common. Sheep farming systems operating in these countries include rangeland sheep production in Australia, pastoral sheep farming in New Zealand, and the stratified sheep management system in the UK. Sheep management in these countries is characterized by an abundance of land, low human population, and several hundred to some thousands of sheep in a flock. Sheep are farmed in hills or in large pasture lands and are rarely housed. Stratification in Britain is the development of a highly structured sheep industry (hill system, upland system, and lowland system) based on the natural resources of different areas of the country and the sale and movement of livestock between these different areas. Typically, sheep flow from the most naturally disadvantaged areas, the hills and mountains, to the more favored areas, the lowlands. In the hills and mountains, sheep are bred to produce animals, which are sold to producers in more favorable areas for use as breeding stock or for further grazing and sale as finished livestock. The intensive system is rarely practiced for sheep production around the world except for dairy sheep production in southern Europe as well as the Near East and Middle East countries. In semiarid open rangelands, pastoralism is the main production system for sheep, as they are managed in nomadic or transhumance pastoral systems in many regions of the world. These are the traditional system of rearing sheep. It is preferred to rear sheep in these systems, as a large array of products can be obtained from sheep. Further, they can reproduce more quickly than other livestock species to make up for any loss in the flock. Pastoralism can be categorized by the degree of movement into three main classes (FAO, 2001): (1) Nomadic: This is a highly mobile form of pastoralism with no established home base. There is flexibility in the timing of seasonal migration. As movements are opportunistic, depending on pasture and water availability, therefore there is a high dependability on the growth cycles of different plant species. Nomadic routes are more or less defined, as nomads prefer established migration routes. Also, sometimes along the way there is an agreement between nomads and the point of trade. Their routes may be changed depending on climatic conditions such as drought or disease outbreaks. (2) Transhumance: Seasonal migration between the established points is the main feature of this system. Further, transhumant pastoralists have a permanent home base. Herd-splitting, leaving some of the animals of the herd at the base and moving with the rest of the animals in search of pasture, is another feature of transhumance. However, with the advent of modern transport systems, the movement of animals between grazing areas is assisted by trucks or rail. For example, during the winter, sheep flocks are moved by trucks to lowlands in Britain (Kilgour et al., 2008). (3) Agropastoralism (ranching): This is a more settled type of pastoralist system mainly found in Australia and the United States, and on a smaller scale in Europe. Agropastoralists own land and grow staple crops. Further, the flock size is smaller compared with nomads or the transhumant system. They do not move farther from their settlements in search of grazing. Rather, supplementary feeding is also practiced in this system.

Goats are the among the first domesticated farm animals and their domestication traces back to at least 10,000 years ago (Ensminger and Parker, 1986). Goats are distributed worldwide and are mainly reared for meat, milk, hides, and hair. At present, the global goat population is roughly 1 billion. More than 90% of these are in Asia and Africa and only 1.6% in Europe (FAO, 2018). Among different countries, China, India, Pakistan, and Bangladesh have the highest goat populations across the world. Goats are raised in a wide range of production systems around the world. They are usually associated with traditional, low-technology production systems (Devendra, 2015). The majority of the world's goat population is reared under smallholder, mixed farming systems. However, goats are an important component of pastoralist (nomadic or transhumance system) herds. Goats are preferred by production systems using the poorest land, shrubland, and forest areas where other species cannot survive (El Khidir et al., 1998). Commonly mixed herds of sheep and goats are kept under extensive systems for the optimum use of natural resources. Goat meat is the primary product obtained in these systems. Unlike other ruminants, goats can derive their nutritional needs primarily by browsing shrubs and trees. Further, owing to their high adaptive capacity (Silanikove, 2000) and level of fitness, goats are kept under a wide range of production and farming systems throughout the world. The intensive system of rearing is limited to only dairy goats in certain temperate regions of the world such as Europe, where high-value animals are raised for specialized dairy production. There has been a simultaneous increase in goat milk production and consumption across the world (Miller and Lu, 2019). Most of the world's goat milk is produced and consumed locally in India, Bangladesh, Pakistan, and Turkey; a negligible amount reaches the market. On the contrary, the most organized market for goat milk is found in Europe. Having only 1.6% of the world's goat population, Europe produces about 15% of the world's goat milk. Moreover, France, Spain, and Greece are the leading producers of cheese from goat milk, which has high demand in the international market.

The monogastric livestock (pigs and chickens) production system across the world presents a rather contrasting picture than ruminant production. More than half of pork and poultry meat currently originates from industrial systems and about half of this production originates from developing countries. More than 95% of the world's industrial pork is produced in the developed world and the Asian region.

Raising pigs and the consumption of pork are largely concentrated in Asia, Europe, and the United States. Out of nearly 1 billion pigs in the world, Asia has 57%, and in particular China has the highest pig population (46%) (FAO, 2018). Moreover, China is a leading pork producer (46% of world pork) as well as consumer of pork worldwide. The European Union is the second largest pork producer and is the largest net exporter of pork, followed by the United States (Koopman et al., 2014). Pigs are raised in a variety of settings, going from small family units of scavenging pigs through small to medium commercial semiintensive units to very large intensive units (FAO, 2020d). Highly intensive and large-scale pig production systems are in practice mostly in developed countries. These intensive farms are found in the major arable areas of Europe, the corn belt states of the United States, the Canadian prairie provinces, the Netherlands, and North Carolina. For example, most US pigs are raised by producers with more than 5000 swine and most are owned by firms that each own more than 50,000 swine (Koopman et al., 2014). Smallholder pig farming is an important livestock activity in the rest of the developing world. The majority of rural pig farmers practice backyard pig farming while those in the periurban areas rear pigs semiintensively with pig sties. Most of the pork in developing countries is produced by smallholder pig farmers.

Chickens are the most omnipresent of all livestock species, as they are found more or less everywhere that is inhabited by people. Over the past 40–50 years, the global stocks of chickens increased about five times and the average carcass weights more than doubled, resulting in a nearly 12-fold rise in chicken over a 50-year period. Growth trends are mostly seen in intensive units with specialized production of meat or eggs, rather than in backyard productions mainly meant for home consumption or sale in a local market.

Other perspectives on livestock production system classifications

Various approaches to classify livestock production systems have been adopted, such as the integration of livestock production with crop production, the relation to land available, the agroecological zone (AEZ), the intensity of production, the type of products and services, etc. Like previously discussed classifications, other attempts to classify livestock production systems could also be appreciated, which were formulated using combinations of criteria (Jahnke, 1982; Mortimore, 1991; Otte and Chilonda, 2003; Seré and Steinfeld, 1996; Wilson, 1986). Certain reports provide more detailed information on specific farming systems. For instance, Devendra (1995) gives a more detailed definition of livestock production in mixed farming systems. Another way to represent livestock systems of the world is by including a combination of production factors. For livestock production systems, which include landless production, the livestock biomass is more relevant than the land area, and the land production factor is better represented by the average biomass of livestock per agricultural person than by agricultural land area (FAO, 2018a).

Livestock can be broadly divided into ruminants (which through enteric fermentation have the capacity to digest rough plant material) and monogastric species (unlike ruminants, their dietary needs are more similar to our own). The main large ruminant species are cattle and buffalo, which may be multipurpose or specialized for beef, dairy, or draft power; and camelids, such as camels, llamas, and alpacas, which play important roles for traction and fiber but also provide meat and milk to many societies. Small ruminants are sheep and goats, which are important sources of meat, milk, wool, and hides. Livestock monogastric species include pigs, which are mainly raised for meat, and a range of poultry species such as chickens, ducks, and turkeys raised for meat, eggs, and feathers. The livestock sector largely depends on the cost of the classic production factors–land, labor, and capital. In line with this, the FAO (FAO, 2018a) discussed land, labor, and capital in a way that for livestock systems, the land factor is replaced by livestock biomass. Labor refers to the agricultural population and capital refers to the amount of gross domestic product (GDP) available per person. A combination of these variables could help to develop a representative map. Using this approach, livestock systems were classified into three broad categories: (1) Extensive livestock systems, (2) labor-intensive livestock systems, and (3) capital-intensive livestock systems (FAO, 2018a).

Extensive livestock systems

Extensive systems are characterized by low labor and capital inputs, and they generally occur in marginal habitats that are typically pastoralist, with ruminant livestock grazing large expanses of marginal rangelands (FAO, 2018a). This system encompasses and directly benefits about 180 million pastoralists, whereas many more linked to export markets, such as the small ruminant trade from the Horn of Africa, also benefit. These extensive livestock systems are efficient but of low productivity which seeks few inputs like basic animal health care. Extensive systems are found in lightly managed areas that are unsuitable for crop growth. Therefore, unlike intensive systems, they do not compete for human food, instead making use of natural resources that would otherwise not contribute to feeding the world's population. Here, livestock are integral to the social fabric of pastoralist societies. Hence, the utility of livestock goes well beyond food and livelihoods and caters to many social and financial roles. These regions face challenges of climate extremes, harsh winters, or extended droughts that cause high levels of mortality among livestock. Although these systems are considered inefficient in terms of GHG emissions, they account for only a small proportion of the overall agricultural emissions. These systems hold great potential to sequester soil organic carbon, if degraded rangelands could be redeemed to health (FAO, 2018a).

Labor-intensive livestock systems

Labor-intensive livestock systems are typically smallholder-based and mostly occur as part of mixed crop-livestock farms (FAO, 2018a). These systems are occupied by arable crops, tree crops, and aquaculture, and are typified by the smallholder systems of Central America, Africa, and Asia. The majority of the labor-intensive systems are run by families with a focus on producing staple foods for subsistence, with surpluses sold or exchanged locally. However in some cases–such as smallholder dairy production in East Africa and South Asia–they are well organized and linked to national and international markets. A number of poor people who rely on livestock globally raise them in these systems. Livestock not only provide nutritious food but also hold sociocultural importance in these farm families. Notably, in developing countries the largest contribution in milk production originates from smallholder dairy farms. However, in terms of efficiency, livestock production under these systems is inefficient compared to extensive or capital-intensive systems. However, there are other efficiencies that must be accounted for, particularly in relation to nutrient cycling, giving value to crop residues, and providing draft power (FAO, 2018a).

Capital-intensive livestock systems

Capital-intensive livestock systems are usually associated with highly modified environments where land and labor inputs have been substituted by capital investment through intensification and mechanization (FAO, 2018a). Capital-intensive livestock systems mainly produce beef, dairy, pork, and poultry products. Intensive beef production is characterized by the feedlots of the southwest United States. Dairy farms of these systems are modern and large. They are also highly mechanized and extremely productive. In fact, half of America's milk comes from farms with more than 1000 cows, and this proportion is continuously increasing each year. Pigs and chickens in particular are part of intensive livestock production. These are the mainstays in high-income countries but are becoming much more prevalent in low- and middle-income countries as well, where they are providing animal protein and contributing to feed-growing urban populations. Capital-intensive livestock systems are highly efficient and extremely productive despite their dependence on large amounts of inputs in terms of feed and fodder, shelter, and operational requirement of unit. Intensive livestock systems are known to impact the environment in many ways, including deforestation, disruption of nutrient cycles, and pollution with pesticides and chemical fertilizers. Another emerging issue in capital-intensive livestock systems is antimicrobial resistance due to its large amounts use which foster ideal conditions for the emergence of microbes (FAO, 2018a).

Drivers of livestock production systems

Economic growth

The demand and consumption of different livestock products have a close relationship with per capita income. Therefore, with the rise in income, the consumption of meat, milk, and eggs increases until it reaches saturation. For example, in developed countries, the per capita consumption of meat is around 80–130 kg per year; there is practically no further increase beyond that level (Steinfeld et al., 2006). In developing countries, compared to cereals, the demand for livestock-related products is more elastic. With rapidly increasing incomes in many developing countries over the past 20–30 years, livestock product consumption levels have also increased. The economies of low- and middle-income countries achieved an average annual growth of 4.5% (3.0% per capita) from 1990 to 2019. Developing countries in South Asia, in particular, experienced very strong economic growth with gross domestic product growth rates of 6.0% over the same period (Table 2). In high-income countries, the average GDP growth was 2.2% at a given period.

Table 2 Economic development indicators in different regions of the world (average for the period 1990–2019).

Regions	GDP growth (annual %)	GDP per capita growth (annual %)	Population growth (annual %)	Population growth in 2019 (%)
East Asia and Pacific	4.2	3.3	0.9	0.5
European Union	1.8	1.5	0.2	0.2
Latin America and Caribbean	2.6	1.2	1.3	0.9
Middle East and North Africa	4.0	1.9	2.1	1.7
South Asia	6.0	4.2	1.7	1.2
Sub-Saharan Africa	3.6	0.8	2.7	2.7
OECD members	2.1	1.4	0.7	0.5
High-income countries	2.2	1.5	0.6	0.4
Low- and middle-income	4.5	3.0	1.5	1.2
Least-developed countries (UN classification)	4.6	2.0	2.5	2.3
World	2.8	1.5	1.3	1.1

OECD, Organization for Economic Cooperation and Development.
Least-developed countries (UN classification) include 47 countries: 33 in Africa, nine in Asia, four in Oceania, and one in Latin America and the Caribbean.
The country classification by income level is based on the 2016 GNI per capita from the World Bank.
From World Bank, 2020. World Development Indicators online DataBank [WWW Document]. World Bank. https://databank.worldbank.org/source/world-development-indicators# (accessed 8.27.20).

Increasing human population

In addition to higher incomes, increases in the human population add to the demand for livestock-related products. The world population is around 7.79 billion (UN, 2020a) and it is increasing at 1.1% annually. This indicates that each year, around 85 million people are added to the existing population. The populations of several developing countries are growing fast compared to developed countries. However, there are variations among the developing world with respect to population growth. Population growth has slowed to 0.5% in East Asia whereas it remains high in sub-Saharan Africa (2.7%). Furthermore, there has been a decline in the total population younger than 15 in many developed and developing countries, which further impacts the demand scenario for livestock products.

Urbanization

Urbanization is another phenomenon that drives livestock production. The societal structure is also changing rapidly in tune with population growth. More than half of the worldwide population (55%) lives in urban areas (UN, 2020b). Further, it has been projected that by 2050, more than two-thirds of the world population will live in urban areas. Urban people embrace new eating habits, consume more animal protein, and eat a higher proportion of their food away from home. The consumption pattern reveals that the rural population on average consumes fewer livestock products than the urban population.

Changes that have occurred in the average human diet across world regions are clearly evident. People in Europe and American countries derive more than 50% of their dietary protein intake from food of livestock origin (the figures do not include fish and other seafood), and little change occurred between 1990 and 2013 (Table 3). Changes have been the most dramatic in Asia, where the total protein supply from livestock for human diets increased by 184%. There was a slight change in African countries (FAO, 2018). Dietary transition is evident toward the increasing share of livestock products in the diet of people in the developing world at the expense of staple foods such as cereals and tubers.

The rapidly increasing demand for livestock products pushes the livestock sector, which needs to transform promptly in an effort to cope with such demand. Hence, a transition in the consumption pattern for livestock products would drive reforms in existing livestock production systems.

Table 3 Daily protein supply from livestock and from all sources in 1990 and 2013 (in grams per capita).

	Total protein from livestock (% of total protein)		Total protein	
Region	**1990**	**2013**	**1990**	**2013**
Africa	12.97 (22.06)	16.06 (23.24)	58.79	69.1
Americas	44.42 (54)	52.13 (55.90)	82.24	93.25
Asia	14.41 (23.5)	26.62 (34.3)	61.17	77.57
Europe	58.33 (56)	57.91 (57)	104.17	102.06
Oceania	68.29 (64.9)	66.18 (65.15)	105.23	101.58
World	25.23 (35.8)	32.13 (39.5)	70.47	81.23

From FAO, 2018. FAOSTAT: Food and Agriculture Data [WWW Document]. Food Agric. Organ. http://www.fao.org/faostat/en/#data/QA (accessed 8.26.20).

Technological advances in livestock production

The availability and use of technology have massive roles in driving livestock production in association with other factors. In the past, several technological interventions helped to meet demand for livestock products. Emerging technologies in this field hold potential to drive changes in the livestock sector. Further, future changes in demand will likely be addressed using advanced technologies in every field of livestock production such as breeding, nutrition, or health care. In the livestock breeding sector, the widespread use of artificial insemination, more efficient statistical tools for genetic merit estimation, and the selection of objective traits have already increased the rate of genetic gain in recent decades. In the poultry and pig sectors, these gains are more visible, particularly in developed regions. However, in the developing world, there are still considerable opportunities to increase livestock productivity. Molecular genetics tools may have impact in the future. Examples include DNA-based tests for genes or marker-assisted selection for traits such as disease resistance, meat quality, etc. At present, the genome sequences of poultry and cattle have been studied entirely. This paves the way for possible advances in evolutionary biology, animal breeding, and animal models for human diseases. Several other techniques such as multiple ovulation and embryo transfer (MOET), transgenic livestock production, and cloning, which is still developing, hold enormous potential. In the livestock nutrition field, recent advances in genomics, transcriptomics, proteomics, and metabolomics will continue to contribute to the field of animal nutrition (Thornton, 2010).

Climate change

Global climate change is primarily caused by greenhouse gas (GHG) emissions that result in the warming of the atmosphere (Stocker et al., 2013). Interactions between livestock and climate happen two ways. Livestock production influences the climate and changes in climate affect the production of livestock. The livestock sector, particularly ruminants, with emissions of 14.5% of global GHG (Gerber et al., 2013) contributes to climate change. In turn, climate change will affect the livestock directly (heat stress, diseases, loss of productivity, etc.) or indirectly (feed and fodder quantity and quality, water scarcity, biodiversity loss, land degradation, extreme weather events, etc.). The environmental temperature critically affects production; per the IPPCC's 2013 (The Intergovernmental Panel on Climate Change) estimates, it is likely to increase in the range of 0.3–4.8°C (Gerber et al., 2013). Such changes will alter the water availability, quality and quantity of fodder, animal production, reproduction, and health. Livestock diseases are also affected by an increase in temperature and precipitation variation. In order to adapt to changing climatic scenarios, several changes are needed in different livestock production systems. In intensive systems, GHG mitigation can be achieved through increasing production efficiency using the available technology (Thornton, 2010); this may involve some shifting toward monogastric species. Under a land-based system, changes in grazing and manure management hold the potential to mitigate GHG emissions. The diversification of livestock production and the integration of livestock with forestry and crop production have been suggested to cope with climate change. Most of the world's livestock population (about two-thirds) is reared in mixed livestock systems, and provides the majority of the milk, meat, and several staple crops (Herrero et al., 2012). Changes in mixed crop-livestock systems with respect to resource use efficiency are inevitable in the coming days.

Livestock welfare concerns

In general, animal welfare can be defined as the ability of animals to cope with their environment (Broom, 1996). Moreover, the five freedoms concept of animal welfare (FAWC, 1993) developed by the UK's Farm Animal Welfare Council is now internationally accepted for defining and assessing animal welfare. It has been used as a framework in guidelines, recommendations, codes, and legislation prepared by several countries and organizations.

Continuous increase in attention toward animal issues and concerns over animal welfare have been seen in the latter half of the 20th century. This was also when livestock production was shifting toward intensification to meet the growing demand for livestock products. In particular, intensive livestock production systems have been clashing with the cherished moral ideals of people related to animal welfare. The major welfare concerns in the livestock industry are focused on the intensive management of dairy calves, the use of gestation and farrowing crates for sows, and the housing of layers in battery cages, among others (Madzingira, 2018). There are differences between different countries in relation to animal welfare legislation. Several countries and organizations have already made livestock welfare mandatory by introducing legislation while others have prepared guidelines, codes, and recommendations in this respect. For example the European Union (EU) established a wide range of legislative provisions concerning animal welfare for the farming of poultry, calves, and pigs and the transport and slaughter operations of all species. In particular, the EU has banned traditional cages for laying hens and requires group housing for pregnant sows (Simonin and Gavinelli, 2019). Further, OIE (Office International des Epizooties) has produced guidelines (not mandatory) on animal welfare to smooth trade between OIE member states (OIE, 2019). Similarly, other nations are moving in the same direction to meet animal welfare needs. Ethical concerns will play an increasing role in the production and consumption of livestock products in the coming days. In most developed countries, consumers apart from concerns for the price and safety of livestock products, are also interested in production systems, particularly the management, transport, and slaughter of livestock. However, improvements in animal welfare do not need to necessarily penalize livestock production; indeed, they may increase profits.

Conclusion

Livestock production systems, largely being a subset of agriculture farming, are result of agro-ecological forces coupledwith the consistent and conscious endeavor of human beings to harness various goods and services toward fulfilling their several type of needs. This has evolved over time and is primarily governed by the agroecology of that region; nevertheless, many driving forces also provide velocity to keep it dynamic. The ever-increasing human population, urbanization coupled with income gain, and more inclination toward including animal origin food into their food bowl continue to fuel the structural changes in the livestock sector. Revolutionizing livestock production with emerging scientific tools and techniques brings not only opportunity but also several challenges. Therefore, the intensification of livestock production should be navigated and promoted in such a way that it should not come at the expense of human, animal, and environmental health.

References

Andreae, B., 1964a. Types of Farming Systems: Development and Transformation of Land Use, Animal Husbandry, and Farming Systems in Europe and Overseas as Well as New Methods of their Demarcation. Stuttgart.

Andreae, B., 1972. Agricultural Forms of Operation in the Tropics: Land Use and Animal Husbandry in the Field of Tension between Tradition and Progress. Parey, Hamburg.

BAHS, 2019. Animal Husbandry Statistics Division. Department of Animal Husbandry & Dairying, Govt. of India.

Broom, D.M., 1996. Animal welfare defined in terms of attempts to cope with the environment. Acta Agric. Scand. Sect. Anim. Sci. Suppl. 27, 22–28.

Bruinsma, J., 2003. World Agriculture: Towards 2015/2030: An FAO Perspective. Earthscan, FAO, Rome, Italy.

Delgado, C., 2005. Rising demand for meat and milk in developing countries: implications for grasslands-based livestock production. Grassl. Glob. Resour., 29–39.

Devendra, C., 1995. Livestock development strategies for low income countries. In: Proceedings of the Joint FAO/ILRI Roundtable on Livestock Development Strategies for Low Income Countries, Addis Abada, 27 February–2 March, FAO and ILRI, Rome and Nairobi, pp. 133–144.

Devendra, C., 2015. Dynamics of goat meat production in extensive Systems in Asia: improvement of productivity and transformation of livelihoods. Agrotechnology 04, 131.

Dixon, J., Gulliver, A., Gibbon, D., Hall, M. (Eds.), 2001. Summary: Farming Systems and Poverty: Improving Farmer's Livelihoods in a Changing World. FAO and World Bank, Rome, Italy and Washington, DC.

El Khidir, I.A., Babiker, S.A., Shafie, S.A., 1998. Comparative feedlot performance and carcass characteristics of Sudanese desert sheep and goats. Small Rumin. Res. 30, 147–151.

Endres, M.I., Schwartzkopf-Genswein, K., 2018. Overview of cattle production systems. In: Tucker, C.B. (Ed.), Advances in Cattle Welfare. Woodhead Publishing, pp. 1–26.

Ensminger, M.E., Parker, R.O., 1986. Sheep & Goat Science, fifth ed. The Interstate Printers & Publishers, Inc., Danville, IL.

FAO, 2001. Pastoralism in the New Millennium [WWW Document]. FAO. http://www.fao.org/3/y2647e/y2647e02.htm#P0_0. (accessed 8.25.20).

FAO, 2018a. Shaping the future of livestock. In: The 10th Global Forum for Food and Agriculture (GFFA), Berlin, 18–20 January 2018. Food and Agricultural Organisation of the United Nation.

FAO, 2018b. FAOSTAT: Food and Agriculture Data [WWW Document]. FOOD Agric. Organ. http://www.fao.org/faostat/en/#data/QA. (accessed 8.26.20).

FAO, 2020a. Domestic Animal Diversity Information System (DAD-IS) [WWW Document]. Food and Agriculture Organization of the United Nations. http://www.fao.org/dad-is/en/. (accessed 8.25.20).

FAO, 2020b. Livestock Systems [Online]. Food and Agriculture Organization of the United Nations. Available at http://www.fao.org/livestock-systems/en/. (Accessed 23 August 2020).

FAO, 2020c. Production Systems | Livestock Systems [Online]. Food and Agriculture Organization of the United Nations. Available at http://www.fao.org/livestock-systems/production-systems/en/. (Accessed 23 August 2020).

FAO, 2020d. Livestock Systems: Food and Agriculture Organization of the United Nations [WWW Document]. http://www.fao.org/livestock-systems/production-systems/en/. (accessed 8.25.20).

FAWC, 1993. Second Report on Priorities for Research and Development in Farm Animal Welfare. FAWC, Tolworth Tower, Surbiton, Surrey.

Gerber, P.J., Steinfeld, H., Henderson, B., Mottet, A., Opio, C., Dijkman, J., Falcucci, A., Tempio, G., 2013. Tackling Climate Change Through Livestock – A Global Assessment of Emissions and Mitigation Opportunities. Food and Agriculture Organization of the United Nations (FAO), Rome.

Herrero, M., Thornton, P.K., Notenbaert, A., Msangi, S., Wood, S., Kruska, R., Dixon, J., Bossio, D., van de Steeg, J., Freeman, H.A., Li, X., Rao, P.P., 2012. Drivers of Change in Crop–Livestock Systems and their Potential Impacts on Agro-Ecosystems Services and Human Wellbeing to 2030: A Study Commissioned by the CGIAR System Wide Livestock Programme. International Livestock Research Institute, Nairobi, Kenya.

Herrero, M., Grace, D., Njuki, J., Johnson, N., Enahoro, D., Silvestri, S., Rufino, M.C., 2013. The roles of livestock in developing countries. Anim. Int. J. Anim. Biosci. 7 (Suppl 1), 3–18.

Hoffmann, I., Boerma, D., Scherf, B., 2011. The global plan of action for animal genetic resources — the road to common understanding and agreement. Livest. Sci., Animal Breeding for Poverty Alleviation: Harnessing Science for Greater Impact 136, 7–14.

Jahnke, H.E., 1982. Livestock Production Systems and Livestock Development in Tropical Africa. 35 Kieler Wissenschaftsverlag Vauk, Kiel.

Kilgour, R.J., Waterhouse, T., Dwyer, C.M., Ivanov, I.D., 2008. Farming systems for sheep production and their effect on welfare. In: Dwyer, C.M. (Ed.), The Welfare of Sheep. Springer Netherlands, Dordrecht, pp. 213–265.

Koopman, R.B., Laney, K., Giamalva, J., 2014. Pork and Swine: Industry & Trade Summary. United States International Trade Commission (No. Publication ITS-011). U.S. International Trade Commission, Washington, DC.

Leakey, R., Kranjac-Berisavljevic, G., Caron, P., Craufurd, P., Martin, A.M., McDonald, A., Abedini, W., Afiff, S., Bakurin, N., Bass, S., 2009. Impacts of AKST on Development and Sustainability Goals.

Lewin, H.A., 2009. It's a Bull's market. Science, 478–479.

Madzingira, O., 2018. Animal welfare considerations in food-producing animals. In: Abubakar, M., Manzoor, S. (Eds.), Animal Welfare. IntechOpen Ltd, Londan, UK, pp. 99–122.

Miller, B.A., Lu, C.D., 2019. Current status of global dairy goat production: an overview. Asian-Australas. J. Anim. Sci. 32, 1219–1232.

Morris, S.T., 2017. Overview of sheep production systems. In: Ferguson, D.M., Lee, C., Fisher, A. (Eds.), Advances in Sheep Welfare. Woodhead Publishing, pp. 19–35.

Mortimore, M., 1991. A review of mixed farming systems in the semi-arid zone of sub-Saharan Africa. ILCA LED Working Document.

Noordhuizen, J., Gunnarsson, S., 2006. Animal production systems in the industrialised world. Rev. Sci. Tech. Off. Int. Epizoot. 25, 493–503.

OIE, 2019. Terrestrial Code: OIE - World Organisation for Animal Health [WWW Document]. https://www.oie.int/standard-setting/terrestrial-code/. (accessed 8.30.20).

Otte, J., Chilonda, P., 2003. Classification of cattle and small ruminant production systems in sub-Saharan Africa. Outlook Agric. 32 (3), 183–190.

Rosegrant, M.W., Fernández, M., Sinha, A., Alder, J., Ahammad, H., de Fraiture, C., Eickhout, B., Fonseca, J., Huang, J., Koyama, O., 2009. Looking into the Future for Agriculture and AKST.

Ruthenberg, H., 1971. Farming Systems in the Tropics, first ed. Clarendon Press, Oxford, UK.

Ruthenberg, H., 1980. Farming Systems in the Tropics, third ed. Clarendon Press, Oxford, UK.
Seré, C., Steinfeld, H., 1996. World Livestock Production Systems: Current Status, Issues and Trends (No. 127), FAO Animal Production and Health Paper. FAO, Rome, Italy.
Silanikove, N., 2000. The physiological basis of adaptation in goats to harsh environments. Small Rumin. Res. 35, 181–193.
Simonin, D., Gavinelli, A., 2019. The European Union legislation on animal welfare: state of play, enforcement and future activities. In: Hild, S., Schweitzer, L. (Eds.), Animal Welfare: From Science to Law, pp. 59–70.
Steinfeld, H., Wassenaar, T., Jutzi, S., 2006. Livestock production systems in developing countries: status, drivers, trends. Rev. Sci. Tech. 25, 505–516.
Stocker, T.F., Qin, D., Plattner, G.-K., Tignor, M.M.B., Allen, S.K., Boschung, J., Nauels, A., Xia, Y., Bex, V., Midgley, P.M., 2013. Climate Change 2013: The Physical Science Basis. Contribution of Working Group I to the Fifth Assessment Report of the Intergovernmental Panel on Climate Change. Cambridge University Press, Cambridge, United Kingdom and New York.
Teufel, N., Markemann, A., Kaufmann, B., Zárate, A.V., Otte, J., 2010. Livestock Production Systems in South Asia and the Greater Mekong Sub-Region: A Quantitative Description of Livestock Production in Bangladesh, Cambodia, India, Lao PDR, Nepal, Pakistan, Sri Lanka, Thailand, and Vietnam (Working Paper No. 48), Pro-Poor Livestock Policy Initiative (PPLPI). FAO.
Thornton, P.K., 2010. Livestock production: recent trends, future prospects. Philos. Trans. R. Soc. B Biol. Sci. 365, 2853–2867.
Thornton, P.K., Kruska, R.L., Henninger, N., Kristjanson, P.M., Reid, R.S., Atieno, F., Odero, A.N., Ndegwa, T., 2002. Mapping Poverty and Livestock in the Developing World. ILRI, Nairobi, Kenya.
Thornton, P.K., Jones, P.G., Owiyo, T., Kruska, R.L., Herrero, M.T., Kristjanson, P.M., Notenbaert, A.M.O., Bekele, N., Omolo, A., 2006. Mapping Climate Vulnerability and Poverty in Africa.
UN, 2020a. World Population Prospects - Population Division - United Nations [WWW Document]. https://population.un.org/wpp/DataQuery/. (accessed 8.27.20).
UN, 2020b. World Urbanization Prospects - Population Division - United Nations [WWW Document]. https://population.un.org/wup/DataQuery/. (accessed 8.27.20).
UNPD, 2017. World Population Prospects: The 2017 Revision, Key Findings and Advance Tables. Working Paper No. ESA/P/WP/248.
Van Vuuren, D.P., Ochola, W.O., Riha, S., Giampietro, M., Ginzo, H., Henrichs, T., Hussain, S., Kok, K., Makhura, M., Mirza, M., 2009. Outlook on Agricultural Change and its Drivers.
Wanapat, M., Kang, S., 2013. World Buffalo production: challenges in meat and milk production, and mitigation of methane emission. Buffalo Bull. 32, 1–21.
Wilson, R.T., 1986. Livestock Production in Central Mali: Long-term Studies on Cattle and Small Ruminants in the Agropastoral System. 14 ILRI (aka ILCA and ILRAD).
Zygoyiannis, D., 2006. Sheep production in the world and in Greece. In: Small Rumin. Res., Keynote Lectures of the 6th International Sheep Veterinary Congress. 62, pp. 143–147.

Further reading

Andreae, B., 1964b. Types of Farming Systems. Development and Transformation of Land Use, Animal Husbandry, and Farming Systems in Europe and Overseas as Well as New Methods of Their Demarcation. Types Farming Syst. Dev. Transform. Land Use Anim. Husb. Farming Syst. Eur. Overseas Well New Methods Their Demarcation.

CHAPTER FOUR

Sustainable livestock production and food security

Akash[a], Mozammel Hoque[a], Sukanta Mondal[b], and Satish Adusumilli[c]

[a]Division of Surgery, ICAR-Indian Veterinary Research Institute, Izatnagar, Uttar Pradesh, India
[b]Animal Physiology Division, ICAR-National Institute of Animal Nutrition and Physiology, Adugodi, Bengaluru, Karnataka, India
[c]Freimann Life Science Center, University of Notre Dame, Notre Dame, IN, United States

Introduction

There are about 7.08 billion people inhabiting the world today. The world needs 2800 million tons of cereals to properly feed everyone, of which global production amounts to just 2100 million tons. This gap in need and development leaves more than 868 million undernourished citizens worldwide, with 850 million of them living in developing countries (Tiwari et al., 2014). Also, rapid urbanization, increased purchasing power, and dietary change drives demand for richer diets and animal-origin proteins. The population is expected to reach around nine billion by 2050 (Kharas, 2010). Global demand for livestock products is forecast to double by 2050, based on projections of world population growth, increasing revenue, and more urbanization. Feeding this large population will be very challenging with daily changing climatic conditions. The sustainable intensification of livestock production must largely meet this rising demand. Intensified livestock development is followed by a change in the human population from rural to urban settings, with a corresponding rise in the size of the farm and animal housing densities. Food security, as described by the World Food Security Committee of the United Nations, is the situation under which all citizens always have physical, social, and economic access to sufficient safe and nutritious food that meets their dietary needs and food preferences for an active and healthy life. A large population can be guaranteed food security by implementing livestock farming and promotion to sustain food demand. Depending on resource quality, environmental factors, and social and economic contexts, there are various types of livestock production systems, and they vary considerably in sustainability. These livestock systems include extensive grassland systems, intensive landless systems,

Emerging Issues in Climate Smart Livestock Production
https://doi.org/10.1016/B978-0-12-822265-2.00011-9

and mixed farming systems. These systems make a significant contribution to human nutrition and livelihoods, and provide important services to the ecosystem (Padmakumar et al., 2019). They may also cause nutrient and environmental pollution as well as land degradation if not properly managed. Current animal production practices put serious environmental pressure through air, water, and soil pollution and the use of natural resources, including land, fossil fuels, and water. The livestock sector currently accounts for approximately 18% of global greenhouse gas emissions (Steinfeld et al., 2006). Food protection must be incorporated and coordinated with government policies on nutrition, food security, poverty alleviation, environmental health, and infrastructure creation within intensified livestock production systems. The intensification of livestock production without coordination can have negative effects, including facilitating disease transmission between animals and from animals to humans. Failure to implement public and environmental health policies will worsen the negative impacts of intensifying livestock (McAllister, 2019). Certain species of livestock such as goats, backyard poultry, pigs, rabbits, etc., can easily be reared without using many inputs and their products lead to food safety (Ahmad et al., 2018).

Livestock farming systems provide food security (Ahmad et al., 2018)

1. Dairy farming
2. Poultry farming
3. Pig farming
4. Sheep farming
5. Goat farming
6. Rabbit farming

1. **Dairy farming**: Dairy farming is the main and most important source of livelihood for the majority of the small and landless farmers. These small farmers rear the low-milk-producing cattle and buffalo. Such farmers may be educated to turn to modern dairy methods by purchasing elite dairy animals and improving husbandry techniques to maximize milk yield, which would help boost their livelihoods and serve as a complete food for other people. Better-quality breeds can be produced by practicing techniques such as artificial insemination from nondescript animals. Educated unemployed young people can also take

up dairy farming as a job through institutional support. They can then deliver clean and healthy milk at economic prices that the poor can afford. Cattle, buffalo, and goats are animals that can be reared as livestock for milk production.

2. **Poultry farming**: Backyard poultry farming can be done with limited inputs, so they need just a night shelter. They need some care in the early days of their lives and are built to endure the harshness of weather. With limited effort, they can be easily reared in groups and can be comfortably managed by the house's children and women. Backyard poultry farming can serve as the main pillar for a "food security model" for rural people. The broilers are the birds that are reared for meat production. The broilers grow to 1 kg of weight in just 28 days under proper management conditions, so that profits can be made within a month. Unemployed villagers can be employed throughout the year by the adoption of broiler farming. It also plays a major role in boosting the nutritional status. Broiler farming along with agriculture together can raise the social and economic status and lead to food security.
3. **Pig farming**: The pig is the most beneficial livestock species among all and can be started with minimum investment by constructing a low-cost shed. A farmer with good management will be able to produce 3–4 crops each year and then can sell them at a reasonable cost. Low gestation and high litter size make this extremely profitable. They are not only producing food but also producing other specialized products such as hams and sausages. They have a high feed conversion rate and can be reared on kitchen waste. Pig farming helps the poor farmer to increase liability and offers a sustainable food source ensuring food security.
4. **Sheep farming**: Sheep offer a steady source of revenue for farmers in the form of mutton, wool, and livestock purchases, thereby creating a healthy economic climate for farmers. This livestock species is best suited for hilly areas and much less maintenance is required to rear them. Farmer living conditions can be improved by selling wool. The income thus obtained supports the farmers' livelihoods.
5. **Goat farming**: Goat is known as the "poor man's cow" due to its numerous advantages despite a low investment. Goats are reared for milk and chevon by poor sections of society. Goats can be reared easily because they can survive on weeds, trees, tree tops, and leaves to feed their appetite. Goats give birth to kids in 5 months and twinning is also common in goats, so a better flock can be maintained by good managemental

practices. To maintain economy and nutrition, goats can prove to be an excellent source of milk, meat, and fiber for the rural poor.

6. **Rabbit farming (Cuniculture)**: Animal husbandry of breeding and raising domestic rabbits as livestock for their meat, fur, or wool. Globally, rabbits are reared for meat production in a variety of settings. Commonly, backyard operations with small holdings are practiced in many countries and large-scale commercial operations are in existence in Europe and Asia. Rabbit farming has significant potential to improve food security and nutrition in developing countries. Rabbits are highly productive, thanks to a short gestation and great prolificacy. A female rabbit can produce up to 80 kg of meat per year, that is, 2900%–3000% of its own weight in meat. Rabbit meat is highly nutritious, low in fat and cholesterol, and rich in proteins, vitamins, and minerals. Backyard rabbit farming provides additional income to small and marginal farmers and enhances the diet quality of the poor rural and urban population. Rabbits adapt well in household production due to their low investment and labor costs. Being herbivores, rabbits do not compete with humans for food, making this a sustainable practice. Rabbits can turn 20% of the proteins they eat into edible meat, compared to 8%–12% for beef, 16%–18% for pigs, and 22%–23% for broiler chickens under efficient production systems. Therefore, rabbit meat is more economical in terms of feed energy than beef. In 1990, the world's annual production of rabbit meat was estimated to be 1.5 million tons. In 2014, the number was estimated at 2 million tons. China is among the world's largest producers and consumers of rabbit meat, accounting for some 30% of the world's total consumption. Rabbit farming also has some constraints, including management of diseases that includes vaccination, the lack of production management and technical training, and taboos about eating rabbit meat. An FAO study in 64 developing countries found that 30% of people surveyed believed that social, religious, or other reasons would not favor the development of rabbit production. The FAO has supported and developed rabbit production projects in various parts of the world (FAO, 2015; Mohan, 2019).

Important consideration about sustainable livestock production

1. **Biosecurity measures**: Housing with higher densities is the main threat for food safety by livestock intensification. Greater animal-to-animal

contact can promote disease transmission. In intensive ruminant systems, animals from a large geographic region are congregated in feedlots or barns at a single site. Pathogens can reach the production system from across this geographic region and be transmitted among individuals. Biocides are used to minimize the incidence and prevalence of pathogens, but caution must be taken to ensure that biocides do not contaminate meat, milk, or the environment. With proper biosecurity measures, disease transmission can be prevented in livestock and production and food security can be maintained.

2. **Health management**: Transportation and changes in nutrition are the main factors that make livestock animals susceptible to diseases. Antibiotics used to treat infections promote resistance and make it difficult to treat the disease. Vaccines reduce the use of antibiotics, but they are useless if the immune system of the animal is compromised. So it is extremely important to promote nutritional management and reduce stress levels to boost the immune system and reduce disease in livestock species. Rules and regulations regarding the feeding and transportation of livestock can reduce the occurrence of infectious diseases and promote production and food safety.
3. **Feeding strategies**: Feeding strategies for livestock must ensure that they improve the nutritional status and decrease the release of harmful byproducts and pathogens. Balanced nutrition, which improves immunity and production, should be followed as a feeding strategy.
4. **Waste management**: To increase livestock production, there is a need to increase the livestock density, but this leads to increased pathogens. Improper waste management can contaminate the soil, water, and feed. So, the proper management of livestock waste is very necessary to restrict microbial contamination in the production system.
5. **HACCP and food safety**: Livestock from a large geographical region are taken to a single place, which increase the pathogen overload. Hazard analysis and critical control point (HACCP) is a systematic preventive approach to food safety from biological, chemical, and physical contaminants in livestock production processes that may make the end product unsafe. This approach designs measures to decrease these risks to a safe level. The implementation of HACCP during slaughter and meat processing is important. Such frameworks must be harmonized in such a way that food protection standards are consistent across countries, but sufficiently versatile to be regionally applicable to different livestock production activities. Regulations on food safety that support public

health should be based on science rather than market forces or international trade.

6. **Awareness about food safety systems**: Educating customers on food health practices is important because 30%–40% of foodborne illnesses occur in the home. Equally important is awareness of the health benefits and risks of meat and milk products. Consumers should be conscious of the effect of various production processes on the environmental factors of livestock goods. Various extension programs and social networking play important roles in promoting and evaluating the effects of food safety education and understanding the environmental impact of various forms of production systems.

Livestock farming with care

This concept of "livestock farming with care" is based on four principles: one health, customized care, no nuisance, and credible performance (Scholten et al., 2013).

A. **One health (health and safety of animals and humans)**: Safeguarding human and animal health related to the production of livestock includes more than food safety. This also means the eradication of infectious animal and zoonotic diseases that arise from livestock farming and transportation as well as mitigation of the effects of antibiotic use, such as the development of antibiotic resistance to microbes.

B. **Customized care (from the perspective of the integrity of the animal)**: Ensuring animal production robustness, dignity, and integrity requires more than regard for human values in terms of animal welfare issues. It also includes concern for the unique biological traits and conditions of particular animals in the herd, resulting in good animal health being observed on the farm.

C. **No nuisance (with regard for environmental and societal perspectives)**: This means avoiding environmental exposure to vital waste material emissions such as dust, noise, and pathogens resulting from livestock production systems; it includes sustainable natural resource management, biodiversity, and land degradation prevention.

D. **Credible performance (from a socioeconomic perspective)**: This means ensuring a safe and efficient livestock production system with good opportunities for farmers in local, state, international, or global food supply chains.

Precision livestock farming

To be able to produce healthy, reliable, inexpensive, environmentally friendly, and welfare-friendly food products and sell such products in an increasingly competitive foreign agricultural sector, livestock farmers must have access to knowledge relevant to the timely production of livestock. Significant advances in the fields of nanotechnology, genomics, and information technology are awaited. Such new technologies have also become available for livestock farming and are the driving forces in information and communication technology (ICT) implementation. Electronically regulated livestock production systems with an emphasis on ICT are needed to ensure that information is obtained in a cost-effective and timely manner and is readily accessible on farms (Banhazi et al., 2012). On the basis of precision agriculture (Gebbers and Adamchuk, 2010), the practice of precision livestock farming (PLF) is currently under development (Wathes et al., 2008; Berckmans, 2008). PLF allows real-time livestock management and complies with the livestock farming with care guidelines by enabling individual animal variations within a herd, spatial and temporal variations in circumstances, and integrative environmental and health performance on the farm. PLF allows farmers to observe the requirements of each and every animal. Advanced electronic and ICT-related technology deployed on farms as part of PLF programs would promote the management of livestock that is more sensitive to market demand. PLF technology includes methods for the electronic calculation of essential components of the production system that demonstrates the quality of resource use as well as analyses the knowledge collected and monitors processes to maintain the optimal performance of both resource use and animal productivity. The international development of PLF technology will provide the tools needed to acquire and manage data in real time. IT methods can be distinguished between sensing, data processing and storage techniques, dynamic decision support modeling, and manual or robotic implementation of routine decision making. The key objective of PLF is to increase farm productivity, efficiency, and sustainability by enhancing the collection, management, and use of on-farm data to enhance the nutritional, environmental, and other management aspects of various livestock species (Niemi et al., 2010; Schmoldt, 2001; Zhang et al., 2002). PLF can also provide a mechanism to "enforce" the implementation of best managemental practices in farms and thus reduce the variability observed in inputs and outputs, such as the varying quality and quantity of different livestock products (Schulze et al., 2007).

Sustainable livestock production and environment

Increased global understanding of climate change and research showing that livestock is a contributor to greenhouse gases, environmental destruction, and biodiversity depletion have contributed to a number of coordinated initiatives aimed at improving or maintaining the viability of livestock systems that provide economic and ecological resources without sacrificing future health and environment. Food production and processing as well as enteric feed fermentation contribute 45% and 39% of total agricultural emissions, respectively (Steinfeld et al., 2006). Ruminants produce about 90% of livestock emissions through enteric fermentation (188 million tons), and the remaining 10% from manure (Swamy and Bhattacharya, 2006). Eutrophication, overgrazing, and soil degradation are some adverse environmental effects from inadequately managed livestock production systems. For example, a frequently cited statistic is that livestock contribute 18% of greenhouse gases worldwide (Steinfeld et al., 2006). A recent metaanalysis of 86 studies that analyzed diverse agroforestry systems showed that when grassland was transformed into silvo-pastures mixing plants, forage, and livestock, the net accumulation of soil carbon or the sink of greenhouse gases was greater (Feliciano et al., 2018). Nutritional, genetic, health, and management strategies have been developed by animal scientists to reduce greenhouse gas emission intensities by as much as 30% (Gerber et al., 2013a,b). One of the main strategies for ensuring the sustainability of livestock production systems is the idea of sustainable diets that are competitive, ethically and socioculturally appropriate, and environmentally safe. In consideration of competitiveness among feed and food systems, the principle of sustainable diets stipulates that potential feed systems will concentrate on the efficiency of the conversion of fibrous foods, such as residue of crops with a large content of poorly digestible structural carbohydrates (lignin and cellulose) into human-consuming animal products. Sustainable food and feed systems thus have the capacity to preserve the productivity of feed systems while at the same time reducing their detrimental environmental and social effects (Bocquier and González-García, 2010).

Sustainable livestock production and human nutrition

According to the World Bank (2016), 30% of the Indian population lives below the poverty line and more than 70% of them live in rural areas with inadequate nutritional means and thus food insecurity. By promoting livestock farming among them and helping them support their food demands through livestock farming, these masses can be targeted and taken under the umbrella of food security. Livestock is a vital and very crucial component of India's agricultural fiber. This not only provides agricultural rural households with security when crop failures occur but also meets their food demands. Food dependent on livestock is complete in every respect, being rich in calories, protein, and lipids, which are important for the ultimate growth of the human body. The selling of value-added goods paves the way for the farmer to gain additional profit from which to support his economy. Even so, heavy focus must be put on the quality management and health of these foods. This can be achieved effectively by increasing consciousness among livestock farmers of good management practices (GMPs) and good farming practices (GHPs) that can eventually contribute to clean, nutritious, and balanced food production for both producers and consumers.

Sustainable livestock production and economy

Livestock produce capital through the sale of animals and livestock products for farmers of all categories. Livestock products are among the top 10 commodities exchanged globally with a value of around $6.5 million (FAOSTAT, 2017). Millions of farmers in low- and middle-income countries hold livestock as a status mark with a higher status or as emergency cover and sell it to fulfill their needs. It also provides opportunity to women who particularly depend on livestock animals as a major source of income. Livestock also helps to motivate women with essential and varied positions in various low- and middle-income countries. The livestock sector is an important part of agriculture, adding 60%–70% of the total jobs in low- and middle-income countries, especially in Africa and Asia, according to the International Labor Organization. It is not restricted to production but includes processing, transportation, storage, and marketing. Aggregate direct and indirect job creation and socioeconomic impacts of the livestock industry in low- and middle-income countries have been noticed.

Effects of animal products on human health

Animal products not only help in human growth but also provide essential micronutrients for the human body. Cow milk is a major source of calcium, which is a main component of bone growth, blood clotting, and muscle contraction. Dairy products promote bone strength during infancy and puberty and decrease the likelihood of osteoporosis and type 2 diabetes (Rozenberg et al., 2016). Dairy consumption was also linked with decreased blood pressure, arterial stiffness, coronary disease, and rickets (Fekete, 2016). Meat is a good source of iron, which is essential to motor growth and neurological processing, including thoughts and memory (Nyaradi et al., 2013). Iron-containing supplemental foods such as meat are especially essential for infants with inadequate iron reserves or inadequate intake (Obbagy et al., 2019). Essential improvement and awareness about livestock production will help to improve the nutritional and health status of humans.

Contribution of livestock in the world

Livestock production provides a major contribution to improved food security and nutrition (FSN) by increasing the quantity of highly nutritious animal source food (ASF) and livestock perishable products that are a valuable source of income for the world's poor to allow them to buy food. It is also a good source of manure that can be used as fertilizer to enhance crop growth. ASF is a rich source of essential micronutrients such as zinc, iron, and vitamin A. The same micronutrients are present in plant source food but their amount and bioavailability are greater in ASF that supports women and young children (Gibson, 2011). ASF is an energy-rich food that is beneficial for undernourished people but bad for overweight people; it provides around 16% of total calories worldwide according to FAOSTAT for 2010. Livestock affect human safety and the climate, too. Infectious agents such as bacteria or viruses can be spread directly from livestock to humans by vectors or food chains and may trigger disease. It has been estimated that approximately 60% of human infectious diseases are zoonotic (Taylor et al., 2001). The latest figures indicate that 14.5% of greenhouse gas emissions come from livestock production systems, directly or indirectly (Gerber et al., 2013a,b; Herrero et al., 2016). Livestock has become the fastest-growing field globally and its development has been accelerated by increasing wages, demand for ASF, and increasing populations. It is currently estimated that the livestock industry contributes 30%–40% to global agricultural GDP and that the survival of 1.3 billion people relies on livestock.

Global challenges in sustainable livestock production

The viewpoint on livestock and related opportunities or challenges as defined in the SOFA (FAO, 2009) differs between postindustrial low agricultural growth and emerging economies. Overall, the emphasis on livestock sustainability issues in high-income countries including Northern and Western Europe is on the environment, including greenhouse gas reduction as well as animal health and welfare.

1. **Environmental challenges**: Environmental challenges mainly focus on the management of climate change and natural resources as well as the reduction of greenhouse gases. The annual figures of the livestock industry greenhouse gas emissions are 14.5% of all anthropogenic emissions. There are significant variations between livestock species and farming systems, with higher production usually providing lower emissions per generated product. As a general rule, intensive systems emit fewer greenhouse gases than extensive systems. Improved productivity and biodiversity as well as the reduction of greenhouse gases can be accomplished by proper animal heath as well as better nutrition and management.
2. **Economic challenges**: Economic challenges mainly relate to trade, markets, and less investment in small-scale farms. The functionality and accessibility of markets for livestock and livestock products are very essential for food security. The basic reasons for poorly developed markets are information gaps and poor supply chains, the negative environmental and social externalities are not priced, and poorly constructed taxes and trade policies. Trade-related issues include unfair competition from subsidized imports, disparities in national production requirements for antimicrobial use, animal health and environmental effects, disease transmission by livestock or agricultural goods, and economic discussions on the pros and cons of food security in relation to foreign trade integration.
3. **Social challenges**: Social challenges are about child labor and gender discrimination. Pastoralist societies often prefer men for work and child labor is very common in these societies. In intensive systems, the conditions for farmers and other workers are unsatisfactory and they have a low social status. There is also a significant proportion of migrant labor in intensive livestock production systems, and this labor is widely established to be synonymous with bad working conditions.

4. **Health challenges**: Health challenges effectively focus on the "one world one health" concept and deal with human as well as animal health. Productivity-reducing animal diseases are a direct threat to food security. The risk of livestock diseases may be immense and is a big reason for poor production in low-income countries. Not only do well-known infectious diseases lead to low productivity, but also animal health is commonly affected by endemic diseases. In addition to the threat of reduced production and productivity, infectious animal diseases may directly affect human health. The risk of transmitting zoonotic agents to humans is greater when humans and animals live in close proximity.
5. **Animal welfare challenges**: There are very different views on animal welfare across countries. This is therefore a major challenge to reconcile increasing productivity and welfare in the developing, and often uncontrolled, intensive livestock systems around the world. Even so, retailers and consumers in a range of high-income countries are rapidly seeking animal-friendly livestock production.

Livestock production and climatic changes

The environmental costs of conventional pig meat production are estimated at EUR 1.9 kg^{-1}, whereas the private costs are estimated at EUR 1.4 (Nguyen et al., 2010). The production of greenhouse gases from milk is 1.2 kg CO2 eqv kg^{-1} relative to 0.8 kg CO2 eqv kg^{-1} from wheat (Garnett, 2009). Agricultural land use generates 35% of industrial greenhouse gases (Stern, 2007). Most of this pollution arises from agricultural processing, which accounts for around 18% of global greenhouse gas emissions, rendering it one of the top two main contributors to the most severe environmental issues (Steinfeld et al., 2006). Much of the livestock-generated greenhouse gases can be mitigated through proper feeding, managing manure, and improving productivity (Steinfeld and Gerber, 2010). Improving the roughage digestibility (Gerber et al., 2010), supplementing prebiotics (Mwenya et al., 2004; Dhama et al., 2008), organic acids (Ungerfeld et al., 2007), tannins (Hess et al., 2006), saponins (Agarwal et al., 2006), essential oils (Calsamiglia et al., 2007) and ionophores (Tedeschi et al., 2003) that minimize enteric methane in ruminants. Greenhouse gas emissions can be mitigated by 496 million tons of CO2 equivalent per year if all the cattle dung collected in the country is used in biogas production (Pathak et al., 2009).

Livestock and biodiversity loss

The global food supply is dependent on genetic diversity (Hoffmann, 2011). India has possibly developed about 30,000 indigenous rice varieties in the last 50 years. It will be reduced to 50 varieties in 20 years, the top 10 of which account for more than 75% of the rice area of the country (Sudarshan, 2002). Of the 8262 breeds reported worldwide, 628 are classified as extinct and 1881 are classified as at-risk (CGRFA, 2012). A large part of this genetic erosion stems from global livestock production practices. The risk of biodiversity loss is impossible to measure; losses in terms of culture values, eco logical resources, loss of genes for disease, adaptation to the environment, and other beneficial characteristics of both plants and animals cannot be overlooked. India is home to 167 species of essential crops and has 320 wild relatives and 298 varieties of 14 livestock species. However, the cross-breeding of cattle with foreign types has contributed to the dilution of the gene pool of local breeds for decades (Mathias and Mundy, 2005). It is advised to preserve 25%–50% local breed inheritance in cross-breeds in order to retain livestock biodiversity at the genetic level. This issue can be solved with the use of automated detection and data collection techniques such as radio frequency recognition in artificial insemination centers and veterinary clinics.

Future demand of sustainable livestock production

The global world food system has been estimated to provide sufficient nutrition to 85% of the population (Black, 2003), that is, only 15% of the world's population is not properly fed. Increasing income rates and demographic increases have contributed to further shifts in lifestyles, such as a transition from rice to wheat as well as increased intake of vegetables, berries, meat, and beverages (Popkin, 2002). Through 2050, it is estimated that the intake of meat and dairy products will rise by 173% and 158%, respectively. The typical Indian food habit is also turning to livestock products. This increased demand for animal protein is hard to reconcile with the present method of livestock farming, as India is already facing shortages in feed and fodder as well as dwindling of grazing space. More than 35% of the cereals produced worldwide are currently fed to livestock (Trostle, 2008). In India, the total consumption of concentrated feed was only 47 million tons in 2003, during which food grain production amounted to 174.7

million tons. However, the demand for concentrated feed is estimated to increase marginally to 55.7 million tons in 2050 with a food and feed grain demand of 377 million tons (Dikshit and Birthal, 2010; Amarasinghe et al., 2007).

Maneuvers to achieve sustainable livestock production for food security

1. **Good governance**: Food security for everyone is the product of the joint efforts of multiple stakeholders, thereby maintaining balanced governance in this sector and allowing every stakeholder to participate. The need for innovation in research and development in agriculture in general is another dimension of governance. The private sector is already the world's pioneer in agricultural research and development (FAO, 2012). Another governance problem is whether to internalize the externalities of the livestock sector. All these issues can be resolved by good governance policies.
2. **Better market access**: Increased market access for smallholders is a way for the majority of the stakeholders to maximize efficiency, output, and profits while contributing to improved food security for everyone. Connections to international trade have also been shown to minimize hunger and increase food security (Anderson et al., 2011; FAO/OECD, 2014). However, when it comes to animal products from low-income nations, they do not reach foreign markets because some of these nations have not been seen to be safe from infectious animal diseases (WTO, 1994; OIE, 2016).
3. **Better resource use efficiency**: Increased resource efficiency in the livestock sector is a policy that encourages environmental sustainability, which always goes side by side with economic sustainability and, if handled carefully, may even encourage social sustainability. A major obstacle, of course, is the expense of the improvements required for this development, which must be taken into consideration to reach especially resource-poor farmers. Yield gap analysis is a valuable method to enhance resource efficiency (Sumberg, 2012). This strategy is focused on the idea of reducing the gap between the best and worst performers in the field by integrating current strategies and techniques and implementing new technology.

4. **Improvement in animal health**: Improved animal health would lead to a healthy livestock sector and food security by improved production as well as reduced environmental footprint and greenhouse gas emissions. The development strategy is to expand animal health programs at all stages, with an emphasis on prevention rather than curing diseases. This method often involves low-cost control of livestock on the field, such as improved reproductive management and the implementation of biosecurity procedures.
5. **Improvement in genetics**: Animal genetics have traditionally been used to boost production. Breeding schemes can take into account other aspects of livestock sustainability, such as disease resistance, heat-stress resilience, and reproductive health. There are still tragic examples of very high-produced cattle breeds in the climate, disease and feed environments they cannot cope with, resulting in very low yields. One may, therefore, look forward to a substantial contribution to efforts to close the yield gap from a genetic approach, taking into account a multitude of animal characteristics.
6. **Opposing social and gender discrimination**: Social and gender equality and responsibility are two aspects of life that are given a very different priority in different cultures and countries. Social accountability is much more important as there are systemic shifts occurring, as in the case with the growth of the livestock sector; more employees would be engaged in large-scale farms with often very dangerous working conditions. As agriculture is traditionally uncontrolled in many countries, the implementation of socially acceptable legislation can be a problem (Gerber et al., 2013a,b). However, social equality must be measured and safeguarded through means or measures that lead to a sustainable livestock sector. Applying a gender viewpoint to the livestock sector could have some significant transformative effects linked to food security. Empowerment of women and innovation in women's education have led to a dramatic change in household diets and nutrition (Smith and Haddad, 2015). In a variety of societies, women in smallholder structures are responsible for the day-to-day control of food, egg processing, and poultry meat products that are eaten in the home or marketed for revenue. To increase the productivity of this ASF, women must be given priority. So social and gender equality must be given importance to improve the sustainable livestock sector.

Conclusion and future prospective

Livestock is a vital and most important component of agriculture globally. The livestock sector is a significant contributor to the livelihoods of the world's poor. It is also the fastest-growing field of agricultural output worldwide and is expected to continue to expand. This not only gives rural farming households security when crop failures occur, but also meets their food requirements. Food based on livestock is complete in all respects, being rich in calories, proteins, and lipids that are essential for the ideal growth of the human body. Even now, a heavy emphasis needs to be put on the quality control and well-being of these products. This can be done effectively by growing knowledge among livestock farmers of good management practices and good farming practices that will eventually lead to clean, healthy, and balanced food production. However, the production and consumption of ASF varies greatly across countries as well as within countries. That, in effect, implies that the problems and strategies for a productive livestock sector that lead to food security can be somewhat different across countries or systems of farming. Food security can be accomplished by narrowing "yield gaps," growing the productivity of crops and livestock, eliminating waste in the food supply chain, the diversification and incorporation of crops and livestock, maintaining agrobiodiversity, and introducing greenhouse gas management and advanced production technology in the agriculture and livestock sectors. Scaling up and connecting smallholders with corporate bodies and global food supply chains require immediate attention, as they own the majority of the livestock in the country and can play a key role in food security. Humanity must agree that a sustainable rise in livestock production is a requirement for fulfilling the future demand for livestock products. As a consequence of climate change and market factors, food security policies would need to be complex and compatible with increasing livestock production practices. Food security policies will be enforced from a systemic viewpoint, taking into consideration livestock security, human health, environmental sustainability, societal gains, and the allocation of resources. Different genetic, nutritional, and health-related approaches exist to reduce the environmental effects of livestock and make them a meaningful contributor to sustainable livestock production and food security.

References

Agarwal, N., Kamra, D.N., Chaudhary, L.C., Patra, A.K., 2006. Effect of Sapindus mukorossi extracts on in vitro methanogenesis and fermentation characteristics in buffalo rumen liquor. J. Appl. Anim. Res. 30, 1–4.

Ahmad, S.F., Singh, A., Kour, G., Arora, R., 2018. Food security through sustainable livestock farming. Krishi Jagran. https://krishijagran.com/animal-husbandry/food-security-through-sustainable-livestock-farming/.

Amarasinghe, U.A., Shah, T., Turral, H., Anand, B.K., 2007. India's water future to 2025-2050: business-as-usual scenario and deviations. Research Report 123, International Water Management Institute, Colombo, Sri Lanka.

Anderson, K., Cockburn, J., Martin, W., 2011. Would freeing up world trade reduce poverty and inequality? The vexed role of agricultural distortions. Policy Research Working Papers, https://doi.org/10.1596/1813-9450-5603.

Banhazi, T., Babinszky, L., Halas, V., Tscharke, T., 2012. Precision livestock farming: precision feeding technologies and sustainable livestock production. Int. J. Agric. Biol. Eng. 5 (4), 54–61.

Berckmans, D., 2008. Precision livestock farming (PLF). Comput. Electron. Agric. 62 (1), 1.

Black, R., 2003. Micronutrient deficiency: an underlying cause of morbidity and mortality. Bull. World Health Organ. 81, 79–87.

Bocquier, F., González-García, E., 2010. Sustainability of ruminant agriculture in the new context: feeding strategies and features of animal adaptability into the necessary holistic approach. Animal 4, 1258–1273.

Calsamiglia, S., Busquet, M., Cardozo, P.W., Castillejos, L., Ferret, A., 2007. Essential oils as modifiers of rumen microbial fermentation. J. Dairy Sci. 90, 2580–2595.

CGRFA, 2012. Status and trends of animal genetic resources-2012. In: Item 4.2 of the Provisional Agenda Fourteenth Regular Session, April 15-19, 2013. Commission on Genetic Resources for Food and Agriculture, Rome.

Dhama, K., Mahendran, M., Tomar, S., Chauhan, R.S., 2008. Beneficial effects of probiotics and prebiotics in livestock and poultry: the current perspectives. Intas Polivet 9, 1–12.

Dikshit, A.K., Birthal, P.S., 2010. India's livestock feed demand: estimates and projections. Agric. Econ. Res. Rev. 23, 15–28.

FAO, 2009. The State of Food and Agriculture 2009—Livestock in the Balance. FAO, Rome. http://www.fao.org/docrep/012/i0680e/i0680e.pdf.

FAO, 2012. World agriculture towards 20130/20150: the 2012 revision, by N. Alexandratos & J. Bruinsma. ESA Working Paper No. 12-03 http://www.fao.org/economic/esa/esag/en/.

FAO Report, 2015. The Rabbit—Husbandry, Health and Production. Food and Agricultural Organization of the United Nations (FAO), Rome (Italy).

FAO/OECD, 2014. Food security and nutrition opportunities for economic growth and job creation in relation to food. Report by FAO and the OECD (with inputs by ADB, IFAD, IFPRI and WTO) to the G20 Development Working Group.

FAOSTAT, 2017. Statistics Database of the Food and Agricultural Organization of the United Nations (FAO). FAO, Rome (Italy).

Fekete, Á.A., 2016. The Comparative Effects of Milk Proteins on Blood Pressure, Vascular Function and Other Cardiovascular Risk Markers (Doctoral Dissertation). University of Reading, UK.

Feliciano, D., Ledo, A., Hillier, J., Nayak, D.R., 2018. Which agroforestry options give the greatest soil and above ground carbon benefits in different world regions? Agric. Ecosyst. Environ. 254, 117–129.

Garnett, T., 2009. Livestock-related greenhouse gas emissions: impacts and options for policy makers. Environ. Sci. Policy 12, 491–503.

Gebbers, R., Adamchuk, V.I., 2010. Precision agriculture and food security. Science 327 (5967), 828–831.

Gerber, P., Vellinga, T., Dietze, K., Opio, C., Sironi, D., Thieme, O., Weiler, V., 2010. Greenhouse gas emissions from the dairy sector: a life cycle assessment. A Report Prepared by Food and Agriculture Organization of the United Nations, Animal Production and Health Division, Rome, Italy.

Gerber, P.J., Henderson, B., Makkar, H.P., 2013a. Mitigation of Greenhouse Gas Emissions in Livestock Production: A Review of Technical Options for Non-CO2 Emissions (No. 177). Food and Agriculture Organization of the United Nations (FAO), Rome.

Gerber, P.J., Steinfeld, H., Henderson, B., Mottet, A., Opio, C., Dijkman, J., Falcucci, A., Tempio, G., 2013b. Tackling Climate Change Through Livestock—A Global Assessment of Emissions and Mitigation Opportunities. Food and Agriculture Organization of the United Nations (FAO), Rome. http://www.fao.org/docrep/018/i3437e/i3437e.pdf.

Gibson, R.S., 2011. Strategies for preventing multi-micronutrient deficiencies: a review of experiences with food-based approaches in developing countries. In: Thompson, E., Amoroso, L. (Eds.), Combating Micronutrient Deficiencies: Food-Based Approaches. FAO, Rome.

Herrero, M., Henderson, B., Havlík, P., Thornton, P.K., Conant, R.T., Smith, P., Wirsenius, S., Hristov, A.N., Gerber, P., Gill, M., Butterbach-Bahl, K., Valin, H., Garnett, T., Stehfest, E., 2016. Greenhouse gas mitigation potentials in the livestock sector. Nat. Clim. Chang. https://doi.org/10.1038/nclimate2925.

Hess, H.D., Tiemann, T.T., Noto, F., Carulla, J.E., Kreuzer, M., 2006. Strategic use of tannins as means to limit methane emission from ruminant livestock. Int. Congr. Ser. 1293, 164–167.

Hoffmann, I., 2011. Livestock biodiversity and sustainability. Livest. Sci. 139, 69–79.

Kharas, H., 2010. The emerging middle class in developing countries. Working Paper No. 285, OECD Development Centre, Paris.

Mathias, E., Mundy, P., 2005. Herd Movements: The Exchange of Livestock Breeds and Genes Between North and South. League for Pastoral Peoples and Endogenous Livestock Development, Ober-Ramstadt, Germany.

McAllister, T.A., 2019. Safe and sustainable livestock production. In: The First FAO/WHO/AU International Food Safety Conference Addis Ababa.

Mohan, C., 2019. How to do profitable commercial rabbit farming in India? Krishi Jagran. 15 June.

Mwenya, B., Zhou, X., Santoso, B., Sar, C., Gamo, Y., Kobayashi, T., Takahashi, J., 2004. Effects of probiotic-vitacogen and β1-4 galacto-oligosaccharides supplementation on methanogenesis and energy and nitrogen utilization in dairy cows. Asian Aust. J. Anim. Sci. 17, 349–354.

Nguyen, T.L.T., Hermansen, J.E., Mogensen, L., 2010. Environmental costs of meat production: the case of typical EU pork production. J. Clean. Prod. 30, 1–9.

Niemi, J.K., Sevón-Aimonen, M.L., Pietola, K., Stalder, K.J., 2010. The value of precision feeding technologies for grow-finish swine. Livest. Sci. 129, 13–23.

Nyaradi, A., Li, J., Hickling, S., Foster, J., Oddy, W.H., 2013. The role of nutrition in children's neurocognitive development, from pregnancy through childhood. Front. Hum. Neurosci. 7, 97.

Obbagy, J.E., English, L.K., Psota, T.L., Wong, Y.P., Butte, N.F., Dewey, K.G., Fox, M.K., Greer, F.R., Krebs, N.F., Scanlon, K.S., et al., 2019. Complementary feeding and micronutrient status: a systematic review. Am. J. Clin. Nutr. 109 (Supplement_7), 852–871.

OIE, 2016. OIE-Listed Diseases, Infections and Infestations in Force in 2016. http://www.oie.int/animal-health-in-the-world/oie-listed-diseases-2016.

Padmakumar, V.P., Mckune, S., Miller, L., Hendrickx, S., Balehegn, M., Dahl, G.E., Adesogan, A.T., 2019. Sustainable livestock systems to improve human health, nutrition, and economic status. Anim. Front. 9 (4), 39–50.

Pathak, H., Jain, N., Bhatia, A., Mohanty, S., Gupta, N., 2009. Global warming mitigation potential of biogas plants in India. Environ. Monit. Assess. 157, 407–418.

Popkin, B.M., 2002. The dynamics of the dietary transition in the developing world. In: Caballlero, B., Popkin, B.M. (Eds.), The Nutrition Transition: Diet and Disease in the Developing World. Academic Press, Amsterdam, USA, pp. 111–121. ISBN-13: 9780080492346.

Rozenberg, S., Body, J.J., Bruyère, O., Bergmann, P., Brandi, M.L., Cooper, C., Devogelaer, J.P., Gielen, E., Goemaere, S.K., J.M., et al., 2016. Effects of dairy products consumption on health: benefits and beliefs—a commentary from the Belgian Bone Club and the European Society for Clinical and Economic Aspects of Osteoporosis, Osteoarthritis and Musculoskeletal Diseases. Calcif. Tissue Int. 98, 1–17.

Schmoldt, D.L., 2001. Precision agriculture and information technology. Comput. Electron. Agric. 30 (1–3), 5–7.

Scholten, M.C.T., de Boer, I.J.M., Gremmen, B., Lokhorst, C., 2013. Livestock farming with care: towards sustainable production of animal-source food. NJAS-Wagen J. Life Sci. 66, 3–5.

Schulze, C., Spilke, J., Lehner, W., 2007. Data modeling for precision dairy farming within the competitive field of operational and analytical tasks. Comput. Electron. Agric. 59 (1–2), 39–55.

Smith, L.C., Haddad, L., 2015. Reducing child undernutrition: past drivers and priorities for the post-MDG era. World Dev. 68, 180–204.

Steinfeld, H., Gerber, P., 2010. Livestock production and the global environment: consume less or produce better? Proc. Natl. Acad. Sci. 107, 18237–18238.

Steinfeld, H., Gerber, P., Wassenaar, T., Castel, V., Rosales, M., De Haan, C., 2006. Livestock's Long Shadow: Environmental Issues and Options. FAO, Rome, Italy.

Stern, N., 2007. The Economics of Climate Change: The Stern Review. Cambridge University Press, Cambridge, p. 692. ISBN-13: 9780521700801.

Sudarshan, H., 2002. Foreword. In: Ramprasad, V. (Ed.), Hidden Harvests: Community Based Biodiversity Conservation. Green Foundation, Bangalore, India.

Sumberg, J., 2012. Mind the (yield) gap(s). Food Security 4 (4), 509–518.

Swamy, M., Bhattacharya, S., 2006. Budgeting anthropogenic greenhouse gas emission from Indian livestock using country-specific emission coefficients. Curr. Sci. 91, 1340–1353.

Taylor, L.H., Latham, S.M., Woolhouse, M.E.J., 2001. Risk Factor for Human Disease Emergence. The Royal Society, https://doi.org/10.1098/rstb.2001.0888.

Tedeschi, L.O., Fox, D.G., Tylutki, T.P., 2003. Potential environmental benefits of ionophores in ruminant diets. J. Environ. Qual. 32, 1591–1602.

Tiwari, R., Kumar, H.D., Dutt, T., Singh, B.P., Pachaiyappan., Dhama, K., 2014. Future challenges of food security and sustainable livestock production in India in the changing climatic scenario. Asian J. Anim. Vet. Adv. 9 (7), 367–384.

Trostle, R., 2008. Global agricultural supply and demand: factors contributing to the recent increase in food commodity prices. Outlook No. WRS-0801/May 2008, United States Department of Agriculture/Economic Research Service, USA, pp. 1–30.

Ungerfeld, E.M., Kohn, R.A., Wallace, R.J., Newbold, C.J., 2007. A meta-analysis of fumarate effects on methane production in ruminal batch cultures. J. Anim. Sci. 85, 2556–2563.

Wathes, C.M., Kristensen, H.H., Aerts, J.M., Berckmans, D., 2008. Is precision livestock farming an engineer's daydream or nightmare, an animal's friend or foe, and a farmer's panacea or pitfall? Comput. Electron. Agric. 64 (1), 2–10.

World Bank, 2016. Drug-Resistant Infections: A Threat to Our Economic Future (Discussion Draft). World Bank, Washington, DC. License: Creative Commons Attribution CC BY 3.0 IGO http://pubdocs.worldbank.org/en/527731474225046104/AMR-Discussion-Draft-Sept18updated.pdf.

WTO, 1994. Agreement on the Application of Sanitary and Phytosanitary Measures. https://www.wto.org/english/docs_e/legal_e/15-sps.pdf.

Zhang, N., Wang, M., Wang, N., 2002. Precision agriculture a worldwide overview. Comput. Electron. Agric. 36 (2–3), 113–132.

CHAPTER FIVE

Sustainable livestock production and biodiversity

Akash[a], Mozammel Hoque[a], Sukanta Mondal[b], and Dipanshu Bisht[a]
[a]Division of Surgery, ICAR-Indian Veterinary Research Institute, Izatnagar, Uttar Pradesh, India
[b]Animal Physiology Division, ICAR-National Institute of Animal Nutrition and Physiology, Adugodi, Bengaluru, Karnataka, India

Introduction

The main challenge of the 21st century is to satisfy rising demands for agricultural productivity and food security while maintaining ecological processes that are critical to long-term agricultural development and human well-being (Johnson et al., 2014). By 2050, the agriculture sector will face a massive challenge of producing 60% more grain, feed, and fiber than is currently available (8.5 billion t/year) to sustain a global population estimated to be 9.3 billion (FAO, 2014a). While agriculture is the world's largest land user, it would be impossible to sustain this level of productivity while still protecting the planet's natural resources for future generations unless we make drastic improvements to our agriculture production and food-processing systems.

Food production can be increased by intensification (i.e., higher yields per hectare) and extensification (i.e., using more hectares). In the past, technical advancements and innovations such as high-yielding plant varieties, irrigation, and high levels of chemical inputs increased global agricultural productivity more than threefold in 50 years with just a 12% rise in farmed land (FAO, 2014a). This green revolution, which raised crop yields, enabled farmers to boost food supply for poor consumers at reasonable rates, saved an unquantifiable quantity of ecological resources, and prevented the release of an estimated 590 billion tons of carbon dioxide (CO_2) into the environment by saving millions of hectares of grasslands and forests from conversion to agricultural land (Burney et al., 2010).

According to recent estimates, intensification will account for 80% of potential increases in global agricultural demand, with extensification accounting for 20% (Alexandratos and Bruinsma, 2012). Closing yield

Emerging Issues in Climate Smart Livestock Production
https://doi.org/10.1016/B978-0-12-822265-2.00012-0

gaps on underperforming lands by raising cropping productivity, changing dietary priorities, and decreasing waste could potentially double food production while significantly reducing agricultural environmental impacts by halting further agricultural expansion into woodlands, grasslands, and wetlands (Foley et al., 2011). However, political instability, a shortage of capital, and poor farmers' inability to invest in fertilizer, equipment, and other inputs all currently impede the impact of intensification in low-yielding areas (Johnson et al., 2014). As a result, the FAO forecasts that the average annual growth in global wheat, rice, and maize yields between 2007 and 2050 would be less than half the increase recorded between 1961 and 2007 (Alexandratos and Bruinsma, 2012).

Although emerging technology such as molecular genetics and genetic engineering would inevitably help increase the productivity of crop improvement projects to raise potential global crop performance in response to climate change, the focus of global food production is moving away from cereals as staples and toward the efficient supply of a protein-rich diet dependent on livestock products, especially as household incomes rise (Raghuvanshi and Singh, 2015). Demand for red meat, which is expected to more than double over the next 20 years (Thornton, 2010), provides opportunities for grass-based ruminant farmers, especially the world's 1 billion smallholders, to use their livestock enterprises as a way out of poverty and to contribute significantly to global food security (Herrero and Thornton, 2013). Attempting to meet potential meat and milk demand using a business-as-usual development model, on the other hand, would only intensify the increasingly significant environmental harm incurred by agricultural systems, as shown by land destruction, biodiversity loss, soil erosion, and greenhouse gas emissions (FAO, 2006a; Havlík et al., 2014). To achieve a sustainable livestock sector, a refashioning of the livestock sector is needed to produce improved nutritional results and more productive processing using the same resources but at a lower environmental expense (Garnett, 2014). In general, achieving sustainable livestock production necessitates measures to increase productivity efficiently; preserve, maintain, and optimize natural ecosystems; change population growth and resource-intensive consumption patterns; strengthen governance systems; protect and improve rural livelihoods and social well-being; and minimize waste (FAO, 2014a; Charles et al., 2015).

Sustainability and its dimensions

Individuals, organizations, and nations are using the concept of sustainability to assess and monitor human impacts on the natural environment and resource base. Sustainable development was defined by the World Commission on Environment and Development (UN, 1987) as "development that meets the needs of current generations without jeopardizing future generations' ability to meet their own needs." This means that resources are limited and should be managed conservatively and carefully, with long-term goals and consequences in mind. Sustainable livestock production should protect land, water, plant, and animal genetic resources while still being environmentally nondegrading, technically suitable, commercially sustainable, socially acceptable, and adhering to responsible animal ethics.

The aim of sustainable development is to build a more robust and environmentally friendly system that creates more food with quality, sufficiency, and consistency (Allievi et al., 2015; Huber, 2000; Garnett, 2014). Efficiency is a supply-side concept, characterized as the use of resources to achieve the best possible economic production. Sustainable intensification of livestock production by effective technology transfer (increasing forage yield or optimizing grazing use of forage) to low-producing nations improves global meat production productivity and can substantially reduce pollution per unit of production (Tilman et al., 2011; Garnett, 2011). A sufficiency, demand restriction viewpoint places the onus on customers and businesses who encourage unhealthy consumption habits of high-impact foods or services to modify their behavior (Garnett, 2014). To achieve environmental and societal sustainability goals, sufficiency objectives that determine the required amount for optimal human health have been proposed for sustainable global meat consumption (Allievi et al., 2015). In fact, the overall meat supply *per capita* is a valuable metric for measuring progress toward sufficiency, taking into account the diverse consumption habits that exist in countries such as the United States and India. The ability of human-made processes to mimic natural systems is measured by consistency or system transformation. If the man-made mechanism strongly mimics the natural system, then this will lead to sustainability, recognizing that natural systems have boom/bust cycles (Allievi et al., 2015). Consistency as a metric will focus on green technologies such as sowing viable pasture plants to replace or supplement degraded grassland. Allievi et al. (2015) provided animal welfare ethics in consistency; because animals will struggle in factory farms, humans may transform to vegetarianism, which will dramatically alter sustainable development goals expressed in the main predictor of the number of animals slaughtered *per capita*. All three techniques are interconnected and relevant to sustainability. Increased inefficiency, for example, may lead to greater consistency because increases in the amount of useful meat provided by each animal would result in fewer animals being required to meet the same degree of demand. All three techniques are interconnected and relevant to sustainability. Increased inefficiency may lead to greater consistency because increases in the amount of useful meat provided by each animal would result in fewer animals being required to meet the same degree of demand (Allievi et al., 2015). This suggests that a combined strategy including all three viewpoints is needed to accurately measure progress toward achieving a globally sustainable livestock sector. Tight enforcement, though, is known to be impossible in free economies.

Challenges in global meat production and consumption

From 1963 to 2012, global meat demand nearly quadrupled, rising from 78 to 311 Mt. Cattle, sheep, and goats account for about 25% of the total (FAO, 2014b). According to the OECD/FAO (2014), global meat demand will increase 19% by 2023, with developing countries accounting for 78% of the additional 57 Mt output, with rising household incomes driving most of this growth. The majority of this expected growth is accounted for by poultry (49%) and pork (29%); developed countries will contribute just one-third of the combined poultry and pork rise. Poultry is the fastest-growing segment of global poultry production because it is less expensive, more effective to feed than other livestock, consumes less land than other meats, and has few religious or cultural restrictions. Poultry production will surpass pork production by 2020 (Heinrich Böll Foundation, 2014). The developing world would account for nearly all of the expected rise in beef, sheep, and goat meat demand. Beef demand in the United States and Europe is stagnant, but production in India, Brazil, and Australia, which account for 59% of all beef exports, is projected to increase. Global meat demand is expected to reach 455 million tons by 2050, with growing contributions from developing countries (Alexandratos and Bruinsma, 2012). Milk production has more than doubled from 340 to 792 Mt between 1963 and 2012 (Gerosa and Skoet, 2012). Milk production has become a significant livestock practice in developing countries, thanks to the rapid growth of smallholder milk production and the introduction of small-scale rural processors. India's milk production has risen nearly sixfold since 1970 (21–117 Mt). India surpassed the United States in dairy production in 1997, becoming the world's top milk producer (FAO, 2006b). Milk consumption increased by an average of 3%–4% per year in South Asia (1995–2005), more than double the growth rates reported for staple foods (Hemme and Otte, 2010). Previously, increased demand for milk was solely motivated by demographic growth, but demand is now primarily being driven by increasing *per capita* milk intake in developing countries. The increase in meat and milk production is being propelled by increasing demand for livestock products, which is concentrated in developing countries experiencing rapid population growth, urbanization, lifestyle changes, and increasing disposable income (Thornton, 2010). Consumers reduce their use of staples while increasing their consumption of animal protein as *per capita* income rises. In Asia, where the total protein supplied from livestock products has increased by 140% since 1980, income-driven dietary changes have been most dramatic (FAO, 2006a). This trend is expected to continue, especially in China and India, due to the enormous demands of their new middle classes. Because of the sheer scale of China's population, the power of their economy, and the vast disparity in demand between rural poor and wealthier urban consumers, the Chinese food sector will continue to rapidly grow as the poor's income level rises (Zhou et al., 2012). South and East Asia are facing the same rapid transition from agriculture to market-oriented livestock processing regimes that many developed countries did decades earlier (Heinrich Böll Foundation, 2014). China, as the world's biggest meat producer, is using its economic clout to boost productivity and quality in the same way that the developed world has done in the past. However, it is unlikely that China will be able to satisfy all the increased demand from domestic production, which opens up opportunities for exporting countries to fill their "diet gap," with unavoidable consequences on global food prices. However, it is unclear if China will be able to meet its the demand for meat by imports because the large amounts expected are beyond the capability of the major exporting countries.

Consumption patterns

Between 1961 and 2012, global meat intake rose from 22 to 40 kg *per capita* (FAOSTAT, 2014). Since the mid-1990s, the consumption of beef, veal, pork, and poultry has risen at a rate of about 3% per year in developing and emerging economies; consumption growth in developed countries has been just around 0.4% per year (Westcott and Hansen, 2015). Developed countries consume 64 kg *per capita* per year on average while emerging countries consume 24 kg *per capita* per year (FAOSTAT, 2014). The most significant disparity is the actual *per capita* meat intake in the United States (117 kg/year), which is twice that of China and 30 times that of India (4 kg/year), which has the world's lowest *per capita* meat consumption (FAOSTAT, 2014). China currently eats more than twice as much meat as the United States, with a gross meat intake of 71 Mt. Africa's meat production and demand are not increasing at the same rate as the rest of the world; Africans consume just 20 kg of meat each year, with no improvement anticipated in the next decade (FAO, 2014b). Overall, the consumption habits of chicken, pork, beef, and sheep in developing countries are unlikely to adjust over the next decade, despite a continued weakening trend among high-income earners in developed countries. In 2023, developing countries will consume 83% of the additional meat eaten, with Asia accounting for half (OECD/FAO, 2014). Consumer preferences across the globe support higher poultry growth than other meats while high-income households in developed countries, especially China, are expected to substitute poultry consumption with beef consumption. In the United States, however, poultry consumption is expected to rise by 12% to 57 kg/person/year. Depending on the price of beef and sheep meat, poultry consumption in countries where pork is not eaten (Malaysia, Israel, Middle East countries) may increase. Pork, which currently ranks first in terms of consumption, would lose market share, especially in Asia, where consumers are shifting their preferences to other meat products (OECD/FAO, 2014). Global beef consumption will remain steady, with a slight decrease in developed country *per capita* consumption and a slight increase in developing country *per capita* consumption. Sheep meat would account for a minor portion of the global meat market. Africa, China, and countries in the Middle East will see the most important increases in sheep meat intake (OECD/FAO, 2014). Globally, the *per capita* consumption of staples is decreasing as household incomes rise, but increased grain production will be expected as the human population grows, and there will be no crop surplus to redirect to animal production. Grain and forage production would need to be increased to feed livestock, especially pigs, chickens, dairy cattle, and some beef, sheep, and goats. Animal feed requirements are less stringent than human feed standards, making it easier to boost crop yields for livestock, but there is limited "new" land suitable for crop production, especially in Asia. This will most likely result in less land being used for cattle grazing, with the more profitable land being used for crops. Just about 10% of the overall global meat supply (30.2 Mt) is traded globally at the moment (FAO, 2014b). Though poultry and pigs will be a large part of the potential increased supply and demand (due to their reliance on crops), the growing trend toward more red meat consumption *per capita* will put significant strain on pastoral lands. As a result of various trade agreements and health and safety issues, developing countries such as China can limit imports to some countries (Australia, New Zealand, Canada, Brazil, and the European Union (EU)), putting additional strain on their resources.

Global distribution of grassland resources

Grasslands occupy more ground on the planet than any other large plant type and are vital to human populations, offering essential ecological products and services (for example, fodder for grazing animals such as cattle, sheep, goats, and camels; wildlife habitats; medicinal resources; climate change regulation; genetic resources; erosion control; water provision; air purification; and cultural and amenity services). The word "grassland" has grown to now include "all land dedicated to forage use," which can include pastureland, forestland, cropland, and rangeland as grassland resources (Blair et al., 2014). This makes grassland resources remarkably complex and difficult to describe (Allen et al., 2011). Grassland vegetation comprises grasses, legumes, other forbs, and woody species that are either native or exotic, and can be found as a monoculture (forage crop), a combination of two or more species (sown pasture), or a rich and varied plant population of more than 400 species of forbs and woody plants (Blair et al., 2014). Grasslands may be either temporary sown pastures used as leys in crop rotation schemes or permanent natural habitats used for grazing by livestock and wildlife, such as the African savannahs; Australian grasslands; South American cerrado, campo, llanos, and pampa; North American prairies; and Eurasian steppes (Archibold, 1995). Natural grasslands are nearly nonexistent in most of Europe; remnant reserves are anthropogenic and need maintenance to prevent reversion to woodland (Tscharntke et al., 2005). The 5.4 billion hectares of grasslands (40% of the world's land area; White et al., 2000) are found on every continent, at all ice-free latitudes, and at elevations ranging from coastal to montane. Grassland climates range from temperate to tropical, with annual rainfall ranging from less than 250 mm in arid grasslands to well more than 1000 mm in mesic grasslands. The average annual temperature ranges from about 0°C to more than 25°C. Precipitation seasonality, soil conditions, herbivory, and fire are all essential factors in deciding the composition and function of local vegetation. Grasslands have a wide variety of climatic and topographic conditions, resulting in a varied biome. The overall areas of grassland in sub-Saharan Africa and Asia are 14.5 and 8.9 Mkm^2, respectively. Five countries (Australia, the Russian Federation, China, the United States, and Canada) have more than 3 million square kilometers of grassland each (White et al., 2000). There are 28 smaller countries (25 in Africa) that have grassland covering more than 60% of their total land area. Five of the 28 countries have more than 1 million square kilometers of grassland (Mongolia, Kazakhstan, Angola, South Africa, Ethiopia).

Conversion of grassland ecosystem into agriculture

Human activity has altered the size, composition, and health of all major habitats on the planet. The conversion of grasslands to croplands and other agro-habitats has been the primary cause of grassland destruction around the world. Temperate grasslands with improved soils and more abundant rainfall were largely cleared for crops by the 1950s (Millennium Ecosystem Assessment, 2005). Lower-quality semiarid grasslands were left for grazing livestock until recently, when emerging innovations such as minimal tillage and drought-resistant plants made it profitable to grow crops in drier conditions. As agriculture industries in South America (Gavier-Pizarro et al., 2012), southern Africa (Maeda et al., 2010), North America (Landis and Weling, 2010), and Asia (Qiu et al., 2010) grow to feed a rapidly expanding human population, the present and potential risks to grassland habitats are large (Foley et al., 2011). Töpfer et al. (2000) found that conversion to croplands, mixed cultivation, and sown pastures have affected 3.3 billion hectare, or 26% of the world's land area, and has resulted in the depletion of one-third of the temperate and tropical forests and one-quarter of the natural grasslands. By 2003, more than 97% of tall grass prairies, 71% of mixed prairies, and 48% of low grass prairies had disappeared in the United States (Robertson et al., 1997; Samson et al., 2004). Tall grass prairies, North America's most endangered prairie (Samson et al., 2004), have been reduced to just 100 km^2 of their former 6000 km^2 in Manitoba (Joyce and Morgan, 1989) and 820 km^2 in Ontario (Samson et al., 2004) in Canada (Ontario Ministry of Natural Resources, 2009). In selected grassland ecoregions, conversion to croplands has been as high as 76% in South America and 20%–40%

in other regions (Töpfer et al., 2000). According to the Millennium Ecosystem Assessment (2005), a further 10%–20% of grassland and woodland lands will be transferred to agriculture, mainly in developed countries, before 2050, encompassing both dry and tropical areas. As part of greenhouse gas (GHG) abatement policies, developing countries are expected to see a modest rise in forest cover.

Challenges and opportunities for livestock production systems

Domestic grazers have access to a large and easily exploited advantage in grasslands (Blair et al., 2014). Because the fate of smallholder pastoralists in these areas is inextricably linked to livestock development, any major change in grassland condition can have a direct impact on subsistence. Biophysical (dust clouds, GHG pollution, and geographic climate change) and social effects cause indirect damage to a greater community.

Managing biomass production

The ability to map the effect of overgrazing on grassland quality requires the ability to inventory ground cover and standing biomass across vast landscapes. According to Wolf et al. (2015), global livestock feed consumption in 2011 was 4.844 Pg DM, with 52% of it coming from grassland and pasture supplies and the rest coming from grain residue (12%) and feed-crop (36%) to provide 0.26 Gt of meat in our modern global world. The productivity of grasslands and forage crops that underpin livestock production systems will be affected by the combined effects of intensified use and climate change in the coming decades (Ghahramani and Moore, 2015). Weindl et al. (2015) expect a global decrease in grassland yield of 2% at constant CO_2 levels, with the highest negative impacts occurring in Australia (−11%) and the Middle East and North Africa (28%). Brookshire and Weaver (2015) found a >50% decrease in aboveground net plant productivity in a natural subalpine *Festuca idahoensis* ecosystem over the last 44 years, with an average rate of $-2.5\,g/m^2/year$. They blamed it on the global temperatures and changing precipitation trends, which limited the ability of grassland habitats to supply goods and services. Additional losses from pests and weeds are expected to rise from the current average of 13% of forage output (Bajželj et al., 2014). It's critical to figure out what's behind the decline in biomass productivity and what biophysical mechanisms are at work before coming up with remedial management strategies. Remote sensing data in the form of time series can be used to assess grassland dynamics at larger scales, particularly now that Moderate Resolution Imaging Spectroradiometer (MODIS) provides near-daily coverage at 250 m to 1 km resolution. Using these instruments, Hilker et al. (2014) discovered that changes in livestock account for about 80% of the decrease in biomass in Mongolia at a landscape level while precipitation accounted for 30% of variation across the country as a whole, but up to 50% in areas with denser vegetation cover. Gang et al. (2014) used common approaches to identify climate change as the primary factor explaining 46% of global grassland destruction, with southern hemisphere grasslands (Australia) becoming more vulnerable and susceptible to climate change. This ensures that keeping a balance between grassland and forage supply as well as livestock demand is critical for long-term livestock sustainability.

For all grassland types, preserving both green and dry biomass within boundaries defined by livestock success is the primary criteria for managing grasslands (Kemp and Michalk, 2007). For example, the general relationship between sheep or cattle growth and green herbage suggests that to reach 90%–95% of full live weight gain, biomass must be sustained at 1.5–2 t DM/ha (Kemp et al., 2015a). In practice, this means that stocking rates must be adjusted on a regular basis to maximize livestock productivity and other environmental benefits provided by grasslands. Degradation is evidence of a failure by management to balance livestock feed demands with the capacity of forage resources to meet these demands. Climate change, which is expected to increase the incidence of droughts and heavy rainfall events, would make it more difficult to control pastures and maintain livestock feed supplies.

In grasslands with adequate productive capacity, pasture intensification interventions such as sowing enhanced and deeper-rooted grass and legume species will dramatically increase forage productivity and quality compared to natural grasslands. However, considerable losses are incurred in lost yield during establishment, and if insufficient stocking rates are used, forage productivity and quality will suffer. Improved grazing management, fertilization, and herbicides to combat weeds are both useful and cost-effective ways to maintain the perennial grass composition at a more desirable level (Kemp et al., 2000). To best balance livestock demand and feed supply, these activities can be combined with tactical management. Simple grazing management strategies have proven to be beneficial, but they must be tailored to individual main species due to differences in phenology, grazing reaction, and competitiveness with related pasture species (Kemp et al., 2015b).

Biodiversity loss

Agricultural biodiversity encompasses all aspects of biological biodiversity that are important to the production of food and the maintenance of the agro-core ecosystem's functions. Agricultural biodiversity thus encompasses the genetic, species, and ecosystem diversity of mammals, plants, and microorganisms as well as biotic, socioeconomic, and cultural influences (Hoffmann, 2011). Grasslands are among the world's most diverse habitats (Wilson et al., 2012), but maintaining biodiversity and habitat structure while producing enough biomass to support livestock is a global challenge. Because of the failure to maintain this balance by continued increases in livestock densities, livestock production is responsible for 30% of the biodiversity loss through grassland destruction caused by overgrazing or deforestation and conversion of land for forage-crop cultivation (Westhoek et al., 2011). A total of 300 of the 825 terrestrial ecoregions and 23 of the 35 global biodiversity hotspots are now threatened by livestock-related processes that trigger habitat degradation, desertification, and alien species rivalry (FAO, 2006a). The functional makeup of the vegetation has changed as a result of the extinction of this order's species, as shown by the widespread substitution of perennial grasses by fast-growing forbs, annual grasses, and woody shrubs, which cascade into soil food webs (density and diversity of soil organisms). Numerous ecological processes are disrupted, and natural resources are depleted as a result (Lavorel and Grigulis, 2012). Maintaining desirable perennial species biomass in grassland ecosystems by effective grazing management is the ideal way to improve population stability, preserve biodiversity, and provide various ecological services that are strongly dependent on grassland production and management (Petz et al., 2014). Because livestock graze selectively, certain plants may invariably be defoliated more heavily, reducing their productivity against less-suitable species and invasive weeds. Because a few diverse species typically account for a large proportion of grassland biomass in both natural and sown grassland ecosystems, it's common to believe that the many uncommon species present have little effect. It is important to note, however, that the relationships between diversity and function are not linear, and that there is a species richness level at which ecological function decreases and no change occurs (Vitousek and Hooper, 1993). Rather than overall species numbers, biodiversity must understand plant function (species richness). Kemp et al. (2003) found that as the number of species in a grassland rose, productivity decreased and variability in productivity increased significantly (Kemp et al., 2003). Many of the species included in this sample were minor forbs, annual monocotyledons, and others that together produced a few plant functional types, implying a high level of species redundancy. Over the course of a year, these classes usually contributed just 1%–5% of overall net primary production. Just a few plants, either native or imported grasses, influenced productivity. There were some minor species that could be called invasive weeds within the minor species. The number of insects, often small ones, increased as grazing pressures increased. The researchers came to the conclusion that biodiversity should be measured on the basis of plant functional types. Further research revealed that the attractive native grasses with the greatest impact on plant production could be kept inside the sward if the herbage mass was kept at 2t DM/ha or higher during the year. More native species lived in grasslands with a low total species number, meaning that they were not as successful as species that invaded grasslands under higher grazing pressures. The optimum number of species or plant functional types isn't always obvious, although studies such as that from Kemp et al. (2003) have shown that in order to put competitive pressure on invasive weed species, the more efficient functional type (typically the preferable perennial grasses) needs to account for at least 60% of the total biomass. The corollary is that invasive plant species should be kept to less than 10%–15% of the total. To help production and retain an optimal composition, the majority (25%–30%) should be legumes and other forbs as well as "gap fillers" (short-lived annuals). A mixture of plant functional groups provides equilibrium and has no negative effects on soil micro- and mesoorganisms (Kemp et al., 2003). The domination of a few highly productive specialized breeds (Friesian-Holsteins, Angus) in the global cattle and sheep industries has reduced the diversity of livestock breeds used for breeding, endangering indigenous ecotypes. The FAO (2007) reports 9% of the original global farm animals are already extinct and more than 20% of the remaining germplasm is under threat. This is the result of poorly engineered cross-breeding schemes that rely on exotic imports and pay no attention to protecting the integrity of local genetic resources, the vast majority of which are managed by smallholder farmers using conventional methods (Ayalew et al., 2003). This has led to the extinction of native breeds that have evolved to thrive in low-input mixed farming and pastoral development systems common in developed countries (ILRI, 1999).

Management of greenhouse gases and climate change

Globally, livestock eat 17 Gt CO_2e/year of feed, with 4.4 Gt CO_2e/year from grain processing, 2.6 Gt CO_2e/year from crop residues, and 10.0 Gt CO_2e/year from grassland and forage resources (Bajželj et al., 2014), accounting for 14.5% of the overall anthropogenic GHG emissions of 49 Gt CO_2e/year (Ripple et al., 2014; Tubiello et al., 2013). Ruminants account for 11.6% of global GHG emissions from all anthropogenic sources, 90% of which are contributed by cattle. The bulk of the GHG input from livestock comes from CH_4 emitted from enteric fermentation and manure (44%), CO_2 emitted from land use transition and fossil fuel use (27%), and N_2O from fertilizer added to feed-crop fields and manure (2%) (Ripple et al., 2014). Global CO_2e emissions from the livestock industry rose by 51% between 1960 and 2010 (Caro et al., 2014). Emissions increased by 117% in developing countries due to growth in beef and dairy cattle numbers, but dropped by 23% in developed economies. Mostly in tropical areas due to the extension of pasture and arable land for animal feed crops, the destruction of native grazing lands, and conversion to cropland (palm oil in Southeast Asia), deforestation accounts for a large portion of global GHG emissions from the livestock industry, with losses of 450,800 Gt CO_2 from biomass and soil carbon pools. Lowering beef demand or improving the current cattle herd's processing productivity would reduce the need for more tropical deforestation while still producing significant benefits from other biomes and processes (Ripple et al., 2014). For example, the total eradication of deforestation by 2030 could potentially result in a CO_2e mitigation capacity of 2.3–5.8 Gt/year. Sowing pasture plants that are more suited to the local climate, more resilient to grazing, more drought-tolerant, and able to improve soil productivity (e.g., N-fixing species) will contribute to improved development and forage quality in both temperate and tropical habitats. Soil C sequestration is reversible, and C can be lost quickly due to a variety of processes linked to climate variability and poor management, including soil disruption, plant destruction due to overgrazing, burning, flooding, nutrient shortages, and water shortages (Soussana et al., 2014). By 2020, it is reasonable to predict that 10% of the global grazing lands will be managed for carbon sequestration (Conant, 2010). While grazing livestock contribute greatly to climate change, through improving management, livestock production systems may become part of the mitigation solution. Improved forage quality and feeding practices can help minimize

GHG emissions (Sirohi, 2015), but their effectiveness is strongly dependent on the a priori collection of areas that can benefit (Henderson et al., 2015). Although livestock emits significant GHG amounts, transitioning from comprehensive to more productive and effective livestock offers appealing mitigation opportunities to minimize GHG emissions per unit of livestock output while enhancing efficiency through improving animal husbandry, improving grassland management practices, and/or rehabilitating degraded grasslands.

Conclusion

Demand for livestock products will continue to rise for at least the next 30 years, fueled primarily by human population development, income growth, and urbanization, with rising demand from the developing world. It is clear that continued dependence on current livestock systems, which rely on land diversion and overexploitation of grassland and forage supplies to satisfy current demand for livestock products, would not lead to sustainability. Reduced productivity, lack of biodiversity, increased soil degradation, reduced water yield and efficiency, and high GHG emissions are all signs of this approach's failure. To achieve sustainability, both increasing livestock production and reducing environmental impacts are needed. The barriers to achieving sustainability in the livestock sector are sociological, fiscal, and political in nature.

References

Alexandratos, N., Bruinsma, J., 2012. World Agriculture Towards 2030/2050: The 2012 Revision. Food and Agriculture Organization of the United Nations, Rome.

Allen, V.G., Batello, C., Berretta, E.J., Hodgson, J., Kothmann, M., Li, X., McIvor, J., Milne, J., Morris, C., Peeters, A., Sanderson, M., 2011. An international terminology for grazing lands and grazing animals. Grass Forage Sci. 66, 2–28.

Allievi, F., Vinnari, M., Luukkanen, J., 2015. Meat consumption and production-analysis of efficiency, sufficiency and consistency of global trends. J. Clean. Prod. 92, 142–151.

Archibold, O.W., 1995. Ecology of World Vegetation. Chapman and Hall, London/New York.

Ayalew, W., King, J.M., Burns, E., Rischkowsky, B., 2003. Economic evaluation of smallholder subsistence livestock production: lessons from an Ethiopian goat development program. Ecol. Econ. 45, 473–485.

Bajželj, B., Richards, K.S., Allwood, J.M., Smith, P., Dennis, J.S., Curmi, E., Gilligan, C.A., 2014. Importance of food demand management for climate mitigation. Nat. Clim. Chang. 4, 924–929.

Blair, J., Nippert, J., Briggs, J., 2014. Grassland ecology. In: Monson, R.K. (Ed.), Ecology and the Environment. Springer Science+Business Media, New York, NY, pp. 389–423 (Chapter 14).

Brookshire, E.N.J., Weaver, T., 2015. Long-term decline in grassland productivity driven by increasing dryness. Nat. Commun. 6, 7148. https://doi.org/10.1038/ncomms8148.
Burney, J.A., Davis, S.J., Lobell, D.B., 2010. Greenhouse gas mitigation by agricultural intensification. Proc. Natl. Acad. Sci. U. S. A. 107, 12052–12057.
Caro, D., Davis, S.J., Bastianoni, S., Caldeira, K., 2014. Global and regional trends in greenhouse gas emissions from livestock. Clim. Chang. 126, 203–216.
Charles, H., Godfray, J., Garnett, T., 2015. Food security and sustainable intensification. Philos. Trans. R. Soc. B 369, 20120273.
Conant, R.T., 2010. Challenges and opportunities for carbon sequestration in grassland systems: a technical report on grassland management and climate change mitigation. In: Integrated Crop Management 9. Plant Production and Protection Division, Food and Agriculture Organization of the United Nations, Rome.
FAO, 2006a. Livestock's Long Shadow. Environmental Issues and Options. Food and Agriculture Organization of the United Nations, Rome, Rome, Italy.
FAO, 2006b. Livestock Report 2006. Food and Agriculture Organization of the United Nations, Rome.
FAO, 2007. The State of the world's Genetic Resources for Food and Agriculture. Food and Agriculture Organization of the United Nations, Rome.
FAO, 2014a. Building a Common Vision for Sustainable Food and Agriculture: Principles and Approaches. United Nations Food and Agriculture Organisation, Rome.
FAO, 2014b. Food Outlook: Biannual Report on Global Food Markets. Food and Agriculture Organisation of the United Nations, Rome. October, 2014.
FAOSTAT, 2014. FAO Statistical Database. Food and Agriculture Organisation of the United Nations, Rome.
Foley, J.A., Ramankutty, N., Brauman, K.A., Cassidy, E.S., Gerber, J.S., Johnston, M., Mueller, N.D., O'Connell, C., Ray, D.K., West, P.C., Balzer, C., Bennett, E.M., Carpenter, S.R., Hill, J., Monfreda, C., Polasky, S., Rockström, J., Sheehan, J., Siebert, S., Tilman, D., Zaks, D.P., 2011. Solutions for a cultivated planet. Nature 478, 337–342.
Gang, C.C., Zhou, W., Chen, Y.Z., Wang, Z.Q., Sun, Z.G., Li, J.L., Qi, J.G., Odeh, I., 2014. Quantitative assessment of the contributions of climate change and human activities on global grassland degradation. Environ. Earth Sci. 72, 4273–4282.
Garnett, T., 2011. Where are the best opportunities for reducing greenhouse gas emissions in the food system (including the food chain)? Food Policy 36, S23–S32.
Garnett, T., 2014. Three perspectives on sustainable food security: efficiency, demand restraint, food system transformation. What role for life cycle assessment? J. Clean. Prod. 73, 10–18.
Gavier-Pizarro, G.I., Calamari, N.C., Thompson, J.J., Canavelli, S.B., Solari, L.M., Decarre, J., Goijman, A.P., Suarez, R.P., Bernardos, J.N., Zaccagnini, M.E., 2012. Expansion and intensification of row crop agriculture in the Pampas and Espinal of Argentina can reduce ecosystem service provision by changing avian density. Agric. Ecosyst. Environ. 154, 44–55.
Gerosa, S., Skoet, J., 2012. Milk availability: trends in production and demand and medium-term outlook. In: ESA Working paper No. 12-01. Agricultural Development Economics Division, Food and Agriculture Organisation of the United Nations, Rome.
Ghahramani, A., Moore, A.D., 2015. Systemic adaptations to climate change in southern Australian grasslands and livestock: production, profitability, methane emission and ecosystem function. Agric. Syst. 133, 158–166.
Havlík, P., Valin, H., Herrero, M., Obersteiner, M., Schmid, M., Rufino, M.C., Mosnier, A., Thornton, P.K., Böttcher, H., Conant, R.T., Frank, S., Fritz, S., Fussa, S., Kraxner, F., Notenbaert, A., 2014. Climate change mitigation through livestock system transitions. Proc. Natl. Acad. Sci. U. S. A. 111, 3709–3714.

Heinrich Böll Foundation, 2014. Meat Atlas: Facts and Figures About the Animals We Eat. Heinrich Böll Foundation, Berlin, Germany.
Hemme, T., Otte, J., 2010. Status and Prospects for Small Holder Milk Production: A Global Perspective. Food and Agriculture Organisation of the United Nations, Rome, p. 186.
Henderson, B., Falcucci, A., Mottet, A., Early, L., Werner, B., Steinfeld, H., Gerber, P., 2015. Marginal costs of abating greenhouse gases in the global ruminant livestock sector. Mitig. Adapt. Strateg. Glob. Chang., 1–26.
Herrero, M., Thornton, P.K., 2013. Livestock and global change: emerging issues for sustainable food systems. Proc. Natl. Acad. Sci. U. S. A. 110, 20878–20881.
Hilker, T., Natsagdorj, E., Waring, R.H., Lyapustin, A., Wang, Y.J., 2014. Satellite observed widespread decline in Mongolian grasslands largely due to overgrazing. Glob. Chang. Biol. 20, 418–428.
Hoffmann, I., 2011. Livestock biodiversity and sustainability. Livest. Sci. 139, 69–79.
Huber, J., 2000. Towards industrial ecology: sustainable development as a concept of ecological modernization. J. Environ. Policy Plan. 2, 269–285.
ILRI, 1999. Economic valuation of animal genetic resources. In: Proceedings of an FAO: ILRI Workshop Held at FAO Headquarters, Rome, Italy, 15–17 March. International Livestock Research Institute, Nairobi, Kenya, p. 80.
Johnson, J.A., Runge, C.F., Senauer, B., Foley, J., Polasky, S., 2014. Global agriculture and carbon trade-offs. Proc. Natl. Acad. Sci. U. S. A. 111, 12342–12347.
Joyce, J., Morgan, J.P., 1989. Manitoba's tall-grass prairie conservation project. Prairie pioneers: ecology, history and culture. August, 1988. In: Bragg, T.B., Stubberndieck, J. (Eds.), Proceeding of the Eleventh North American Prairie Conference, held August 7-11, 1988. University of Nebraska Printing, Lincoln, NE.
Kemp, D.R., Michalk, D.L., 2007. Towards sustainable grassland and livestock management. J. Agric. Sci. 145, 543–564.
Kemp, D.R., Michalk, D.L., Virgona, J.M., 2000. Towards more sustainable pastures: lessons learnt. Aust. J. Exp. Agric. 40, 343–356.
Kemp, D.R., King, W.M.G., Lodge, G.M., Murphy, S.R., Quigley, P., Sanford, P., 2003. SGS biodiversity theme: the impact of plant biodiversity on the productivity and stability of grazing systems. Aust. J. Exp. Agric. 43, 962–975.
Kemp, D.R., Badgery, W.B., Michalk, D.L., 2015a. Sustainable grasslands: resolving management options for livelihood and environmental benefits. In: Proceedings of the XXIII International Grassland Congress, New Delhi, India.
Kemp, D.R., Badgery, W.B., Michalk, D.L., 2015b. Principles for grassland systems research for livelihoods and environmental benefits. In: See Special Volume Produced for 23rd International Grassland Congress, India November 2015.
Landis, D.A., Weling, B.P., 2010. Arthropods and biofuel production systems in North America. Insect Sci. 17, 220–236.
Lavorel, S., Grigulis, K., 2012. How fundamental plant functional trait relationships scale-up to trade-offs and synergies in ecosystem services? J. Ecol. 100, 128–140.
Maeda, E.E., Pellikka, P.K.E., Siljander, M., Clark, B.J.F., 2010. Potential impacts of agricultural expansion and climate change on soil erosion in the Eastern Arc Mountains of Kenya. Geomorphology 123, 279–289.
Millennium Ecosystem Assessment, 2005. Ecosystems and Human Well-Being: Synthesis. Island Press, Washington, DC.
OECD/FAO, 2014. OECD-FAO Agricultural Outlook 2014. OECD Publishing, Paris.
Ontario Ministry of Natural Resources, 2009. Natural Heritage Information Center (NHIC) Database. Ontario Ministry of Natural Resources, Peterborough, ON, Canada.
Petz, K., Alkemade, R., Bakkenes, M., Schulp, C.J.E., van der Velde, M., Leemans, R., 2014. Mapping and modelling tradeoffs and synergies between grazing intensity and

ecosystem services in rangelands using global-scale datasets and models. Glob. Environ. Chang. 29, 223–234.

Qiu, H.G., Huang, J.K., Yang, J., Rozelle, S., Zhang, Y.H., Zhang, Y.H., Zhang, Y.L., 2010. Bioethanol development in China and the potential impacts on its agricultural economy. Appl. Energy 87, 76–83.

Raghuvanshi, G., Singh, Y., 2015. Agriculture biotechnology for sustainable food security: a review. Indian Res. J. Genet. Biotechnol. 7, 311–319.

Ripple, W.J., Smith, P., Haberl, H., Montzka, S.A., McAlpine, C., Boucher, D.H., 2014. Ruminants, climate change and climate policy. Nat. Clim. Chang. 4, 1–3.

Robertson, K.R., Anderson, R.C., Schwartz, M.W., 1997. The tallgrass prairie mosaic. In: Schwartz, M.W. (Ed.), Conservation in Highly Fragmented Landscapes. Chapman and Hall, New York, NY, pp. 55–87.

Samson, F.B., Knopf, F.L., Ostlie, W.R., 2004. Great Plains ecosystems: past, present and future. Wildl. Soc. Bull. 32, 6–15.

Sirohi, S., 2015. Opportunities and challenges for carbon trading from livestock sector. In: Sejian, V., Gaughan, J., Baumgard, L., Prasad, C. (Eds.), Climate Change Impact on Livestock: Adaptation and Mitigation. Springer India, pp. 239–252.

Soussana, J.-F., Klumpp, K., Ehrhardt, F., 2014. The role of grassland in mitigating climate change. In: Hopkins, A., Collins, R.P., Fraser, M.D., King, V.R., Lloyd, D.C., Moor, J. M., Robson, P.R.H. (Eds.), The Future of European Grasslands. Proceedings of the 25th General Meeting of the European Grassland Federation, Aberystwyth, Wales 7-11 September 2014, pp. 76–87.

Thornton, P.K., 2010. Livestock production: recent trends, future prospects. Philos. Trans. R. Soc. B 365, 2853–2867.

Tilman, D., Balzer, C., Hill, J., Befort, B.L., 2011. Global food demand and the sustainable intensification of agriculture. Proc. Natl. Acad. Sci. U. S. A. 108, 20260–20264.

Töpfer, K., Wolfensohn, J.D., Lash, J., 2000. People and ecosystems: the fraying web of life. In: World Resources 2000–2001. World Resources Institute, Washington DC.

Tscharntke, T., Klein, A.M., Kruess, A., Steffandewenter, I., Thies, C., 2005. Landscape perspectives on agricultural intensification and biodiversity – ecosystem service management. Ecol. Lett. 8, 857–874.

Tubiello, F.N., Salvatore, M., Rossi, S., Ferrara, A., Fitton, N., Smith, P., 2013. The FAOSTAT database of greenhouse gas emissions from agriculture. Environ. Res. Lett. 8, 1–10.

UN, 1987. Our Common Future. Report of the World Commission on Environment and Development, World Commission on Environment and Development, 1987. Published as Annex to General Assembly document A/42/427, Development and International Co-operation: Environment August 2, 1987.

Vitousek, P.M., Hooper, D.U., 1993. Biological diversity and terrestrial ecosystem biogeochemistry. In: Schulze, E.-D., Mooney, H.A. (Eds.), Biodiversity and Ecosystem Function. Springer-Verlag, Berlin, Germany, pp. 3–14.

Weindl, I., Lotze-Campen, H., Popp, A., Müller, C., Havlík, P., Herrero, M., Schmitz, C., Rolinski, S., 2015. Livestock in a changing climate: production system transitions as an adaptation strategy for agriculture. Environ. Res. Lett. 10, 094021.

Westcott, P., Hansen, J., 2015. USDA Agricultural Projections to 2024. USDA Agricultural Projections No. (OCE-151). 97 pp, February 2015.

Westhoek, H., Rood, T., Berg, M.v.d., Janse, J., Nijdam, D., Reudink, M.A., Stehfest, E., 2011. The Protein Puzzle: The Consumption and Production of Meat, Dairy and Fish in the European Union. PBL (Netherlands Environmental Assessment Agency), The Hague/Bilthoven.

White, R., Murray, S.M., Rohweder, M., 2000. Pilot Analysis of Global Ecosystems (PAGE): Grassland Ecosystems. World Resources Institute (WRI), Washington, DC, USA.

Wilson, J.B., Peet, R.K., Dengler, J., Pärtel, M., 2012. Plant species richness: the world records. J. Veg. Sci. 23, 769–802.

Wolf, J., West, T.O., Le Page, Y.L., Kyle, G.P., Zhang, X.S., Collatz, G.J., Imhoff, M.L., 2015. Biogenic carbon fluxes from global agricultural production and consumption. Glob. Biogeochem. Cycles. https://doi.org/10.1002/2015GB005119.

Zhou, Z.Y., Tian, W.M., Wang, J.M., Liu, H.B., Cao, L.J., 2012. Food Consumption Trends in China. April 2012. Report Submitted to the Australian Government Department of Agriculture, Fisheries and Forestry, Canberra, Australia.

Further reading

Clay, D.E., Clay, S.A., Reitsma, K.D., Dunn, B.H., Smart, A.J., Carlson, G.G., Horvath, D., Stone, J.J., 2014. Does the conversion of grasslands to row crop production in semiarid areas threaten global food supplies? Glob. Food Sec. 3, 2–30.

Kemp, D.R., Dowling, P.M., 2000. Towards sustainable temperate perennial pastures. Aust. J. Exp. Agric. 40, 125–132.

Olofsson, J., Hickler, T., 2008. Effects of human land-use onthe global carbon cycle during the last 6,000 years. Veg. Hist. Archaeobotany 17, 605–615.

Shevliakova, E., Pacala, S.W., Malyshev, S., Hurtt, G.C., Milly, P.C.D., Caspersen, J.P., Sentman, L.T., Fisk, J.P., Wirth, C., Crevoisier, C., 2009. Carbon cycling under 300 years of land use change: importance of the secondary vegetation sink. Glob. Biogeochem. Cycles 23, 1–16.

Smith, P., Haberl, H., Popp, A., Erb, K.-H., Lauk, C., Harper, R., Tubiello, F.N., de Siqueira Pinto, A., Jafari, M., Sohi, S., Masera, O., Böttcher, H., Berndes, G., Bustamante, M., Ahammad, H., Clark, H., Dong, H.M., Elsiddig, E.A., Mbow, C., Ravindranath, N.H., Rice, C.W., Robledo Abad, C., Romanovskaya, A., Sperling, F., Herrero, M., House, J.I., Rose, S., 2013. How much land based greenhouse gas mitigation can be achieved without compromising food security and environmental goals? Glob. Chang. Biol. 19, 2285–2302.

Stoll-Kleemann, S., O'Riordan, T., 2015. The sustainability challenges of our meat and dairy diets. Environ. Sci. Policy Sustain. Dev. 57, 34–48.

Suttie, J.M., Reynolds, S.G., Batello, C., 2005. Grasslands of the world. In: Food and Agriculture Organization of the United Nations, Plant Production and Protection Series No 34. Food and Agriculture Organization, Rome, Italy.

CHAPTER SIX

Climate change impact on livestock production

N.P. Soumya[a], Ramanuj Banerjee[b], Mamoni Banerjee[c], Sukanta Mondal[d], R.L. Babu[e], Mozammel Hoque[f], I.J. Reddy[d], S. Nandi[d], P.S.P. Gupta[d], and P.K. Agarwal[g]

[a]University of Kerala, Thiruvananthapuram, Kerala, India
[b]Department of Scientific and Industrial Research, Ministry of Science and Technology, Government of India, New Delhi, India
[c]Rajendra Mishra School of Engineering Entrepreneurship, IIT, Kharagpur, West Bengal, India
[d]Animal Physiology Division, ICAR-National Institute of Animal Nutrition and Physiology, Adugodi, Bengaluru, Karnataka, India
[e]Department of Bioinformatics and Biotechnology, Karnataka State Akkamahadevi Women's University, Jnanashakthi Campus, Vijayapura, Karnataka, India
[f]Division of Surgery, ICAR-Indian Veterinary Research Institute, Izatnagar, Uttar Pradesh, India
[g]NASF, ICAR, New Delhi, India

Introduction

Livestock production is an essential component of world agriculture, contributing to food security, nutrition, poverty alleviation, and economic growth. About 30% of the Earth's terrestrial areas are occupied with livestock systems, sustaining the livelihood of marginal farmers. Demand for livestock products is also increasing due to population growth, income increases, lifestyle changes, and fast urbanization, especially in developing countries (Delgado et al., 2001). Globally, the human population is anticipated to rise to 9.2 billion by 2050. However, Earth's climate is changing, causing severe threats to livestock production and the world economy (Rosenzweig et al., 2008). The primary culprit for global climate change is said to be the emission of greenhouse gases (GHG), which results in atmospheric warming (IPCC, 2013). GHG emissions from human activity and livestock are a significant driver of climate change, trapping heat in the Earth's atmosphere and triggering global warming, which is the ongoing rise of the average temperature of the Earth's climate. The significant consequences of climate change are increased atmospheric temperature, alterations in rainfall, reduced water availability, changes in the quality and quantity of feed, and unexpected extreme weather events such as elevating sea surface temperature, rising sea levels, and melting of glaciers and ice caps

Emerging Issues in Climate Smart Livestock Production
https://doi.org/10.1016/B978-0-12-822265-2.00010-7

in the polar region (Stocker et al., 2013). Increased atmospheric temperature leads to heat stress, which reduces animal production and reproduction, reduces feed and fodder production, and increases susceptibility to disease (JGCRI &BMI, 2009). The susceptibility of livestock to climate change also depends on factors such as the duration, frequency, and severity of the exposure. Moreover, sensitivity varies with the breed, immunity status, and other management practices such as housing and feeding (ICEM, 2013). It has been noted that outbreaks of foot and mouth disease increase with temperature, humidity, and rainfall; similarly, mastitis increases in dairy animals during hot and humid weather, which is also associated with increases in flies and cattle ticks (JGCRI &BMI, 2009). Grazing systems may be more vulnerable to climate change due to their dependence on natural resources and climatic conditions (Aydinalp and Cresser, 2008). Confined livestock production systems have a better opportunity for adaptation, as they have more control over the microclimate as well as improved husbandry practices and mitigation strategies (Rötter and Van de Geijn, 1999; FAO, 2009; Thornton and Gerber, 2010).

The effect of climatic change on livestock can be either direct or indirect. as shown in Fig. 1. The direct effects of climate change on health include temperature-related illnesses and death. Indirect impacts can be changes in the crop yields with variations in quantity and quality, changes in water quality and availability, shifts in the availability of feed and fodder crops, greater competition for resources, distribution of vector-borne diseases, loss of livestock biodiversity, and overall modifications in ecosystems (FAO, 2009; Thornton and Gerber, 2010; Lacetera et al., 2013). Those outcomes will possibly continue for the future, affecting both animal and human populations (IPPC, 2007). Competition for natural resources will adversely affect livestock production and food security (Thornton, 2010). Therefore, the challenge is to maintain a balance between productivity, household food security, and environmental preservation.

To combat this adverse effect of climate change on animals, humans, plants, and the environment, the government of India formulated and launched several national action plans, strategies, and policies (JGCRI &BMI, 2009). The prime objective is conserving environmental resources as well as securing livelihoods and the well-being of all, including fauna and flora. It primarily includes animals, that is, livestock production and health, wildlife, and predominantly human health. The policies are formulated to stimulate partnerships of different stakeholders, that is, public agencies, private agencies, communities at the local level, academicians, scientific institutions, investment agencies, and international development partners in

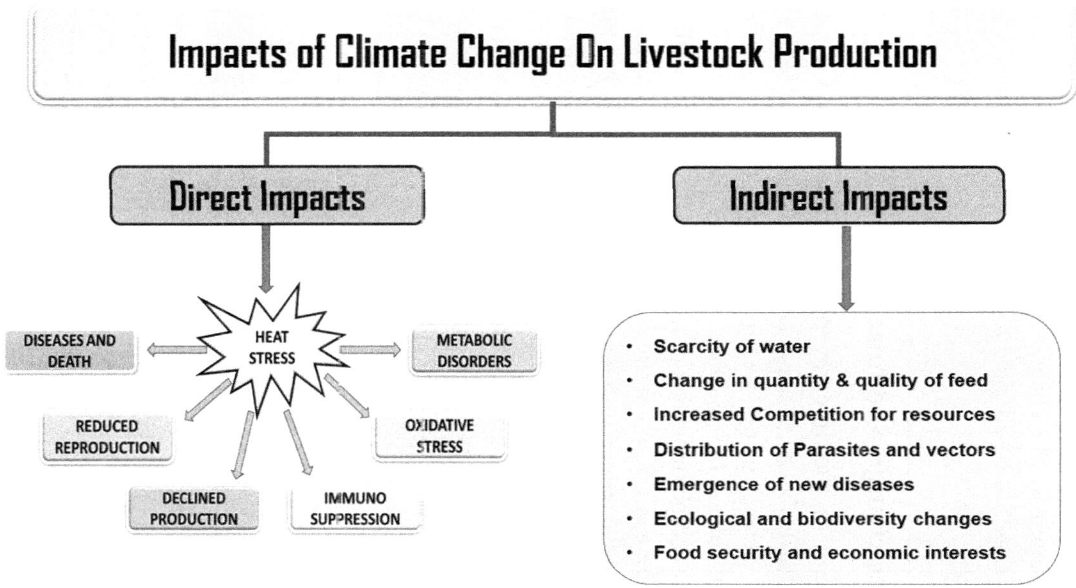

Fig. 1 Impacts of climate change on livestock production.

harnessing their respective resources and strengths for environmental management to protect animals, plants, and humans against the adverse effects of climate change (ECOW&PA, 2017).

This review is targeted at collating the impacts of climate change on livestock in terms of production, reproduction, and food security, with a specific focus on dairy animals. The chapter also summarizes various adaptation and mitigation strategies to be adopted to sustain livestock in the changing climate scenario.

Climate change and heat stress

According to the Intergovernmental Panel on Climate Change (IPCC), the Earth's average temperature could increase to 1.4–5.8°C by 2100, and an increase in the average global temperature above 2°C may be alarming and beyond the bearable limit (Richardson et al., 2009). Places where there is a combined effect of high relative humidity with high ambient temperature can create a more significant challenge for livestock. The global mean annual temperature over land shows 2016 as the warmest year, followed by the second highest in 2019. The average temperature change in the current decade (2011–2019) was 1.26°C. It was 1.01°C in the previous decade (2001 – 2010) and 0.58°C in the decade before that (1991–2000). Already, 8 of the 9 years in the most-recent decade (2011–2019) were among the 10 warmest years on record in terms of global mean annual temperatures (FAO, 2020).

Heat stress is the significant direct impact of climate change that strongly affects livestock production. Each species of animal has its comfort zone in which the energy expenditure is minimum. Thermoregulation is the underlying mechanism by which an animal maintains its body temperature. The thermoneutral zone (TNZ) of dairy animals is reported to range from 16 to 25°C, within which they maintain a physiological body temperature of 38.4–39.1°C (Yousef, 1985). Above this comfort zone, livestock starts to experience heat stress, as extra energy is spent on maintaining homeothermy. An environmental temperature above 35°C is enough to activate the stress response mechanisms in lactating dairy cows. There are variations among livestock for susceptibility to heat stress, depending on the species, genetic potential, life stage, and nutritional status. Therefore, the magnitude reduction in their performance differs. Compared to monogastric animals, ruminants have a higher degree of thermal tolerance. Among dairy animals, goats are the most adapted species to heat stress concerning production, reproduction, and disease resistance (Silanikove and Koluman, 2015). There exists a significant variation in thermal tolerance between different breeds of a species as well. Tropical cattle breeds (*Bos indicus)* are more resistant to heat stress and survive and perform better in a hotter climate due to their adaptive genetic build-up. However, high-producing exotic/temperate cattle breeds () often fail to adopt to the tropical climate due to the inability of expressing thermo-tolerant genes (Kumar et al., 2014). Thus, high-yielding breeds originating from temperate regions are more vulnerable to heat stress because of increased metabolic heat production (Hoffmann, 2008), due to vital body activities that are essential for survival. For instance, Holstein cattle can maintain ambient temperatures up to 25–26°C, but above 25°C, management practices should be modified to alleviate heat stress (Berman et al., 1985).

Impact of heat stress on animal health

The impacts of heat stress on livestock can be on feed intake, feed nutrient utilization, animal production, and reproduction, ultimately affecting the well-being of animals (Fig. 2). Above the thermoneutral zone, animals activate adaptive mechanisms of acclimatization through genetic and phenotypic modifications. To avoid hyperthermia, animals try to reduce internal heat production as well as increase their heat loss from the body, resulting in reduced feed intake, increased water intake, enhanced respiratory and sweating rates, and other altered physiological functions (Nardone et al., 2010).

Effect on feed intake and digestive physiology

Reducing feed intake is an adaptive mechanism to decrease heat production in warm environments. Accordingly, animals will be in a negative energy balance (NEB) due to a depressed appetite, resulting in a reduced body condition score and weight gain (Lacetera et al., 1996). Rhoads et al. reported that feed intake in lactating cows declines by 40% at air temperatures above 40°C, whereas the decline is around 22%–35% in dairy goats and 8–10% in buffalo heifers (Hamzaoui et al., 2013; Hooda and Singh, 2010). But water intake will rise by 1.2kg/°C above the minimum ambient temperature (West, 2003).

When the core body temperature rises, blood flow increases toward the extremities and reduces at the gastrointestinal tract to support the cooling mechanism (Baumgard and Rhoads, 2012). Hence, gut motility rumination

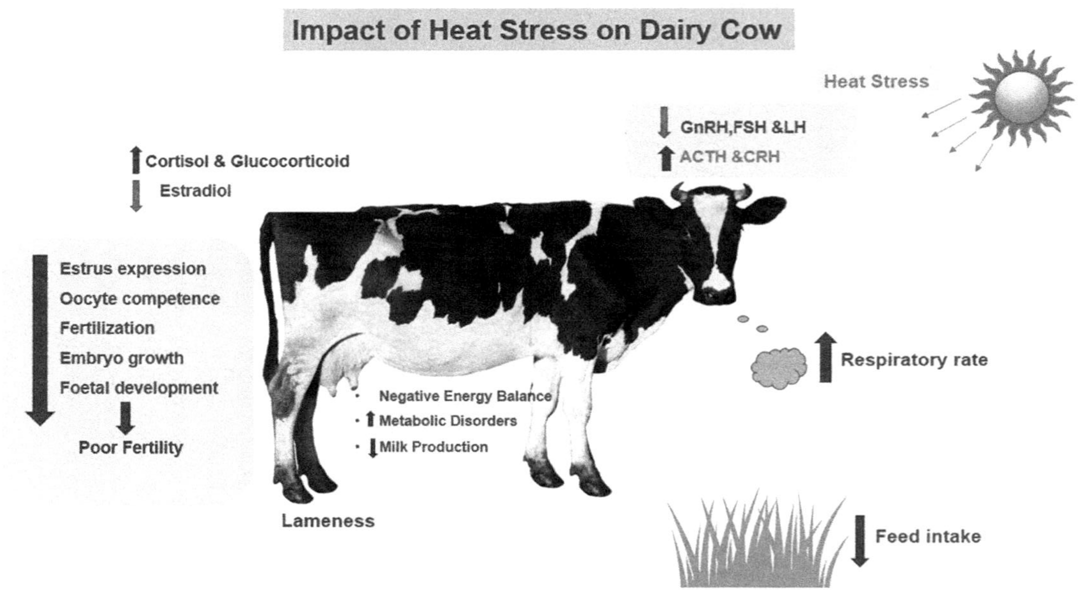

Fig. 2 Impact of heat stress on dairy cow.

and ruminal contractions are reduced in heat load conditions (Beede and Collier, 1986). At the same time, the production of ruminal volatile fatty acids (VFA), the primary source of metabolizable energy, is lowered due to decreased dry matter intake (DMI) or inadequate absorption (Gengler et al., 1970). Heat stress and poor forage quality result in the production of lower acetate and higher propionate and butyrate in the rumen. Thus, changes in the VFA ratio and an altered buffering mechanism during heat stress can result in changes in ruminal pH. Similarly, high temperature affects the movement and retention of digesta, causing changes in digestibility (Yadav et al., 2013). The ruminal microflora and fermentation pattern also changes due to heat stress, resulting in variations in digestibility (Zhao et al., 2019). Moreover, decreased feed consumption during day time and increased feeding during colder hours can cause rumen acidosis and laminitis in dairy cows (Shearer, 1999). The increased respiration and elevated exhalation of CO2 can also contribute to the incidence of rumen acidosis. It is reported that enteric methane production is higher when the ruminal temperature increases (King et al., 2011).

Effect on acid–base balance and oxidative stress

The altered metabolisms in heat-stressed animals are due to increased oxidative stress leading to mitochondrial damage, which is the source of reactive oxygen species (ROS). Altan et al. reported that thermal stress causes increased lipid peroxidation of polyunsaturated fatty acids, resulting in the accumulation of free radicals and other ROS. This leads to oxidative stress and associated cellular damage. Oxidative stress is an imbalance between free radicals and antioxidants. When more free radicals are formed than can be kept in balance by antioxidants, the free radicals start damaging fatty tissue, DNA, and proteins in the body, making it more vulnerable to disease. During the summer, lactating cows showed a higher reactive oxygen metabolite level in plasma, whereas the vitamin E and total carotene levels were decreased. Also, increased oxidant and decreased antioxidant molecule levels were reported from dairy cattle and buffalos with increased activity of the antioxidant enzyme (Lacetera, 2019). Increased water intake is one of the major physiological responses to heat stress. It aids in cooling digestive systems (Bianca, 1964) and acts as a vehicle for heat dissipation through panting and sweating. The increased water loss through sweat, urine, and via the respiratory tract changes the acid–base and electrolyte balance in heat-stressed cows. Because sweat in cattle is high in potassium and low in sodium, the increased sweating rate of heat-stressed cows increases potassium requirements by as much as 12%. The condition can be aggravated if there is a deficiency of forage in diets (Schneider et al., 1984; Collier et al., 2006).

Effect on immunity and incidence of disease

The animal's state of health plays a significant role in its ability to deal with heat-induced stresses. The prevalence of infectious and metabolic diseases may be greater in hot weather, as the cell-mediated immune response decreases during heat stress (Sunil Kumar et al., 2011). Severe heat stress reduces colostral immunoglobulins (IgG and IgA) in dairy cows, causing harmful effects on immunization and the survival of newborn calves (Nardone et al., 1997). During heat stress, cows' lymphocyte activity declines before and after parturition, potentially reducing the efficiency of vaccines and increasing their vulnerability to infections. Besides, there is a considerable reduction in neutrophil function during high temperatures, weakening the mammary gland's protection against infections (Lecchi et al., 2016) and leading to a higher occurrence of mastitis during the summer months (Morse et al., 1988). It has been found that increased ambient temperature causes significant cellular damage, leading to the activation and expression of heat shock factors (HSF) and heat shock proteins (HSPs) (Pirkkala et al., 2001). HSPs are a family of proteins produced by cells in response to stressful stimuli. They can bind with whole proteins and peptides apart from stimulating immune receptors and regulating the proper folding of proteins involved in proinflammatory signaling pathways.

There can be higher incidences of nutrient deficiencies, ketosis, ruminal acidosis, and respiratory alkalosis accompanying heat shock in dairy cattle (Kadzere et al., 2002). Cows in early lactation are prone to subclinical or clinical ketosis, a metabolic disease that occurs when the animal is in a severe state of negative energy balance (Lacetera et al., 1996). Liver function is also compromised in heat-stressed cattle, causing reduced liver enzyme activities and albumin secretion (Bernabucci et al., 2010). Further, lipid mobilization from adipose tissue during NEB can lead to liver lipidosis and the incomplete oxidation of leading fat accumulation of ketone bodies and ketoneurea. As ambient temperatures rise, the respiratory rate increases, resulting in panting, open-mouth breathing, and subsequent respiratory alkalosis due to the rapid loss of carbon dioxide. It will be compensated by the increased urinary output of bicarbonate and decreased salivary bicarbonate level, affecting healthy rumen buffering (Collier et al., 1982). The occurrence of lameness is also reported in dairy cattle in association with heat stress (Shearer, 1999), which is due to ruminal acidosis or loss of bicarbonate (Cook and Nordlund, 2009).

Effect of heat stress on production performance

The increased environmental temperature and the heat generated from the metabolism results in enhanced heat gain in animals. Srivastava (2010) reported that increased ambient temperature adversely affects production and reproduction performances in livestock.

Effect on milk production

Thermal stress adversely affects milk production, causing substantial economic loss to dairy farmers. Increased air temperature along with a high temperature-humidity index (THI) causes reduced DMI. This causes declined milk yield, reduced body weight, poor reproductive performance, and altered immunity (West et al., 1991; Berman, 2005; Vale, 2007). Animals with high genetic merit were found to be more susceptible to heat stress and its deleterious effects, disturbing the milk yield and composition (Wheelock et al., 2010). Many studies reported that dairy cattle's sensitivity to heat stress increases with enhanced milk production, which might be due to a rise in metabolic heat output through high production levels. For the same reason, Holstein cows were found to suffer from a more prominent effect of heat stress compared to Jersey cows, resulting in decreased milk yield (Kadokawa et al., 2012).

There exists a complex neuro-endocrine interaction to maintain homeostasis during heat stress in livestock. It has been proved that neuroendocrine regulation by the hypothalamic–pituitary–adrenal (HPA) axis plays a significant role in livestock tolerating and surviving a specific environment (Afsal et al., 2018). The HPA axis controls the thermoregulatory mechanism through the release of various neurotransmitters and hormones. High levels of epinephrine and norepinephrine ensure the blood supply to vital organs and regulate cardiovascular function in heat-stressed animals (Fuller et al., 2020). Hepatic gluconeogenesis is also under the control of the HPA axis, which supports life-sustaining activities in stressed animals. The primary hormones involved in the metabolic pathways of heat generation are the thyrotropin-releasing hormone (TRH), the thyroid-stimulating hormone (TSH), and the thyroid hormones (T3 and T4). It has been found that the levels of the growth hormone somatotropin were increased during heat stress. However, the levels of prolactin, the lactogenic hormone, were decreased in heat stress, which will adversely affect milk production and growth. During heat stress, the satiety center located in the ventromedial hypothalamus is stimulated, and the lateral "feeding center" is inhibited, leading to reduced feed intake and a consequent drop in milk production (Albright and Alliston, 1971). With every 1°C rise in air temperature, the DMI will reduce by 0.85 kg, which accounts for around a 36% decrease in milk production (West et al., 1991). A reduced milk yield in dairy animals of up to 50% might be due to reduced feed intake, and the remaining could be due to altered metabolisms as a part of reduced feed intake (Baumgard and Rhoads, 2013).

The results of Bouraoui et al. showed an increase in THI value from 68 to 78 reduced DMI by 9.6% and milk production by 21%. The effect of heat stress can vary depending on the stage of lactation. It has been found that the severity of heat stress is more in mid-lactation compared to early and late lactation. The decline in milk production was 14% in early lactation and 35% in mid-lactation (Basiricò et al., 2009). The heat stress on the dry period reduces mammary cell proliferation, reducing yield in the next lactation and having adverse effects on the immunity of dairy cows (Tao and Dahl, 2013). Singh et al. reported that heat stress would negatively affect the lactation length, dry period, calving interval, milk constituents, and milk yield in Murrah buffaloes. High temperature and high humidity during calving had a detrimental effect on the lactation yield, lactation length, and milk quality of cross-bred cattle (Kumar et al., 2014). Kadzere et al. reported that the milk fat, solids-not-fat (SNF), and milk protein percentages drop by 39.7, 18.9, and 16.9%, respectively, due to heat stress. Changing environments have significant effects on the organic and inorganic components of milk (Summer et al., 2019). The levels of milk fat and milk protein are lower in the summer season compared to other seasons (Bouraoui et al., 2002). Usually, about 60% of the VFAs produced in the rumen are acetic acid, a precursor of milk fat predominant in a high roughage diet (Schingoethe et al., 1976). The decreased availability and limited intake of forage during a hotter climate cause low acetate and high propionate levels, leading to milk fat depression (MFD) in lactating animals. The lactose component of milk is not much affected by heat stress (Cowley et al., 2015). Similarly, the milk protein content and casein fraction of milk tend to decrease when the THI goes above 75 (Abeni et al., 1993). As per Mariani et al. (1993), the lower content of milk ash and phosphorus in summer could be attributed to heat stress. Even though there is a significant effect of environmental factors on milk production, quantifying such environmental effects is often tricky, as milk production is also influenced by other factors such as nutrition and management that may or may not be directly linked to environmental factors (Collier et al., 2017).

Effect on meat production

Climate change will affect the quality and quantity of meat produced by livestock. Unlike daily production metrics such as milk or eggs, the impact of heat stress on the meat sector cannot be directly measured. Even then, there exists a high risk of mortality and adverse effects on animal health and welfare. The effect of heat stress on physiological and biochemical pathways will reflect on meat quality. The growth performance of beef calves will be adversely affected by a reduction in weight gain, feed intake, and feed conversion efficiency. Heat stress in beef cattle is visible by adjustments in homeostatic mechanisms such as panting, sweating, urination, and behavioral alterations such as a reduction in activity, increased water intake, and reduced feed intake (Summer et al., 2019). The effect will be rigorous for those maintained at grazing systems. Heat stress can cause a slow growth rate as well as lower slaughter weightless muscle mass and fat content. Hence, the quality and palatability of meat will decrease with altered tenderness and juiciness (Keane and Allen, 1998). Because changed climates lead to a shift in the availability of feed and fodder crops, there will be greater competition for resources. It has been reported that meat from grass-fed cattle contains more n-3 fatty acids and conjugated linoleic acid than those fed with concentrate (Scollan et al., 2006). These fatty acids are known for their health benefits in preventing cardiovascular and other lifestyle diseases. Hence, to keep the meat quality as per standards, supplemental nutrition has to be provided. Furthermore, heat stress and extreme weather conditions can negatively affect the health status, as mentioned above, in lactating cows. Heat stress affects meat quality parameters and meat safety. The occurrence of pale, soft, oxidative (PSE) meat in pork increases with high temperatures (Dalla Costa et al., 2007).

A high environmental temperature on or before slaughter decreases the glycolytic potential of meat with altered biochemical and histochemical characteristics resulting in darker meat (Spencer et al., 2005). Upon exposure to heat stress, *B. indicus* cattle led to dark, firm, and dry (DFD) beef (Wheeler et al., 2001). Moreover, it has been found that the risk of carcass contamination with *Escherichia coli* and Salmonella in poultry and pigs increased during heat stress (Gregory, 2010). An experiment on heat stress with indigenous sheep showed significant changes in carcass characteristics and meat quality (Rana et al., 2014). Intense heat provokes an adrenergic stress response, and adrenaline stimulates peripheral vasodilatation and muscle glycogenolysis, leading to high pH and darker meat. Also, hyperthermia along with an anaerobic metabolism due to exercise leads to early and robust rigor mortis. High temperature and dehydration in slaughtering animals can cause shrinkage of the myofibers, ultimately leading to dryness and dark-colored meat. In broilers, chronic heat stress from 4 to 6 weeks of age resulted in a decreased proportion of breast muscle (Zhang et al., 2012). In broilers, a high environmental temperature harms carcass composition and meat yield, notably in a reduction in breast meat. Also in response to heat stress, broilers were observed to have decreased protein synthesis and increased protein breakdown (Lin et al., 2006) with lower body protein and muscle tissue protein as well as higher fat levels (Gonzalez-Esquerra and Leeson, 2005; Akşit et al., 2006). Thus, the quality of meat is depreciated with the mobilization of minerals and vitamins from tissues (Sahin et al., 2009), reducing the nutritive value of eggs and meat. Hence, the addition of extra vitamins, minerals, and fats along with a reduction of excess protein is recommended in broiler feed to counter the effects of heat stress (Daghir, 2009).

Effect of heat stress on egg production

Climate change and associated heat stress are major constraints in the poultry industry. High temperature and reduced feed intake will adversely affect broilers and layers. Birds have a high rate of metabolism as well as intense heat production and low heat dissipation due to the presence of feathers and the absence of sweat glands. The comfort zone for poultry is between 18 and 22°C, and above 30°C, the feed intake reduces, leading to declined production performance and increased mortality depending on the severity and duration of exposure. It has been found that severe heat stress in layers reduces feed intake by 30–50%.

Moreover, the digestibility of nutrients is found to be reduced with high environmental temperature (Hai et al., 2000). As a result of insufficient nutrient supply, egg production drops. In addition, increased panting during heat stress causes a loss of bicarbonates from the body; these are essential for eggshell formation. Consequently, nutrient deficiency together with bicarbonate loss can result in compromised eggshell formation, leading to more broken eggs and decreased egg production (Mashaly et al., 2004), aggravating losses during the warm season. In a nutshell, high ambient temperature harms egg production by decreasing the egg weight, number, and eggshell quality.

Effect of heat stress on reproduction

High environment temperatures may compromise the reproductive efficiency of farm animals in both sexes (Fig. 3). It has been reported that heat stress has negative consequences on various reproductive functions. Many reproductive processes, such as spermatogenesis, oocyte competence, embryonic growth, gonadotropin secretion, ovarian follicular growth, steroidogenesis, development of the corpus luteum, and uterine endometrial responses, are impaired by heat stress (Wolfenson and Roth, 2019; Mondal et al., 2015a,b; Mondal et al., 2017). Hyperthermia and corresponding physiological modifications generated by heat stress are responsible for these changes. Increased ambient temperature and high-intensity radiant heat can alter normal reproductive physiology through the hypothalamic–hypophyseal–ovarian axis. The consequences are found to be more severe in high producers compared to low yielders. (Wolfenson et al., 2000; Hansen, 2009; Nardone et al., 2010). In males, thermal stress causes spermatic injury as a result of increased testicular metabolism. Thermal stress can be more deleterious to lactating dairy cows than nonlactating dairy cows or beef cattle, as metabolic heat production in lactating dairy cows is relatively high. Moreover, beef cattle breeds were found to be genetically resistant to heat stress, limiting the impact of heat stress on reproduction and fertility (Gaughan et al., 2010).

Effects of heat stress on female reproduction

Effects of heat stress on estrus and sexual behavior

Many studies show that heat stress reduces the duration and intensity of estrus in dairy cattle (Younas et al., 1993; Gwazdauskas et al., 1981). Because reproduction in females is influenced by the photoperiod, estrus cycles are further affected by seasonal variations. It has been observed that the frequency of anestrus and silent ovulation is higher during summer, leading to hidden estrus and failed insemination. Heat stress alters motor activities in dairy cows, such as decreased walking time during estrus resulting in poor estrus detection in the summer (Kadokawa et al., 2012). Sexual behavior reflects the proper functioning of reproductive systems in livestock. Thermal stress can induce a rise in the ACTH-regulated cortisol level, which blocks estradiol-induced sexual behavior (Hein and Allrich, 1992). Likewise, low

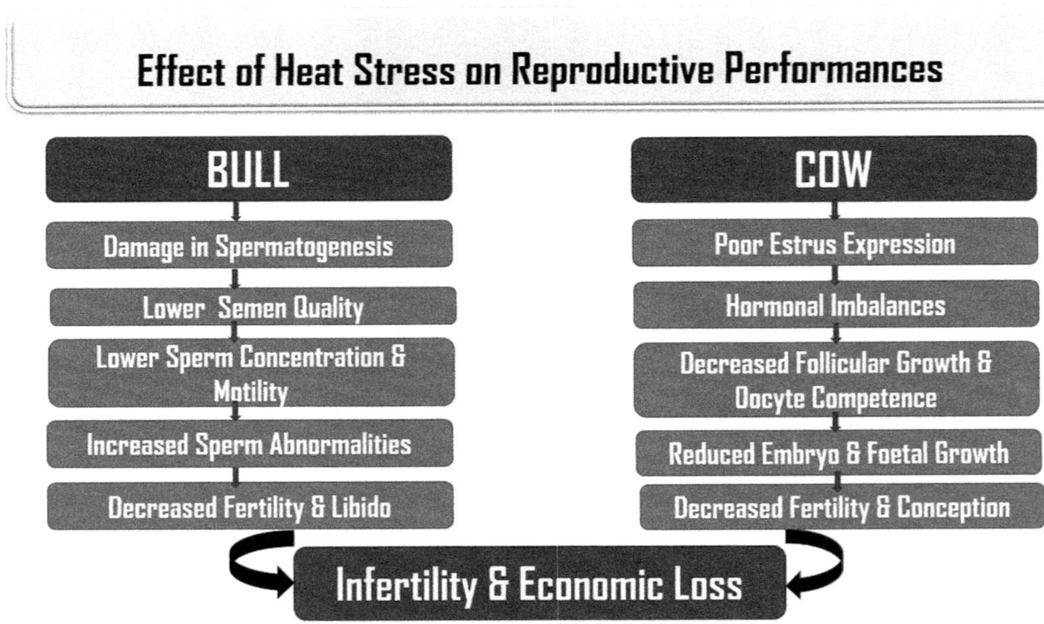

Fig. 3 Effect of heat stress on reproductive performances.

estradiol in Indian buffaloes on the day of estrus is responsible for the reduced expression of estrus in summer (Upadhyay et al., 2009). Also, physical lethargy associated with heat stress is believed to be a reason for weak estrus expression as an adaptive mechanism to reduce internal heat production (Hansen and Aréchiga, 1999). An investigation on the effect of heat stress on the day of estrus in cows under field conditions showed a significant reduction in estrus sign intensity and a reduction in the size of follicles. Moreover, vaginal discharge and mounting behavior were reduced with increased THI (Schüller et al., 2017). In summer seasons, cows prefer to exhibit standing heat during night hours when the atmospheric temperature is comparatively cold (West, 2002). Moreover, a study on Japanese black cows reported a longer estrus cycle with a subsequent delay in the postestrus progesterone level rise in the summer months (Sakatani et al., 2012).

Effect on fertility and pregnancy

Heat stress is the leading cause of poor fertility and conception rate in dairy cattle. The fertility of dairy cows and buffalos can be affected by various factors, including genetic, nutritional, hormonal, pathological, management, and environmental factors. However, as the heritability of fertility traits is low, most of the variations are determined by environmental effects rather than genetic factors (Thiruvenkadan et al., 2010). The reproductive efficiency of dairy cattle declines with exposure to heat stress (Fig. 2). High ambient temperatures cause poor detection of estrus and limit the possibility of a cow becoming pregnant. Altered follicular dynamics and poor oocyte quality led to low fertility and interrupted milk production throughout the year. Damage to oocytes and embryos can decrease fertility in summer, with an almost 20% reduction in pregnancy rates (Hansen and Aréchiga, 1999) and a 20–30% reduction in conception rates (Ingraham et al., 1974., Badinga et al., 1985, Chebel et al., 2004)). Also, a 90-day nonreturn of estrus to the first service in lactating dairy cows was reported in the summer months (Al-Katanani et al., 1999). Due to lowered conception rates, more artificial insemination (AI) is required to achieve a successful pregnancy. Furthermore, delayed fetal growth and increased fetal loss have been reported from thermally stressed pregnant animals.

The effect of heat stress on the hypothalamic–hypophyseal–ovarian axis

The reproductive hormones play a significant role in different phases of the function and development of the reproductive system (Mondal and Prakash, 2002; Mondal et al., 2003, 2004a, 2007, 2008a,b,c). Heat stress caused by high ambient temperature affects the reproductive cycles via the hypothalamic- hypophyseal–ovarian axis (Reddy and Rao, 2014). Several studies show an inverse relationship between a high environmental temperature and reproductive hormone concentration, consequently impairing reproductive efficiency in farm animals (Palta et al., 1997). The primary reproductive hormones are the gonadotropin-releasing hormone from the hypothalamus, gonadotropins, luteinizing hormone (LH), and follicle-stimulating hormone (FSH) from the anterior pituitary gland. Heat stress causes an altered secretion of LH and FSH at the hypophyseal level and androstenedione, progesterone, and estradiol at an ovarian level in livestock (Ronchi et al., 2001; Mondal et al., 2004b). The rise in environmental temperature is capable of increasing the cortisol and adrenocorticotropic hormone (ACTH) levels. Reduced circulating estradiol and prolactin levels result in diminished sexual behavior and irregular estrus cycles, leading to infertility (Singh et al., 2013).

Similarly, low levels of tonic LH secretion hamper luteal development by inhibiting follicular growth in cyclic cows (Wolfenson et al., 2000). The altered production of progesterone, LH, and FSH during heat stress impairs oocyte growth. Decreased estradiol production along with low LH surge frequency affect the maturation and ovulation of follicles. Consequently, the plasma inhibin level diminishes as a considerable quantity of plasma inhibin comes from small and medium follicles. The reduced preovulatory release of LH causes delayed ovulation and reduced expression of estrus behavior. Due to these discrepancies in the endocrine profile, buffaloes often failed to display estrus signs or show silent heat, owing to reduced ovarian activity and anestrous during heat stress (Das and Khan, 2010). On exposure to a higher temperature, the level of GnRH-induced FSH was also significantly reduced (Gilad et al., 1993) while tonic FSH secretion was elevated because of the inhibition of negative feedback from small follicles (Ozawa et al., 2005). This ultimately affected the reproductive efficiency of dairy animals. Likewise, heat stress reduces estradiol secretion from the ovarian follicle due to reduced theca cell androstenedione production and diminished granulosa cell aromatase activity. Decreased follicular size also possibly results in lower serum estradiol and decreased steroidogenesis within the theca and granulosa cells. The plasma progesterone concentration from the dominant follicle is also reduced during heat stress, limiting endometrial function and subsequent embryo development (Wakayo et al., 2015). Studies report that an increased level of circulating prolactin during heat results leads to the suspension of estrous cycles, whereas the increased secretion of endometrial PGF-2α during heat stress results in threatened pregnancy and infertility (Bilby et al., 2008; Singh et al., 2013).

Effect on follicular growth and oocyte development

The most prominent outcome of heat stress, limiting the reproductive function, is its detrimental effect on the ovarian follicles. It has been found that developed follicles undergo permanent damage whenever the core body temperature exceeds 40°C, becoming nonviable. Heat stress alters follicular growth and development along with reduced steroid hormone secretions, resulting in an extended follicular wave. All these issues disrupt oocyte growth and quality, adversely affecting the ovarian follicular dynamics (Wolfenson et al., 1995; Mondal et al., 2015a,b). Further, the degeneration of the premature dominant follicle before reaching the appropriate size leads to a significant drop in ovulation percentage. It has been found that low LH levels, high tonic FSH, and negative animal energy balance during summer prevent the maturation and ovulation of dominant follicles (Wakayo et al., 2015). Heat stress in summer not only lessens the number of dominant follicles, but also more medium subordinate follicles will survive. The reduction in individual follicular dominance and the development of more than one dominant follicle are believed to be the reasons for increased twinning in summer (Ryan et al., 1991). The. development and function of the corpus luteum are also affected by heat stress (Howell et al., 1994), and progesterone secretion by the luteal cells is suppressed (Wolfenson et al., 2000). Heat stress causes delayed luteolysis in cows, triggering deprivation in follicular estradiol secretion, which initiates endocrine events resulting in luteolysis. As a result, longer luteal phases are observed in heat-stressed cows (Wilson et al., 1998).

The growth and maturation of oocytes are adversely affected during heat stress due to oxidative damage and apoptotic cell death. It induces irreversible damage on the cytoskeleton and meiotic spindle and impairs mitochondrial function, compromising the oocyte (Hansen, 2009). The damage to the cells is due to changes in progesterone, LH, and FSH secretions during the estrous cycle. It has been found that the cytoplasm of the oocyte is more vulnerable to heat shock than the nucleus (Paula-Lopes et al., 2013). Heat stress can compromise the oocyte as early as 105 days before ovulation (de Torres-Júnior et al., 2008) or as late as the periovulatory period (Putney et al., 1989). Reports show that the deleterious effect of heat stress on oocyte competence is delayed in *B. indicus* cattle compared to *Bos taurus* (de Torres-Júnior et al., 2008). Heat stress can also harm the oocyte development, protein synthesis, and transcription formation needed for subsequent embryonic development (Edwards and Hansen, 1997; Mondal et al., 2015a,b). Moreover, reduced mRNA content changed the membrane integrity while inadequate protein storage needed for early embryonic growth altered the signal transmission and protein transport. During summer, heat stress damages developing small follicles, causing ovulation of the infantile oocyte (Howell et al., 1994). Early meiosis and aged oocytes with poor competency occur as a result of prolonged follicular dominance. Incompetent oocytes become transcriptionally inactive and lose the ability to respond to heat by synthesizing heat shock protein 70 (HSP70). Therefore, cows exposed to thermal stress lose their oocyte competence for fertilization and development to the blastocyst stage, decreasing the fertility rate.

Effect on fertilization and embryo growth

Early embryonic mortality is a major cause of post fertilization losses in domestic ruminants (Ghosh and Mondal, 2006; Suresh et al., 2009; Mondal et al., 2009; Mondal et al., 2010; Nicholas et al., 2013; Nandi et al., 2012; Mondal et al., 2013; Mondal et al., 2015a,b). Thermal stress affects healthy embryonic growth and survival in dairy animals. The spermatozoa formations need a temperature below the average body temperature. Similarly, the development of oocytes is also susceptible to temperature variations. Thus, heat stress leads to lower fertilization and the development of deficient embryos (Wolfenson et al., 2000). Besides, a compromised intrauterine environment in heat-stressed cows increases the chance of early embryonic loss and inhibits healthy embryonic development, reducing the percentage of successful inseminations (Rivera and Hansen, 2001). An increase in the uterine temperature of 0.5°C showed a decline in the conception rate (Naqvi et al., 2012).

The maximum embryonic losses due to heat stress occur during the early embryonic period, and the effect is critical during the first cleavage divisions when most of the embryonic genome is inactive (Ahmad Para et al., 2018). Heat stress causes embryonic death via interfering with protein synthesis, oxidative cell damage, and the activation of stress-related genes associated with apoptosis. Embryo survival during heat shock depends on its genotype, the stage of development, exposure to regulatory molecules, and redox status. It has been observed that the bovine embryo becomes more resistant to heat shock as growth progresses. *B. indicus* embryos are more resistant to heat shock than embryos from nonadapted *B. Taurus* (Hansen, 2007).

Effect on fetal development

Heat stress during pregnancy delays fetal growth due to reduced nutrient and hormone supply to the endometrium owing to decreased uterine blood supply (Savasani et al., 2015). Studies show that embryos heat stressed at the postimplantation period are more prone to fetal malnutrition and various other teratologies (Kadokawa et al., 2012). The weak growth of the embryo restricts the production of an adequate amount of the maternal recognition of pregnancy factor (interferon tau–IFN T in ruminants), which makes the endometrial environment suitable for a successful pregnancy. The consequent rise in endometrial prostaglandin F2 alpha (PGF2α) elicits luteolysis, resulting in a threatened pregnancy (Upadhyay et al., 2009). Heat exposure studies in ewes after 120 days of gestation showed a reduction in feed intake by 25% and placental weight by 54%. Heat stress in ewes was found to be positively correlated with fetal weight and negatively correlated with fetal/placental weight ratio and fetal brain/liver weight ratio. Thus, fetal growth retardation occurs due to a primary reduction in placental growth caused by the reduced activity of thyroid hormones, progesterone, and placental lactogen (Bell et al., 1989). Early et al. (1991) reported that fetal and placental weights and the total content of protein, RNA, and DNA were reduced in heat-stressed ewes. A reduction in placental function and a decline in progesterone secretion affect fetal growth and aggravate the chances of teratological defects. A recent study on the Murciano-Granadina breed of dairy goats revealed a shortened gestation length and altered investigative behavior of the offspring in their early life when heat stress hit the dam during the first third of pregnancy (Coloma-García et al., 2020). Heat stress may have severe effects during early and mid-gestation, as angiogenesis is more extensive during that period (Hansen, 2009). Heat stress during the late gestation period in Holstein cows resulted in calves with a lower birth weight and impaired postpartum productive and reproductive performance (Reyes et al., 2006).

Effect of heat stress on the postpartum period

Heat stress delays the postpartum return of estrus and reduces the conception rate and subsequent pregnancies in cows (Nardone et al., 2010). A negative energy balance in dairy cows as a result of reduced DMI may be the primary reason for the prolonged postpartum period and reduced fertility during summer. Further, heat stress-associated hormone imbalances in cattle such as high levels of prolactin, low levels of luteinizing hormone and estradiol, and consequent impairment in follicle maturation result in ovarian inactivity

(Wolfenson et al., 2000). Thus, poor folliculogenesis and delayed ovulation result in more extended service periods and calving intervals along with reduced calf birth weight and milk yields (Singh et al., 2011; Savasani et al., 2015). Similarly, Dash et al. reported that high ambient temperature combined with high relative humidity in the summer season results in more prolonged service periods in cows and buffalos. Furthermore, the postpartum negative energy balance of dairy cows due to reduced DMI under heat stress will also extend the interval between the calving to the next conception. Thus, a delay in the postpartum breeding of cows reduces the pregnancy rate in the summer. Some studies show that first lactation cows are more vulnerable to heat stress in summer with a significantly longer postpartum period than cows with multiple lactations (Ray et al., 1992). All this leads to extended calving intervals, poor birth rates, and reduced milk yield per year, significantly affecting the economic outcome. Hence, a cooler region will favor optimum reproductive performance in livestock.

Effects of heat stress on male reproduction

Heat stress is an essential factor that affects the reproductive competence of bulls. Because bulls comprise more than half the herd, bull fertility is crucial for successful artificial insemination and to produce a good, viable, and genetically potential young one. Exposure of bulls to high ambient temperatures was found to have an undesirable effect on testicular size, libido, and semen quality (Meyerhoeffer et al., 1985; Söderquist et al., 1996). Spermatogenesis occurs in testicles, and for successful spermatogenesis in mammals, the temperature of the testis must be 2–8°C below body temperature (Senger, 2003). The optimum temperature for spermatogenesis is between 15 and 30°C (Sahni and Roy, 1967). The scrotal skin temperature has a strong negative association with serum testosterone, libido, sperm motility, sperm concentration, and rate of conception; it is positively linked with abnormal and dead sperm. Scrotal circumference, testicular consistency, tone, size, and weight are inversely related to higher ambient temperatures. However, the semen quality of *B. indicus* bulls is less susceptible to defects attributed to high temperatures than those of *B. taurus* bulls (Skinner and Louw, 1966).

Effect on spermatogenesis

Spermatogenesis is the formation of spermatozoa from primordial germ cells. Through sequential stages of differentiation, spermatogonia are transformed into spermatozoa. Significant events during spermatogenesis are spermatocytogenesis, spermatidogenesis, and spermiogenesis. Spermatocytogenesis is the formation of spermatocytes with half the usual complement of genetic material, whereas spermatidogenesis is the formation of spermatids from secondary spermatocytes. Spermiogenesis refers to the differentiation of spermatids to mature spermatozoa, which are released into the lumen of seminiferous tubules and stored in the epididymis. The total duration of spermatogenesis in bulls is 61 days (Staub and Johnson, 2018). Sertoli cells are the main somatic component of seminiferous tubules, and their number is an essential factor determining the daily production of spermatozoa in bulls. Leydig cells, located in the interstitial compartment outside the tubules, performs the endocrine function.

Heat stress causes degeneration in both Sertoli and Leydig cells in the testis in addition to the collapse of the germinal epithelium and the partial atrophy of seminiferous tubules. The consequent degeneration of germ cells in seminiferous tubules damages spermatogenesis (Naqvi et al., 2012) and lowers testosterone levels (Murugaiyah, 1992). The proliferation of spermatogonia is interrupted by high relative humidity and ionizing irradiation (Amundson et al., 2005). Elevated temperatures and increased testicular metabolism without an equivalent rise in blood supply induce local hypoxia and testicular injures (Reyes et al., 2012). Also, high temperature induces mitochondrial dysfunction in sperm cells and interferes with glucose oxidative metabolism, resulting in germ cell apoptosis and DNA damage (Paul et al., 2008). The subsequent increase in lipid peroxidation and the accumulation of ROS produce primary sperm defects (Nichi et al., 2006). This can also bring about secondary defects in sperm following interventions in sperm quality and semen production, reducing the fertilizing ability of spermatozoa. Heat stress in the form of 48-h scrotal insulation in Holstein bulls showed adverse effects on the chromatin of epididymal and testicular sperm (Karabinus et al., 1997). Thus, exposure of bulls to extreme ambient temperature seems to damage the primary spermatocytes, spermatids, and spermatozoa and reduces the breeding efficiency of males. Spermatozoa are more susceptible to heat stress during postmeiotic stages of development, as histone modifications and hyperacetylation cause the chromatin condensation of spermatozoa. This will disrupt the DNA methylation reprogramming in the paternal pronucleus, causing disordered active DNA demethylation followed by de novo methylation patterns. Both paternal and maternal pronuclei tend to decrease in size after fertilization with heat-stressed spermatozoa, resulting in lower fertilization rates (Rahman et al., 2018).

Effect on semen quality and characteristics

Heat stress affects the endocrine system of the bull, compromising sperm quality and concentration and resulting in reduced fertilization rates and high embryonic mortality in females. However, some reports show that stress has the least effect on testicular endocrinology in bulls (Barth and Bowman, 1994) compared to altered steroidogenesis in boars (Wettemann and Desjardins, 1979). Summer heat stress significantly reduced sperm quality in Simmental, Limousin, and Yanbian bulls. Simmental bulls were found to be most sensitive to heat stress, and the severity of heat stress depends on the bull's breed. With heat stress, the expression levels of HSP60 and HSP70 increased while HSP90 decreased. Alterations in heat shock protein expressions may have a deleterious effect on sperm quality in bulls (Cheng et al., 2016). THI beyond 60 can elicit detrimental effects on semen volume, semen output, number of semen doses per ejaculate, and semen concentration. Semen production losses were more for adult bulls than young bulls at extremely high THI (Al-Kanaan et al., 2015). The impact of heat stress on the reproductive performances of dairy cows and bulls is shown in Fig. 3.

A recent study reported that bull semen collected at a high THI compromised sperm viability and in vitro embryo development with a decrease in blastocyst development as well as delayed hatching (Luceño et al., 2020). Similarly, in boars exposed to thermal stress, a significant reduction in sperm motility and total sperm count as well as a significant increase in primary and secondary morphologically abnormal spermatozoa were observed (McNitt and First, 1970). Many studies report a decrease in sperm motility and semen volume along with an increase in dead sperm, abnormal sperm, or defective sperm in semen samples from heat-stressed bulls (Pinto et al., 2001; Kastelic et al., 2001). In commercial AI centers, only bulls with more than 75% normal spermatozoa in the ejaculate will be selected. If abnormal spermatozoa are due to heat stress, the chances of rejecting good bulls by mistake can lead to higher economic losses.

Other indirect impacts of climate change

The indirect effects of climate changes mainly arise due to changes in the quality and availability of feed and water. Other impacts are the survival and allocation of pathogens and their vectors, spreading vector-borne diseases and food-borne diseases, changes in host resistance, and feed and water scarcity (Nardone et al., 2010; Thornton et al., 2009).

Change in the quality and quantity of feed and water

Warm environments and extended droughts, together with higher atmospheric CO2 levels, can alter the availability of fodder and water. This will elicit livestock movement to areas where feed and fodder are available, enhancing the risk of disease transmission among domestic animals and between domestic and wild animals (Pinto et al., 2008). There can be modifications in ecosystems and changes in the quantity, quality, variability, and availability of feed and fodder crops, generating greater competition for resources among species (FAO, 2009; Thornton, 2010). Poor forage quality is thought to be a significant reason for increased methane emissions per unit of gross energy consumed (Rötter and Van de Geijn, 1999; Benchaar et al., 2001). The impacts can be more severe in grazing systems. In nongrazing systems, indirect impacts can be reflected in higher prices for feed and fodder, with reduced crop yields and reduced feed availability.

Global agriculture uses 70% of freshwater resources, making it the world's largest consumer (Thornton et al., 2009). However, water scarcity is a severe issue for the livestock sector, which uses water for animal drinking, feed crops, and other production processes (Thornton et al., 2009). The livestock sector accounts for about 8% of global human water use, and global warming causes a shortage of water (Nardone et al., 2010). Global warming and the associated rise in sea levels have increased salinity, and high concentrations of heavy metals may influence livestock production (Nardone et al., 2010). It has been reported that healthy metabolism, digestion, fertility, etc., are affected by the water's high salinity. Also, chemical contaminants and heavy metals in water can harm cardiovascular, excretory, skeletal, nervous, and respiratory systems (Nardone et al., 2010). This demands the cultivation of low-water crop varieties and heat-tolerant livestock species.

Disease epidemics

The indirect effects of climate change on disease epidemics are due to changes in the temporal and spatial distribution of microbial communities (pathogens or parasites), vector-borne diseases, and food-borne diseases as well as host resistance (Nardone et al., 2010; Thornton et al., 2009). Poor feeding and watering can make livestock more vulnerable to infectious diseases. There can be an altered distribution and transmission of animal pathogens (Baylis and Githeko, 2006), and heat stress can improve the survival capability of pathogens or their vectors (Chirico et al., 1997). Climate change and associated warm environments are capable of disturbing animal health by altering the host–pathogen–environment cycle. Higher temperatures and changing rainfall patterns can alter the abundance, distribution, and transmission of animal pathogens (Baylis and Githeko, 2006) as well as modify the number and spread of vector-borne pests and insects (Thornton et al., 2009). Therefore, there can be shifts in the spread of diseases along with the emergence of new infections in which livestock are not usual hosts and could even act as a mixing vessel between humans and animals. Many of these diseases can be zoonotic, causing severe consequences for public health, biodiversity conservation, and the livestock sector (Pinto et al., 2008). For instance, increased tick infestations were found to cause weight loss by up to 18% in livestock (White et al., 2003). Also, Bluetongue virus that mainly affects sheep is widespread with increased environmental temperature due to the shift in the distribution of the Culicoides species (Wittmann et al., 2001). Similarly, vector-borne diseases such as Rift Valley fever (RVF), West Nile virus (WNV), and trypanosomosis may occur more frequently in the future as an effect of global climate change (FAO, 2015). Precipitation variation can make livestock and wildlife prone to diseases like anthrax, when grazing (WHO, 2008).

Impact on biodiversity

Fluctuations in crop production due to elevated temperature, high CO2 levels, and water scarcity may indirectly affect biodiversity. It may change the migration and reproduction patterns of livestock. Those species with limited habitats, smaller populations, restricted mobility, and reduced breeding rates will be more susceptible (Steinfeld et al., 2006). Deforestation and overgrazing can destroy many unique livestock habitats. Besides, an increase in livestock rearing can also limit the availability of natural resources, causing ecological loss. Moderately grazed areas are more species-rich than under- or overgrazed areas (FAO, 2016). Moreover, GHG emissions play a significant role in the loss of biodiversity (Alkemade et al., 2013).

The Ministry of Environment, Forest, and Climate Change of the Indian government made several plans, including nationally determined contribution (NDC), to protect national biodiversity, forests, and wildlife; ensure the welfare of animals, birds, and wildlife; and prevent and abate pollution and climate change (MoEF&CC, 2018–2019, 2019–2020).

Impact on GHG emissions

The livestock sector plays a significant role in GHG emissions such as CO2, NO2, and CH4, and contributes about 14.5% of anthropogenic GHG emissions (FAO, 2006; Gerber et al., 2013). The enteric fermentation, respiration, and excretions of livestock can cause direct emissions of these gases (Jungbluth et al., 2001). However, emissions derived from feed crops, manure application, farm operations, livestock product processing, transportation, and land-use allocation for livestock production are responsible for indirect emissions (IPCC, 1997). The livestock segment contributes 44% of the global anthropogenic methane (CH4) emissions through enteric fermentation and manure production. The dairy sector accounts for 65% of the total livestock GHG emissions (Gerber et al., 2013). Beef cattle are first with 41% of the sector's emissions, followed by dairy cattle (20%), swine (9%), buffalo (8%), poultry (8%), and small ruminants. However, the digestive process of pigs does not produce as much methane as in ruminants. The reduced efficiency and productivity of livestock are responsible for the higher concentration of these gases (Gerber et al., 2013). Enteric fermentation is the major source of methane emissions, which can be reduced through improved nutrition and careful genetic selection and breeding of animals (US-EPA, 1999). Various management practices to mitigate enteric fermentation include:

- Supplementation of dietary fat content (Beauchemin et al., 2008; Martin et al., 2010).
- Provision of good-quality forage (Hristov et al., 2013).
- Increased protein content in the diet (ICF International, 2013).
- Provision of feed supplements and antibiotics (Boadi et al., 2004).
- Use of antimethanogens and vaccines to suppress methane emissions (EPA, 2013).

There is a need to develop a policy that differentiates between poor emissions from the breeding of farm animals and rich emissions from the usage of automobiles. The distinction is necessary because the first is referred to as survival emissions, while the second is referred to as luxury emissions. (Narain et al., 2009).

Adaptation and mitigation strategies

Adaptation and mitigation strategies are critical to preserving the agriculture and livestock sector to be sustainable, productive, and resilient, as the impact of climate change will seriously compromise food security. Adaptation is a vital response to the adverse effects of climate and preparation for future impacts. There are a great variety of adaptation options available, including technological (heat-tolerant breeds, development and promotion of new crop varieties, livestock feed improvements in water and soil management, and improved animal health technology), behavioral (changes in dietary choice), managerial, (different farm management practices), and policy options (planning regulations and infrastructural development) (Kurukulasuriya and Rosenthal, 2003). Adequate policy adaptation is required to provide environmental stability, preserve ecological balance, and address the international climate change challenges. The policy is expected to address the impact of climate change on animals, birds, and human-animal conflicts, mainly through public-private participation models (Sharma, 2018). Mitigation strategies are human interventions to ameliorate the negative effects of climate change on living things. Different management practices are needed to optimize the production of livestock in hot and humid areas (Beede and Collier, 1986). Important management tactics to minimize the effects of heat stress include:

Physical modification of the environment

Alterations in housing and management strategies can be attempted to alleviate the negative impact of heat stress. Modification of the microenvironment with the provision of shade and cooling can enhance heat dissipation, thus helping to relieve heat stress in the hot environment. Cooling ponds or technologies such as fans, misters, sprinklers, and showers can be used to cool the environment. Cooling can improve milk production, reproduction performance, and feed conversion efficiency in dairy cows (Wolfenson, 2009). Natural or artificial shade is a cost-effective method to minimize heat stress in grazing animals from solar radiation. Likewise, planting more trees surrounding the cowshed will provide natural shade and freshening effects to animals. Elevated roofs with ideal roofing material can reduce solar radiation and heat load inside the barn. Similarly, barn orientation can also benefit by lessening heat stress by reducing the ambient temperature.

Improved nutritional management practices

Enhanced nutrition can support livestock to get through the adverse effects of climate change. Because DMI is reduced during heat stress, the concentration of absorbable nutrients per unit of blood volume must be increased to satisfy the animal's daily requirements (Beede and Collier, 1986). As the buffering capacity of the rumen is affected during thermal stress, supplementation of yeast products can improve the digestibility of nutrients. It enriches the rumen microbial population and the VFA ratio in the rumen with decreased rumen ammonia production. Fungal cultures in the diet have also been shown to reduce the respiration rate and body temperature during thermal stress (Atrian and Shahryar, 2012). It is beneficial to supplement rumen-active fat to lactating cows to increase the energy density during the summer months. The addition of dietary fat in the feed and augmenting rumen undegraded protein increase milk yield in hot climates.

The oxidative cell damage during heat stress could be minimized through supplementation with vitamins and minerals, particularly vitamins A, C, and E, and mineral zinc (McDowell, 1989). Vitamins E and C prevent lipid peroxidation and associated cellular damage. Vitamins E and C as well as zinc are known as the scavengers of ROS during oxidative stress. Additionally, vitamin C acts as an antioxidant and helps in the absorption of folic acid. Also, supplementation of electrolytes to the heat-stressed animal will better regulate the acid–base balance in the blood. Supplementation of vitamin C with other electrolytes can improve cell-mediated immunity (Kumar et al., 2010).

Animal breeding and genetic selection

Even though environmental adaptations and nutritional management can lessen the impact of heat stress to some extent, some long-term policies have to be adopted to cope with climate change. Climate change projections suggest that many livestock succumb to heat stress and breeds with effective thermoregulatory control may be needed for the future. Thermoregulation is an evolutionary adaptation that allows mammals to maintain homeothermy irrespective of environmental fluctuations (Silanikove, 2000). There is a difference in thermal tolerance between species, so livestock species that have natural heat tolerance can thrive better in the future. For instance, *B. indicus* breeds show evolutionary adaptation to heat stress due to morphological, physiological, and cellular traits (Finch, 1986; Hansen, 2004), enabling them to cope with heat stress conditions. On the other hand, *B. taurus* evolved in temperate zones and lacks suitable heat-managing mechanisms (Landaeta-Hernández et al., 2011). Morphological features such as a higher ratio of skin surface to mass, better skin pigmentation, shorter hair, lighter-colored coats, larger and more numerous sweat glands, and increased skin vascularity improve thermoregulation efficiency in *B. indicus* (Landaeta-Hernández et al., 2011; Riley et al., 2012). Regardless of greater thermoregulatory adaptabilities, *B. indicus* breeds are also susceptible to the adverse effects of heat stress and can experience compromised welfare (Srikandakumar and Johnson, 2004). At the same time, for temperate breeds such as the Holstein Friesian, the impact can be more severe and harmful.

Cross-breeding and selective breeding programs can provide concurrent adaptation, mitigation, and food security benefits. To enhance the thermoregulatory ability in Holstein genetics, attempts have been made to introduce the slick hair gene (SLICK) through selective breeding. This gene is responsible for controlling hair length, which is a major factor for evaporative heat loss and efficient heat transfer to surroundings (Dikmen et al., 2008). Composite cattle breeds are expected to have higher heat tolerance, better disease resistance and fitness, and greater reproductive traits than pure temperate breeds (Bentley et al., 2008). Thus, cross-breeding can make use of locally adapted breeds, which are heat-tolerant, thrive on poor nutrition, and are resistant to parasites diseases (Hoffmann, 2008). Furthermore, the conservation of locally adapted but threatened breeds is essential to maintain genetic diversity.

Government of India policy on climate change vis a vis animal production

Of several policy initiatives from the Indian government, Rashtriya Krishi Vikas Yojana (RKVY) under the National Mission for Sustainable Agriculture (NMSA) is an important scheme. It is incentivized through the National Adaptation Fund for Climate Change and encourages growth in agriculture and allied sectors toward crop husbandry, animal husbandry, research and development for animal production, forestry and wildlife, food storage, and infrastructure development (Johnson and Kenchington, 2019). The NMSA scheme has been implemented under the National Action Plan on Climate Change (NAPCC), delivering several milestones such as: (i) Crop diversification, (ii) the identification and release of climate-resilient varieties (CRV), (iii) the identification of genotypes of crops with enhanced CO2 fixation potential and lower consumption of water and nutrients, (iv) climate-resilient genotypes with greater adaptation to drought, flood, salinity, and high temperature, (v) coverage of milch animals under a ration balancing program, and (vi) the establishment of a bypass protein fee-making unit (MoEF&CC, 2019–2020). In the era of climate change vis-à-vis sustainable livestock production, there is an extreme need for impact assessments related to the use of adaptation and mitigation measures tailored to the location and livestock production system in practice globally, and stringent policies that support and facilitate the implementation of climate change adaptation and mitigation measures vis-à-vis livestock production (Rojas-Downing et al., 2017).

Conclusion

Climate change generates a variety of challenges for livestock production through elevated temperature and water scarcity. This chapter discussed in detail the impact of heat stress on livestock performances in terms of growth, production, and reproduction in addition to mitigation strategies. The impacts of growing temperature and increased climate variability on livestock are reflected in animal health and growth. The adverse effects can be due to the changed quantity and quality of fodder, the high price of feed production, and changing patterns of pests and diseases. Heat stress on livestock reduces feed intake and nutrient utilization, ultimately affecting the well-being of animals. At the same time, the livestock sector contributes to climate change through enteric emissions of GHG. Reducing enteric methane can bring immediate improvements in climate change mitigation and food security. Raising more productive animals is a strategy to enhance efficiency and, thus, lower methane emission concentrations. Because the demand for animal products in the upcoming decades will increase massively, the sector has to explore opportunities to raise sustainable production with the efficient use of natural resources. Hence, it is necessary to adopt climate-smart livestock production strategies. Adaptation to climate change can be done by substituting climate-resilient livestock species and modifying animal husbandry practices such as improved nutrition and assisted reproductive strategies as well as proper animal health. The risk of emerging complex diseases can be best tackled by advanced veterinary care with improved animal husbandry practices. More research and stringent policies are needed in this area to enhance further practical options for adaptation and mitigation strategies.

Acknowledgments

The second author acknowledges that writing this chapter would not be possible without working experience at the Department of Scientific and Industrial Research, the Ministry of Science and Technology, Government of India. The authors also acknowledge the contributions and initiatives of different scientists, including officials of the Ministry of Environment, Forest and Climate Change, Ministry of Agriculture and Farmers Welfare, etc., in formulating different rules, acts, guidelines, and strategies on climate change, which helps to frame this chapter in a structured manner.

Disclaimer

The present chapter has been written in the personal capacity of all authors and the views do not necessarily represent those of any ministries, government departments, public, private, society, trust or other organizations nationally or internationally.

Conflicts of interest

The authors declare that there are no conflicts of interest regarding publication of this paper as a chapter.

References

Abeni, F., Calamari, L., Maianti, M.G., Cappa, V., Stefanini, L., 1993. Effetti dello stress termico sulle bovine in lattazione ed accorgimenti alimentari miranti ad attenuarne l'impatto su quantità e qualità del latte prodotto. Annali Fac. Agric. UCSC33, 151–170.

Afsal, A., Sejian, V., Bagath, M., Krishnan, G., Devaraj, C., Bhatta, R., 2018. Heat stress and livestock adaptation: neuro-endocrine regulation. Int. J. Vet. Anim. Med. 1 (2), 108.

Ahmad Para, I., Ahmad Dar, P., Ahmad Malla, B., Punetha, M., Rautela, A., Maqbool, I., Mohd, A., Ahmad Shah, M., Ahmad War, Z., Ishaaq, R., Akram, M.W., 2018. Impact of heat stress on the reproduction of farm animals and strategies to ameliorate it. Biol. Rhythm. Res., 1–7.

Akşit, M., Yalcin, S., Özkan, S., Metin, K., Özdemir, D., 2006. Effects of temperature during rearing and crating on stress parameters and meat quality of broilers. Poult. Sci. 85 (11), 1867–1874.

Albright, J.L., Alliston, C.W., 1971. Effects of varying the environment upon the performance of dairy cattle. J. Anim. Sci. 32 (3), 566–577.

Al-Kanaan, A, König, S, Brügemann, K, 2015. Effects of heat stress on semen characteristics of Holstein bulls estimated on a continuous phenotypic and genetic scale. Livestock Science 177, 15–24.

Al-Katanani, Y.M., Webb, D.W., Hansen, P.J., 1999. Factors affecting seasonal variation in 90-day nonreturn rate to first service in lactating Holstein cows in a hot climate. J. Dairy Sci. 82 (12), 2611–2616.

Alkemade, R., Reid, R.S., van den Berg, M., de Leeuw, J., Jeuken, M., 2013. Assessing the impacts of livestock production on biodiversity in rangeland ecosystems. Proc. Natl. Acad. Sci. 110 (52), 20900–20905.

Amundson, J.L., Mader, T.L., Rasby, R.J., Hu, Q.S., 2005. Temperature and temperature–humidity index effects onpregnancy rate in beef cattle. In: Proceedings of 17th International Congress on Biometeorology, Deutscher Wetterdienst, Offenbach, Germany.

Atrian, P., Shahryar, H.A., 2012. Heat stress in dairy cows (a review). Research in Zoology 2 (4), 31–37.

Aydinalp, C., Cresser, M.S., 2008. The effects of climate change on agriculture. Agric. Environ. Sci. 5, 672–676.

Badinga, L., Collier, R.J., Thatcher, W.W., Wilcox, C.J., 1985. Effect of climatic and management factors on conception rate of dairy cattle in subtropical environment. J. Dairy Sci. 68, 78–85.

Barth, A.D., Bowman, P.A., 1994. The sequential appearance of sperm abnormalities after scrotal insulation or dexamethasone treatment in bulls. Can. Vet. J. 35 (2), 93.

Basiricò, L., Bernabucci, U., Morera, P., Lacetera, N., Nardone, A., 2009. Gene expression and protein secretion of apolipoprotein B100 (ApoB100) in transition dairy cows under hot or thermoneutral environments. Ital. J. Anim. Sci. 8 (sup2), 492–594.

Baumgard, L.H., Rhoads, R.P., 2012. Ruminant nutrition symposium: ruminant production and metabolic responses to heat stress. J. Anim. Sci. 90 (6), 1855–1865.

Baumgard, L.H., Rhoads, R.P., 2013. Effects of heat stress on postabsorptive metabolism and energetics. Annu. Rev. Anim. Biosci. 1, 311–337.

Baylis, M., Githeko, A.K., 2006. The effects of climate change on infectious diseases of animals. Report for the Foresight Project on Detection of Infectious Diseases. Department of Trade and Industry, UK Government. 35 pp.

Beauchemin, K.A., Eriksen, L., Norgaard, P., Rode, L.M., 2008. Nutritional management for enteric methane abatement: a review. Aust. J. Exp. Agric. 48, 21–27.

Beede, D.K., Collier, R.J., 1986. Potential nutritional strategies for intensively managed cattle during thermal stress. J. Anim. Sci. 62, 543–554. https://doi.org/10.2527/jas1986.622543x.

Bell, A.W., McBride, B.W., Slepetis, R., Early, R.J., Currie, W.B., 1989. Chronic heat stress and prenatal development in sheep: I. Conceptus growth and maternal plasma hormones and metabolites. J. Anim. Sci. 67 (12), 3289–3299.

Benchaar, C., Pomar, C., Chiquette, J., 2001. Evaluation of dietary strategies to reduce methane production in ruminants: a modeling approach. Can. J. Anim.Sci. 81, 563–574.

Bentley, D., Hegarty, R.S., Alford, A.R., 2008. Managing livestock enterprises in Australia's extensive rangelands for greenhouse gas and environmental outcomes: a pastoral company perspective. Aust. J. Exp. Agric. 48 (2), 60–64.

Berman, A., 2005. Estimates of heat stress relief needs for Holstein dairy cows. J. Anim. Sci. 83 (6), 1377–1384.

Berman, A., Folman, Y., Kaim, M., Mamen, M., Herz, Z., Wolfenson, D., Arieli, A., Graber, Y., 1985. Upper critical temperatures and forced ventilation effects for high-yielding dairy cows in a subtropical climate. J. Dairy Sci. 68, 1488–1495.

Bernabucci, U., Lacetera, N., Baumgard, L.H., Rhoads, R.P., Ronchi, B., Nardone, A., 2010. Metabolic and hormonal acclimation to heat stress in domesticated ruminants. Animal 4 (7), 1167–1183.

Bianca, W., 1964. Thermoregulatory responses of the dehydrated ox to drinking cold and warm water in a warm environment. Res. Vet. Sci. 5, 75.

Bilby, T.R., Baumgard, L.H., Collier, R.J., Zimbelman, R.B., Rhoads, M.L., 2008. Heat stress effects on fertility: consequences and possible solutions. In: Proceedings of the 2008 South Western Nutritional Conference.

Boadi, D., Benchaar, C., Chiquette, J., Massé, D., 2004. Mitigation strategies to reduce enteric methane emissions from dairy cows: update review. Can. J. Anim. Sci. 84 (3), 319–335.

Bouraoui, R., Lahmar, M., Majdoub, A., Belyea, R., 2002. The relationship of temperature-humidity index with milk production of dairy cows in a Mediterranean climate. Anim. Res. 51 (6), 479–491.

Chebel, R.C., Santos, J.E., Reynolds, J.P., Cerri, R.L., Juchem, S.O., Overton, M., 2004. Factors affecting conception rate after artificial insemination and pregnancy loss in lactating dairy cows. Anim. Reprod. Sci. 84 (3–4), 239–255.

Cheng, Y., Liu, S., Zhang, Y., Su, D., Wang, G., Lv, C., Zhang, Y., Yu, H., Hao, L., Zhang, J., 2016. The effect of heat stress on bull sperm quality and related HSPs expression. Anim. Biol. 66 (3–4), 321–333.

Chirico, J., Jonsson, P., Kjellberg, S., Thomas, G., 1997. Summer mastitis experimentally induced by Hydrotaea irritans exposed to bacteria. Med. Vet. Entomol. 11 (2), 187–192.

Collier, R.J., Beede, D.K., Thatcher, W.W., Israel, L.A., Wilcox, C.J., 1982. Influences of environment and its modification on dairy animal health and production. J. Dairy Sci. 65 (11), 2213–2227.

Collier, R.J., Dahl, G.E., VanBaale, M.J., 2006. Major advances associated with environmental effects on dairy cattle. J. Dairy Sci. 89 (4), 1244–1253.

Collier, R.J., Renquist, B.J., Xiao, Y., 2017. A 100-year review: stress physiology including heat stress. J. Dairy Sci. 100 (12), 10367–10380.

Coloma-García, W., Mehaba, N., Llonch, P., Caja, G., Such, X., Salama, A.A., 2020. Prenatal heat stress effects on gestation and postnatal behavior in kid goats. PLoS One 15 (2), e0220221.

Cook, N.B., Nordlund, K.V., 2009. The influence of the environment on dairy cow behavior, claw health and herd lameness dynamics. Vet. J. 179 (3), 360–369.
Cowley, F.C., Barber, D.G., Houlihan, A.V., Poppi, D.P., 2015. Immediate and residual effects of heat stress and restricted intake on milk protein and casein composition and energy metabolism. J. Dairy Sci. 98 (4), 2356–2368.
Daghir, N.J., 2009. Nutritional strategies to reduce heat stress in broilers and broiler breeders. Lohmann Inform. 44 (1), 6–15.
Dalla Costa, O.A., Faucitano, L., Coldebella, A., Ludke, J.V., Peloso, J.V., Dalla Roza, D., da Costa, M.P., 2007. Effects of the season of the year, truck type and location on truck on skin bruises and meat quality in pigs. Livest. Sci. 107 (1), 29–36.
Das, G.K., Khan, F.A., 2010. Summer anoestrus in buffalo–a review. Reprod. Domest. Anim. 45 (6), e483–e494.
de Torres-Júnior, J.R.S., de FA Pires, P.M., de Sá, W.F., de Ferreira, A.M., Viana, J.H., Camargo, L.S., Ramos, A.A., Folhadella, I.M., Polisseni, J., de Freitas, C., Clemente, C.A., de Sá Filho, M.F., Paula-Lopes, F.F., Baruselli, P.S., 2008. Effect of maternal heat-stress on follicular growth and oocyte competence in Bos indicus cattle. Theriogenology 69, 155–166.
Delgado, C., Rosegrant, M., Steinfeld, H., Ehui, S., Courbois, C., 2001. Livestock to 2020: the next food revolution. Outlook Agric. 30 (1), 27–29.
Dikmen, S., Alava, E., Pontes, E., Fear, J.M., Dikmen, B.Y., Olson, T.A., Hansen, P.J., 2008. Differences in thermoregulatory ability between slick-haired and wild-type lactating Holstein cows in response to acute heat stress. J. Dairy Sci. 91 (9), 3395–3402.
Early, R.J., McBride, B.W., Vatnick, I., Bell, A.W., 1991. Chronic heat stress and prenatal development in sheep: II. Placental cellularity and metabolism. J. Anim. Sci. 69 (9), 3610–3616. https://doi.org/10.2527/1991.6993610x.
Edwards, J.L., Hansen, P.J., 1997. Differential responses of bovine oocytes and preimplantatin embryos to heat shock. Mol. Reprod. Dev. 46, 138–145.
ENVIS Centre on Wildlife & Protected Areas (ECOW&PA), 2017. Wildlife Institute of India, Dehradun, Ministry of Environment, Forests & Climate Change, Govt of India. Climate Change: Impacts on Wildlife. pp. 1–6. http://www.wiienvis.nic.in/Database/Climate_Change_8631.aspx.
EPA (U.S. Environmental Protection Agency), 1999. Livestock Manure Management. http://www.epa.gov/methane/reports/05-manure.pdf. (Accessed 20 October 2017).
EPA (U.S. Environmental Protection Agency), 2013. Global Mitigation of Non-CO_2 Greenhouse Gases. http://www.epa.gov/climatechange/Downloads/EPAactivities/MAC_Report_2013.pdf. (acessed 12.10.17).
FAO, 2006. Livestock's Long Shadow: Environmental Issues and Options. Rome.
FAO, 2009. State of Food and Agriculture—Livestock in the Balance. Rome.
FAO, 2015. In: Scherf, B.D., Pilling, D. (Eds.), The Second Report on the State of the World's Animal Genetic Resources for Food and Agriculture. FAO Commission on Genetic Resources for Food and Agriculture Assessments, Rome.
FAO, 2016. Principles for the Assessment of Livestock Impacts on Biodiversity. Livestock Environmental Assessment and Performance (LEAP) Partnership. FAO, Rome, Italy.
FAO, 2020. Environment Statistics, FAOSTAT Temperature Change Data: 1961–2019. Rome.
Finch, V.A., 1986. Body temperature in beef cattle: its control and relevance to production in the tropics. J. Anim. Sci. 62 (2), 531–542.
Fuller, A., Maloney, S.K., Blache, D., Cooper, C., 2020. Endocrine and metabolic consequences of climate change for terrestrial mammals. Curr. Opin. Endocr. Metabol. Res. 11, 9–14.
Gaughan, J.B., Mader, T.L., Holt, S.M., Sullivan, M.L., Hahn, G.L., 2010. Assessing the heat tolerance of 17 beef cattle genotypes. Int. J. Biometeorol. 54, 617–627.

Gengler, W.R., Martz, F.A., Johnson, H.D., Krause, G.F., Hahn, L., 1970. Effect of temperature on food and water intake and rumen fermentation. J. Dairy Sci. 53 (4), 434–437.
Gerber, P.J., Steinfeld, H., Henderson, B., Mottet, A., Opio, C., Dijkman, J., Falcucci, A., Tempio, G., 2013. Tackling Climate Change Through Livestock: A Global Assessment of Emissions and Mitigation Opportunities. Food and Agriculture Organization of the United Nations (FAO).
Ghosh, J., Mondal, S., 2006. Nucleic acids and protein content in relation to growth and regression of buffalo corpora lutea. Anim. Reprod. Sci. 93, 316–327.
Gilad, E., Meidan, R., Berman, A., Graber, Y., Wolfenson, D., 1993. Effect of heat stress on tonic and GnRH-induced gonadotrophin secretion in relation to concentration of oestradiol in plasma of cyclic cows. J. Reprod. Fertil. 99 (2), 315–321.
Gonzalez-Esquerra, R., Leeson, S., 2005. Effects of acute versus chronic heat stress on broiler response to dietary protein. Poult. Sci. 84 (10), 1562–1569.
Gregory, N.G., 2010. How climatic changes could affect meat quality. Food Res. Int. 43 (7), 1866–1873.
Gwazdauskas, F.C., Thatcher, W.W., Kiddy, C.A., Pape, M.J., Wilcox, C.J., 1981. Hormonal pattern during heat stress following PGF2alpha-tham salt induced luteal regression in heifers. Theriogenology 16, 271–285.
Hai, L., Rong, D., Zhang, Z.Y., 2000. The effect of thermal environment on the digestion of broilers. J. Anim. Physiol. Anim. Nutr. 83 (2), 57–64.
Hamzaoui, S.A., Salama, A.A., Albanell, E., Such, X., Caja, G., 2013. Physiological responses and lactational performances of late-lactation dairy goats under heat stress conditions. J. Dairy Sci. 96 (10), 6355–6365.
Hansen, P.J., 2004. Physiological and cellular adaptations of zebu cattle to thermal stress. Anim. Reprod. Sci. 82, 349–360.
Hansen, P.J., 2007. To be or not to be—determinants of embryonic survival following heat shock. Theriogenology 68, S40–S48.
Hansen, P.J., 2009. Effects of heat stress on mammalian reproduction. Philos. Trans. Roy. Soc. B: Biol. Sci. 364 (1534), 3341–3350.
Hansen, P.J., Aréchiga, C.F., 1999. Strategies for managing reproduction in the heat-stressed dairy cow. J. Anim. Sci. 77 (Suppl 2), 36–50. https://doi.org/10.2527/1997.77suppl_236x.
Hein, K.G., Allrich, R.D., 1992. Influence of exogenous adrenocorticotropic hormone on estrous behavior in cattle. J. Anim. Sci. 70 (1), 243–247.
Hoffmann, I., 2008. Livestock genetic diversity and climate change adaptation. Livestock Global Clim. Change 76.
Hooda, O.K., Singh, G., 2010. Effect of thermal stress on feed intake, plasma enzymes and blood biochemicals in buffalo heifers. Ind. J. Anim. Nutr. 27 (2), 122–127.
Howell, J.L., Fuquay, J.W., Smith, A.E., 1994. Corpus luteum growth and function in lactating Holstein cows during spring and summer. J. Dairy Sci. 77 (3), 735–739.
Hristov, T.A., Yang, W., Lee, C., Gerber, P.J., Henderson, B., Tricarico, J.M., 2013. Mitigation of methane and nitrous oxide emissions from animal—Special Topics. J. Anim. Sci. 91, 5045–5069.
ICF International, 2013. Greenhouse Gas Mitigation Options and Costs for Agricultural Land and Animal Production Within the United States. http://www.usda.gov/oce/climate_change/mitigation_technologies/GHG_Mitigation_Options.pdf.
Ingraham, R.H., Gillette, D.D., Wagner, W.D., 1974. Relationship of temperature and humidity to conception rate of Holstein cows in subtropical climate. J. Dairy Sci. 57, 476–481.
International Centre for Environmental Management (ICEM), 2013. USAID Mekong ARCC Climate Change Impact and Adaptation on Livestock. Prepared for the United States Agency for International Development by ICEM.

IPCC (Intergovermental Panel on Climate Change), 1997. IPCC/OECD/IEA Programme on National Greenhouse Gas Inventories. Intergovernmental Panel on Climate Change.
IPCC (Intergovermental Panel on Climate Change), 2013. Climate change 2013: the physical science basis. In: Stocker, T.F., Qin, D., Plattner, G.-K., Tignor, M., Allen, S.K., Boschung, J., Midgley, P.M. (Eds.), Contribution of Working Group I to the Fifth Assessment Report of the Intergovernmental Panel on Climate Change. Cambridge University Press, Cambridge, United Kingdom and New York, NY, p. 1535.
IPPC (Intergovernmental Panel on Climate Change), 2007. Climate Change 2007: The Physical Science Basis. IPPC, Geneva. 996p.
Johnson, D.E., Kenchington, E.L., 2019. Should potential for climate change refugia be mainstreamed into the criteria for describing EBSAs? Conserv. Lett. 12 (4), e12634.
Joint Global Change Research Institute and Battelle Memorial Institute (JGCRI &BMI), 2009. Pacific northwest division. India: the impact of climate change to 2030 a commissioned research report. In: National Intelligence Council (NIC) Workshops and Research With Nongovernmental Experts, 2009 April, pp. 1–50.
Jungbluth, T., Hartung, E., Brose, G., 2001. Greenhouse gas emissions from animal houses and manure stores. Nutr. Cycl. Agroecosyst. 60 (1–3), 133–145.
Kadokawa, H., Sakatani, M., Hansen, P.J., 2012. Perspectives on improvement of reproduction in cattle during heat stress in a future Japan. Anim. Sci. J. 83, 439–445.
Kadzere, C.T., Murphy, M.R., Silanikove, N., Maltz, E., 2002. Heat stress in lactating dairy cows: a review. Livest. Prod. Sci. 77 (1), 59–91.
Karabinus, D.S., Vogler, C.J., Saacke, R.G., Evenson, D.P., 1997. Chromatin structural changes in sperm after scrotal insulation of Holstein bulls. J. Androl. 18 (5), 549–555.
Kastelic, J.P., Cook, R.B., Pierson, R.A., Coulter, G.H., 2001. Relationships among scrotal and testicular characteristics, sperm production, and seminal quality in 129 beef bulls. Can. J. Vet. Res. 65 (2), 111–115.
Keane, M.G., Allen, P., 1998. Effects of production system intensity on performance, carcass composition and meat quality of beef cattle. Livest. Prod. Sci. 56 (3), 203–214.
King, C.C., Dschaak, C.M., Eun, J.S., Fellner, V., Young, A.J., 2011. Quantitative analysis of microbial fermentation under normal or high ruminal temperature in continuous cultures. Profess. Anim. Sci. 27 (4), 319–327.
Kumar, S., Mote, S., Singh, D., Chauhan, S.S., Ghosh, N., 2014. Effects of environmental factors on lactation yield and lactation length of Holdeo crossbred cattle. Indian J. Appl. Res. 4 (10), 4–7.
Kumar, B.V., Singh, G., Meur, S.K., 2010. Effects of addition of electrolyte and ascorbic acid in feed during heat stress in buffaloes. Asian-Australas. J. Anim. Sci. 23 (7), 880–888.
Kurukulasuriya, P., Rosenthal, S., 2003. Climate Change and Agriculture: A Review of Impacts and Adaptations; Climate Change Series. World Bank, Washington, DC. Paper No. 91.
Lacetera, N., 2019. Impact of climate change on animal health and welfare. Anim. Fron. 9 (1), 26–31. https://doi.org/10.1093/af/vfy030.
Lacetera, N., Bernabucci, U., Ronchi, B., Nardone, A., 1996. Body condition score, metabolic status and milk production of early lactating dairy cows exposed to warm environment. Riv. Agric. Subtrop. Trop. 90 (1), 43–55.
Lacetera, N., Segnalini, M., Bernabucci, U., Ronchi, B., Vitali, A., Tran, A., Guis, H., Caminade, C., Calvete, C., Morse, A., Baylis, M., 2013. Climate induced effects on livestock population and productivity in the Mediterranean area. In: Regional Assessment of Climate Change in the Mediterranean. Springer, Dordrecht, pp. 135–156.
Landaeta-Hernández, A., Zambrano-Nava, S., Hernández-Fonseca, J.P., Godoy, R., Calles, M., Iragorri, J.L., Añez, L., Polanco, M., Montero-Urdaneta, M., Olson, T., 2011. Variability of hair coat and skin traits as related to adaptation in Criollo Limonero cattle. Trop. Anim. Health Prod. 43 (3), 657–663.

Lecchi, C., Rota, N., Vitali, A., Ceciliani, F., Lacetera, N., 2016. In vitro assessment of the effects of temperature on phagocytosis, reactive oxygen species production and apoptosis in bovine polymorphonuclear cells. Vet. Immunol. Immunopathol. 182, 89–94.

Lin, H., Sui, S.J., Jiao, H.C., Buyse, J., Decuypere, E., 2006. Impaired development of broiler chickens by stress mimicked by corticosterone exposure. Comp. Biochem. Physiol. A Mol. Integr. Physiol. 143 (3), 400–405.

Luceño, N.L., Angrimani, D.D., de Cássia, B.L., Szymańska, K.J., Van Poucke, M., Demeyere, K., Meyer, E., Peelman, L., Mullaart, E., Broekhuijse, M.L., Van Soom, A., 2020. Exposing dairy bulls to high temperature-humidity index during spermatogenesis compromises subsequent embryo development in vitro. Theriogenology 141, 16–25.

Mariani, P., Zanzucchi, G., Blanco, P., Masoni, M., 1993. Variazioni stagionali del contenuto in fosforo del latte di massa di singoli allevamenti. Ind. Latte 29, 39–53.

Martin, C., Morgavi, D.P., Doreau, M., 2010. Methane mitigation in ruminants: from microbe to the farm scale. Animal 4 (3), 351–365.

Mashaly, M.M., Hendricks 3rd, G.L., Kalama, M.A., Gehad, A.E., Abbas, A.O., Patterson, P.H., 2004. Effect of heat stress on production parameters and immune responses of commercial laying hens. Poult. Sci. 83 (6), 889–894.

McDowell, L.R., 1989. In: McDowell, L.R. (Ed.), Analítico: Vitamins in Animal Nutrition; Comparative Aspects to Human Nutrition. Academic Press, London, pp. 10–52. 93–131.

McNitt, J.I., First, N.L., 1970. Effects of 72—hour heat stress on semen quality in boars. Int. J. Biometeorol. 14 (4), 373–380.

Meyerhoeffer, D.C., Wettemann, R.P., Coleman, S.W., Wells, M.E., 1985. Reproductive criteria of beef bulls during and after exposure to increased ambient temperature. J. Anim. Sci. 60 (2), 352–357.

Ministry of Environment, Forest & Climate Change (MoEF&CC), 2018–2019. Government of India. Climate Change. Annual Report. pp. 67–95. 208–220.

Ministry of Environment, Forest & Climate Change (MoEF&CC), 2019–2020. Government of India. Climate Change. Annual Report. pp. 70–100. 180–196.

Mondal, S., Prakash, B.S., 2002. Comparison of luteal function between cows and buffaloes during estrous cycle. Indian J. Dairy Sci. 55 (3), 142–144.

Mondal, S., Prakash, B.S., Palta, P., 2003. Relationship between peripheral plasma inhibin and progesterone concentrations in Sahiwal cattle (Bos indicus) and Murrah buffaloes (Bubalus bubalis). Asian-Australasian J. Anim. Sci. 16 (1), 6–10.

Mondal, S., Kumar, V., Reddy, I.J., Singh, K., 2004a. Progesterone and nucleic acid contents of buffalo corpus luteum in relation to stages of estrous cycle. Ind. J. Anim. Sci. 74 (6), 710–712.

Mondal, S., Palta, P., Prakash, B.S., 2004b. Influence of season on peripheral plasma progesterone concentrations in Murrah buffaloes (Bubalus bubalis). Buffalo J. 1, 95–100.

Mondal, S., Prakash, B.S., Palta, P., 2007. Endocrine aspects of estrous cycle in buffalo : an overview. Asian Australas. J. Anim. Sci. 20 (1), 124–131.

Mondal, S., Minj, A., Tiwari, A.K., Sharma, B., Varshney, V.P., 2008a. Molecular characterization of FSH receptor (FSHR) gene in Indian river buffalo. Gen. Comp. Endocrinol. 158, 147–153.

Mondal, S., Praksh, B.S., Palta, P., 2008b. Peripheral plasma FSH concentrations in relation to expression of oestrus in Murrah buffaloes (Bubalus bubalis). Buffalo Bull. 27 (4), 258–262.

Mondal, S., Praksh, B.S., Palta, P., 2008c. Relationship between peripheral plasma inhibin and FSH concentrations in buffaloes (Bubalus bubalis) and Sahiwal cows (Bos indicus) during estrous cycle. Trop. Anim. Health Prod. 40, 403–406.

Mondal, S., Nandi, S., Reddy, I.J., Suresh, K.P., 2009. Isolation, culture and characterization of endometrial epithelial cells in buffalo (*Bubalus bubalis*). Buffalo Bull. 28 (2), 101–106.

Mondal, S., Nandi, S., Reddy, I.J., Suresh, K.P., 2010. Isolation, culture and characterization of endometrial stromal cells in buffalo (*Bubalus bubalis*). Ind. J. Anim. Sci. 79 (12), 24–26.

Mondal, S., Nandi, S., Reddy, I.J., 2013. Isolation and characterization of luteal cells in buffalo (*Bubalus bubalis*). Indian J. Physiol. Pharmacol. 57 (1), 1–6.

Mondal, S., Mor, A., Reddy, I.J., Nandi, S., 2015a. Impact of exposure of elevated temperature (42.5°C) on in vitro maturation, metabolic and ionic contents of sheep oocytes. Biomedicine 35 (1), 65–71.

Mondal, S., Mor, A., Reddy, I.J., Nandi, S., Gupta, P.P., 2017. Heat stress induced alterations in prostaglandins, ionic and metabolic contents of sheep endometrial epithelial cells in vitro. Biomed. J. Sci. Tech. Res. 1 (4), 1–5.

Mondal, S., Mor, A., Reddy, I.J., Soumya, N.P., 2015b. Genes regulating maternal recognition of pregnancy in domestic animals: an update. Braz. Arch. Biol. Technol. 58 (6), 854–863.

Morse, D., DeLorenzo, M.A., Wilcox, C.J., Collier, R.J., Natzke, R.P., Bray, D.R., 1988. Climatic effects on occurrence of clinical mastitis. J. Dairy Sci. 71, 848–853. https://doi.org/10.3168/jds.S0022-0302(88)79626-5.

Murugaiyah, M., 1992. Changes in the semen characteristics of Kambing Katjan crossbreed buck under hot and humid environmental temperatures. In: Recent Advances in Goat Production. New Delhi: Proc. 5th Int. Conf. Goats, pp. 1126–1129.

Nandi, S., Mondal, S., Reddy, I.J., 2012. Effect of prostaglandin producing modulators on *in vitro* growth characteristics in buffalo endometrial epithelial cells. Theriogenology 77, 1014–1020.

Naqvi, S.M., Kumar, D., Paul, R.K., 2012. Environmental stresses and livestock reproduction. In: Environmental Stress and Amelioration in Livestock Production. Springer, Berlin, Heidelberg, pp. 97–128.

Narain, S., Ghosh, P., Saxena, N.C., Parikh, J., Soni, P., 2009. Climate Change Perspectives From India. United Nations Development Programme, India (Lasting Solutions for Development Challenges). pp. 1–68.

Nardone, A., Lacetera, N., Bernabucci, U., Ronchi, B., 1997. Composition of colostrum from dairy heifers exposed to high air temperatures during late pregnancy and the early postpartum period. J. Dairy Sci. 80 (5), 838–844.

Nardone, A., Ronchi, B., Lacetera, N., et al., 2010. Effects of climate changes on animal production and sustainability of livestock systems. Livest. Sci. 130, 57–69.

Nichi, M., Bols, P.E., Zuche, R.M., Barnabe, V.H., Goovaerts, I.G., Barnabe, R.C., Cortada, C.N., 2006. Seasonal variation in semen quality in *Bos indicus* and *Bos taurus* bulls raised under tropical conditions. Theriogenology 66 (4), 822–828.

Nicholas, C.R., Chen, J., Tang, Y., Southwell, D.G., Chalmers, N., Vogt, D., Arnold, C. M., Chen, Y.J., Stanley, E.G., Elefanty, A.G., Sasai, Y., 2013. Functional maturation of hPSC-derived forebrain interneurons requires an extended timeline and mimics human neural development. Cell Stem Cell 12 (5), 573–586.

Ozawa, M., Tabayashi, D., Latief, T.A., Shimizu, T., Oshima, I., Kanai, Y., 2005. Alterations in follicular dynamics and steroidiogenic abilities induced by heat stress during follicular recruitment in goats. Reproduction 129, 621–630.

Palta, P., Mondal, S., Prakash, B.S., Madan, M.L., 1997. Peripheral inhibin levels in relation to climatic variations and stage of estrous cycle in buffaloes (Bubalus bubalis). Theriogenology 47 (5), 989–995.

Paul, C., Murray, A.A., Spears, N., Saunders, P.T., 2008. A single, mild, transient scrotal heat stress causes DNA damage, subfertility and impairs formation of blastocysts in mice. Reproduction 136 (1), 73.

Paula-Lopes, F.F., Lima, R.S., Satrapa, R.A., Barros, C.M., 2013. Physiology and endocrinology symposium: influence of cattle genotype (Bos indicus vs. Bos taurus) on oocyte and preimplantation embryo resistance to increased temperature. J. Anim. Sci. 91 (3), 1143–1153.

Pinto, M.E., de Alencar-Araripe, M.A., Alencar, A.A., 2001. Effects of scrotal insulation on testis size and semen criteria in Santa Ines hairy sheep raised in the State of Ceara, Northeast of Brazil. Rev. Bras. Zootec. 30 (6), 1704–1711.

Pinto, J., Bonacic, C., Hamilton-West, C., Romero, J., Lubroth, J., 2008. Climate change and animal diseases in South America. Rev. Sci. Tech. 27 (2), 599–613.

Pirkkala, L., Nykanen, P., Sistonen, L., 2001. Roles of the heat shock transcription factors in regulation of the heat shock response and beyond. FASEB J. 15, 1118–1131.

Putney, D.J., Mullins, S., Thatcher, W.W., Drost, M., Gross, T.S., 1989. Embryonic development in superovulated dairy cattle exposed to elevated ambient temperatures between the onset of estrus and insemination. Anim. Reprod. Sci. 19, 37–51.

Rahman, M.B., Schellander, K., Luceño, N.L., Van Soom, A., 2018. Heat stress responses in spermatozoa: mechanisms and consequences for cattle fertility. Theriogenology 113, 102–112.

Rana, M.S., Hashem, M.A., Akhter, S., Habibullah, M., Islam, M.H., Riswas, R.C., 2014. Effect of heat stress on carcass and meat quality of indigenous sheep of Bangladesh. Bangladesh. J Anim Sci. 43.

Ray, D.E., Halbach, T.J., Armstrong, D.V., 1992. Season and lactation number effects on milk production and reproduction in dairy cattle in Arizona. J. Dairy Sci. 75, 2976–2983.

Reddy, S., Rao, K.A., 2014. Effects of climate change on livestock production and mitigation strategies. Int. J. Innov. Res. Rev. 2 (4), 124–144.

Reyes, L., Alvarez-Valenzuela, F.D., Correa-Calderón, A., Saucedo-Quintero, J.S., Robinson, P.H., Fadel, J.G., 2006. Effect of cooling Holstein cows during the dry period on postpartum performance under heat stress conditions. Livest. Sci. 105 (1–3), 198–206.

Reyes, J.G., Farias, J.G., Henríquez-Olavarrieta, S., et al., 2012. The hypoxic testicle: physiology and pathophysiology. Oxidative Med. Cell. Longev. 2012, 929285. 15 pages.

Richardson, K., Steffen, W., Schellnhuber, H., et al., 2009. Synthesis Report from Climate Change: Global Risks, Challenges and Decisions, Conference, 10–12 March. University of Copenhagen, Copenhagen.

Riley, D.G., Chase, C.C., Coleman, S.W., Olson, T.A., 2012. Genetic assessment of rectal temperature and coat score in Brahman, Angus, and Romosinuano crossbred and straight bred cows and calves under subtropical summer conditions. Livest. Sci. 148, 109–118.

Rivera, R.M., Hansen, P.J., 2001. Development of cultured bovine embryos after exposure to high temperatures in the physiological range. Reproduction 121, 107–115.

Rojas-Downing, M.M., Nejadhashemi, A.P., Harrigan, T., Woznicki, S.A., 2017. Climate change and livestock: impacts, adaptation and mitigation. Clim. Risk Manag. 16, 145–163. https://doi.org/10.1016/j.crm.2017.02.001.

Ronchi, B., Stradaioli, G., Supplizi, A.V., Bernabucci, U., Lacetera, N., Accorsi, P.A., Nardone, A., Seren, E., 2001. Influence of heat stress or feed restriction on plasma progesterone, oestradiol-17β, LH, FSH, prolactin and cortisol in Holstein heifers. Livest. Prod. Sci. 68 (2–3), 231–241.

Rosenzweig, C., Karoly, D., Vicarelli, M., Neofotis, P., Wu, Q., Casassa, G., Menzel, A., Root, T.L., Estrella, N., Seguin, B., Tryjanowski, P., 2008. Attributing physical and biological impacts to anthropogenic climate change. Nature 453 (7193), 353–357.

Rötter, R., Van de Geijn, S.C., 1999. Climate change effects on plant growth, crop yield and livestock. Clim. Chang. 43 (4), 651–681.

Ryan, D.P., Kopel, E., Boland, M.P., Godke, R.A., 1991. Pregnancy rates in dairy cows following the administration of a GnRH analogue at the time of artificial insemination or at mid-cycle post insemination. Theriogenology 36 (3), 367–377.

Sahin, N., Tuzcu, M., Ozercan, I., Sahin, K., Prasad, A.S., Kucuk, O., 2009. Zinc picolinate in the prevention of leiomyoma in Japanese quail. J. Med. Food 12 (6), 1368–1374.

Sahni, K.L., Roy, A., 1967. A note on summer sterility in Romney Marsh rams under tropical conditions. Indian J. Vet. Sci. 37, 335–338.

Sakatani, M., Alvarez, N.V., Takahashi, M., Hansen, P.J., 2012. Consequences of physiological heat shock beginning at the zygote stage on embryonic development and expression of stress response genes in cattle. J. Dairy Sci. 95 (6), 3080–3091.
Savasani, H.H., Padodara, R.J., Bhadaniya, A.R., Kalariya, V.A., Javia, B.B., Ghodasara, S. N., Ribadiya, N.K., 2015. Impact of climate on feeding, production and reproduction of animals. Agric. Rev. 36 (1), 26–36.
Schingoethe, D.J., Voelker, H.H., Beardsley, G.L., Parsons, J.G., 1976. Rumen volatile fatty acids and milk composition from cows fed hay, haylage, or urea-treated corn silage. J. Dairy Sci. 59 (5), 894–901.
Schneider, P.L., Beede, D.K., Wilcox, C.J., Collier, R.J., 1984. Influence of dietary sodium and potassium bicarbonate and total potassium on heat-stressed lactating dairy cows. J. Dairy Sci. 67 (11), 2546–2553.
Schüller, L.K., Michaelis, I., Heuwieser, W., 2017. Impact of heat stress on estrus expression and follicle size in estrus under field conditions in dairy cows. Theriogenology 102, 48–53. https://doi.org/10.1016/j.theriogenology.2017.07.004.
Scollan, N., Hocquette, J.F., Nuernberg, K., Dannenberger, D., Richardson, I., Moloney, A., 2006. Innovations in beef production systems that enhance the nutritional and health value of beef lipids and their relationship with meat quality. Meat Sci. 74 (1), 17–33.
Senger, P., 2003. The organization and function of the male reproductive system. In: Pathways to Pregnancy and Parturition, Pullman: Current Conceptions, fourth ed, pp. 44–79.
Sharma, N., 2018. India's new National Forest Policy to address the challenge of climate change. The Economic Times Bureau 16, 1–2 (March).
Shearer, J.K., 1999. Foot health from a veterinarian's perspective. In: Proc. Feed and Nutritional Management Cow College. Virginia Tech, pp. 33–43.
Silanikovc, N., 2000. Effects of heat stress on the welfare of extensively managed domestic ruminants. Livest. Prod. Sci. 67 (1–2), 1–8.
Silanikove, N., Koluman, N., 2015. Impact of climate change on the dairy industry in temperate zones: predications on the overall negative impact and on the positive role of dairy goats in adaptation to earth warming. Small Rumin. Res. 123 (1), 27–34.
Singh, S.V., Upadhyay, R.C., Ashutosh, Hooda, O.K., Vaidya, M.M., 2011. Climate change: impacts on reproductive pattern of cattle and buffaloes. Wayamba J. Anim. Sci. 8, 199–208.
Singh, M., Chaudhari, B.K., Singh, J.K., Singh, A.K., Maurya, P.K., 2013. Effects of thermal load on buffalo reproductive performance during summer season. J. Biol. Sci. 1 (1), 1–8.
Skinner, J.D., Louw, G.N., 1966. Heat stress and spermatogenesis in Bos indicus and Bos taurus cattle. J. Appl. Physiol. 21 (6), 1784–1790.
Söderquist, L., et al., 1996. Influence of season, age, breed and some other factors on the variation in sperm morphological abnormalities in Swedish dairy AI bulls. Anim. Reprod. Sci. 44 (2), 91–98.
Spencer, J.D., Gaines, A.M., Berg, E.P., Allee, G.L., 2005. Diet modifications to improve finishing pig growth performance and pork quality attributes during periods of heat stress. J. Anim. Sci. 83 (1), 243–254.
Srikandakumar, A., Johnson, E.H., 2004. Effect of heat stress on milk production, rectal temperature, respiratory rate and blood chemistry in Holstein, Jersey and Australian milking zebu cows. Trop. Anim. Health Prod. 36 (7), 685–692.
Srivastava, A.K., 2010. Impact of climate change on animal health and performance proc. In: Upadhyay, R.C., Singh, S.V., Ashutosh, Ashutosh, M., Aggrawal, A. (Eds.), National Symposium on Climate change and Livestock Productivity in India held at N.D.R.I., Karnal from October 7–8, pp. 3–9.
Staub, C., Johnson, L., 2018. Spermatogenesis in the bull. Animal 12 (s1), s27–s35.

Steinfeld, H., Gerber, P., Wassenaar, T., Castel, V., Rosales, M., Haan, C., 2006. Livestock's Long Shadow: Environmental Issues and Options. FAO, Rome.
Stocker, T.F., Qin, D., Plattner, G.K., Tignor, M., Allen, S.K., et al., 2013. IPCC: The Physical Science Basis. Contribution of Working Group I to the Fifth Assessment Report of the Intergovernmental Panel on Climate Change., p. 1535.
Summer, A., Lora, I., Formaggioni, P., Gottardo, F., 2019. Impact of heat stress on milk and meat production. Anim. Front. 9 (1), 39–46.
Sunil Kumar, B.V., Kumar, A., Kataria, M., 2011. Effect of heat stress in tropical livestock and different strategies for its amelioration. J. Stress Physiol. Biochem. 7 (1), 45–54.
Suresh, K.P., Nandi, S., Mondal, S., 2009. Factors affecting laboratory production of buffalo embryos: a meta-analysis. Theriogenology 72, 978–985.
Tao, S., Dahl, G.E., 2013. Invited review: heat stress effects during late gestation on dry cows and their calves. J. Dairy Sci. 96 (7), 4079–4093.
Thiruvenkadan, A.K., Panneerselvam, S., Rajendran, R., Murali, N., 2010. Analysis on the productive and reproductive traits of Murrah buffalo cows maintained in the coastal region of India. Appl. Anim. Husb. Rural Dev. 3, 1–5.
Thornton, P.K., 2010. Livestock production: recent trends, future prospects. Philos. Trans. Roy. Soc. B: Biol Sci. 365 (1554), 2853–2867.
Thornton, P.K., Gerber, P.J., 2010. Climate change and the growth of the livestock sector in developing countries. Mitig. Adapt. Strateg. Glob. Chang. 15 (2), 169–184.
Thornton, P.K., van de Steeg, J., Notenbaert, A., Herrero, M., 2009. The impacts of climate change on livestock and livestock systems in developing countries: a review of what we know and what we need to know. Agric. Syst. 101 (3), 113–127.
Upadhyay, R.C., Ashutosh, Singh, S.V., 2009. Impact of climate change on reproductive functions of cattle and buffalo. In: Aggarwal, P.K. (Ed.), Global Climate Change and Indian Agriculture. ICAR, New Delhi, pp. 107–110.
Vale, W.G., 2007. Effects of environment on buffalo reproduction. Ital. J. Anim. Sci. 6 (sup2), 130–142.
Wakayo, B.U., Brar, P.S., Prabhakar, S., 2015. Review on mechanisms of dairy summer infertility and implications for hormonal intervention. Open Vet. J. 5 (1), 6–10.
West, J.W., 2002. Physiological effects of heat stress on production and reproduction. In: The Proceedings of Tri-State Dairy Nutrition Conference; April 16 and 17, 2002; Grand Wayne Center Fort Wayne, Indiana, pp. 1–9.
West, J.W., 2003. Effects of heat-stress on production in dairy cattle. J. Dairy Sci. 86 (6), 2131–2144.
West, J.W., Mullinix, B.G., Sandifer, T.G., 1991. Changing dietary electrolyte balance for dairy cows in cool and hot environments1. J. Dairy Sci. 74 (5), 1662–1674.
Wettemann, R.P., Desjardins, C., 1979. Testicular function in boars exposed to elevated ambient temperature. Biol. Reprod. 20 (2), 235–241.
Wheeler, T.L., Cundiff, L.V., Shackelford, S.D., Koohmaraie, M., 2001. Characterization of biological types of cattle (cycle V): carcass traits and longissimus palatability. J. Anim. Sci. 79 (5), 1209–1222.
Wheelock, J.B., Rhoads, R.P., VanBaale, M.J., Sanders, S.R., Baumgard, L.H., 2010. Effects of heat stress on energetic metabolism in lactating Holstein cows. J. Dairy Sci. 93 (2), 644–655.
White, N., Sutherst, R.W., Hall, N., Whish-Wilson, P., 2003. The vulnerability of the Australian beef industry to impacts of the cattle tick (Boophilus microplus) under climate change. Clim. Chang. 61 (1–2), 157–190.
WHO, 2008. Anthrax in Humans and Animals, fourth ed. World Health Organization, World Organisation for Animal Health, Food and Agriculture Organization of the United Nations, Geneva, Switzerland.

Wilson, S.J., Marion, R.S., Spain, J.N., Spiers, D.E., Keisler, D.H., Lucy, M.C., 1998. Effects of controlled heat stress on ovarian function of dairy cattle. 1. Lactating Cows1. J. Dairy Sci. 81 (8), 2124–2131.

Wittmann, E.J., Mellor, P.S., Baylis, M., 2001. Using climate data to map the potential distribution of Culicoides imicola (Diptera: Ceratopogonidae) in Europe. Rev. Sci. Techn.-Off. Int. Epizooties 20 (3), 731–740.

Wolfenson, D., 2009. Impact of heat stress on production and fertility of dairy cattle. In: Proceedings of the 18th Annual Tri-State Dairy Nutrition Conference, Fort Wayne, IN, pp. 21–22.

Wolfenson, D., Roth, Z., 2019. Impact of heat stress on cow reproduction and fertility. Anim. Front. 9 (1), 32–38.

Wolfenson, D., Thatcher, W.W., Badinga, L., Savio, J.D., Meidan, R., Lew, B.J., Braw-Tal, R., Berman, A., 1995. Effect of heat stress on follicular development during the estrous cycle in lactating dairy cattle. Biol. Reprod. 52 (5), 1106–1113.

Wolfenson, D., Roth, Z., Meidan, R., 2000. Impaired reproduction in heat-stressed cattle: basic and applied aspects. Anim. Reprod. Sci. 60, 535–547.

Yadav, B., Singh, G., Verma, A.K., Dutta, N., Sejian, V., 2013. Impact of heat stress on rumen functions. Vet. World 6 (12), 992.

Younas, M., Fuquay, J.W., Smith, A.E., Moore, A.B., 1993. Estrus and endocrine responses of lactating Holsteins to forced ventilation during summer. J. Dairy Sci. 76, 430–434.

Yousef, M.K., 1985. Stress Physiology in Livestock. Volume I. Basic Principles. CRC Press.

Zhang, Z.Y., Jia, G.Q., Zuo, J.J., Zhang, Y., Lei, J., Ren, L., Feng, D.Y., 2012. Effects of constant and cyclic heat stress on muscle metabolism and meat quality of broiler breast fillet and thigh meat. Poult. Sci. 91 (11), 2931–2937.

Zhao, S., Min, L., Zheng, N., Wang, J., 2019. Effect of heat stress on bacterial composition and metabolism in the rumen of lactating dairy cows. Animals 9, 925.

CHAPTER SEVEN

Nanotechnology in animal production

Kingshuk Poddar[a] and Anyam VVNGSV Kishore[b]
[a]New Materials Business, Tata Steel Limited, Kolkata, India
[b]Department of Surgery and Radiology, West Bengal University of Animal and Fishery Sciences, Kolkata, India

Introduction

Nanotechnology is no longer a new concept in today's world but has become a new empowering technology over the years, with so much potential to improve agriculture and livestock worldwide. The term commonly defines the science and technology of small sizes, largely associated with the use and manipulation of matter at a scale of one-billionth of a meter (Regis, 1995). The description includes an additional attribute for a domain of integrative and multidisciplinary science that already existed, but at a tinier scale. Essentially, nanotechnology also inculcates any and every material at the atomic, molecular, and supramolecular scales across all dimensions (Varnek et al., 2005).

Nanotechnology enables new implementations for molecular and cellular biology, veterinary physiology, and animal production and reproduction. Although the science essentially originated from physics in recent times, there has been incremental usage in integrative scientific disciplines, primarily in life sciences applications and endeavors; this matches what Feynman said about "plenty of room" while defining the science (Feynman, 1960; Peterson, 2004). Nanotechnology has largely been used for disease diagnosis along with definitive therapeutics, thus making it essentially accommodative to the science and engineering of agriculture and animal and food systems (Scott and Chen, 2003). Nanotechnology correlates to length scales of roughly 1–100 nm, limiting not only to natural, incidental, or manufactured or synthetic material containing particles, in an unbound free state or in an aggregate or agglomerate form whereby at least 50% of the particles in the number size distribution mentioned (EU Commission, 2011). This source-independent but size limiting dimensional attribute of nanotechnology entities makes them even more application in a life sciences context, in specific

Emerging Issues in Climate Smart Livestock Production
https://doi.org/10.1016/B978-0-12-822265-2.00009-0

livestock. At this scale of less than a micrometer, all the properties of materials differ fundamentally and usually cause unexpectedly integrated. Detecting, monitoring, and regulating systems could sense disease and report that to farmer and veterinarian to activate a targeted treatment delivery system. Using nanotechnology could permit a wide range of advances in animal and veterinary sciences by improving reproductive science, breeding management, early diagnosis of disease, and the prevention and treatment of disease in animals, thereby improving public health.

The word nanotechnology was initially used in 1974 to describe scientific tools used for the manipulation of materials at their nanoscale. The use of nanotechnology in biological science is called nanobiology (Troncarelli et al., 2013). Similarly, nanomedicine is defined as the application of different tools based on nanotechnology to develop faster and more efficient solutions for medical problems or disease management. When applied in medicine, nanotechnology helps to understand several physiological and pathological processes. Such acquired knowledge gives new therapeutic solutions and concepts for already existing problems (Mohantya et al., 2014). Nanotheranostics are new formulations that contain both diagnostic and therapeutic properties in a single application. They provide us with useful information about drug delivery, the site where it's released, and the efficiency. These data can be used for designing therapeutic interventions.

Growth in these fields can be effectively transferred to poultry and livestock production systems with the aim to improve animal welfare, increase the efficacy of production systems, and produce necessary high-quality poultry and animal products. Diagnosing sensors as nanochips have found application in close monitoring and diagnosing of livestock, animal husbandry, poultry health, disease physiology and behavior, generic pathology, animal health, and reproductive health (Kannaki and Verma, 2006b). These tracking mechanisms inculcate smaller details in animal, bird, and fish tracking systems to improve the quality and commerce of the final product as per international standards. This tracking and sensing facilitation through nanotechnology-based sensors not only helps monitor the quality of protein produced, but also helps track possible contaminants such as chemicals, additives, fertilizers, poisons, and pesticides. These synthetic toxins can often lead to allergic reactions with animal health as well as end users. This trackable sensor application using nanotechnology helps monitor the quality of protein, thus preventing spoilage in a real-time and affordable process (Rajkumar et al., 2006). This potential nanotechnological application can revolutionize emerging domains such as artificial meat production or the nonanimal-sourced meat industry (Moraru et al., 2003). Among other nonmeat-based animal produce industries and ancillary applications, nanotechnology can help improve milk production by preventing bacterial spoilage, using pasteurization techniques, and retaining essential nutrients and milk. This is significant both in economy and volume of market share (Kuzma and Verhage, 2006). In addition, for poultry and the next sizeable market share of egg production, the widespread application of novel nanodelivery-mediated nutrient fortification and reduced pathogen infection has found widespread applications, including nanocarriers and nanooptics (Kannaki and Verma, 2006a). This would also entail defining the parameters of nanoparticles (NPs), a common determining factor for such targeted delivery applications entailing nanotechnology in livestock management and improvement. These NPs can largely be naturally self-assembled and synthetically engineered without any specific reassembly but within specific parameters (Monica and Cremonini, 2009). The <100 nm diameter NPs are known to possess special physical, chemical, biological, and electrical properties across one dimension, thus making them significantly different from the bulk material (>500 nm). These enable larger applications of the same material, albeit at much smaller sizes and dimensions, largely because of its larger surface area-to-volume ratio. An intermediate-sized group of NPs, ranging between 100 and 500 nm, shows distinctly different physical, chemical, and biological surface properties, but not essentially attributing to an average property of both the extreme classes. Attributed to the type and structure of the nanomaterial molecular assembly, the properties and applications in a biological system of NPs vary greatly. One of the major applications of nanotechnology in medicine is the use of nanoparticles for targeted drug delivery, heat, light, or other substances to specific types of cancerous cells. Worldwide, researchers are developing nanoparticles to fight viruses, which is more relevant in recent times with the pandemic affecting the entire world. These nanoparticles destroy viruses by delivering certain enzymes that stop the reproduction cycle of virus molecules in the host's bloodstream. Even oral administration of these drugs is being developed (Dilbaghi et al., 2013).

A larger market-share of livestock-based commerce where significant nanotechnological advances has been is accounted for, is aquaculture, both freshwater and marine. The effective and targeted dosage of essential proteins and oils make these produce more marketable in quality and consumable parameters makes nanotechnological application an integral part of sustainable practices. In congruence with agricultural practices (Poddar et al., 2018) of mixed farming integrated with fish culture and concurrent water conservation and ground water replenishment paired have led to significant runoff prevention. Smart nanosensors and nano drug delivery vehicles help coordinate such practices, promote health and nutrition thereby preventing pathological anomalies and track effectively on growth parameters. These nanoentity-based applications do not limit themselves to fish alone, but also to water-based microbial flora and fauna such as certain algae, plankton, etc., that often carry a commercial ticket of more than the fish or seafood itself. On the other hand, as an actual application, there have been reports of lanthanum-based NPs capable of preventing algae populations in water though the removal of phosphate compounds, among other contaminants (Zhang et al., 2011). This process, as a part of sustainable farming practices, can contribute to eutrophication remedies (Copetti et al., 2016), thus preventing fouling or cross-contamination of groundwater levels in the long run. Nanoparticle applications mitigated by phosphorous and other nutrient binding clay have also promoted flora and fauna growth of water bodies in addition to basic water quality, including potability for enclosed resources, tanks, and bunds (Van Oosterhout and Lurling, 2013). These nanoclay-based applications with impregnated positively charged ions as the matrix or mesh support materials have also effectively controlled algae and other allied pathogens and weeds (Ivanova et al., 2008).

As part of a growing fish feed industry, NP-encapsulated pelleted fish feed has resulted in significant growth improvement and a feed-to-volume ratio with zero side effects in physiology, protein content, or taste. This pelleted formulation using nanotechnological advances has also been employed in packaged domesticated and companion animal feed, which is a big part of the urban economy, a sector where rearing livestock was commonly never practiced. Thus, nanotechnology has ushered in avenues of animal feed applications using nanoencapsulation, nanoadditive-based pelleting, and obvious targeted nanocarrier delivery in sectors of animal business and commerce that did not actually exist before. In addition, the cyanobacterial population, a common issue in such ecosystems, has also been controlled effectively using nano-based approaches. The following details offer further information about these animal production techniques.

Nanotechnology in boosting animal production

When used in both medical and food industries, nanotechnology gives the best results in improving animal productivity. To maximize profitability, the livestock need to reach their ideal slaughter weights rapidly. In order to achieve this, antibiotics are used as feed additives to protect the herds from illness and increase growth so that the animal production cycle will be shortened (Wang et al., 2016). The regular practice of antibiotics in the animal production sector has led to major problems such as drug-resistant bacteria, which have caused devastating damage in livestock and produced meat with antibiotic residue (Schwarz et al., 2001). The majority of the antibiotics are used in agriculture compared to human use (Silbergeld et al., 2008). This has caused many countries to pass legislation restricting the use of certain antibiotics in animal production (Casewell et al., 2003). Nanoparticles can be the solution for this problem as they can be used effectively without causing antibiotic resistance in microbes. Other than this, nutrient delivery, biocides, veterinary medicine, and reproduction are areas of the animal production industry that could benefit from nanotechnology to increase the productivity indirectly.

Nutrition is the major portion ($\approx$60%–65%) of animal production system inputs. Feed efficiency can be improved by using nanofeed additives, which help in reducing feed cost and in turn increasing the quantity and quality of animal products. The most common cause for feed contamination in storage silos and bulk storage is by fungi or toxin-producing insects. Nanobiosensors can be used for the early detection of such organisms causing contamination. Fats and amino acids get degraded in the rumen, making them less available for the animal. So, nanoparticles that are engineered to protect fats in ruminant diets from the fermentation process in the rumen and to protect essential amino acids and make them more available for the host animal might be advantageous to increase the yield. Currently, there are many studies showing the successful use of different nanomaterials in poultry and livestock feeding. The use of nanofeed additives and new detoxifying nanomaterials in animal production could provide added value in feeding practices due to their positive effect on stimulating productivity and liveability.

Minerals play a major role in animal nutrition. But the form in which these minerals are available in nature influences their bioavailability. Therefore, if the bioavailability is less, they will not be of much nutritional use to the animal. An example of this is iron, which is a major essential mineral and its deficiency causes major problems with both animals and humans, especially in the neonatal stages of life, during pregnancy, and in worm toxicity (Pond et al., 2003). Though ferrous sulfate is the highly bioavailable form (McDowell et al., 1983), it gives food a metallic taste that makes it less palatable; it also hastens the oxidation process of fats in cereals, thus causing rancidity (Hurrell, 2002). The alternative is to use ferric phosphate (FePO4) as even though it has less bioavailability, it is more stable. Regarding this, Rohner et al. (2007) developed nanoparticles of ferric phosphate that are highly available, proving that at the nanoscale, this source can increase its nutritional value. So, the production of such nanominerals provides many advantages for the animal feed industry. They are economically cheaper, needed in less volume, help in growth, and have immune-stimulating effects. They can also be used to overcome common reproductive problems in the herds (Swain et al., 2015). Currently, many nanominerals are available commercially such as nano-ZnO, which helps improve the growth rate, immunity, and reproductive status of farm animals. It also decreases the incidence of diarrhea among young piglets (Mishra et al., 2014; Yang and Sun, 2006). The use of nano-Zn in dairy cattle improved milk production and reduced the somatic cell count (SCC) in cows suffering from subclinical mastitis (Rajendran, 2013). Vitamins in a liquid state prepared by nanotechnology are available for poultry feed. These nanosized nutrients are designed to pass through the gastrointestinal tract (GIT) and deliver vitamins or other nutrients directly in the bloodstream so that the bioavailability will be increased. They do not alter the taste but improve the nutrient dispersibility and durability of the feed. In addition, the use of preservatives can be reduced due to the use of nanotechnology (Thulasi et al., 2013). The microencapsulation of feed ingredients aids in protecting the feed particles from being inactivated by light and oxidation. It also prevents degradation by proteases and other digestive enzymes, keeps them relatively stable at variable pH values and temperature. Thus enabling better dispersing of them and enables good mixing of fat-soluble additives in feed while increasing their shell life time during storage (Meena et al., 2018). Mycotoxicosis is a serious disease that affects both animals and humans. Mycotoxins are usually detected in about 25% of animal feeds, with a higher prevalence in developing countries. With the help of nanotechnology, a nanomycotoxin binder of MgO-SiO2 is manufactured that binds aflatoxins efficiently and helps in reducing the risk of mycotoxins (Moghaddam et al., 2010).

Nanotechnology in disease management and preventing pathogenesis

Nanomedicine is an interesting field in nanotechnology that has shown significant progress in both diagnostics and therapeutics. It has a great scope to play a leading role in global veterinary practice (Feneque, 2003; Scott, 2005). Metallic and nanostructured particles are used as diagnostic tools in biomedical research, which helps to visualize the position of a cell or the distribution of a drug in the body. Functionalized NPs conjugated to monoclonal antibodies have been used to instantly and specifically detect the presence of respiratory syncytial virus, both in vitro and in vivo. The results were promising and suggested that functionalized NPs can be used very effectively for the rapid and sensitive detection of viruses; they could replace the current less-sensitive detection methods (Tripp et al., 2007).

In the veterinary field, nanotechnology has been used in the treatment of African animal trypanosomiasis (Kroubi et al., 2010). It helped in developing a new drug delivery system against diaminazene (DMZ), a trypanocide drug registered for veterinary use. The porous cationic nanoparticles considerably increased the potential targeting of trypanosomes. In another study, mice pretreated with protein cage nanoparticles (PCN) irrespective of any specific viral antigens, showed improved tolerance against both sublethal and lethal doses of two different influenza viruses, a mouse-adapted SARS coronavirus and a mouse pneumovirus. PCN treatment increased survival and even showed a marked viral clearance rate, the early production of viral-specific antibodies, and less morbidity and lung damage (Wiley et al., 2009). Vaccination against the foot-and-mouth disease virus (FMDV) has been a major problem, as currently there is no way to differentiate the infected from the vaccinated. Moreover, the large-scale production of inactivated virus has its own risks. To counter this problem, a study was conducted using inert nanobeads, which can target antigen to dendritic cells (DCs) to trigger immune responses against FMDV-specific synthetic peptides in sheep. In single peptide-induced responses in sheep species, the preferred combination of more than one peptide, either conjugated on its own to individual nanobeads or conjugated as a combination or mixture, also triggered significant cell-mediated and humoral immune responses. These nanodrugs offered many advantages over the conventional drugs in many aspects. For example, when gentamicin is bound to hydrogel by a peptide linker, it helps gentamicin to retain the active form as long as the linker is intact. A protease enzyme produced by *Pseudomonas aeruginosa* is the only source that can break down the peptide link, that is, the gentamicin will only be liberated and activated in the presence of *P. aeruginosa* (Soppimath et al., 2002; Suzuki et al., 1998). The development of NPs combined with antibodies or nucleic acid enables the development of sensitive diagnostic assays. The development of nano-based biochips helped not only in pathogen diagnosis but also led to a better understanding of the genetic predisposing factors. The simultaneous detection of thousands of genes, antigens, or disease biomarkers can be possible by using high-density nanoarray chips. DNA and protein microarrays were also used to assess the efficiency of drug and gene expression. With the introduction of the lab-on a-chip (LOC), even nano/picoliters of a sample can be used to detect target DNA or proteins (Manuja et al., 2012). NPS can be used as an imaging agent for MRI in cats (Kim et al., 2010; Underwood and Van Eps, 2012). So, it is proven that nanotechnology offers revolutionary solutions for the many serious problems that veterinarians commonly face such as tuberculosis, brucellosis, methicillin-resistant *Staphylococcus aureus* (MRSA), foot-and-mouth disease, even intracellular infectious or blood pathogens.

Nanotechnology in animal breeding and growth

The management of animal breeding is an expensive and laborious process for livestock owners. Currently, nanotubes are being studied as a solution to provide real-time measurement of changes in the level of estradiol in the blood. These nanotubes can be implanted under the skin of animals. These nanotubes are absolute tools for tracking estrus in animals because these tubes have the capability to specifically bind and detect the estradiol antibody present during estrus by near-infrared fluorescence. The signal from this sensor will be incorporated as part of central monitoring, so this information is used to detect estrus at the right time and can lead to effective breeding practices (O'connell et al., 2002).

Some nanoparticles have the ability to improve fertility and protect spermatozoa through the functional groups they carry. Artificial insemination is the most-preferred method in animal production as an alternative to natural mating due to the lower risk for animals and producers. This helps in boosting the selection of livestock traits and improves genetic diversity. Nanotechnology can also be incorporated in artificial insemination techniques and can to lead to improve chances of conception. The noninvasive bioimaging of gametes, nanopurification, and protectants in cryopreservation are some of the aspects where artificial insemination can be improved using nanotechnology. Quantum dots have been used to improve the study of mammalian spermatozoon and oocyte movement and their interactions in a normal physiological state. These inorganic nanoparticles are better than organic fluorescent molecules as they are more self-illuminating, biocompatible, photo-stable, and have greater signal intensity, which gives better imaging of gametes and other cell types in vivo (Druart et al., 2009; Feugang et al., 2015; Feugang et al., 2012; Long et al., 2005). Quantum dots can be used to produce targeted or non-targeted imaging as a function of their size, emitted wavelengths, and conjugation possibilities (Druart et al., 2009; Feugang et al., 2015; Feugang et al., 2012; Hasuwa et al., 2010; Long et al., 2005). This engineered NP is used to visualize the molecular and cellular events during fertilization, just as the fluorescent proteins but at greater depths of tissue (Feugang et al., 2015; Hasuwa et al., 2010). A higher concentration may be required for in vivo imaging on larger animals as the signal strength of quantum dots increases with the increase in the dose. So the dose has to be optimized for improved biocompatibility, as many of the current ones include heavy metals such as cadmium and lead that can cause cytotoxicity when used at high levels (Vasquez et al., 2016). However, cytotoxicity may be decreased or eliminated by the careful selection of the concentration and surface chemistries of quantum dots (Lin et al., 2014). Damaged sperm from undamaged sperm can be separated using nanopurification. Ubiquitin is a surface marker found on damaged sperm. Magnetic NPs with antibodies against ubiquitin are used for a protein-based removal strategy (Odhiambo et al., 2014). A lectin-based strategy is also used that features magnetic NPs coated with lectins that can specifically bind to glycan, which is exposed at the surface of the sperm that has acrosomal damage (Petruska et al., 2014). A half-concentration of nanopurified bull spermatozoa achieved conception rates equal to those of full-concentration unpurified semen without any recorded negative impacts to the inseminated cows or calves (Odhiambo et al., 2014). Thus, one sample of nanopurified, extended semen can be used to inseminate a lot of female animals. Antibody and lecithin strategies can be used for increased selection ability and fertility improvement. Nanoprotectant additives used in extenders during cryopreservation can increase the survivability of sperm. Extenders are buffering agents that provide sperm with the nutrients required for prolonged storage. They serve to protect and contain antibiotics as well as to prevent bacterial growth from affecting sperm quality and infecting inseminated females (Bryla and Trzcińska, 2015). NPs with antimicrobial properties can replace the antibiotics used in the extender, as many antibiotics have been shown to inhibit sperm motility and viability depending on the dose (Hargreaves et al., 1998). NPs are also known to facilitate the addition of natural products in extenders to increase sperm motility. This promises future advancements in reproductive biotechnology with the plausible greater inclusion of NPs in molecular biology techniques. Sperm-mediated gene transfer is a novel approach where mesoporous silica NPs are loaded with nucleic acids and proteins (Barkalina et al., 2014). Such NPs can form strong links with spermatozoa in vitro and show no adverse effects on sperm function or quality. Transfections with polymeric NPs such as PDMAEMA, chitosan, and polyethyleneimine have proven to be advantageous over traditional viral approaches, given that low concentrations of polymers are used (Agarwal et al., 2012). The toxicity and efficacy of transfection highly depend on the molecular weight of the nanopolymer. 60 kDa is the optimal molecular weight of PDMAEMA for transfection (Agarwal et al., 2012). With further exploration and advances, NPs could play a major role in animal reproduction. However, the spermatotoxic effects of some NPs should be noted, as they may have serious ill effects if breeder reproduction is altered. Zinc oxide and titanium oxide NPs showed diminishing effects in in vitro sperm viability in a dose- and time-dependent manner by weakening the membrane and DNA fragmentation (Barkhordari et al., 2013; Pawar and Kaul, 2014). Human sperm when incubated with zinc oxide NPs showed a significant increase in cell death after 45 min at a concentration of 500 μg/mL; a concentration of 100 μg/mL would significantly increase cell death after 180 min (Barkhordari et al., 2013). Pawar and Kaul found that buffalo sperm (*Bubalus bubalis*) when incubated with 100 μg/mL of titanium oxide NPs reduced the viability of spermatozoa. At 10 μg/mL, titanium oxide increased the premature sperm capacitation, which is the final required step in sperm maturation for the penetration of oocyte and fertilization. The use of microfluidics in animal science significantly simplifies traditional in vitro fertilization procedures used in animal breeding. It helps in physically sorting the sperm and eggs. Just as a computer circuit sorts data through wires and logic gates, microfluidic and nanofluidic systems also analyze by controlling the flow of liquids or gases through a series of tiny channels and valves. With the help of human genome mapping, geneticists are rapidly sequencing the genomes of cattle, sheep, poultry, pigs, and other livestock, hoping to identify gene sequences that relate to commercially valuable traits such as disease resistance and leanness of meat. Including probes for these traits on biochips helps the breeders to speedily identify champion breeders and screen out genetic diseases. NPs may be the source of advancement for the animal production industry, but precautions should also be taken when considering the application of NPs for assisting reproduction. More detailed studies are needed to determine the long-term effects of these NPs.

Nanotechnology in treating reproductive problems

There are numerous applications of nanotechnology in animal reproduction that optimize the general reproductive performance at all stages, including the diagnosis and treatment of reproductive disorders, estrus detection, sorting and freezing sperm, the direct interference of nanodevices during calving, and also managing reproductive problems such as a retained placenta (Swain et al., 2015). The NPs can also be used for the sustained release of reproductive hormones. They protect the hormones and vitamins from getting inactivated and degraded by oxidation (e.g., vitamins and steroid hormones) or hydrolysis (e.g., gonadotropic hormones) (Joanitti and Silva, 2014). Nanosensors are very sensitive devices supplied with mobile probe biomolecules. The probes are made of nanomaterials and are usually used for diagnostic purposes. They may be used to diagnose genital tract infectious diseases as well as metabolic and hormonal disorders and for the detection of estrus (Saragusty and Arav, 2011; Scott, 2007). Choosing the sex of the fetus (fetus sexing) could be made possible by the use of biochips (Patil et al., 2009).

The use of microfluidics in animal science significantly simplifies traditional in vitro fertilization procedures used in animal breeding. It helps in physically sorting sperm and eggs. Just as a computer circuit sorts data through wires and logic gates, microfluidic and nanofluidic systems also analyze by controlling the flow of liquids or gases through series of tiny channels and valves. With the help of human genome mapping, geneticists are rapidly sequencing the genomes of cattle, sheep, poultry, pig and other livestock hoping to identify gene sequences that relate to commercially valuable traits such as disease resistance and leanness of meat. By including probes for these traits on biochips, helps the breeders to speedily identify champion breeders and screen out genetic diseases.

Nanotechnology in sterilization of animals

Despite all the advantages associated with the use of NP systems in improving reproductive performance, it is well documented that some NPs (e.g., metal NPs) can be toxic and cause deleterious side effects to living organisms. This property of a few metal NPs can be used for reproduction technologies as contraceptive agents. Several metals, including cadmium, cause dose-dependent sterility, especially when used in low to moderate levels. The delivery of metals as NPs to reproductive and related organs like pituitary by targeting molecules such as antibodies or using magnetic field-based delivery of magnetic NPs minimizes the systemic toxicity (Jha et al., 2014).

Nanotechnology in reducing transport and storage of animal protein (food packaging)

The major reason for food spoilage in the food industry is oxygen. It can cause discoloration and spoil the food. Nanotechnology can be applied in the food industry to counter this problem. With the help of nanotechnology, new plastic can be developed for food packaging. NPs that are used in the production of these plastics have the ability to prevent oxygen from entering the package. In other words, this packaging material acts as a barrier for oxygen and protects the products from spoiling during long transport and storage. Nanosilver, nanozinc oxide, and nanotitanium dioxide with functional properties such as a lighter weight with stronger packaging barriers are also used in new packaging systems (Bumbudsanpharoke and Ko, 2015). Even fruits can be covered using nanocoatings, which prevent the fruit from weight loss and shrinkage (Predicala, 2009). Packaging plays a major role in improving the product shelf life. so many companies are trying to develop smart packaging systems. These packaging systems are capable of self-repairing small holes/tears, react to the external environmental conditions such as changes in the temperature and moisture, and can alert the customer if the food gets contaminated. Nanotechnology can be used to change certain properties of an already-existing packaging material such as increasing the mechanical, thermal, chemical, and antimicrobial properties. These enhanced packaging systems can sense and signal microbiological and biochemical changes (Joseph and Morrison, 2006; Moraru et al., 2003). With enzymes coated by nanotechnology, we can protect them from the environment, delaying nutrient corruption and increasing shelf life. Ethylene absorbent is one of the most important materials produced by nanotechnology. Ethylene gas produced by fruits can lead to decay and absorbing it can keep fruit fresh for long periods. Biological components such as a cell, enzyme, or antibody linked to a tiny transducer are used in biosensors. They are powered by one system that then supplies power to a second system. Even the slightest changes in cells and molecules can be detected by biosensors. This is then used to measure and identify the test substance, even if there is a very low concentration of the tested material. So, the larger the concentration of bacteria in a particular food, the stronger the signal will be, indicating that the food is unsafe to eat. With the help of this technology, large amounts of food can be easily checked for safety (Johnson, 2005).

Food processing nanoapplications

Food processing is an important part of both poultry and livestock production systems because this step adds value to the product. The most common use of nanotechnology in the food-processing industry is nanoenabled water treatment technologies, antimicrobial surfaces, and quality monitoring of food products with the help of nanosensors. Water is considered the necessary element for all animal species as well as in food processing. A nanofiltration process is used to purify water. A filtration membrane is used to remove solids, bacteria, and parasites from the water (Sekhon, 2014). This technology can be used in providing high-quality water for food processing as well as recycling the water from processing plants, especially if the water resources are limited. The accurate detection of antibiotic residues in meats can be done with the use of nanobiosensors (Mungroo and Neethirajan, 2014; Sekhon, 2014). This application is very important for providing live birds and animals on farms with fresh water free from contamination and residues and also at the time of processing to ensure that the products for consumers are of good quality and safe. Antimicrobial materials play a major role in providing safe products that are free from pathogens while also increasing shelf life. Low and medium molecular weight chitosan showed higher antimicrobial activity than normal chitosan on different bacteria such as *E. coli* (NCIMB 11943), *S. aureus* (NCIMB 13062), *Bacillus cereus* (NCIMB 9373), and *P. fluorescens* (NCIMB 9046) that were obtained from raw chicken fillets. Nanomaterials will lead to new applications in the fields of food preservation and packaging.

Nanotechnology in livestock

Meat and egg quality

Several studies were conducted to evaluate the potential use of NPs in enhancing the quality of meat and eggs. Chromium NPs (200 μg/kg) in feed to pigs destined for market resulted in 14.06% leaner body fat at slaughter than the control pigs, which were fed a basic diet of corn and soybean meal. Increases in both skeletal muscle mass and quality were achieved; the same effects were found when finishing pigs were fed chitosan nanoparticle supplements loaded with chromium (Wang et al., 2014; Wang et al., 2012).

These chromium-loaded chitosan NPs increased the activity of hormone-sensitive lipase in adipose tissue while decreasing fatty acid synthase activity and boosting blood serum immune components (Wang et al., 2014, Wang et al., 2012). These data show how these NPs work and how they affect meat quality. The nanomaterials are often found beneficial if added in livestock feed or water. Chromium NPs added to poultry feed not only improved breast and thigh muscle protein content while simultaneously lowering cholesterol, but also raised the average daily gain and feed efficiency of the broilers in the experimental group fed 500 μg/kg $Cr3+$ (Zha et al., 2009). The advantages of these additives result in significantly shorter production cycles and superior quality of meat with much less requirement of feed to have broilers arrive at market weight. Conversely, when the same chromium NPs were supplied to layers, there was no significant effect on body weight or egg production (Sirirat et al., 2013). However, Sirirat et al. did find that higher chromium and calcium levels increased the quality of yolks and shells, respectively. The bioaccumulation of NPs in the liver was noted for the experimental group, an observation shared with Chauke and Siebrits (2012) in a study that replaced an antibiotic against coccidiosis with silver NPs in water (0.083 mg/kg of silver compared to 0.001 mg/kg in the control). More information on NPs and their mechanisms of action is needed to ensure that quality is not sacrificed after long-term exposure (Panea et al., 2014) The introduction of nutrient supplements in livestock feed, irrespective of particle size, will benefit the producer if there is good demand for the product. If meat and eggs obtained from an animal fed nanoparticle supplements are enhanced from the original product, they are then likely to still be favorable to consumers. However, it is crucial to understand the role of the nanoparticle additive in a given biological system and byproducts from that system to ensure that it is safe for consumption before its application in animal production.

Dairy and milk

Mastitis is a very common problem among dairy cows with a variety of inciting factors, often bacterial, that can require the use of antibiotics to treat the condition. As an example, a drug used in mastitis cases, tilmicosin, has been found to have negative side effects at a significantly high concentration. But the release of tilmicosin can be controlled by using hydrogenated castor oil–solid lipid NP carriers (Han et al., 2009). In was observed that a lower dose was sufficient in controlling a *S. aureus*-induced murine mastitis model

(10 vs 20 mg/kg) (Wang et al., 2012). The proper use of therapeutic nanocarriers with the desired concentration and half-life of the drug could benefit producers by minimizing milk expiration times and the amount of milk wasted. Nanotechnology can also help to detect food-borne pathogens and ensure product safety. An interesting feature of these nanocomposites is the antibody, whose specificity and selectivity could be modified to capture a variety of pathogens (Sung et al., 2013). Though the main goal has been to remove potentially harmful contaminants from milk, there has also been some interest in mixing nanoparticle supplements directly into cow milk for human consumption. A nanopowdered oyster shell is added to milk to increase the calcium content from 100 to 120 mg/mL, a level more suitable for growing children and postmenopausal women. This introduction of nanopowdered oyster shell into milk did not negatively alter its sensory or physicochemical qualities after 16 d of storage at 4°C (Lee et al., 2015).

Biological waste management through nanotechnology for sustainable animal husbandry and livestock production

One of the major problems in livestock management is contamination from multiple solid wastes. This can lead to microbial infections and has effects on water management and soil contamination at large. Solid waste management hence has been one effective domain of nanotechnological advances. The currently available material choices can be effectively substituted by new age nanobiomaterials and nanobiomimics, thereby creating sustainable and effective waste management solutions. These naturally inspired nanofibers can offer nanodelivery and support scaffold structures. To cite an example, nanofibers inspired from natural fibers such as silk, hemp, and spider web have found applications in readily absorbable and faster degradable materials with minimal toxic remnants or residues. The material choices include cellulose and collagen substitutes as well as biorenewable substitutes such as green tea, which offer additional free radicle quenching theranostic properties (Kim et al., 2015). These effective approaches would culminate in a new age of nanocomposite material sources based on nanotechnology applications that can effectively improve quality biomass procurement from animal remains and fibrous excretes.

Common livestock diseases such as trypanosomiasis, helminthiasis, tick-borne diseases, influenza, and clostridial infections, just to name a few, have caused considerable economical losses all over the world because of their fatal effects on livestock population and the potential for zoonosis (Engels et al., 2002; Thontiravong et al., 2012; Useh et al., 2012). A lack of advances in nanotechnology has made it very difficult to differentiate animal and human pathogens that cause similar diseases (Nagano et al., 2008; Weatherhead and Tweardy, 2012).

Toxicological and environment safety concerns

As the science of nanotechnology is relatively new and interdisciplinary, the sudden growth in research and application has also brought concerns about safety, particularly in usage in biological and living systems. Many nanoentities are known to be toxic compared to micro-sized living creatures, thereby raising concerns on their uncontrolled applications as nanoadditives and chemicals. Thus limiting usage in nanopesticides, nanofungicides, and nanovitamin formulations for use on livestock with possible build-up on residual environmental pollution and decelerated degradation. The new materials in nanoform might enter the soil and groundwater with the potential for food chain entry, thereby raising concerns to design and allocate safety protocols for wide-scale applications of nanomaterial-based animal production technologies. There have been reports of nanoforms of copper oxides, carbon nanotubes (Nair et al., 2010), zinc-based oxides, aluminum oxides (Burklew et al., 2012), and cesium oxides (Priester et al., 2012) in food crops used as fodder for commercial animal production processes (Lin and Xing, 2007). Thus, geography-specific documented and periodically reviewed environmentally safe dosage of nanoparticle applications based on time-resolved residual studies highlighting retention and degradation must be carried out and maintained (Vishwakarma, 2010). This is particularly relevant for certain size exclusions of nanomaterials as they can cross barriers and affect the uptake of water, minerals, sugar and its intermediates and derivatives in animal health like vasculature and blood vessel and hygiene like pathological infestation and organ-based build up on foreign entities. In the long run, these nanoentities can also potentially affect the genome and integrate with DNA, leading to inflammatory responses apart from respiratory and hematological side effects resulting from nanoparticle toxicity. Hence, nanotechnological use as biomolecules in sustainable animal production requires extreme caution with adequate risk assessment and gap analysis at all stages as per global guidelines (FAO/WHO, 2013).

Conclusion and future outlook

Nanotechnological advances have shown high efficacy and application potential in sustainable animal production techniques, yielding improved commercial avenues. Once empowered by suitable regulatory, ethical, and legal guidelines and framework would enable an environmentally safe and economic viable model leading to a concerted effort to this regard thus enabling effective and redefinedsustainable and affordable animal production techniques. Concerted attention and interest in nanotechnology in animal production procedures can possibly lead to sustainable livestock and pisciculture techniques. A focussed approach enabled by policy support would enable the discipline of nanotechnology to evolve effectively. As an integrative scientific domain, the size exclusivity of nanotechnology can potentially lead to this otherwise improved efficacy entity into a toxic substance. Thus, all regulatory concerns around nanotechnological approaches would need careful consideration, thus enabling the discipline to evolve much more comprehensively. Hence, improved research and commercialization, ably supported by funding support from both the government and venture firms, should effectively work to remove the potential bottlenecks of nanotechnology applications by developing environmentally safe nanotechnology-based animal production agricultural systems.

References

Agarwal, S., Zhang, Y., Maji, S., Greiner, A., 2012. PDMAEMA based gene delivery materials. Mater. Today 15, 388–393.

Barkalina, N., Jones, C., Kashir, J., Coote, S., Huang, X., Morrison, R., Townley, H., Coward, K., 2014. Effects of mesoporous silica nanoparticles upon the function of mammalian sperm in vitro. Nanomedicine 10, 859–870.

Barkhordari, A., Hekmatimoghaddam, S., Jebali, A., Khalili, M.A., Talebi, A., Noorani, M., 2013. Effect of zinc oxide nanoparticles on viability of human spermatozoa. Iran. J. Reprod. Med. 11, 767.

Bryla, M., Trzcińska, M., 2015. Quality and fertilizing capacity of boar spermatozoa during liquid storage in extender supplemented with different antibiotics. Anim. Reprod. Sci. 163, 157–163.

Bumbudsanpharoke, N., Ko, S., 2015. Nano-food packaging: an overview of market, migration research, and safety regulations. J. Food Sci. 80, R910–R923.

Burklew, C.E., Ashlock, J., Winfrey, W.B., Zhang, B., 2012. Effects of aluminum oxidenanoparticles on the growth, development, and microRNA expression of tobacco (*Nicotiana tabacum*). PLoS One 7 (5), e34783.

Casewell, M., Friis, C., Marco, E., McMullin, P., Phillips, I., 2003. The European ban on growth-promoting antibiotics and emerging consequences for human and animal health. J. Antimicrob. Chemother. 52, 159–161.

Chauke, N., Siebrits, F.K., 2012. Evaluation of silver nanoparticles as a possible coccidiostat in broiler production. S. Afr. J. Anim. Sci. 42, 493–497.

Copetti, D., Finsterle, K., Marziali, L., Stefani, F., Tartari, G., Lurling, M., 2016. Eutrophication management in surface waters using lanthanum modified bentonite: a review. Water Res. 97, 162–174.

Dilbaghi, N., Kaur, H., Kumar, R., Arora, P., Kumar, S., 2013. Nanoscale device for veterinary technology: trends and future prospective. Adv. Mater. Lett. 4, 175–184.

Druart, X., Cognié, J., Baril, G., Clément, F., Dacheux, J.-L., Gatti, J.-L., 2009. In vivo imaging of in situ motility of fresh and liquid stored ram spermatozoa in the ewe genital tract. Reproduction 138, 45–53.

Engels, D., Chitsulo, L., Montresor, A., Savioli, L., 2002. The global epidemiological situation of schistosomiasis and new approaches to control and research. Acta Trop. 82, 139–146.

EU Commission, 2011. EU Commission Recommendation of 18 October 2011 on the Definition of Nanomaterial. http://eur-lex.europa.eu/legal-content/EN/TXT/PDF/?uri5CELEX:32011H0696&from5EN. Accessed 16.06.20.

FAO/WHO. Food and Agriculture Organization of the United Nations and World HealthOrganization, 2013. Technical Paper on the "State of the Art on the Initiatives and Activities relevant to Risk Assessment and Risk Management of Nanotechnologies in the Food and Agriculture Sectors". Available at http://www.fao.org/fileadmin/templates/agns/pdf/topics/Nano_2.pdf. Accessed 16.06.20.

Feneque, J., 2003. Brief introduction to the veterinary applications of nanotechnology. Nanotechnology. Now. On line: http://www.nanotech-now.com/Jose-Feneque/Veteri-nary-Applications-Nanotechnology.htm.

Feugang, J.M., Youngblood, R.C., Greene, J.M., Fahad, A.S., Monroe, W.A., Willard, S. T., Ryan, P.L., 2012. Application of quantum dot nanoparticles for potential non-invasive bio-imaging of mammalian spermatozoa. J. Nanobiotechnol. 10, 45.

Feugang, J.M., Youngblood, R.C., Greene, J.M., Willard, S.T., Ryan, P.L., 2015. Self-illuminating quantum dots for non-invasive bioluminescence imaging of mammalian gametes. J. Nanobiotechnol. 13, 38.

Feynman, R., 1960. There's plenty of room at the bottom: an invitation to enter a new field of physics. Eng. Sci. 23, 22.

Han, C., Qi, C.M., Zhao, B.K., Cao, J., Xie, S.Y., Wang, S.L., Zhou, W.Z., 2009. Hydrogenated castor oil nanoparticles as carriers for the subcutaneous administration of tilmicosin: in vitro and in vivo studies. J. Vet. Pharmacol. Ther. 32, 116–123.

Hargreaves, C.A., Rogers, S., Hills, F., Rahman, F., Howell, R.J., Homa, S.T., 1998. Effects of co-trimoxazole, erythromycin, amoxycillin, tetracycline and chloroquine on sperm function in vitro. Hum. Reprod. 13, 1878–1886.

Hasuwa, H., Muro, Y., Ikawa, M., Kato, N., Tsujimoto, Y., Okabe, M., 2010. Transgenic mouse sperm that have green acrosome and red mitochondria allow visualization of sperm and their acrosome reaction in vivo. Exp. Anim. 59, 105–107.

Hurrell, R.F., 2002. Fortification: overcoming technical and practical barriers. J. Nutr. 132, 806S–812S.

Ivanova, J., Toncheva-Panova, T., Chernev, G., Samuneva, B., 2008. Effect of Ag21, Cu21 and Zn21 containing hybrid nano matrixes on the green algae Chlorella keissleri. J. Appl. Plant Physiol. 34, 339–348.

Jha, R.K., Jha, P.K., Chaudhury, K., Rana, S.V., Guha, S.K., 2014. An emerging interface between life science and nanotechnology: present status and prospects of reproductive healthcare aided by nano-biotechnology. Nano Rev. 5, 22762.

Joanitti, A.G., Silva, L.P., 2014. The emerging potential of by-products as platforms for drug delivery systems. Curr. Drug Targets 15, 478–485.

Johnson, A., 2005. Agriculture and Nanotechnology. University of Wisconsin. Madison Student-Produced Report.
Joseph, T., Morrison, M., 2006. Nanotechnology in Agriculture and Food: A Nanoforum Report. Nanoforum.org.
Kannaki, T.R., Verma, P.C., 2006a. The challenges of 2020 and the role of nanotechnology in poultry research. In: Proceedings of the National Seminar on Poultry Research Priorities to 2020. Central Avian Research Institute, India.
Kannaki, T.R.V., Verma, P.C., 2006b. Poultry research priorities to (2020). In: Sasidhar, P. V.K. (Ed.), Proceedings of National Seminar. Keynote Addresses and Lead Papers. Central Avian Research Institute, Izatnagar, India, pp. 273–277.
Kim, B.Y., Rutka, J.T., Chan, W.C., 2010. Nanomedicine. N. Engl. J. Med. 363, 2434–2443.
Kim, K.S., Lee, D., Song, C.G., Kang, P.M., 2015. Reactive oxygen species-activated nanomaterials as theranostic agents. Nanomedicine 10, 2709–2723.
Kroubi, M., Daulouede, S., Karembe, H., Jallouli, Y., Howsam, M., Mossalayi, D., Vincendeau, P., Betbeder, D., 2010. Development of a nanoparticulate formulation of diminazene to treat African trypanosomiasis. Nanotechnology 21, 505102.
Kuzma, J., Verhage, P., 2006. Nanotechnology in Agriculture and Food Production: Anticipated Applications (Report). Woodrow Wilson International Center.
Lee, Y.K., Ahn, S.I., Chang, Y.H., Kwak, H.S., 2015. Physicochemical and sensory properties of milk supplemented with dispersible nanopowdered oyster shell during storage. J. Dairy Sci. 98, 5841–5849.
Lin, D., Xing, B., 2007. Phytotoxicity of nanoparticles: inhibition of seed germination androot growth. Environ. Pollut. 150, 243–250.
Lin, G., Ding, Z., Hu, R., Wang, X., Chen, Q., Zhu, X., Liu, K., Liang, J., Lu, F., Lei, D., 2014. Cytotoxicity and immune response of CdSe/ZnS quantum dots towards a murine macrophage cell line. RSC Adv. 4, 5792–5797.
Long, J.Z., Lackan, C.S., Hadjantonakis, A.-K., 2005. Genetic and spectrally distinct in vivo imaging: embryonic stem cells and mice with widespread expression of a monomeric red fluorescent protein. BMC Biotechnol. 5, 1–11.
Manuja, A., Kumar, B., Singh, R.K., 2012. Nanotechnology developments: opportunities for animal health and production. Nanotechnol. Dev. 2, e4.
McDowell, L.R., Conrad, J.H., Ellis, G.L., Loosli, J.K., 1983. Minerals for Grazing Ruminants in Tropical Regions. University of Florida Gainesville.
Meena, N.S., Sahni, Y.P., Thakur, D., Singh, R.P., 2018. Applications of nanotechnology in veterinary therapeutics. J. Entomol. Zool. Stud. 6, 167–175.
Mishra, A., Swain, R.K., Mishra, S.K., Panda, N., Sethy, K., 2014. Growth performance and serum biochemical parameters as affected by nano zinc supplementation in layer chicks. Indian J. Anim. Nutr. 31, 384–388.
Moghaddam, S.H., Jebali, A., Daliri, K., 2010. The use of MgO-Sio2 nano-composite for adsorption of aflatoxin in wheat flour samples. Nano 10, 12–14.
Mohantya, N., Palaib, T., Prustyc, B., Mohapatrad, J., 2014. An overview of nanomedicine in veterinary science. Vet. Res. 2, 90–95.
Monica, R.C., Cremonini, R., 2009. Nanoparticles and higher plants. Caryologia 62, 161–165.
Moraru, C.I., Panchapakesan, C.P., Huang, Q., Takhistov, P., Liu, S., Kokini, J.L., 2003. Nanotechnology: a new frontier in food science understanding the special properties of materials of nanometer size will allow food scientists to design new, healthier, tastier, and safer foods. Nanotechnology 57.
Mungroo, N.A., Neethirajan, S., 2014. Biosensors for the detection of antibiotics in poultry industry—a review. Biosensors 4, 472–493.

Nagano, N., Isomine, S., Kato, H., Sasaki, Y., Takahashi, M., Sakaida, K., Nagano, Y., Arakawa, Y., 2008. Human fulminant gas gangrene caused by clostridium chauvoei. J. Clin. Microbiol. 46, 1545–1547.

Nair, R., Varghese, S.H., Nair, B.G., Maekawa, T., Yoshida, Y., Kumar, D.S., 2010. Nanoparticulate material delivery to plants. Plant Sci. 179 (3), 154–163.

O'connell, M.J., Bachilo, S.M., Huffman, C.B., Moore, V.C., Strano, M.S., Haroz, E.H., Rialon, K.L., Boul, P.J., Noon, W.H., Kittrell, C., 2002. Band gap fluorescence from individual single-walled carbon nanotubes. Science 297, 593–596.

Odhiambo, J.F., DeJarnette, J.M., Geary, T.W., Kennedy, C.E., Suarez, S.S., Sutovsky, M., Sutovsky, P., 2014. Increased conception rates in beef cattle inseminated with nanopurified bull semen. Biol. Reprod. 91, 97–101.

Panea, B., Ripoll, G., González, J., Fernández-Cuello, Á., Albertí, P., 2014. Effect of nanocomposite packaging containing different proportions of ZnO and Ag on chicken breast meat quality. J. Food Eng. 123, 104–112.

Patil, S., Kore, K., Kumar, P., 2009. Nanotechnology and its applications in veterinary and animal science. Vet. World 2, 475.

Pawar, K., Kaul, G., 2014. Toxicity of titanium oxide nanoparticles causes functionality and DNA damage in buffalo (Bubalus bubalis) sperm in vitro. Toxicol. Ind. Health 30, 520–533.

Peterson, C.L., 2004. Nanotechnology: from Feynman to the grand challenge of molecular manufacturing. IEEE Technol. Soc. Mag. 23, 9–15.

Petruska, P., Capcarova, M., Sutovsky, P., 2014. Antioxidant supplementation and purification of semen for improved artificial insemination in livestock species. Turk. J. Vet. Anim. Sci. 38, 643–652.

Poddar, K., Vijayan, J., Ray, S., Adak, T., 2018. Nanotechnology for sustainable agriculture. Biotechnology for Sustainable Agriculture. Elsevier.

Pond, W.G., Church, D.C., Pond, K.R., 2003. Fundamentos de nutrición y alimentación de animales. Summary of Foundations of Nutrition and Animal Feeding, 2nd. World Wide Books.

Predicala, B., 2009. Nanotechnology: Potential for Agriculture. Prairie Swine Centre Inc, University of Saskatchewan, Saskatoon, SK, pp. 123–134.

Priester, J.H., Ge, Y., Mielke, R.E., et al., 2012. Soybean susceptibility to manufactured nanomaterialswith evidence for food quality and soil fertility interruption. Proc. Natl. Acad. Sci. 109, E2451–E2456.

Rajendran, D., 2013. Application of nano minerals in animal production system. Res. J. Biotechnol. 8, 1–3.

Rajkumar, R.S., Kandeepan, G., Prejit Susitha, K., 2006. Poultry research priorities to 2020. In: Sasidhar, P.V.K. (Ed.), Proceedings of National Seminar. Contributory Papers. Central Avian Research Institute, Izatnagar, India.

Regis, E., 1995. Nano: The Emerging Science of Nanotechnology. Little Brown, New York, NY.

Rohner, F., Ernst, F.O., Arnold, M., Hilbe, M., Biebinger, R., Ehrensperger, F., Pratsinis, S. E., Langhans, W., Hurrell, R.F., Zimmermann, M.B., 2007. Synthesis, characterization, and bioavailability in rats of ferric phosphate nanoparticles. J. Nutr. 137 (3), 614–619. https://doi.org/10.1093/jn/137.3.614.

Saragusty, J., Arav, A., 2011. Current progress in oocyte and embryo cryopreservation by slow freezing and vitrification. Reproduction 141, 1.

Schwarz, S., Kehrenberg, C., Walsh, T.R., 2001. Use of antimicrobial agents in veterinary medicine and food animal production. Int. J. Antimicrob. Agents 17, 431–437.

Scott, N.R., 2005. Nanotechnology and animal health. Rev. Sci. Tech. 24, 425.

Scott, N.R., 2007. Nanoscience in veterinary medicine. Vet. Res. Commun. 31, 139–144.

Scott, N.R., Chen, H., 2003. Nanoscale science and engineering or agriculture and food systems. In: Roadmap Report of National Planning Workshop, pp. 18–19. Washington, DC, http://www.nseafs.cornell.edu/web.roadmap.pdf.

Sekhon, B.S., 2014. Nanotechnology in agri-food production: an overview. Nanotechnol. Sci. Appl. 7, 31.

Silbergeld, E.K., Graham, J., Price, L.B., 2008. Industrial food animal production, antimicrobial resistance, and human health. Annu. Rev. Public Health 29, 151–169.

Sirirat, N., Lu, J., Hung, T., Lien, T., 2013. Effect of different levels of nanoparticles chromium picolinate supplementation on performance, egg quality, mineral retention, and tissues minerals accumulation in layer chickens. J. Agric. Sci. 5, 150–159.

Soppimath, K.S., Aminabhavi, T.M., Dave, A.M., Kumbar, S.G., Rudzinski, W.E., 2002. Stimulus-responsive "smart" hydrogels as novel drug delivery systems. Drug Dev. Ind. Pharm. 28, 957–974.

Sung, Y.J., Suk, H.-J., Sung, H.Y., Li, T., Poo, H., Kim, M.-G., 2013. Novel antibody/gold nanoparticle/magnetic nanoparticle nanocomposites for immunomagnetic separation and rapid colorimetric detection of Staphylococcus aureus in milk. Biosens. Bioelectron. 43, 432–439.

Suzuki, Y., Tanihara, M., Nishimura, Y., Suzuki, K., Kakimaru, Y., Shimizu, Y., 1998. A new drug delivery system with controlled release of antibiotic only in the presence of infection. J. Biomed. Mater. Res. 42, 112–116.

Swain, P.S., Rajendran, D., Rao, S.B.N., Dominic, G., 2015. Preparation and effects of nano mineral particle feeding in livestock: a review. Vet. World 8, 888.

Thontiravong, A., Wannaratana, S., Tantilertcharoen, R., Prakairungnamthip, D., Tuanudom, R., Sasipreeyajan, J., Pakpinyo, S., Amonsin, A., Kitikoon, P., Oraveerakul, K., 2012. Comparative study of pandemic (H1N1) 2009, swine H1N1, and avian H3N2 influenza viral infections in quails. J. Vet. Sci. 13, 395–403.

Thulasi, A., Rajendran, D., Jash, S., Selvaraju, S., Jose, V.L., Velusamy, S., Mathivanan, S., 2013. Nanobiotechnology in animal nutrition. In: Sampath, K.T., Ghosh, J. (Eds.), Nutrition and Reproductive Physiology (Recent Concepts), first ed. Serial Publishing House, New Delhi, pp. 499–516.

Tripp, R.A., Alvarez, R., Anderson, B., Jones, L., Weeks, C., Chen, W., 2007. Bioconjugated nanoparticle detection of respiratory syncytial virus infection. Int. J. Nanomedicine 2, 117.

Troncarelli, M.Z., Brandão, H.M., Gern, J.C., Guimarães, A.S., Langoni, H., 2013. Nanotechnology and antimicrobials in veterinary medicine. Formatex 13, 543–556.

Underwood, C., Van Eps, A.W., 2012. Nanomedicine and veterinary science: the reality and the practicality. Vet. J. 193, 12–23.

Useh, N.M., Nok, A.J., Ibrahim, N., Esievo, K.A., 2012. Anaemia in clostridium chauvoei infection is masked by haemoconcentration. Vet. Arh. 82, 433–447.

Van Oosterhout, F., Lurling, M., 2013. The effect of phosphorus binding clay (Phoslocks) in mitigating cyanobacterial nuisance: a laboratory study on the effects on water quality variables and plankton. Hydrobiologia 710, 265–277.

Varnek, A., Fourches, D., Hoonakker, F., Solov'ev, V.P., 2005. Substructural fragments: an universal language to encode reactions, molecular and supramolecular structures. J. Comput. Aided Mol. Des. 19, 693–703.

Vasquez, E.S., Feugang, J.M., Willard, S.T., Ryan, P.L., Walters, K.B., 2016. Bioluminescent magnetic nanoparticles as potential imaging agents for mammalian spermatozoa. J. Nanobiotechnol. 14, 20.

Vishwakarma, S.S.A.N.M.V., 2010. Safety and risk associated with nanoparticles—a review. J. Miner. Mater. Charact. Eng. 9, 455–459.

Wang, X.F., Zhang, S.L., Zhu, L.Y., Xie, S.Y., Dong, Z., Wang, Y., Zhou, W.Z., 2012. Enhancement of antibacterial activity of tilmicosin against Staphylococcus aureus by solid lipid nanoparticles in vitro and in vivo. Vet. J. 191, 115–120.

Wang, M.Q., Wang, C., Du, Y.J., Li, H., Tao, W.J., Ye, S.S., He, Y.D., Chen, S.Y., 2014. Effects of chromium-loaded chitosan nanoparticles on growth, carcass characteristics, pork quality, and lipid metabolism in finishing pigs. Livest. Sci. 161, 123–129.

Wang, J.B., Qi, L.L., Han, F., 2016. Antibiotic with different antibacterial spectrum changed intestinal microflora structure and reduced immune response of Lingnan yellow broiler. Res. J. Biotechnol. 11, 121–129.

Weatherhead, J.E., Tweardy, D.J., 2012. Lethal human neutropenic entercolitis caused by clostridium chauvoei in the United States: tip of the iceberg? J. Infect. 64, 225–227.

Wiley, J.A., Richert, L.E., Swain, S.D., Harmsen, A., Barnard, D.L., Randall, T.D., Jutila, M., Douglas, T., Broomell, C., Young, M., 2009. Inducible bronchus-associated lymphoid tissue elicited by a protein cage nanoparticle enhances protection in mice against diverse respiratory viruses. PLoS One 4, e7142.

Yang, Z.P., Sun, L.P., 2006. Effects of nanometre ZnO on growth performance of early weaned piglets. J. Shanxi Agric. Sci 3, 577–588.

Zha, L.-Y., Zeng, J.-W., Chu, X.-W., Mao, L.-M., Luo, H.-J., 2009. Efficacy of trivalent chromium on growth performance, carcass characteristics and tissue chromium in heat-stressed broiler chicks. J. Sci. Food Agric. 89, 1782–1786.

Zhang, J., Shen, Z., Shan, W., Mei, Z., Wang, W., 2011. Adsorption behavior of phosphate on lanthanum(III)-coordinated diamino-functionalized 3D hybrid mesoporous silicates material. J. Hazard. Mater. 186, 7683.

CHAPTER EIGHT

Transgenic animals

Pankaj Singh[a], Pradeep Kumar Singh[b], and Ram Lakhan Singh[c]
[a]Department of Biotechnology, Dr. Rammanohar Lohia Avadh University, Ayodhya, Uttar Pradesh, India
[b]Department of Biochemistry, Dr. Rammanohar Lohia Avadh University, Ayodhya, Uttar Pradesh, India
[c]Nilamber-Pitamber University, Palamu, Jharkhand, India

Introduction

In the past several years, animal breeders have attempted to improve the productive performance of animals for characteristics such as milk yield, growth, feeding efficiency, etc. Phenotypic-selective breeding was the dominant practice to develop advanced genetic traits in livestock. These genetic improvements lead to the increased productivity of many dairy animals. In this method, the regulation and function of the animal genes were based on the observation of inherited characteristics. Selective breeding was a common practice among farmers for the improvement of a particular characteristic (Vithani, 2008). After the evolvement of molecular genetics and recombinant DNA technology, it became possible to control the desired characteristics/traits in animals, and transgenics were produced. A transgenic animal is one whose genome has been deliberately changed, and they carry genes from other related or unrelated species. Recombinant DNA technology involves integrating genetic information from different organisms, which produces genetically modified organisms. Recombinant DNA technology was first used to clone the gene of interest in a bacterial cell for the expression of the desired pure proteins/materials. During transgenesis, it is possible to alter the DNA of any animal for improvement (Baldassarre et al., 2003). Presently, biotechnology has reached a level where it is feasible to make desirable changes in the animal genome. Further, transgenic animals are modified in the laboratory to strengthen the desired characteristics for human welfare (Chrenek and Makarevic, 2008). The first transgenic organism was the *E. coli* bacterium, produced with the help of plasmid DNA (Cohen et al., 1973). Soon after 1974, the transformation technique was extended to produce the first transgenic mouse, which contained DNA

Emerging Issues in Climate Smart Livestock Production
https://doi.org/10.1016/B978-0-12-822265-2.00001-6

fragments of the Simian Virus 40 (SV-40) genome (Jaenisch and Mintz, 1974). Later, several transgenic animals such as pigs, rabbits, sheep, cattle, etc., were produced for various purposes. Beside transgenic animal development, this technology is also used for a number of vertebrate and invertebrate species for transgenic production (Mehta et al., 2017).

If the germ cell line of an animal is modified, characteristics are passed on to their descendants during normal reproduction. The transgenic competence and strict control of gene expression are the main limiting factors in the production of transgenic animals (Kohler et al., 1991).

Techniques for producing transgenic animals

DNA microinjection

This technique was first developed by Lin (1996). In this, a direct and microsurgical procedure is followed to introduce a foreign DNA fragment (transgene) into either the cytoplasm or nucleus of a single cell. Many copies of the foreign DNA fragment are injected directly into one or both of the pronuclei of a newly fertilized egg cell. If the transgene is ligated at this pronuclear stage, then of course each cell of the resulting animal (the founder animal for a transgenic line) has this transgene. Generally, the transgene is not necessarily always ligated at this stage, but it is ligated after one or more cellular divisions. It requires a glass needle (works as a microcapillary pipette), a microinjector, and a positioning device (a micromanipulator) to control the movement of the micropipette. Mostly male pronuclei of embryos are used for injection due to their large size. Generally, in a single embryo, 200–500 copies of the transgene can be injected (Chrenek et al., 2010). After injection, the transgene randomly gets incorporated in the genome of the pronuclei of injected embryos. The resulting embryos with the transgene are cultured in vitro for 24 h, and are then implanted into a pseudopregnant foster mother. The pseudopregnant foster mother is made pseudopregnant either by mating her with vasectomized males or by injecting her with hormones that cause a false pregnancy and allow the uterus to receive the egg. Actually, this process is used to stimulate their reproductive system to prepare the female's body for transplantation (Krimpenfort et al., 1991). Muta Mouse (Vijg and Douglas, 1996) and Big Blue animals (Kohler et al., 1991) have been produced by using this method. In this technique, the transgene integration into the genome of the founder animal is very low, i.e., only 0.5%–3% (Ebert and Schindler, 1993). This technique is not useful for poultry because of the difficulty in gaining access to the fertilized egg when it is still at the single cell stage (Rutovitz and Mayer, 2002). Similarly, in other nonmammalian species, pronuclear is not access, as the embryonic cell is implanted in the opaque vitellus. To overcome this problem, large amounts of DNA fragments are microinjected into the cytoplasm of one-cell embryos in these species. This simple technique is efficient in several fish species (MacLean, 2003). On the other hand, the rate of integration is extremely low in various species such as *Xenopus*, chickens, and insects, for unknown reasons.

This technique has a number of limitations such as low efficiency, random integration, and variable expression patterns. The overall efficiency of this technique in the mouse is superior compared to cattle. The presence of the inserted sequence in embryos is reported by a polymerase chain reaction or a green fluorescent protein (Chauhan et al., 1999). Regardless of these challenges, this technique has been used successfully to generate transgenic animals such as sheep and goats, which produce valuable biopharmaceutical proteins (Ebert et al., 1991). The first recombinant human plasma protein produced by a transgenic goat was ATryn, an anticoagulant antithrombin (Zhou et al. 2005).

Embryonic stem cell-mediated gene transfer

In this method, the first totipotent stem cells derived from early preimplantation embryos are isolated. Then, the insertion of the desired gene into the embryonic stem cells (ES cells) is allowed by a process known as transfection (Gordon, 1996). These stem cells containing the desired gene are incorporated by homologous recombination into the host embryos, resulting in the production of chimeric animals. Thus, this technique allows gene targeting rather than insertion by allowing the directed modification of endogenous genes by the desired gene (Robertson, 1991). Transfected ES cells are identified and selected by using a cloned selectable marker gene. Finally, the transformed ES cells are microinjected into the animal blastocyst embryonic stage. The microinjected embryos are transferred into a surrogate mother where they complete their development. Presently, this method is most successful with mice, as mouse ES cells are pluripotent and when integrated into blastocysts, they can divide and differentiate efficiently in the mouse embryo. Thus, the inserted gene reaches the somatic and germ line tissues from where they are then passed on to successive generations.

The ability to isolate and maintain ES cells in vitro increases the efficiency of the production of transgenic animals. Embryonic stem cells are pluripotent in nature, thus they can differentiate into various cell/tissue types and be genetically modified for the production of animals of desired genotypes. The ES cells and primordial germ cells offer rapid production of transgenic animals through targeted gene modifications.

Retrovirus-mediated gene transfer

A retrovirus is a single-stranded RNA virus with a reverse transcriptase enzyme that, upon transfection, makes DNA from RNA. It has the capacity to integrate into the host DNA and be copied when the cell starts division (Chrenek et al., 2010). After integration, the production of chimeras takes place because not all cells can have the transgene. A pure homozygous transgene may be obtained after 10–20 generations by inbreeding and can be stored and used for subsequent implantation (Chrenek et al., 2010). Transgenic mice have been successfully produced by this method. Male germ line stem cells with the ability to self-renew as well as genetic modifications are transfected with the retrovirus. Furthermore, these cell lines are used to study the biology of their complex self-renewal and differentiation processes as well as to generate a wide range of transgenic animal species (Makoto et al., 2001). This method is suitable to make transgenic poultry and other species of animals (Houdebine, 2002). The most commonly used retroviral vectors are the Moloney murine leukemia virus (causes lymphoid leukemia in hamsters, rats, and mice); the Rous sarcoma virus (cancer formation in humans), and the avian leukosis virus (poultry flocks) (Rutovitz and Mayer, 2002). The major limitation of this method is that the retrovirus is unable to carry a large size of the desired DNA fragment for transfer into the cell (Rutovitz and Mayer, 2002). However, this method has proved highly efficient in several species, including birds (Lillico et al., 2007) and mammals (Pfeifer, 2006).

Use of transposons

Transposons are natural DNA sequence regions (about 1–3 kb) that self-replicate and are randomly integrated at multiple sites within the same genome. Integration is possible because the transposon itself encodes an integrase enzyme. In many species, especially nonmammals, foreign DNA injected in the embryo hardly ever becomes integrated in the genome; examples include the chicken, *medaka*, *Drosophila,* and silk worms. Foreign DNA can be introduced into the transposon in vitro and the recombinant transposon can be microinjected into one-cell embryos along with the transposon integrase enzyme. Now, all transgenic insects are generated by using specific transposon vectors and their efficiency is also proved in the production of transgenic fish, chickens, and mammals (Ding et al., 2005). The potential of transposon-mediated gene transfer was recently enhanced to a great extent, as it transfers foreign DNA as long as 120 kb with high efficiency (Sumiyama et al., 2010).

Nuclear transfer method

This method began in the mid-1980s. After 30 years of successful experiments with frogs, it is now practiced in different species of animals, except humans (Heyman et al., 1998; McCreath et al., 2000; Kuroiwa et al., 2002). The transgenic sheep Dolly was produced by using the somatic cell nuclear transfer technique (Eenennaam, 2008). Pluripotency can be tested by transferring a nucleus from a cultured cell to an enucleated egg and determining if the reconstituted ovum supports development and the formation of viable offspring. A number of laboratories have tested embryonic, fetal, and adult somatic cell lines for pluripotency with varying success. The nuclear transfer method or somatic cell nuclear transfer (SCNT) is a safer and more efficient way of creating transgenic animals in terms of large numbers of embryos being saved. During SCNT, a nucleus is taken from a somatic cell such as a skin cell and a nucleus containing the transgene is then transferred into an enucleated egg. When the egg develops into a blastocyst, it is implanted into a foster mother (Fig. 1). In this technique, the transfer of the somatic cell nucleus to the cytoplasm of the enucleated egg is to be reprogrammed by egg cytoplasmic factors to form a zygote (Wilmut and Whitelaw, 1994; Campbell et al., 1996). This assures the offspring to be 100% transgenic because the newly implanted nucleus already has the transgene in its genome. Somatic cell nuclear transfer has the ability to reprogram cells into an undifferentiated state. This unique oocyte ability permits SCNT to be used to generate stem cells from somatic cells. Once somatic cells are differentiated and returned to the embryonic state, the ICM of the embryo can be isolated and potentially produce cells that are like embryonic stem cells. In theory, by generating stem cells from the somatic cells of a patient, the stem cells could be a source to cure degenerative diseases. SCNT was the main technology in fully reprograming somatic cells until the invention of induced pluripotent (iPS) cells. iPS cells are somatic cells reprogramed into stem cells by introducing specific factors: Oct4, Sox2, Klf4, and c-myc. When the technology was first introduced, many believed that it would

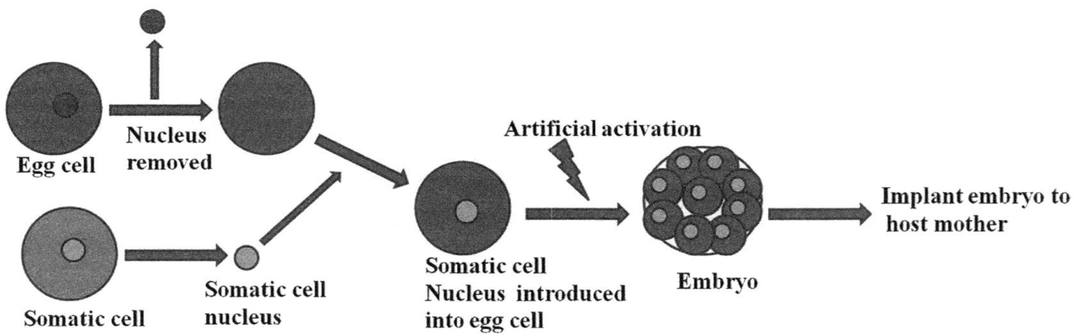

Fig. 1 Somatic cell nuclear transfer process involves the removal of the nucleus from the egg cell, followed by the transfer and fusion of a donor somatic cell nucleus to the enucleated egg cell. The manipulated egg cell is then artificially activated, which divides to form an embryo to be implanted into a host mother for further development.

replace the need for SCNT, as the generation of iPS cells is ethically less controversial (Lee and Prather, 2013). Experimental results suggested by Gong et al. (2004) showed that transgenic calves could be efficiently produced by SCNT using transgenic cells as nuclei donors. Bovine fetal oviduct epithelial cells were transfected with a constructed double marker selective vector (pCE-EGFP-IRES-Neo-dNdB) containing the enhanced green fluorescent protein (EGFP) and neomycin-resistant (Neo^r) genes by electroporation, and a transgenic cell line was obtained.

Sperm-mediated gene transfer

For the production of transgenic animals, a new method called sperm-mediated gene transfer (SMGT) is available that is based on the intrinsic ability of sperm cells to bind and internalize exogenous DNA molecules and to transfer them into the oocyte at fertilization as an alternative to microinjection (Houdebine, 2002; Lavitrano et al., 2006). There are three critical steps for successful sperm-mediated gene transfer to create transgenic animals. The first step is binding DNA on the sperm-cell surface, followed by the internalization of foreign DNA into the nucleus, and finally the integration of exogenous DNA (Fig. 2). Although the mechanism of each individual step is not fully understood, much advancement in the past few years has provided a better understanding of the process (Chan et al., 2000). The first transgenic mouse was created in the early 1990s via the incubation of spermatozoa with exogenous DNA, followed by in vitro fertilization (IVF). This success led to a new interest in transgenic technology because of its simplicity; various methods have been successfully used to create transgenic animals, including mice, pigs, fish, cattle, chickens, and *Xenopus*. The advantages of this method include greater efficiency, low cost, ease of use, the absence of embryo handling, and the need for expensive equipment (Chrenek et al., 2010).

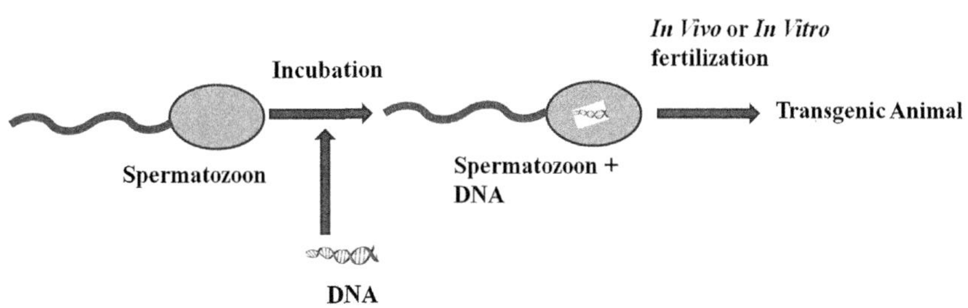

Fig. 2 Sperm-mediated gene transfer technique.

Artificial chromosome-mediated gene transfer

Genes used to express in prokaryotes are generally cDNAs, a smaller gene; cDNAs are poorly expressed in mammalian cells. Hence, larger-sized genomic DNA are used for transgenesis. For transgene expression in mammalian cells, important gene regulatory sequences must be present upstream or downstream of the gene. However, during the transfer of the transgene into the cell, these regulatory sequences are rarely retained as part of the insert. Complete transgene complexes become too large (more than 100 kb) for a conventional vector. For this reason, yeast artificial chromosomes (YACs) that carry genomic DNA ranging from 100 to 1000 kb have been used for transgenesis. Transgenic mice have been produced either by microinjection of the pronucleus of the fertilized egg or the transfection of embryonic stem cells with YACs.

Human artificial chromosomes (HACs) or extra minichromosomes artificially created by either a top-down approach (engineered chromosome) or a bottom-up approach (de novo artificial chromosome) represent a new generation of gene delivery vectors. They have several benefits over other gene delivery vectors, including the ability to carry large gene inserts and stable episomal maintenance in a single copy. HAC-based vector systems may allow the physiological regulation of the introduced gene as a native chromosome. HAC can also carry genomic loci with regulatory elements, which allows for the expression of transgenes in a genetic environment similar to the natural chromosome (Oshimura and Katoh, 2008). Human artificial chromosomes (HAC) replicate and segregate independently from the host genome as natural chromosomes (Grimes et al., 2001; Katoh et al., 2004). HACs can faithfully mimic the normal pattern of gene expression because they can hold complete genomic loci, including upstream and downstream regulatory elements (Ikeno et al., 1998). An advantage of this procedure is that foreign genetic materials can be transferred to mouse embryonic stem cells by using the human artificial chromosome as a vector. It can transfer large segments of genomic DNA greater than the size of the maximum capacity of conventional cloning vectors such as the yeast artificial chromosome (YAC), although the transfer efficiency needs to be improved. During the chromosome transfer process, the accidental fragmentation of an intact chromosome can take place, thus producing human chromosome fragments (hCF). The low-dose irradiation of microcells followed by fusion with recipient cells also induces chromosome breakage, thus resulting in the production of an endogenous functional centromere (Dowdy et al., 1990). HACs allow therapeutic transgenes to be expressed in target cells similar to the endogenous loci (Kouprina et al., 2014). HAC carrying the suicide gene of herpes simplex virus thymidine kinase (HSV-TK) in tumor cells were selectively killed by ganciclovir under in vitro and in vivo conditions. Thus, this novel HAC vector may be useful not only for animal transgenesis but also in gene therapy (Kazuki et al., 2011).

Testis cell transplantation method

This works by transfecting the spermatogonia in situ through the transfer of transgenes into seminiferous tubules, followed by transplantation into the host testis. Testis-derived cells transplanted into the testis of infertile males populate the host testis, generate sperm, and produce transgenic offspring (Chrenek et al., 2010). This technique provided the first functional assay for spermatogonial stem cells. In 2001, it became possible to produce transgenic animals using this transplantation technique (Takashima and Shinohara, 2018). In this technique, we genetically manipulate the donor spermatogonial stem cells before transplantation into the testis of a recipient animal. Genetically manipulated spermatogonial stem cells will colonize and initiate donor-derived spermatogenesis into transgenic animals (Fig. 3). It also has potential applications in fertility preservation in cattle and humans. Recent reports demonstrated that rhesus monkeys recovered spermatogenesis after the

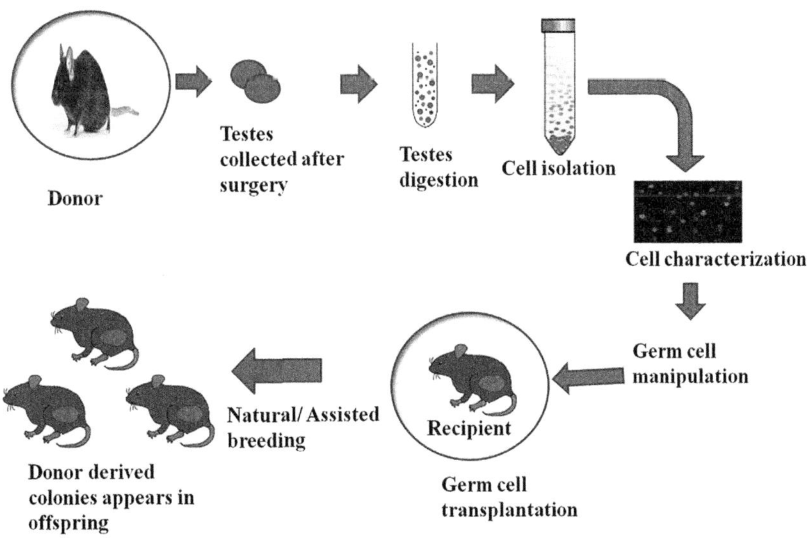

Fig. 3 Schematic overview of germ cell transplantation.

allogenic transplantation of spermatogonial stem cells (McLean, 2008; Medrano et al., 2014).

Different types of transgenic animals

Transgenesis is a powerful new innovation for modifying animal traits by introducing outside hereditary material. Organisms containing the transgenes or specified sequences of cloned DNA transferred by using techniques such as pronuclear microinjection, embryonic stem cell manipulation, the Cre-lox technique, viral vectors, cytoplasmic injection, spermatogonial manipulation, etc., are called transgenic animals. The development of transgenic animals has been part of biomedical research. It has produced better and good-quality breeds, increased milk yield, and produced organs to meet the demand for organ transplantation. Transgenic animals are of even more vital uses in human welfare in many ways, such as agriculture, medicine, food, disease model organisms, industrial purposes, drug development, and the development of new treatments and cures for many serious diseases. The use of transgenic animals indicates that within 5–8 years, the transgenic animal will play a critical and significant role in the biomedical field with the production of important pharmaceutical proteins.

During the 1980s, the idea of genetically manipulating animals by introducing genes into fertilized eggs was converted into reality. The DNA that is introduced is called the transgene and the overall process is called transgenesis or transgenic technology. Transgenesis has become a powerful technique for studying the fundamental problems of mammalian gene expression and development to create animal model systems for human diseases.

The first genetically modified mammal resulted from the early finding in which transgenic mice could be generated with the injection of viral (SV40) DNA into preimplantation mice blastocysts (Jaenisch and Mintz, 1974; Jaenisch et al., 1975). Further, a significant development was achieved with the successful injection of a rat growth hormone gene containing the promoter of the mouse metallothionein-I gene into the pronuclei of fertilized mouse eggs (Palmiter et al., 1982), which resulted in the generation of transgenic mice. By 1981, different researchers revealed the effective implantation of outside DNA into mice, thereby changing the hereditary makeup of the animals. These studies contributed to an understanding of gene regulation, tumor development, immunological specificity, molecular genetics of development, and many other biological processes of fundamental interest

(Van Keuren et al., 2009). Efforts have also been made to use transgenic animals as living bioreactors. The goal of transgenesis is not to produce cows as big as elephants but to produce specific, economically significant traits in livestock. The aim of transgenesis is to produce animals with a hereditable change in their genotype so that the benefits of gene manipulation can be passed on to their offspring.

Mice

There are many human diseases for which animal models are lacking, which greatly hinders the development of therapeutic agents. By transgenesis, a number of mice have been produced that carry genetic lesions identical to those existing in certain human inherited diseases. Not only will such mice allow a full understanding of the biochemical consequences of gene deficiency, but they may also make possible the preliminary screening of potential drugs. Working with large domestic animals is much more difficult than with mice. First, large domestic animals do not produce a large number of eggs. Second, the reimplantation of manipulated embryos is more difficult because sheep and cattle will not give more than two offspring. Third, the eggs of many domestic animals have an opaque cytoplasm, so it is impossible to see their pronuclei or nuclei without resorting to centrifugation to separate the cytoplasmic contents. For experimental purposes, the production of transgenic mice is an appropriate choice. A superovulated mouse can yield up to 40 eggs, reimplantation is relatively easy, and mice can carry up to 20 offspring. Mice are routinely used for this purpose because they are relatively inexpensive, easy to maintain and breed, and large amounts of data are available with regard to the chromosomal mapping and linkage analysis of many mouse genes (Nagy et al., 2003).

Methods for producing transgenic mice

There are two methods that are widely used to produce transgenic mice:

1. Transforming ES cells growing in tissue culture with the desired DNA (Kumar et al., 2009) (Fig. 4).
2. Injecting the desired gene into the pronucleus of a fertilized mouse egg (Becker and Jerchow, 2011) (Fig. 4).

Embryonic stem cell method

ES cells are harvested from the inner cell mass (ICM) of mice blastocysts, i.e., the stage of embryonic development when implantation occurs (Bradley and Van-der Weyden, 2006; Capecchi, 2005; Evans, 2005). They can be

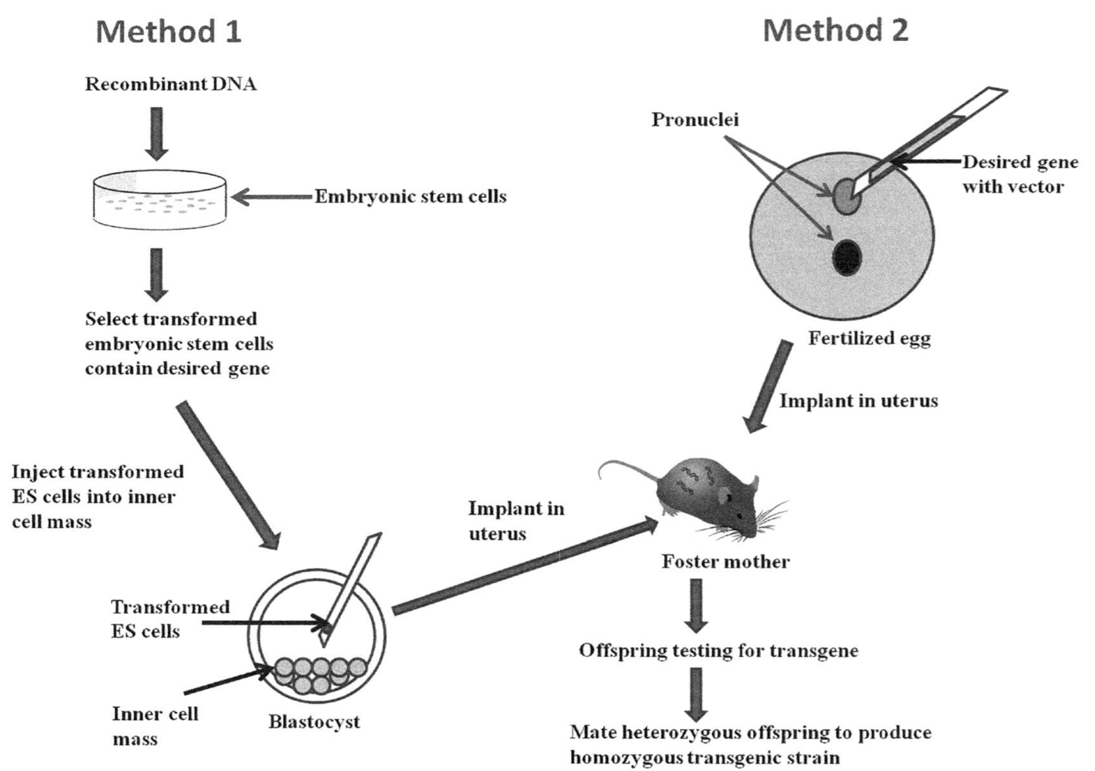

Fig. 4 Methods of producing transgenic mice. Method 1 produces transgenic mice by transforming embryonic stem cells growing in tissue culture with the desired DNA; Method 2 produces transgenic mice by injecting the desired gene into the pronucleus of a fertilized mouse egg.

developed in culture and hold their maximum capacity to create all the cells of the mature animal, including its gametes.

The steps involved in ES cell-mediated transgenesis are:

1. Preparation of recombinant DNA that includes the gene of interest; vector DNA as the carrier for a gene to be inserted into the host DNA molecules; and the promoter and enhancer sequences required for gene expression into host cells.
2. Transformation of ES cells in culture with the exposure of cultured cells to the DNA.
3. Selection of successfully transformed cells.
4. Injection of transformed cells into the ICM of mouse blastocysts.
5. Embryo transfer to a pseudopregnant mouse that is prepared by mating a female mouse with a vasectomized male. The mating process stimulates the hormonal changes needed to make her uterus receptive. The embryos are transferred into her uterus to develop healthy pups.
6. Transgenic offspring testing.
7. Establishment of a transgenic strain.

Pronucleus method

1. Preparation of recombinant DNA as in the previous method.
2. Transformation of fertilized eggs, which includes:
   - Harvesting freshly fertilized eggs before the sperm head has become a pronucleus.
   - Injection of the male pronucleus with recombinant DNA.
   - Two pronuclei are fused to form a diploid zygote nucleus and divided by mitosis to form a two-cell embryo.
3. Transfer of the embryos into a pseudopregnant foster mother.
4. Confirmatory test for the transgene.

Types of transgenic mice

The production of transgenic is much more difficult, as it can take several years to establish a mutation in the ES cells and develop and validate a new transgenic mouse model. Conventional transgenic mice are broadly utilized in preclinical exploration in oncology as well as in other research fields. Transgenic mice can be divided into two groups based on the functional activation and deactivation of the gene.

Knockout mice

Transgenic mice in which the insertion of transgene causes a loss of gene function are called knockout mice. These mice give important clues about the biological function of a normal gene. In translational cancer research, this represents a powerful tool in assessing the potential validity of targeted therapy because the targets can be precisely inactivated in the setting of a developing or developed tumor. Such studies with knockout mice can be applied for the assessment of many gene classes, including oncogenes, tumor-suppressor genes, and metabolic (housekeeping) genes (Walrath et al., 2010). The loss-of-function models may also contain dominant-negative transgenic mice models that carry specific mutations or other types of modifications of gene structure that disrupt the activity of the wild-type gene (Peri et al., 1998). Two types of knockout mice models are frequently used: constitutive and conditional, which can affect a specific target tissue or can occur in a time-controlled manner (Kersten et al., 2017).

In the constitutive knockout model, the target gene is permanently inactivated in the whole animal, in every cell of the organism (Jiang, 2013). The mouse phenotype, such as the physiology, anatomy, behavior, and other observable characteristics, can be accounted for with the help of these types of mouse models. In the observation of cancer development stages, knockout mouse models have played priceless roles in the identification and validation of novel cancer genes. Simple knockouts are frequently intended to lead to a loss of protein function, whereas in cancer, a subset of cancer-causing mutations consistently also results in a gain of function (Maddison and Clarke, 2005).

To remove the limitations of the constitutive knockout model, conditional knockout models emerged. The conditional knockout model can provide better understanding about spontaneous carcinogenesis. In humans, tumors evolve in a wild-type environment and at any time of life. Therefore, the time at which an organism loses its gene function may be a critical factor in disease development. This technology utilizes bacterial Cre and yeast FLP enzymes, which act as site-specific recombinases to catalyze recombination between specific 34-bp loxP and FRT sites, respectively. When Cre or FLP proteins are expressed, homologous recombination is induced between loxP and FRT sites. These sites flank the gene of interest and are oriented in the same direction, which causes the deletion of the gene of interest after the recombination of the flanks of the genetic sequence. The expression of the recombinase can be controlled temporally or spatially. Therefore, we can control the deletion of the gene of interest in temporal and spatial manners, thus overcoming interference due to developmental abnormalities and lethality (Branda and Dymecki, 2004).

Knock-in mice

In this method, an exogenous gene is transferred into the mice genome to analyze the expression pattern of the transgene and/or product function. The insertion of the transgene disrupts the endogenous gene, resulting in knockout/knock-in mice. In knock-in mice, reporter gene (LacZ, green fluorescent protein, etc.) is used to investigate the expression pattern of the exogenous gene. The most common use of knock-in mice is for the creation of a disease model organism. Gain-of-function studies are often used to study oncogenes in mouse models. Oncogene-driven carcinogenesis under in vivo conditions can be studied easily with the help of knock-in models. Gene knock-in has allowed scientists for the first time to conduct a concept-based investigation about gene modifications and resultant phenotypes. Two types of knock-in mice– the constitutive random insertion model and the conditional knock-in model–are produced to understand the gain of function.

The constitutive random insertion model can be produced by the pronuclear injection of the transgene directly into fertilized oocytes or by viral vector-based transfection with early mouse embryos. The exogenous gene is then integrated randomly into the genome of the embryonic cell. Although the procedure is very straightforward and relatively simple, the random incorporation into the genome is the main drawback of this model because it can result in an undesirable expression level or spatiotemporal distribution of transgene activity, or even deleterious effects, thus limiting the usefulness of the model. A study done by Konishi et al. (2007) reported how oncogenes such as *K-ras* drive tumorigenesis in vivo with the use of the random insertion model.

The constitutive gene knock-in model leads to lethality, sterility, and abnormal developmental defects. Hence, conditional knock-in models are generated by using tissue-specific promotors. In vivo models have been generated using knock-in alleles in which missense mutations are introduced. Critically, these studies examine the in vivo and physiological functions of p53. The human p53 gene can be mutated by exposure to benzo(*a*)pyrene (BaP) and then the mutated p53 gene can be inserted into mouse genomes to create the in vivo model. Lung tumors observed in the knock-in mice offer support for the hypothesis of BaP's carcinogenicity (Liu et al., 2005; Iwakuma and Lozano, 2007). Intravital microscopy is a valuable tool for assessing the dynamic changes in the tumor vasculature as well as following the transcriptional targeting of gene expression in various tissues (Kamensek et al., 2013).

Cow

The first transgenic cow was produced in 1997. Cows are an attractive target for transgenesis because the transgene will be expressed in the cow's milk, which will contain proteins of therapeutic benefit (to treat human diseases) and can easily be isolated. Transgenic cows contain a transgene that may come from the same species or different species. When transferred at the embryonic level, every cell in the transgenic cow has the transgene. However, it's expression is only present in mammary tissue. This means that the transgene's products will only be found in the cow's milk and can only be extracted from there. The expression of the transgene is controlled by the specific lactating mammary cell promoter in mammary tissue.

Important steps for the production of the desired protein include the identification of the desired gene, the construction of the recombinant vector, and its expression into a suitable host. The constructed gene is then transferred into female bovine (cow) cells by transfection and transformed cells are selected. The constructed gene contains an antibiotic-resistance gene, a tissue-specific promoter sequence, and a stop sequence. After transfection, the transformed cells were selected with the addition of an antibiotic. Only the transgenic bovine cells that contain an antibiotic-resistance gene will survive. Cells without the antibiotic gene will have no resistance to the antibiotic and will die. Our initial strategies were based on microinjection but recently, the nuclear transfer method has been adopted for transgenesis. This method has a high efficiency of generating transgenic animals as well as product consistency and time of product development (Brink et al., 2000). The transgenic bovine cell is fused with a bovine enucleated oocyte. After fusion, the transgenic cell is reprogrammed to direct development into an embryo, which is implanted into a recipient cow.

One of the most important aspects of transgenic cow production research is that transgenic cows must pass their transgenes to subsequent generations. If a transgenic cow and bull mate, she will have a higher chance of having transgenic offspring. However, if a transgenic cow is mated with a non-transgenic bull, there will be only a 50% chance of having transgenic offspring. Transgenic cows have been produced to fulfill a number of applications in biomedicine, agriculture, animal health, and environmental sustainability. AgResearch gained approval in 2010 to extend its research on transgenic cows. Proteins of therapeutic benefit could be produced in large quantities with low cost in comparison to other transgenic animals.

Experimental results showed that higher concentrations of beta and kappa casein proteins were reported when researchers inserted extra copies of genes responsible for the production of these two proteins into cows to increase the cheese per liter with better quality. The beta lactoglobulin (BLG) protein of milk causes allergic reactions, i.e., milk intolerance, preventing allergic people from consuming milk products. Scientists used antisense technology and inserted DNA sequences that suppress the production of BLG protein in milk. Lactoferrin is an iron-binding protein that plays a significant role in preventing bacterial infections. The human lactoferrin gene has been transferred to cows, and transgenic cows showed expression.

Transgenic fish

Attempts to produce transgenic fish started in 1985. Gene transfers have been successfully done in several fish such as the common carp, rainbow trout, catfish, salmon, goldfish, loach, tilapia, zebrafish, medaka, etc. In fish, the pronuclei are not readily seen under a microscope after fertilization. Therefore, linearized transgene DNA is microinjected into the cytoplasm of either the fertilized egg or embryo that has reached the four-cell stage of development. Unlike, mammalian embryonic development, fish egg development is external so there is no need for implantation. The rate of survival of microinjected embryos varies from 35% to 80%, of which 10%–70% may be transgenic. Electroporation is much less difficult than microinjection and can be used to transfect a large number of eggs within a limited period of time. About 20% or more of the embryos surviving electroporation give rise to transgenic fish. The genes introduced by microinjection in fish include human or rat genes for growth hormone; chicken genes for delta-crystalline protein; the *E. coli* gene β-glactosidase for neomycin resistance; the winter founder gene for antifreeze protein; and the rainbow trout gene for growth hormone. The human growth hormone gene transferred to transgenic fish increased the fish to twice the size of non-transgenic fish. In a similar experiment, the antifreeze protein gene was transferred in several cases but the expression level was too low to protect against freeze.

Goat

The first report of transgenic goat production was published from the United States (Freitas et al., 2007). Nowadays, the isolation of human therapeutic proteins from the milk of transgenic animals is a major route to fulfill the world's needs. The American biotechnological company Genzyme Transgenic Corporation was the first to obtain a commercial pharmaceutical product, the anticoagulant antithrombin III, from the milk of transgenic goats (Goldman et al., 2002). Human breast milk contains valuable antibacterial enzymes that are absent in the milk of dairy animals. One of the most important antibacterial enzymes is lysozyme, which destroys harmful bacteria by tearing open the cell wall. This enzyme is also found in the milk, saliva, and tears of all animals, but human breast milk contains about 3000 times more than goat's milk. Researchers from the University of California at Davis created transgenic goats by injecting the human gene for lysozyme which produces goat milk having 68% as much lysozyme as human milk. Researchers suggest that transgenic goats successfully produced milk containing the enzyme lysozyme, and the milk had an antibacterial effect. The modified milk also increased the amount of coliform gut bacteria, which did not have negative health impacts on the animals themselves (Schrock, 2006). In the future, researchers assume that a transgenic animal's milk will provide an immune boost to children to prevent the onset of various diseases in the developing world, where diarrhea kills more than two million people each year.

Sheep

The production of transgenic sheep has been proven difficult compared to mice and lower animals (Nancarrow et al., 1993). The success rates and procedural work for the production of transgenic sheep are far greater by most criteria. However, due to the enormous benefits to human and animal health, the continuation of the work is assured. The first report of transgenic sheep was published by J.P. Simons in 1988 from Edinburgh in Scotland. Two transgenic ewes were produced, each carrying about 10 copies of the human antihemophilic factor IX gene (cDNA) fused with the BLG gene (β-lactoglobulin) because it is necessary for the specific expression of the gene in mammary glands. Transgenic sheep were produced to explore the possibility of use as a good source of growth hormone and meat production. The fusion gene was constructed with mouse metallothionein-I promotor/regulator and ligated to either gene for human growth hormone-releasing factor (mMThGRF) or the structural gene for bovine growth hormone (mMTbGH). After that, several hundred copies of the constructed gene were injected to the pronuclei of Merino sheep embryos.

It is the only published method to produce transgenic sheep regularly. In the experiment, 842 sheep ova were injected with the mMTbGH gene and transferred into recipient ewes; 47 lambs were born. Out of the 47 lambs, only two were transgenic with mMTbGH, and both had bGH mRNA present in the liver, kidney, and gut. In one lamb, the plasma growth hormone was as high as 700 ng/mL (Caird et al., 1989).

Transgenic sheep have also been produced to achieve high quality and quantity of meat with faster growth rates. Other experimental results also revealed that transgenic sheep express the transgene in the mammary tissue when human genes for blood clotting factor IX and alpha 1 antitrypsin are transferred in sheep. The human growth hormone gene has also been introduced into sheep to promote growth and meat production. The keratin protein is found in wool. The genes involved in the biosynthesis of amino acids and important in keratin biosynthesis are being identified and introduced into sheep to improve the quality of wool. The transgenic sheep do express this gene and have increased wool production with improved wool quality. The transgenic sheep also showed improvements in body weight gain, quantity of meat, wool production, feed efficiency, fat composition, etc. However, they also showed several undesirable side effects such as joint pathology, skeletal defects, gastric ulcers, infertility, etc.

Pig

The generation of a transgenic pig is an entirely new way of breeding. However, the rate of transgenic pig production is much lower (usually $<1\%$) than mice (usually 3%–6%). In pigs, DNA microinjection into the pronuclei of zygotes is the most popular method for the transfer of genetic material. This process involves collection, manipulation, microinjection, cultivation, and implantation of early embryos. Gene transfer in pigs can be employed to increased growth and meat production and serve as bioreactors. Transgenic pigs expressed human growth hormone by using genes encoding hormones of the growth hormone cascade (GHRH, GH, IGF—I), which showed improved growth and meat production. However, they also showed several health problems most likely due to the overproduction of growth hormone (Brem, et al., 1988). Several other human genes have been expressed in pigs and their protein products have been harvested. These protein products are either secreted in milk or in blood serum from where these products are isolated. The human proteins expressed in the blood serum are hemoglobin and specific immunoglobins.

Rabbit

The rabbit is quite promising for molecular farming, which aims to produce large quantities of pharmaceutically or biologically important proteins encoded by the transgene. Transgenes are designed to be expressed in mammary tissue so that protein products are secreted in milk. Protein products can be easily harvested from the milk and purified. Human genes such as interleukin 2, growth hormone, tissue plasminogen activator, alpha lactalbumin, etc., are expressed in the mammary tissue and expressed proteins are harvested from the milk.

Applications of transgenic animals

A transgenic animal carries a known sequence of DNA in its cells and transfers it to its offspring. The term "transgenic" is now extended to gene targeting experiments. An example is the production of chimeric or knockout mice in which a gene has been selectively removed from the mice genome (Beardmore, 1997). Sometimes, the gene under investigation is knocked out to examine its effect on the normal body metabolism; examples include the transgenic monkey, smart mouse, youth mouse, super mouse, and influenza-resistant mouse. Transgenic animals carry foreign DNA to express the same kind of protein no matter which animal or microbe (or even plant) is producing it. Many proteins predominantly used for therapeutic purposes have been produced in transgenic animals (Echelard et al., 2006; Lillico et al., 2007). Some of these proteins provide the animals with better disease resistance than their wild-type relatives and counterparts. The potential applications of transgenic technology are to develop new or altered strains of agriculturally important livestock, such as increased growth rate, improved feed utilization, improved milk production and composition, enhanced reproductive performance and prolificacy, and altered cell and tissue characteristics for biomedical research and manufacturing. The development of animal-derived food products with functional components that improve food safety or nutritional value for improving human health is another promising application of transgenic animals. For example, increasing the dominance of valuable fatty acids in pork tissue or the potential to reduce the amount of deleterious fats and cholesterol in milk has been demonstrated and met with dietary recommendations for cardiovascular health (James et al., 2010). Thus, transgenic technology has the potential to transfer nutritionally beneficial traits to other transgenic products (Wheeler, 2007).

Enhancement of milk production

Recombinant DNA technology has the ability to change the milk composition or to produce entirely new proteins in milk. These changes lead to increased potential uses of milk. The improved livestock growth or survivability via the modified milk composition needs the production of transgenic animals that produce large quantities of milk that contain valuable proteins or that produce milk of higher nutrient content (Noble et al., 2002). The most important nutrient components in milk are protein, fat, and lactose sugar. The growth and health of the developing offspring improve by enriching any of these components. Transgenesis may add certain proteins or growth factors that are deficient in milk (Wall et al., 1991). Some of these proteins are growth factor beta (TGF-β), epidermal growth factor (EGF), insulin-like growth factor 1 (IGF1), and lactoferrin (Zhang et al., 2008). Furthermore, several active proteins have been expressed in the milk of transgenic animals, such as human lysozyme in cattle (Yang et al., 2011) and goats (Jackson et al., 2010), human lactoferrin in mice (Kim et al., 1999) and cattle (Yang et al., 2008), and porcine lactoferrin in mice (Wu et al., 2007). The human lactoferrin expressed in transgenic cows was granted generally recognized as safe (GRAS) status by the USA, FDA, 2005

It is also possible to produce antibodies in the mammary glands of animals. These antibodies are capable of preventing mastitis in sheep, goats, and cattle and mastitis-metritis-agalactia (MMA) in pigs. The antimicrobial peptide lysozyme is also engineered to secrete in milk (Maga et al., 2006). Transgenic animals are considered bioreactors for the expression of the recombinant proteins that are secreted into milk. These animals permit the production of recombinant proteins in fairly large amounts with reduced costs than the production of proteins in eukaryotic cell cultures. Some commercially available drugs (Table 1) such as recombinant human antithrombin III (Atryn) produced from the milk of transgenic goats and the recombinant human C1 esterase inhibitor (Ruconest) obtained from rabbit milk are novel applications of transgenic animals (Maksimenko et al., 2013).

Protection of animals against diseases

Pathogens/infectious diseases have negative effects on livestock production, both in terms of animal welfare and economics. The economic losses due to disease are estimated to be 17% in the developed world and 35%–50% in developing countries. The application of genetic engineering in livestock

Table 1 Mammary gland expression of transgenic proteins.

Protein expressed	Promotor	Host	References
α -lactalbumin	Bovine α –lactalbumin	Pig	Bleck et al. (1998)
Bovine β and k-casein	Bovine β-casein	Cattle	Brophy et al. (2003)
IGF-1	Bovine α –lactalbumin	Pig	Donovan et al. (2001)
IGF-1	Bovine α s1-casein	Rabbits	Wolf et al. (1997)
Lactoferrin	Bovine α s1-casein	Cattle	Krimpenfort et al. (1991)
Lysostaphin	Ovian β-lactoglobulin	Cattle	Wall et al. (2005)
Lysozyme	Bovine α s1-casein	Goat	Maga et al. (2006)

to improve resistance against disease is a very striking idea. Disease resistance is augmented by manipulating natural genes or introducing foreign genes using transgenesis. Natural disease resistance is generally managed by many genes, which are influenced by the environment. There are several stages of infection at which these genes may act, such as initial infection, pathogen multiplication, release, and infection of other hosts. A major objective of genomics in livestock is the identification of endogenous natural resistance genes or genes that augment the immune response (Muller and Brem, 1998). It is now an easy task to transfer resistance genes between species (intraspecies) or to amend disease resistance genes and improve their function. To develop or enhance disease resistance through transgenic approaches, the transfer of major histocompatibility complex (MHC) genes, immunoglobulin (Ig) genes, T cell receptor (TCR) genes, genes that affect lymphokines, or specific disease-resistance genes is very common. The transfer of MHC genes in animals through nuclear transfer or the embryonic cell technique has a useful effect on disease resistance for livestock producers. Transgenic dairy cows that secrete lysostaphin (an antibacterial enzyme) into their milk show higher resistance to mastitis because lysostaphin kills the bacterium *Staphylococcus aureus* in a dose-dependent manner (Dunham and Devlin, 1999). Similarly, the immunoglobulin A (Ig A) gene has been successfully introduced into sheep, pigs, and mice to increase resistance against various oral infections. The production of cattle lacking the prion protein is possible, and this can stop the infection and transmission of spongiform encephalopathies such as bovine spongiform encephalopathy or scrapie (Richt et al., 2007). Influenza-resistant pigs are produced after the transfer of mice fibroblast cell lines that contain the protein that is resistant to influenza. Although attempts to produce disease-resistant animals (Table 2) such as influenza-resistant pigs have been successful, a limited number of genes are presently identified to confer disease resistance in farm animals (Vithani, 2008).

Table 2 Livestock animals with disease-resistant transgenes.

Gene target	Disease/pathogen	Gene transfer method	Target species	References
Mouse SP110	Bovine TB	Somatic cell nuclear transfer	Cattle	Wu et al. (2015)
PrP	Bovine spongiform encephalopathy	Somatic cell nuclear transfer	Cattle	Richt et al. (2007)
Flu 5′ 3′NCR	Influenza A	Lentivector	Chickens	Lyall et al. (2011)
Lysozyme	Mastitis	Pronuclear injection	Goats	Maga et al. (2003)
FMDV3D	Foot and mouth disease virus	Somatic cell nuclear transfer	Goats	Deng et al. (2013)
Porcine Mx1	Influenza A and classical swine fever virus	Somatic cell nuclear transfer	Pigs	Yan et al. (2014)
Murine Mx1	Influenza A	Pronuclear injection	Pigs	Muller et al. (1992)
BVDV	Bovine viral diarrhea virus	Somatic cell nuclear transfer	Sheep	Ni et al. (2015)
1D/VP1	Foot and mouth disease virus	Somatic cell nuclear transfer	Sheep	Hu et al. (2015)

The levels of antimicrobial agents such as lysozyme and lactoferrin in human milk are much higher than in bovine milk. The expression of the human lysozyme gene in mice achieved a significant reduction of bacteria and reduced the frequency of mammary gland infections. Human lactoferrin expressed in the milk of transgenic mice and cattle at high levels is correlated with an increased resistance against mammary gland infections (Niemann et al., 2005). Disease resistance is not only advantageous for livestock producers and their animals but also for consumers, as this offers safer animal products in the market and fewer incidences of human contagious diseases such as avian influenza (Alison and Davis, 2009). A transgenic mouse that secretes recombinant antibodies in milk to neutralize the coronavirus causes contagious gastroenteritis (GEV) in pigs (Castilla et al., 1998). Marek's disease, a disorder in poultry, is believed to be a promising target for transgenesis. Marek's disease is a herpes virus-induced cancer of the lymphoid system. It is a foremost safety problem because it can destroy poultry flocks and give rise to anemia, lameness, and chronic wasting in birds. Transgenic technology was also used to produce sheep resistant to the maedi-visna virus (MVV), an HIV-like virus. This disease is endemic worldwide in sheep, leading to serious production losses due to encephalitis, arthritis, and pneumonia (Clements et al., 1994).

Enhancement of nutritional value

For thousands of years, livestock animals have been farmed for human food production and have been the primary source of nutritional carbohydrates, proteins, lipids, vitamins, and minerals from meat and milk. The henetic alteration of livestock promises to improve public health via increased nutrition. Farmers have better livestock to provide nutritious and cost-effective animal products. Although milk and meat are high-quality food sources for humans, the effects of upgrading and changing farming systems and public lifestyles create new challenges for providing a healthy diet with high nutritional quality. Improved understanding about how human nutritional necessities relate to health and longevity provides new prospects to enhance the nutritional quality of food products and finally improve human health and well-being. Animal fats are relatively low in beneficial unsaturated fats but high in saturated fats. To improve the nutritional quality of animal-derived food products, higher levels of unsaturated fats and lower levels of saturated fats are necessary. Transgenic animal technology is a more targeted approach to introduce specific DNA fragments that are known to impact defined phenotypic traits to improve existing characteristics. Unlike traditional breeding and selection, the manipulation of livestock by genetic engineering is not limited by the species barrier. Transgenic technology can utilize the gene pool of unrelated species to introduce entirely novel and distinctive characteristics. Additionally, transgenic technology is a multipurpose technology that can be used for strategies such as the introduction of a new gene function, the deletion of gene functions, or the replacement of an endogenous gene function with a different one. Thus, it offer great promise as a new tool that can present solutions for several problems related to the efficient production of high-quality, nutrition-rich foods from animals.

The introduction of a novel fatty acid (FA) desaturase activity genetically in mammals with the capacity to synthesize essential *polyunsaturated fatty acid* (PUFA) and get better nutritional characteristics of meat was successful in a transgenic mouse model (Kang et al., 2004). During this experiment, humanized versions of the fat-1 gene were integrated into the mice genome. The expression of the *Caenorhabditis elegans* n-3 FA desaturase in transgenic mice causes muscles to be highly enriched with beneficial omega-3 FAs. Initial transgenic animal studies mainly studied modifying the body composition for the increased production of meat. This was achieved by stimulating muscle growth through the introduction of genes for growth factors such as growth hormone (GH) or insulin-like growth factor I (IGF—I) (Laible, 2012). However, transgenic animals attain only slightly increased growth rates and have a range of harmful side effects due to high systemic GH levels. Poor regulatory control of the transgene activity leads to these consequences (Pursel and Rexroad, 1993). Targeting the foreign gene expression to the skeletal muscle (Pursel et al., 1999) or ensuring a conditional expression approach with the capacity to switch it on or off (Nottle et al., 1999) essentially overcome the problem of harmful side effects in transgenic animals.

Improvement of hair and fiber

The control of the yield, quality, color, and ease of harvesting hair, wool, and fiber for yarn and fabric production is possible due to genetic engineering in livestock. Transgenic methods are used to manipulate the fineness, quality, length, and fold of the wool and hair fiber from goats and sheep . Transgenic methods successfully improve fiber elasticity and strength (Bawden et al., 1998). Presently, the transgenic manipulation of wool focuses on the surface of the fibers. Shrinkage of garments made from these fibers is minimized by decreasing the surface interactions between fibers. Recently, a new approach to produce useful fiber has been achieved using the milk of transgenic goats (Karatzas et al. 1999). Another application of transgenesis is the effort to induce sheep to shed their wool at particular times to ease the need for hand shaving these fiber-producing animals. To achieve this, genes such as epidermal growth factor (EGF) with inducible promoters have been introduced into sheep. When EGF expression is induced, a weak spot is generated in the wool fiber that allows for easy shave off (Wheeler, 2007). These transgenic sheep are produced for superior wool with improved fiber properties. The amino acid cysteine was found to be the limiting amino acid for wool synthesis. To achieve increased production of wool, the transfer of the cysteine biosynthesis bacterial origin gene to the sheep genome is required (Murray et al., 1999).

Spiders that produce orb webs synthesize seven different types of silk for making these webs. Out of these seven types of silk, the most durable one is dragline silk. This silk can be elongated up to 35% and has high tensile properties similar to the synthetic fiber Kevlar. Its energy-absorbing capabilities are superior to those of steel. Medical devices, sutures, ballistic protection, aircraft, automotive composites, tire cords, airbags, and clothing are important potential applications of these fibers.

Enhancement of growth rates

Transgenic animals have been produced that show enhanced growth rate and improved quality of food. Growth hormone (GH) and insulin-like growth factor (IGF) genes are cloned and expressed at different stages in these animals. The production of genetically engineered animals is the key source for understanding the mechanisms of gene action governing growth (Rexroad et al., 1991). Now, it is possible to manipulate growth factors, growth factor receptors, and growth modulators by using transgenic technology. Transgenic livestock and fish have been produced with foreign GH gene that enables us to study the chronic expression of this hormone on growth in mammals and fish. A study has shown that an increase in porcine-produced GH as a result of an exogenous gene contributed to enhanced growth and feed efficiency in pigs. Similarly, a dramatic increase was observed in the growth rate of transgenic Atlantic salmon upon using the gene promoter and growth hormone gene derived from fish species (Hew et al., 1995). The transfer of the chicken ski gene lead to muscular hypertrophy in pigs (Pursel et al., 1999) and cattle (Bowen et al., 1994). The acid meat gene or Rendement Napole gene is responsible for lower yields of pork, as they affect the quality of pig meat. Silencing the expression of this gene in pigs improves the quality of meat. Genes such as GH releasing factor (GHRF) and IGF binding proteins also have significant roles in the modification of growth. Transgenic pigs with a human metallothionein promoter/porcine growth hormone gene construct have been produced that show significant improvements in traits such as growth rate, feed conversion, and body fat/muscle ratio. Similarly, transgenic pigs with human IGF-I had 10% more carcass lean tissue, 30% larger loin mass, and 20% less total carcass fat. Recently, the first transgenic pigs with the spinach desaturase gene were produced that generated increased quantities of unsaturated fatty acids. This transgenic pig contains a higher ratio of unsaturated to saturated fatty acids in their striated muscle. Thus, they have healthier meat due to the presence of unsaturated fatty acids, reducing the risk of stroke and coronary disease (Niemann et al., 2005). Transgenesis shows the path for introducing beneficial fats such as the omega-3 fatty acids from fish or other animals in livestock (Lai et al., 2006). The receptors such as the low-density lipoprotein (LDL) receptor gene and hormones such as leptin are also possible targets that would decrease fat and cholesterol in animal products ().

Reduced environmental impact

Farm animals release waste/pollutants into their environment that adversely affect our environment. Transgenic animals present an alternative way of protecting our environment by improving the efficiency of farming and reducing pollution. Improving efficiency and productivity through gene transfer could offer the use of limited land and water resources, thus protecting the valuable soil and groundwater. Transgenic animals are involved in decreasing phosphorous and nitrogen pollution in the Chesapeake Bay watershed or in the aquifers in hog and poultry producing areas such as Minnesota, North Carolina, and Arkansas. One of the examples is Enviropig, produced by genetic engineering (Golovan et al., 2001). Such transgenic pigs can easily digest plant feed due to an additional gene (Forsberg et al., 2013). Although pigs are omnivorous, they cannot digest a phosphate-rich compound, phytic acid, present in plants. Soil bacteria easily digest phytic acid rejected by pigs, leading to an excessive release of phosphate into the environment. The transfer of a bacterial phytase gene into pigs allows them to digest a larger proportion of the phytic acid. The phytase enzyme is secreted in the saliva of transgenic pigs as early as 7 day of age and they use the phosphate for their growth (Golovan et al., 2001). Thus they reduce their need for dietary supplements and also cut environmental pollution. AquAdvantage, a transgenic salmon with improved feed efficiency, will contribute to a major reduction in the environmental pollution of aquaculture while producing a safe, healthy food.

Ethics in transgenic animal production

In the past three decades, technologies to produce transgenic animals have revolutionized the biomedical field. But advancements in these technologies do not mean that they should be utilized with sacrifice of life. There should be ethics for the use of transgenic animals. The development and increasing requirement of transgenic models has transformed the face of biomedical research and engineering. However, the ability to create new species containing a transgene, whether it is for research or economic advancement, will necessarily raise moral obligations and critical ethical questions to the bioengineer against new genetic lineages. To date, with the use of new research technologies, scientists have successfully created transgenic sheep, cattle, goats, chicken, fish, mice, etc. (Shuldiner, 1996). These transgenic animals serve as modal organisms to study various human diseases such as Alzheimer's disease, Huntington's disease, neuromuscular disease, HIV/AIDS, diabetes mellitus, cardiovascular disease, angiogenesis, cancer, and metabolism, underlying the pathophysiology and developmental process of novel therapeutic interventions (Bagle et al., 2013).

The first document to guide research involving animals was published in 1985. This document was revised by an international cooperative of veterinarians, more than 30 professional societies, scientists, and biomedical experts. In December 2012, the Council for the International Organization of Medical Sciences (CIOMS) and the International Council for Laboratory Animal Science (ICLAS) released a guideline about the use of animals for research, the *International Guiding Principles for Biomedical Research Involving Animals*. The goal of this revision of previous guidelines was to prepare more regulated and ethical guidelines for the welfare of laboratory animals. The social opinions on transgenic research are divided almost in the middle. Opinion surveys in the United States, Japan, and New Zealand reveal that only 42%, 54%, and 58%, respectively, of the people participating in the survey favor such research. The Indian government has set up the Genetic Engineering Approval Committee (GEAC) regarding the validity of research involving genetically modified organisms (GMOs). This organization makes decisions about the use of GMOs in research and addresses the safety of GMOs.

Different kinds of technologies have been improved in the area of animal biotechnology that provide a great opportunity to human beings. With the use of these advanced technologies, now biotechnologists are able to produce more nutritious food and medicine. Despite these contributions to human welfare, the application of animal biotechnology has raised different controversies regarding human health with the use of transgenic products as well as different ethical and environmental challenges. If we concentrate only on the human benefits with animal biotechnology, we may not find any problem in it in terms of the consequences. But, there are also dark sides of biotechnology (as well as animal biotechnology) that cannot be ignored. People feel that the production of oncomice, i.e.mice genetically modified to inevitably develop cancers, is an ethical and moral issue because no one can feel the pain of oncomice due to cancer. In a genetic rearrangement, in the nonsexual or sexual exchange or in the laboratory, unavoidable suffering of animals is beyond description. In a research study, it was found that animals suffer from infections from rotaviruses, damage their intestinal mucosa, and have heavy diarrhea. In a similar experiment, the insertion of additional gene copies of growth hormone into pigs showed faster growth of the offspring, but the offspring also suffered from severe arthritis, which seriously affects their health (BABAS, 1999). An Australian proposal to release transgenic bacteria to minimize the poisonous effect of fluoroacetate was rejected by the Genetic Manipulation Advisory Committee after widespread criticism on environmental grounds.

Nuclear power can contribute positively to the well-being of mankind, but at the same time it can destroy innumerable lives. In a similar way, biotechnology also has some evil purposes that would cause "incalculable risks for human integrity, well-being, and freedom" (Becker, 1996). J. Thomson focuses on the unintended consequences and ethical concerns of modern biotechnology that can be transformed into moral concerns (Thomson, 1996). Recently, a conceptual study has shown that any sort of animal biotechnology, specifically the use of transgenic animals, is morally unacceptable because of its unnaturalness (AEBC, 2002). The European Commission also agreed that animal biotechnology is "unnatural." Somehow, the term "natural" gives satisfaction to the user that it is normal, right, appropriate, and suitable. On the other hand, the term "unnatural" indicates something that is man-made or artificial (BABAS, 1999).

Future prospects

History is filled with the use of animals for the welfare of humanity. Transgenic animals that carry genes from other species have great potential to contribute to human welfare. Transgenic technology is an important tools that will provide future challenges for increased animal-based products (Wall, 1996; Rajoriya et al., 2013). Advancements in the field of genetic engineering allowed scientists to carry out the transfer of transgenes into the animal genome by genetic engineering. In the coming years, genetically modified animals will play a significant role in biomedical engineering and medicine, especially in animal disease models, antibody production, drug development, xenotransplantation, gene pharming, blood replacement, and production of nutritious products. Transgenic technology has revolutionized the formation of biopharmaceutical products with efficiencies far superior than any conventional microbial or cell-culture production system. As the complete genomic sequences of all farm animals become available, it will be possible to improve targeted genetic modification in animal-breeding programs.

There are several techniques for the production of transgenic animals, including DNA microinjection, embryonic stem cell-mediated gene transfer, retrovirus-mediated gene transfer, transposons, nuclear transfer, transfection of gametes, artificial chromosome-mediated gene transfer, etc. But recent results show that these techniques still have low success rates and high rates of prenatal mortality and variable transgenic expression. It will require screening transgenic products before generalizing their applications (Edwards et al., 2003). Considerable effort and time are required to propagate transgenic animal genetics into commercial levels. The transgenic efficiency and precise control of gene expression are the key limiting factors for the production of transgenic animals (Sharfudeen et al., 2014). During genetic engineering procedures, large numbers of embryos do not survive, and of those that do survive, only a small proportion carry the genetic alteration of interest. This means that to obtain genetically engineered animals, large numbers of animals are produced. This creates contradictions to our concepts about the minimal use of animals in research programs. Although there are a number of key limiting factors for the production of transgenic animals, we have successfully transferred transgenes into most species, including cattle, sheep, goats, pigs, rabbits, chickens, fish, etc. Despite further challenges, appropriate lines of transgenic pigs are likely to be available as organ donors within the next 5–10 years (Tadesse and Koricho, 2017). Kozak (2002) revealed that animals produced in this way would help to fight famine and save human health and life. Yongzhong (2001) stated that we can use transgenic animals in the study of gene expression and its control, the production of valuable medicines, the creation of animal models for human disease, breeding new strains of farm animals, and so on.

Until the 1980s, there were no such efficient and cost-effective methods for the production of biopharmaceuticals. In 2006, GTC Biotherapeutics produced the first therapeutic product, ATryn, which a recombinant form of human antithrombin III in the milk of transgenic livestock. It created a milestone in the production of biopharmaceuticals. In the near future, investor confidence and the acceptability of transgenic products in society will establish the production of transgenic animals (Schnieke, 2009). Human welfare and ethical concerns will also determine the acceptability of genome editing and genetic manipulation to consumers. There are some possible risks for human health due to the consumption of transgenic products. Transgenic animal protein products may contain new allergens, which could trigger allergic reactions in humans. They could also produce new toxins that will cause unexpected interactions with other biomolecules.

Conclusion

Nowadays, transgenic animal technology is a leading biotechnology that involve the integration of a transgene into the genome of the targeted animal and the expression of the transgene in the offspring of the genetically modified organism. But when we are going to introduce new, sophisticated techniques such as genetic modification of animals at large scales, moral as well as ethical issues have to be addressed. It becomes necessary to give evidence and clear demonstrations to the general public about the benefits of new products. The recent developments in reproductive technologies such as somatic nuclear transfer, lentiviral transfer of oocytes and zygotes, in vitro production of embryos, sperm sexing, etc., provide a new height to animal breeding programs. Veterinarians as well as animal scientists must be aware of the techniques involved in animal transgenesis and the application of the product so that they can give clear explanations about transgenic products to consumers. All the developments in animal biotechnology will bring great changes in medicine, health, livestock improvement, organ culture for human transplantation, and economic and social benefits from the production of bioreactors. There are several controversies involved in the production of transgenic animals such as whether making them is ethical, whether transgenic food will be safe for humans, and whether it has a positive impact on environment. Scientists are assuming that in the next 8–10 years, transgenic animals will play significant and important roles in human health and well-being.

References

AEBC, 2002. Agriculture and Environment Biotechnology Commission (AEBC). Animals and Biotechnology, AEBC, London. Available in www.aebc.gov.uk/aebc/pdf/aebc0117.pdf.

Alison, V., Davis, U., 2009. Potential Effects of Biotechnology on Animal Health and Well-Being; Applications of Animal Biotechnology in Animal Health, Department of animal science. University of California, California, pp. 1–5.

BABAS, 1999. Ethical Aspects of Agricultural Biotechnology, Bioethical Aspects of Biotechnology in the Agro Food Sector. Cambridge Biomedical Consultants, The Hague.

Bagle, T.R., Kunkulol, R.R., Baig, M.S., More, S.Y., 2013. Transgenic animals and their application in medicine. Int. J. Med. Res. Health Sci. 2 (1), 107–116.

Baldassarre, H., Keefer, C.L., Wang, B., Lazaris, A., Karatzas, C.N., 2003. Nuclear transfer in goats using *in vitro* matured oocytes recovered by laparoscopic ovum pick-up. Cloning Stem Cells 5, 270–285.

Bawden, C.S., Powell, B.C., Walker, S.K., Rogers, G.E., 1998. Expression of a wool intermediate filament keratin transgene in sheep fiber alters structure. Transgenic Res. 7, 273–287.
Beardmore, J.A., 1997. Transgenics: autotransgenics and allotransgenics. Transgenic Res. 6, 107–108.
Becker, G., 1996. Biotechnology—the new ethical frontier: an introduction. In: Gerhold, K. (Ed.), Changing Nature's Course :The Ethical Challenge of Biotechnology. Hong Kong University Press, Hong Kong. Becker associate with James P. Buchanan.
Becker, K., Jerchow, B., 2011. Generation of transgenic mice by pronuclear microinjection. In: Pease, S., Saunders, T. (Eds.), Advanced Protocols for Animal Transgenesis. Springer Protocols Handbooks. Springer, Berlin, Heidelberg.
Bleck, G.T., White, B.R., Miller, D.J., Wheeler, M.B., 1998. Production of bovine α-lactalbumin in the milk of transgenic pigs. J. Anim. Sci. 76, 3072–3078.
Bowen, R.A., Reed, M., Schnieke, A., Seidel, G.E., Stacey, A., Thomas, W.K., Kaijkawa, O., 1994. Transgenic cattle resulting from biopsied embryos: empression of c-ski in a transgenic calf. Biol. Reprod. 8, 664–668.
Bradley, A., Van-der Weyden, L., 2006. Mouse chromosome engineering for modeling human disease. Annu. Rev. Genomics Hum. Genet. 7, 247–276.
Branda, C.S., Dymecki, S.M., 2004. Talking about a revolution: the impact of site-specific recombinases on genetic analyses in mice. Dev. Cell 6 (1), 7–28.
Brem, G., Brenig, B., Müller, M., Kräußlich, H., Winnacker, E., 1988. Production of transgenic pigs and possible application to pig breeding. BSAP Occas. Publ. 12, 15–31.
Brink, M.F., Bishop, M., Pieper, F.R., 2000. Developing efficient strategies for the generation of transgenic cattle which produce biopharmaceuticals in milk. Theriogenology 53, 139–148.
Brophy, B., Smolenski, G., Weeler, T., Wells, D., Huiller, P.L., Labile, G., 2003. Cloned transgenic cattle produced milk with higher levels of b-casein and k-casein. Nat. Biotechnol. 21 (2), 157–162.
Caird, E., Rexroad, J., Hammer, R.E., Bolt, D.J., Mayo, K.E., Frohman, L.A., Palmiter, R. D., Brinster, R.L., 1989. Production of transgenic sheep with growth-regulating genes. Mol. Reprod. Dev. 1 (3), 164–169.
Campbell, K.H., McWhir, J., Ritchie, W.A., Wilmut, I., 1996. Sheep cloned by nuclear transfer from a cultured cell line. Nature 380 (6573), 64–66.
Capecchi, M.R., 2005. Gene targeting in mice: functional analysis of the mammalian genome for the twenty-first century. Nat. Rev. Genet. 6, 507–512.
Castilla, J., Printado, B., Sola, I., Sanchez-Morgado, J.M., Enjuanes, L., 1998. Engineering passive immunity in transgenic mice secreting virus-neutralizing antibodies in milk. Nat. Biotechnol. 16 (4), 349–354.
Chan, A.W.S., Luetjens, C.M., Schatten, G.P., 2000. Sperm-mediated gene transfer. Curr. Top. Dev. Biol. 50, 89–102.
Chauhan, M.S., Nadir, S., Baily, T.L., Pryor, A.W., Butler, S.P., Valander, W.H., Gwazdaukas, F.C., 1999. Bovine follicular dynamics, oocyte recovery and development of oocytes microinjected with green fluorescent protein construct. J. Dairy Sci. 82, 918–926.
Chrenek, P., Makarevic, A., 2008. Transgenic rabbits production and application. Slovak J. Anim. Sci. 41, 113–120.
Chrenek, P., Makarevic, A.V., Pivko, H.J., Bulla, J., 2010. Transgenic farm animal production and application: minireview. Slovak J. Anim. Sci. 43 (2), 45–49.
Clements, J.E., Wall, R.J., Narayan, O., Dauer, D., Schoborg, R., Sheffer, D., Powell, A., Carruth, L.M., Zink, M.C., Rexroad, C.E., 1994. Development of transgenic sheep that express the Visna virus envelope gene. Virology 200 (2), 370–380.

Cohen, S., Chang, A., Boyer, H., Helling, R., 1973. Construction of biologically functional bacterial plasmids in vitro. Proc. Natl. Acad. Sci. U. S. A. 70, 3240–3244.

Deng, S., Yu, K., Li, W.I., Yao, Y., Lu, T., Zhao, Y., Yuan, S., Fu, J., Lian, Z., 2013. RNA interference against FMDV*3D* replication in transgenic goat. Sci. Sin. Vitae 43 (5), 404–410.

Ding, S., Wu, X., Li, G., Han, M., Zhuang, Y., Xu, T., 2005. Efficient transposition of the piggyBac (PB) transposon in mammalian cells and mice. Cell 122, 473–483.

Donovan, S.M., Monaco, M.H., Bleck, G.T., Cook, J.B., Noble, M.S., Hurley, W.L., Wheeler, M.B., 2001. Transgenic over expression of bovineα-lactalbumin and human insulin like such as growth factor-I in porcine mammary gland. J. Dairy Sci. 84 (E. Suppl), E216–E222.

Dowdy, S.F., Scanlon, D.J., Fasching, C.L., et al., 1990. Irradiation microcell mediated chromosome transfer (XMMCT): the generation of specific chromosomal arm deletions. Genes Chromosomes Cancer 2, 318–327.

Dunham, R.A., Devlin, R.H., 1999. Comparison of traditional breeding and transgenesis in farmed fish with implications for growth enhancement and fitness. In: Murray, J.D., Anderson, G.B., Oberbauer, A.M., Mc Gloughlin, M.M. (Eds.), Transgenic Animals in Agriculture. CABI Publishing, New York, pp. 209–230.

Ebert, K.M., Schindler, J.E.S., 1993. Transgenic farm animals: progress report. Theriogenology 39, 121–135.

Ebert, K.M., Selgrath, J.P., DiTullio, P., Denman, J., Smith, T.E., Memon, M.A., Schindler, J.E., Monastersky, G.M., Vitale, J.A., Gordon, K., 1991. Transgenic production of a variant of human tissue-type plasminogen activator in goat milk: generation of transgenic goats and analysis of expression. Biotechnology (NY) 9 (9), 835–838.

Echelard, Y., Ziomek, C.A., Meade, H.M., 2006. Production of recombinant therapeutic proteins in the milk of transgenic animals. BioPharm. Int. 19, 36–46.

Edwards, J.L., Schrick, F.N., McCracken, M.D., Amstel, S.R., Hopkins, F.M., et al., 2003. Cloning adult farm animals: a review of the possibilities and problems associated with somatic cell nuclear transfer. Am. J. Reprod. Immunol. 50 (2), 113–123.

Eenennaam, A.V., 2008. Genetically engineered animals: an overview. In: Uc Davis. The University of California, One Shields Avenue, pp. 1–6.

Evans, M., 2005. Embryonic stem cells: a perspective. Novartis Found. Symp. 265, 98–103.

Forsberg, C.W., Meidinger, R.G., Liu, M., Cottrill, M., Golovan, S., Phillips, J.P., 2013. Integration, stability and expression of the E. coli phytase transgene in the Cassie line of Yorkshire Enviropig™. Transgenic Res. 22 (2), 379–389.

Freitas, V.J., Serova, I.A., Andreeva, L.E., et al., 2007. Production of transgenic goat (*Capra hircus*) with human Granulocyte Colony Stimulating Factor (hG-CSF) gene in Brazil. An. Acad. Cienc. 79 (4), 585–592.

Goldman, I.L., Kadulin, S.G., Razin, S.V., 2002. Transgenic goats in the world pharmaceutical industry of the 21st century. Russ. J. Genet. 38, 1–14.

Golovan, S.P., Meidinger, R.G., Ajakaiye, A., Cottrill, M., Wiederkehr, M.Z., Barney, D.J., Plante, C., Pollard, J.W., Fan, M.Z., Hayes, M.A., Laursen, J., Hjorth, J.P., Hacker, R. R., Phillips, J.P., Forsberg, C.W., 2001. Pigs expressing salivary phytase produce low phosphorus manure. Nat. Biotechnol. 19, 741–745.

Gong, G., Dai, Y., Fan, B., Zhu, H., Wang, L., Wang, H., Tang, B., Liu, Y., Li, R., Wan, R., Huang, Y., Li, N., 2004. Production of transgenic calves by somatic cell nuclear transfer. Chin. Sci. Bull. 49, 161–166.

Gordon, I., 1996. Controlled reproduction in cattle and buffalo: production of transgenic animal. vol. 1 Cab International, Wallingford, UK, pp. 338–372.

Grimes, B.R., Schindelhauer, D., McGill, N.I., Ross, A., Ebersole, T.A., Cooke, H.J., 2001. Stable gene expression from a mammalian artificial chromosome. EMBO Rep. 2 (10), 910–914.

Hew, C.L., Fletcher, G.L., Davies, P.L., 1995. Transgenic salmon: tailoring the genome for food production. J. Fish Biol. 47 (Suppl. A), 1–19.
Heyman, Y., Vignon, X., Chesn, P., Bourhis, D.L., Marchal, J., Renard, J., 1998. Cloning in cattle: from embryo splitting to somatic nuclear transfer. Reprod. Nutr. Dev. 38 (6), 595–603.
Houdebine, L.M., 2002. The methods to generate transgenic animals and to control transgene expression. J. Biotechnol. 98, 145–160.
Hu, S., Qiao, J., Fu, Q., Chen, C., Ni, W., Wujiafu, S., Ma, S., Zhang, H., Sheng, J., Wang, P., Wang, D., Huang, J., Cao, L., Ouyang, H., 2015. Transgenic shRNA pigs reduce susceptibility to foot and mouth disease virus infection. Elife 4. https://doi.org/10.7554/eLife.06951, e06951.
Ikeno, M., Grimes, B., Okazaki, T., Nakano, M., Saitoh, K., Hoshino, H., McGill, N.I., Cooke, H., Masumoto, H., 1998. Construction of YAC-based mammalian artificial chromosomes. Nat. Biotechnol. **16**, 431–439.
Iwakuma, T., Lozano, G., 2007. Crippling p53 activities via knock-in mutations in mouse models. Oncogene 26 (15), 2177–2184.
Jackson, K.A., Berg, J.M., Murray, J.D., Maga, E.A., 2010. Evaluating the fitness of human lysozyme transgenic dairy goats: growth and reproductive traits. Transgenic Res. 1 (6), 977–986.
Jaenisch, R., Mintz, B., 1974. Simian virus 40 DNA sequences in DNA of healthy adult mice derived from preimplantation blastocysts injected with viral DNA. Proc. Natl. Acad. Sci. U. S. A. 71, 1250–1254.
Jaenisch, R., Fan, H., Croker, B., 1975. Infection of preimplantation mouse embryos and of newborn mice with leukemia virus: tissue distribution of viral DNA and RNA and leukemogenesis in the adult animal. Proc. Natl. Acad. Sci. U. S. A. 72 (10), 4008–4012.
James, D., Nuradilla, M., Caitlin, A., Elizabeth, A., 2010. Current status of transgenic animal research for human health applications. Acta Sci. Vet. 38, 627–632.
Jiang, X.C., 2013. Generation of general and tissue-specific gene knockout mouse models. Methods Mol. Biol. 1027, 253–271.
Kamensek, U., Sersa, G., Cemazar, M., 2013. Evaluation of p21 promoter for interleukin 12 radiation induced transcriptional targeting in a mouse tumor model. Mol. Cancer 12 (1), 136.
Kang, J.X., Wang, J.D., Wu, L., Kang, Z.B., 2004. Transgenic mice fat-1 mice convert n-6 to n-3 fatty acids. Nature 427, 504.
Karatzas, C.N., Zhou, J.F., Huang, Y., Duguay, F., Chretien, N., Bhatia, B., Bilodeau, A., Keyston, R., Tao, T., Keefer, C.L., Wang, B., Baldassare, H., Lazaris, A., 1999. Production of recombinant spider silk [BiosteelTM] in the milk of genetically engineered animals. Transgenic Res. 8, 476–477.
Katoh, M., Ayabe, F., Norikane, S., Okada, T., Masumoto, H., Horike, S., Shirayoshi, Y., Oshimura, M., 2004. Construction of a novel human artificial chromosome vector for gene delivery. Biochem. Biophys. Res. Commun. 321 (2), 280–290.
Kazuki, Y., Hoshiya, H., Takiguchi, M., Abe, S., Iida, S., Osaki, M., Katoh, M., Hiratsuka, M., Shirayoshi, Y., Hiramatsu, K., Ueno, E., Kajitani, N., Yoshin, T., Kazuki, K., Ishihara, C., Takehara, S., Tsuji, S., Ejima, F., Toyoda, A., Sakaki, Y., Larionov, V., Kouprina, N., Oshimura, M., 2011. Refined human artificial chromosome vectors for gene therapy and animal transgenesis. Gene Ther. 18, 384–393.
Kersten, K., de Visser, K.E., van Miltenburg, M.H., Jonkers, J., 2017. Genetically engineered mouse models in oncology research and cancer medicine. EMBO Mol. Med. 9 (2), 137–153.
Kim, S.J., Sohn, B.H., Jeong, S., Pak, K.W., Park, J.S., Park, I.Y., Lee, T.H., Choi, Y.H., Lee, C.S., Han, Y.M., Yu, D.Y., Lee, K.K., 1999. High level expression of human lactoferrin in milk of transgenic mice using genomic lactoferrin sequence. J. Biochem. 126 (2), 320–325.

Kohler, S.W., Provost, G.S., Fieck, A., Kretz, P.L., Bullock, W.O., Putman, D.L., Sorge, J.A., Short, J.M., 1991. Analysis of spontaneous and induced mutations in transgenic mice using a lambda ZAP/lacI shuttle vector. Environ. Mol. Mutagen. 18, 316–321.

Konishi, H., Karakas, B., Abukhdeir, A.M., Lauring, J., Gustin, J.P., Garay, J.P., Garay, J.P., Konishi, Y., Gallmeier, E., Bachman, K.E., Park, B.H., 2007. Knock-in of mutant K-ras in nontumorigenic human epithelial cells as a new model for studying K-ras-mediated transformation. Cancer Res. 67 (18), 8460–8467.

Kouprina, N., Tomilin, A.N., Masumoto, H., Earnshaw, W.C., Larionov, V., 2014. Human artificial chromosome based gene delivery vectors for biomedicine and biotechnology. Expert Opin. Drug Deliv. 11 (4), 517–535.

Kozak, M., 2002. Transgenic animal—future prospects and threats. Med. Weter. 58 (2), 83–88.

Krimpenfort, P., Rademakers, A., Eyestone, W., van der Schans, A., van den Broek, S., Kooiman, P., Kootwijk, E., Platenburg, G., Pieper, F., Strijker, R., 1991. Generation of transgenic dairy cattle using 'in vitro' embryo production. Biotechnology 9, 844–847.

Kumar, T.R., Larson, M., Wang, H., McDermott, J., Bronshteyn, I., 2009. Transgenic mouse technology: principles and methods. Methods Mol. Biol. 590, 335–362.

Kuroiwa, Y., Kasinathan, P., Choi, Y.J., Naeem, R., Tomizuka, K., Sullivan, E.J., Robl, J. M., 2002. Cloned trans chromosomic calves producing human immunoglobulin. Nat. Biotechnol. 20 (9), 889–894.

Lai, L., Kang, J., Li, R., Wang, J., Witt, W., Yong, H., Hao, Y., Wax, D., Murphy, C., Rieke, A., Samuel, M., Linville, M., Korte, S., Evans, R., Starzl, T., Prather, R., Dai, Y., 2006. Generation of cloned transgenic pigs rich in omega-3 fatty acids. Nat. Biotechnol. 24, 435–436.

Laible, G., 2012. Transgenic livestock, enhanced nutritional quality in. In: Meyers, R.A. (Ed.), Encyclopedia of Sustainability Science and Technology. Springer, New York, NY.

Lavitrano, M., Busnelli, M., Cerrito, M.G., Giovannoni, R., Manzini, S., Vargiolu, A., 2006. Sperm-mediated gene transfer. *Reprod. Fertil. Dev.* 18 (1–2), 19–23.

Lee, K., Prather, R.S., 2013. Advancements in somatic cell nuclear transfer and future perspectives. Anim. Front. 3 (4), 57–61.

Lillico, S.G., Sherman, A., McGrew, M.J., Robertson, C.D., Smith, J., et al., 2007. Oviduct specific expression of two therapeutic proteins in transgenic hens. Proc. Natl. Acad. Sci. U. S. A. 104, 1771–1776.

Lin, T.P., 1996. Microinfection of mouse eggs. Science 151 (708), 333–337.

Liu, Z., Muehlbauer, Rudolf, K., Schmeiser, H.H., Hergenhahn, M., Belharazem, D., Hollstein, M.C., 2005. p53 mutations in benzo(a)pyrene-exposed human p53 knock-in murine fibroblasts correlate with p53 mutations in human lung tumors. Cancer Res. 65 (7), 2583–2587.

Lyall, J., Irvine, R.M., Sherman, A., McKinley, T.J., Nunez, A., Purdie, A., Outtrim, L., Brown, I.H., Rolleston-Smith, G., Sang, H., Tiley, L., 2011. Suppression of avian influenza transmission in genetically modified chickens. Science 331 (6014), 223–226.

MacLean, N., 2003. Genomic inheritance. In: Hall, B.K., Olsen, W.H. (Eds.), Key concepts in Evolutionary Developmental Biology. Harvard University Press, Cambridge MA, USA, pp. 209–217.

Maddison, K., Clarke, A.R., 2005. New approaches for modelling cancer mechanisms in the mouse. J. Pathol. 205 (2), 181–193.

Maga, E.A., Sargent, R.G., Zeng, H., Pati, S., Zarling, D.A., Oppenheim, S.M., Collette, N. M., Moyer, A.L., Conrad-Brink, J.S., Rowe, J.D., BonDurant, R.H., Anderson, G.B., Murray, J.D., 2003. Increased efficiency of transgenic livestock production. Transgenic Res. 12 (4), 485–496.

Maga, E.A., Cullor, J.S., Smith, W., Anderson, G.B., Murray, J.D., 2006. Human lysozyme expressed in the mammary gland of transgenic dairy goats can inhibit the growth of bacteria that cause mastitis and the cold-spoilage of milk. Foodborne Pathog. Dis. 3, 384–392.

Makoto, N., Clayton, J., Brinster, et al., 2001. Transgenic mice produced by retroviral transduction of male germ-line stem cells. PNAS 98 (23), 13090–13095.

Maksimenko, O.G., Deykin, A.V., Khodarovich, Y.M., Georgiev, P.G., 2013. Use of transgenic animals in biotechnology: prospects and problems. Acta Nat. 5 (1), 33–46.

McCreath, K.J., Howcroft, J., Campbell, K.H., Colman, A., Schnieke, A.E., Kind, A.J., 2000. Production of gene-targeted sheep by nuclear transfer from cultured somatic cells. Nature 405 (6790), 1066–1069.

McLean, D.J., 2008. Spermatogonial stem cell transplantation, testicular function, and restoration of male fertility in mice. Methods Mol. Biol. 450, 149–162.

Medrano, J.V., Martínez-Arroyo, A.M., Sukhwani, M., Noguera, I., Quiñonero, A., Martínez-Jabaloyas, J.M., Pellicer, A., Remohí, J., Orwig, K.E., Simón, C., 2014. Germ cell transplantation into mouse testes procedure. Fertil. Steril. 102 (4), e11–e12.

Mehta, P., Sharma, A., Kaushik, R., 2017. Transgenesis in farm animals-a review. Agric. Rev. 38 (2), 129–136.

Muller, M., Brem, G., 1998. Transgenic approaches to the increase of disease resistance in farm animals. Rev. Sci. Tech. 17, 365–378.

Muller, M., Brenig, B., Winnacker, E.L., Brem, G., 1992. Transgenic pigs carrying cDNA copies encoding the murine Mx1 protein which confers resistance to influenza virus infection. Gene 121 (2), 263–270.

Murray, J., Oberbauer, A., Mcgloughlin, M., 1999. Transgenic Animals in Agriculture. CABI Publishing, Davis, p. 304.

Nagy, A., Gertsenstein, M., Vintersten, K., Behringer, R., 2003. Manipulating the mouse embryo: a laboratory manual, third ed. Cold Spring Harbor Laboratory, New York.

Nancarrow, C.D., Marshall, J.T.A., Ward, K.A., 1993. Production of transgenic sheep. In: Murphy, D., Carter, D.A. (Eds.), Transgenesis Techniques. Methods in Molecular Biology™. vol 18. Humana Press, pp. 273–303.

Ni, W., Qiao, J., Ma, Q., Wang, J., Wang, D., Zhao, X., Cao, Y., Li, Q., Hu, S., Chen, C., 2015. Development of sheep kidney cells with increased resistance to different subgenotypes of BVDV-1 by RNA interference. J. Virol. Methods 218, 66–70.

Niemann, H., Kues, W., Carnwath, J., 2005. Transgenic farm animals: present and future, Department of Biotechnology, Institute for Animal Breeding (FAL), Mariensee, Germany. Rev. Sci. tech. Off. Int. 24, 285–298.

Noble, M., Rodriguez, S., Bleck, G., Cook, J., Hurley, W., Wheeler, M., 2002. Lactational performance of first parity transgenic gilts expressing bovinelactalb min in their milk. J. Anim. Sci. 80, 1090–1096.

Nottle, M.B., Nagashima, H., Verma, P.J., Du, Z.T., et al., 1999. Production and analysis of transgenic pigs containing a metallothionein porcine growth hormone gene construct. In: Murray, J.D., Anderson, G.B., Oberbauer, A.M., McGloughlin, M.M. (Eds.), Transgenic Animals in Agriculture. CABI, Oxon, UK, pp. 145–156.

Oshimura, M., Katoh, M., 2008. Transfer of human artificial chromosome vectors into stem cells. Reprod. Biomed. Online 16 (1), 57–69.

Palmiter, R.D., Brinster, R.L., Hammer, R.E., Trumbauer, M.E., Rosenfeld, M.G., Birnberg, N.C., Evans, R.M., 1982. Dramatic growth of mice that develop from eggs microinjected with metallothionein-growth hormone fusion genes. Nature 300 (5893), 611–615.

Peri, A.K., Wilgenbus, P., Dahl, U., Semb, H., Christofori, G., 1998. A causal role for E-cadherin in the transition from adenoma to carcinoma. Nature 392 (6672), 190–193.

Pfeifer, A., 2006. Lentiviral transgenesis-a versatile tool for basic research and gene therapy. Curr. Gene Ther. 6, 535–542.

Pursel, V.G., Rexroad, C.E., 1993. Recent progress in the transgenic modification of swine and sheep. Mol. Reprod. Dev. 36, 251–254.

Pursel, V.G., Wall, R.J., Mitchell, A.D., Elasser, T.H., et al., 1999. Expression of insulin like such as growth factor-1 in skeletal muscle of transgenic swine. In: Murray, J.D., Anderson, G.B., Oberbauer, A.M., McGloughlin, M.M. (Eds.), Transgenic Animals in Agriculture. CABI, Oxon, UK, pp. 131–144.

Rajoriya, R., Rajoriya, S., Kumar, N., 2013. Transgenic animals: prospects for improving livestock productivity. J. Bio. Innov. 2 (5), 240–259.

Rexroad Jr., C.E., Mayo, K.M., Bolt, D.J., Elsasser, T.H., Miller, K.F., Behringer, R.R., Palmiter, R.D., Brinster, R.L., 1991. Transferrin and albumin directed expression of growth related peptides in transgenic sheep. J. Anim. Sci. 69, 2995–3004.

Richt, J.A., Kasinathan, P., Hamir, A.N., Castilla, J., Sathiyaseelan, T., Vargas, F., Sathiyaseelan, J., Wu, H., Matsushita, H., Koster, J., Kato, S., Ishida, I., Soto, C., Robl, J.M., Kuroiwa, Y., 2007. Production of cattle lacking prion protein. Nat. Biotechnol. 25 (1), 132–138.

Robertson, E.J., 1991. Using embryonic stem cells to introduce mutations into the mouse germ line. Biol. Reprod. 44 (2), 238–245.

Rutovitz, J., Mayer, S., 2002. Genetically Modified and Cloned Animals; All in a Good Cause. GeneWatch, UK. Retrieved from www.genewatch.org.

Schnieke, A., 2009. Animal pharming: past experience and future prospects. In: Genetic Engineering in Livestock. vol. 34. Springer, Berlin, Heidelberg, pp. 49–67.

Schrock, K., 2006. Transgenic Goat's Milk Kicks Up Immunity. Scientific American News Letter.

Sharfudeen, S., Begum, M.C., Deepthi, C.D.N., Lahari, G., Sulthana, M.R., Revathi, A., Tejaswini, S.S.N., 2014. Transgenic technology: an overview, current status & future perspectives. J. Pharm. Res. 8 (4), 474–485.

Shuldiner, A.R., 1996. Molecular medicine transgenic animals. N. Engl. J. Med. 334 (10), 653–655.

Sumiyama, K., Kawakami, K., Yagita, K., 2010. A simple and highly efficient transgenesis method in mice with the T012 transposon system and cytoplasmic microinjection. Genomics 95, 306–311.

Tadesse, T., Koricho, D., 2017. Biomedical application and future prospects of transgenic animal: review. J. Nat. Sci. Res. 7 (23), 82–88.

Takashima, S., Shinohara, T., 2018. Culture and transplantation of spermatogonial stem cells. Stem Cell Res. 29, 46–55.

Thomson, J., 1996. The Realm of Rights. Harvard University Press, Massachusetts.

USA, FDA, 2005. Notice of GRAS Exemption for Human Lactoferrin Derived From the Milk of Transgenic Cows Expressing a Human Gene Encoding Human Lactoferrin. Food and Drug Administration (USA FDA), Maryland, USA. Available at: http://www.accessdata.fda.gov/scripts/fcn/gras_notices/grn000189.pdf.

Van Keuren, M.L., Gavrilina, G.B., Filipiak, W.E., Zeidler, M.G., Saunders, T.L., 2009. Generating transgenic mice from bacterial artificial chromosomes: transgenesis efficiency, integration and expression outcomes. Transgenic Res. 18 (5), 769–785.

Vijg, J., Douglas, G.R., 1996. Bacteriophage lambda and plasmid lacZ transgenic mice for studying mutations in vivo. In: Pfeifer, G.P. (Ed.), Technologies for Detection of DNA Damage and Mutations. Plenum Press, New York, pp. 391–410.

Vithani, N., 2008. Definition and Introduction to the Transgenic System. http//www.bookrags.com/research/transgenicanimalgene. Accessed on 11 March, 2016.

Wall, R.J., 1996. Transgenic livestock: progress and prospects for the future. Theriogenology 45 (1), 57–68.

Wall, R., Pursel, V., Shamay, A., McKnight, C., Pittius, R., Henninhausen, L., 1991. High-level synthesis of a heterologous milk protein in the mammary glands of transgenic swine. Proc. Natl. Acad. Sci. U. S. A. 88, 1696–1700.

Wall, R.J., Powell, A.M., Paape, M.J., Kerr, D.E., Bannerman, D.D., Pursel, V.G., 2005. Genetically enhanced cows resist intra mammary *Staphylococcus aureus* infection. Nat. Biotechnol. 23, 445–451.

Walrath, J.C., Hawes, J.J., Van Dyke, T., Reilly, K.M., 2010. Genetically engineered mouse models in cancer research. Adv. Cancer Res. 106, 113–164.

Wheeler, M., 2007. Cloning and Transgenic Animals for Agriculture. http://www.Geraembryo.com.br/t.technicos/p4/matthew-wheeler.pdf. Accessed on 8 March, 2016.

Wilmut, I., Whitelaw, C.B.A., 1994. Strategies for production of pharmaceutical proteins in milk. Reprod. Fertil. Dev. 6 (5), 625–630.

Wolf, E., Jehle, P.M., Weber, M.M., Sauerwein, H., Daxenberger, A., Breier, B.H., Besenfelder, U., Frenyo, L., Brem, G., 1997. Human insulin-likesuch as growth factor I (IGF-I) produced in the mammary glands of transgenic rabbits: yield, receptor binding, mitogenic activity, and effects on IGF-binding proteins. Endocrinology 138, 307–313.

Wu, S.C., Chen, H.L., Yen, C.C., Kuo, M.F., Yang, T.S., Wang, S.R., Weng, C.N., Chen, C.M., Cheng, W.T.K., 2007. Recombinant porcine lactoferrin expressed in the milk of transgenic mice enhances offspring growth performance. J. Agric. Food Chem. 55, 4670–4677.

Wu, H., Wang, Y., Zhang, Y., Yang, M., Lv, J., Liu, J., Zhang, Y., 2015. TALE nickase-mediated SP110 knockin endows cattle with increased resistance to tuberculosis. Proc. Natl. Acad. Sci. U. S. A. 112 (13), 1530–1539.

Yan, Q., Yang, H., Yang, D., Zhao, B., Ouyang, Z., Liu, Z., Fan, N., Ouyang, H., Gu, W., Lai, L., 2014. Production of transgenic pigs over-expressing the antiviral gene Mx1. Cell Regen. (Lond) 3 (1), 11. https://doi.org/10.1186/2045-9769-3-11.

Yang, P., Wang, J., Gong, G., Sun, X., Zhang, R., Du, Z., Liu, Y., Li, R., Ding, F., Tang, B., Dai, Y., Li, N., 2008. Cattle mammary bioreactor generated by a novel procedure of transgenic cloning for large-scale production of functional human lactoferrin. PLoS One 3 (10), e3453.

Yang, B., Wang, J., Tang, B., Liu, Y., Guo, C., Yang, P., Yu, T., Li, R., Zhao, J., Zhang, L., Dai, Y., Li, N., 2011. Characterization of bioactive recombinant human lysozyme expressed in milk of cloned transgenic cattle. PLoS One 6 (3), e17593.

Yongzhong, Z., 2001. The application of the transgenic animals and a prospect for their future. J. Shandong. Norm. Uni. (Nat. Sci.) 16 (1), 83–87.

Zhang, J., Li, L., Cai, Y., Xu, X., Chen, J., Wu, Y., Yu, H., Yu, G., Liu, S., Chen, A., Cheng, G., 2008. Expression of active recombinant human lactoferrin in the milk of transgenic goats. Protein Expr. Purif. 57, 127–135.

Zhou, Q., Kyazike, Echelard, J., Meade, Y.H.M., Higgins, E., Cole, E.S., Edmunds, T., 2005. Effect of genetic background on glycosylation heterogeneity in human antithrombin produced in the mammary gland of transgenic goats. J. Biotechnol. 117 (1), 57–72.

CHAPTER NINE

Potential implications of gonadotropin-releasing hormone analogues in assisted reproductive technology

Joseph R.D. Fernandes[a,*], Moitreyi Das[b,*], Kavya Chandra[a], Indrashis Bhattacharya[c], and Arnab Banerjee[a]
[a]Department of Biological Sciences, KK Birla, BITS Pilani, Zuarinagar, Goa, India
[b]Department of Zoology, Goa University, Goa, India
[c]Department of Zoology, HNB Garhwal University, Uttarakhand, India

Introduction

During the past few decades, assisted reproductive technology (ART) has made significant advancement in achieving the genetic propagation of farm animals. ART has shown remarkable progress in livestock management and dairy production, with the first successful artificial insemination (AI) in cattle during 1900s. The impelling force of using AI is its potential to amplify the genetic gain by breeding among elite livestock (Perry, 1960). ART is globally practiced to overcome the reproductive incompetence of farm industries that could lead to a potential economic loss in this era. The reproductive efficiency in farm animals may decline due to persistent anovulation, infertility, inaccurate detection of reproductive cyclicity, and conditions of "silent heat." The progressive applications of AI generate innovative methods such as semen extenders, the invention of the electroejaculator, gamete cryopreservation, progeny testing, in vitro fertilization (IVF), embryo transfer, etc., to increase the yield and efficiency of conception. Additionally, there are numerous tools to assist the conception rate in farm animals, such as supplementations of prostaglandin F2alpha (PGF2a), controlled internal drug release (CIDR), and the analogues of the gonadotropin releasing hormone (GnRH). These fascinating technologies collectively

[*]Joint first author.

Emerging Issues in Climate Smart Livestock Production
https://doi.org/10.1016/B978-0-12-822265-2.00004-1

reduce the generation intervals in cattle and have been proven to be effective in restricting the spread of infectious diseases within the livestock community.

GnRH is a decapeptide that consists of 10 amino acids (pGlu-His-Trp-Ser-Tyr-Gly-Leu-Arg-Pro-Gly-NH_2), and is conserved in their lengths across all vertebrates. GnRH was also found in several invertebrates, emphasizing its importance over a million years of evolution (Minakata, 2010). GnRH that stimulates hypophysiotropic gonadotropin release was isolated and designated as GnRH-I (mammal GnRH) (Sealfon et al., 1997). Another highly conserved and early evolved GnRH discovered in the chicken was designated GnRH-II (chicken GnRH) (Miyamoto et al., 1984). Also, a third isoform found in salmons was named GnRH-III (salmon GnRH) (Montaner et al., 1998). The GnRH neuronal population is mainly present in the mediobasal region, the preoptic area (POA), and the infundibular and arcuate nucleus (ARC). Synthesized GnRH binds to the GnRH-receptor present on the gonadotropic cells of the anterior pituitary. Further, it stimulates the secretion of the gonadotropin-like luteinizing hormone (LH) and the follicle-stimulating hormone (FSH). GnRH secretion involves positive and negative feedback regulatory loops. The hypothalamic GnRH neurons act as a pulse generator that eventually drives both the basal (pulsatile) and tonic (surge) release of GnRH to regulate gonadal functions (Herbison, 2016; Plant, 2015). At present, there are two major regulatory mechanisms of GnRH pulse generation. The first proposes a role of the intrinsic pulsatility of the GnRH neuron, whereas the second concept suggests that the neurons in the ARC nuclei are responsible for pulse generation (Herbison, 2016; Plant, 2015). With the discovery of kisspeptin in 2003, this latter has been established as a fundamental contributor of the hypothalamic GnRH pulsatility. After secretion, GnRH is rapidly cleaved by peptidases and thus has a short half-life of 2–4 min. Due to this degradation and massive dilution, the peripheral circulation does not contain biologically active concentrations of GnRH, which is potentially detrimental to various reproductive functions.

The characterization and structural activity analysis of the three-dimensional (3D) model of GnRH variants, including the substitution of amino acids, led to the discovery of the various antagonists and agonists of GnRH (Fig. 1). Positions 1, 2, 3, 6, and 10 of amino acid residues in the decapeptide play important roles in the structure and function of GnRH (Fig. 1A). Positions 5–6 and 6–7 are sensitive for enzymatic cleavage, causing the GnRH native peptide to have a very short half-life. Positions 2 and 3 are important in elucidating the LH-FSH release from the pituitary, whereas

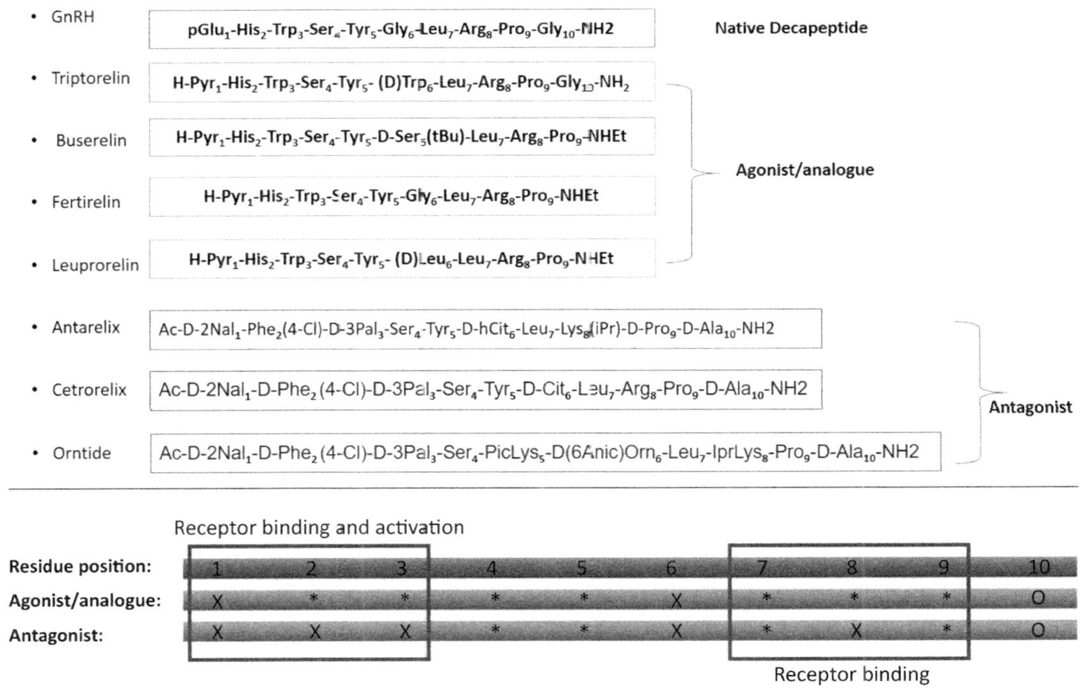

Residue position:	1	2	3	4	5	6	7	8	9	10
Agonist/analogue:	X	*	*	*	*	X	*	*	*	O
Antagonist:	X	X	X	*	*	X	*	X	*	O

Fig. 1 (A) The primary structure of the GnRH native peptide (*blue box* (light gray in the print version)), agonists (*green box* (gray in the print version)), and antagonist (*red box* (dark gray in the print version)). The native decapeptide has a "pyroglycine" at the amino terminus and carboxamide at the carbonyl end. All the amino acids of native GnRH are in an L configuration. The agonist/analogue peptides have all the residues the same as the native, except the first and sixth residues. On the other hand, antagonists only have four residues the same as the native molecule (i.e., the fourth, fifth, seventh, and ninth residues). The agonist molecules have pyroglutamyl instead of pyroglycine as the first amino acid residue. The sixth amino acid is essential for the tertiary structure (bent conformation) of the decapeptide. Also, replacing the sixth amino acid residue with D amino acid aids in binding. Residues 2 and 3 are essential for activation of the GnRH receptor (Sealfon et al., 1997). Proline ethylamide in the agonists helps enhance the potency along with increasing the half-life of the peptide (Karten and Rivier, 1986a, b). (B) Amino acid residue alignment. Amino acid residues in positions 1, 2, and 3 are essential for receptor binding and activation, whereas the residues in positions 7, 8, and 9 are essential for binding the peptide to the receptor (Sealfon et al., 1997). *, same residues as the native decapeptide; X, different residue than the native form; O, not functionally essential. (For interpretation of the references to color in this figure legend, the reader is referred to the web version of this article.)

positions 1, 6, and 10 are important for its 3D structure. Deletion or substitution of the histidine amino acid in position 2 has been demonstrated to be a competitive antagonist of the native GnRH. It can bind to the GnRH receptor, but cannot activate it. Substitutions at positions 1, 3, 6, 8, and 10 lead to an increase in the antagonistic property of the synthesized peptide. Substituting with a D-amino acid for the native L-amino residue at position 6 results in GnRH agonist development. Structurally, the GnRH antagonists differ from the agonists (Coccia et al., 2004). Some basic mechanistic differences are also found in the activity of agonists and antagonists. The agonists produce an initial stimulation of the gonadal response by exerting a positive effect on pituitary gonadotrophs, which results in the secretion of FSH and LH, followed by the downregulation and inhibition of the pituitary-gonadal axis. Unlike the agonist, GnRH antagonists immediately suppress pituitary gonadotropin by GnRH-R competition, thereby reducing the initial stimulatory phase of the agonists.

These variants have wide clinical applications, such as involvement in the induction of ovulation in mammals as well as in the treatment of benign and malignant tumors. A low-dose treatment of the native peptide in a pulsatile manner improves fertility in hypogonadal cases. However, high doses of the native peptide or agonist decrease gametogenesis and steroidogenesis while also decreasing gonadotropin secretion, as the desensitization of the gonadotrope gland has been observed in animal studies (Coccia et al., 2004). The usefulness of GnRH agonists over native GnRH is due to their improved potency and longer half-life (Kumar and Sharma, 2014). Interestingly, GnRH-antagonists decrease the reproductive axis function by competitive inhibition with the native peptide toward the GnRH receptor (Coccia et al., 2004). A study on the GnRH analogue's pharmacokinetic parameters suggested that analogues (e.g., buserelin and dalarelin) in suspension form increased its bioavailability and directly correlated with the hormonal profiles of LH and FSH in the serum (Suszka-Świtek et al., 2017). A few of the approved and available forms of GnRH agonists and antagonists include leuprolide, also called leuprorelin (Lupron: 1985); goserelin (Zoladex: 1989); histrelin (Supprelin, Vantas: 1991 and 2004); triptorelin (Trelstar: 2000); and degarelix (Firmagon: 2008). The inherent issues of peptidic drugs targeting GnRH-R, such as low aqueous solubility and the needs of parenteral administration, have limited their clinical exploitation (Limonta et al., 2012; Limonta and Manea, 2013). This has led to the recent oral bioavailable development of the nonpeptide GnRH receptor antagonist (Tukun et al., 2017).

In this chapter, we will review the use of GnRH agonists (which mimic the GnRH action in the pituitary gonadotrophic cells) and the antagonists (specific GnRH-R blockers) in ART with a particular emphasis in livestock.

GnRH agonists-antagonists in female livestock

ART in modern agriculture is used out of the estrus season to increase reproductive performance and genetic enhancement. It is also used to preserve endangered species and breeds and in eradicating various diseases. In this continuously growing population, livestock animals play a strong in meeting global demand for food and fiber production. Seasonal breeding, ovarian asynchrony, and herd size could hinder their availability and propagation. Therefore, ART helps overcome this intrinsic reproductive restriction in these animals. Our expanding understanding of the female livestock estrous cycle helps in designing new approaches to control both the luteal and follicular functions. This, in turn, permits a high efficacy of ART, thereby ensuring a highly successful pregnancy rate. GnRH occupies a central role in governing the reproductive function in animals. Numerous studies have shown an increase in efficient conception rates using GnRH agonist–antagonist methods. The advantage of using an agonist is its higher affinity toward GnRH receptors as well as its longer half-life in circulation (Karten and Rivier, 1986a, b). However, continuous exposure to a GnRH agonist can lead to blocking LH pulsatile secretion (D'Occhio et al., 1996). This biphasic property allows them to be used in pro- and antifertility applications. A study showed that GnRH treatment induced ovulation in cattle with a growing dominant follicle (Macmillan and Thatcher, 1991). This study further showed that buserelin, a potent GnRH agonist, when injected during the diestrus phase in cyclic cattle, altered ovarian follicle development and induced ovulation. However, this method was more efficacious in lactating dairy cows than in heifers. Studies have shown a higher pregnancy rate in cross-bred cattle using GnRH-based protocols compared to other treatments (Baruselli et al., 2000; Mayor et al., 2008). Pursley et al. developed an ovulation synchronization scheme using GnRH (OvSynch) for fixed time AI in lactating dairy cattle (Pursley et al., 1995). The major advantages of using this technique are the estrous detection rate can be increased to 100% while conception and pregnancy rates show significant increases. A study using a GnRH analogue (GnRH-A, fertirelin acetate) showed a significantly higher increase in the conception rate of dairy cows as compared to other ovulation synchronization methods (Yamada, 2005).

Another application using GnRH agonists is the development of the GnRH agonist–LH protocol for ovarian follicle superstimulation (D'Occhio et al., 1996). Here, the heifers and cows to be superstimulated are implanted with a GnRH-A (deslorelin) bioimplant, which prevents the endogenous preovulatory LH surge. Therefore, the time of ovulation can be controlled by the exogenous administration of LH, which allows fixed-time AI. This technique optimizes the rate of ovulation in response to exogenous LH, further releasing fertilized oocytes, and results in embryos that establish normal pregnancies after transfer to recipients (Nogueira et al., 2002). GnRH antagonists have been successfully used to inhibit the LH surge and ovulation in a range of species. However, the failure to develop a controlled release system has limited studies in developing efficient mechanisms to enhance bovine ART. A study using Nelore cows (*Bos indicus*) showed that the OvSynch protocol using GnRH and PGF2α did not have good reliability or a good conception rate. However, this same study showed that using GnRH and equine chorionic gonadotropin (eCG) together or eCG alone significantly increased the pregnancy rate among Nelore cows in the reproductive season (Gaievski et al., 2015).

ART is especially important in horses. Horses are seasonally polyestrous species, with uncertain success or failure of conception during their limited breeding season that undoubtedly affects productivity. Thus, ART conception is ideal to effectively induce ovulation and increase the pregnancy rates in horses. Studies have shown that human chorionic gonadotropin (hCG) is highly efficient in inducing ovulation in horses. However, there is a risk of twin pregnancies due to multiple ovulation as well as decreased reactivity due to antibody production (Perkins and Grimmett, 2001). Numerous studies have tried using a GnRH agonist and deslorelin acetate implants (Derar et al., 2002; McCue et al., 2002). These studies showed a 3–4 times increase in LH surge after GnRH-agonist treatment, thus improving the breeding management of horses. The administration of a GnRH agonist bioimplant on the beginning of the dominant follicle can start the preovulatory LH surge and advance ovulation with minimal gonadotropin support. A study showed that the length of time to return to estrus was prolonged after treatment with a GnRH agonist in mares (Morehead and Blanchard, 2000). The reason for this is not very well understood, thus requiring more studies on the relationship between the GnRH dosage and follicular development after ovulation and their association with body weight. To improve the efficacy of equine embryo transfer, reliable superovulatory techniques are essential. A report showed that when the GnRH agonist was administered prior to superovulatory treatment, it suppressed follicular development, leading to a higher ovulation rate. This protocol enhanced the embryo recovery rate in superovulatory mares compared to control mares (Dippert et al., 1992). The GnRH antagonist antarelix attenuated progesterone levels, delayed ovulation, and eliminated the large diestrus pulse of LH in mares (Denniston et al., 2006).

The success rate of induced exogenous ovulation significantly depends on the hormone and dose used, the stage of follicle development at the time of treatment, and ensuring that the treatment occurs prior to the LH surge. Studies conducted in pigs suggest that the approach used for induced ovulation varies depending on the maturity of the animal and whether the follicle synchronization protocol was applied. The administration of GnRH agonist treatment too early in follicular development results in the ovulation of more immature cystic follicles. The efficacy of GnRH analogues has been evaluated in swine. A study using gilts showed that when injected, the GnRH agonist induced ovulation. Studies in pigs usually involved the use of hCG and GnRH analogues, individually or in combination. A study using goserelin, a GnRH agonist, increased the LH surge that peaked at 30 min after treatment in barrows and gilts. Prepubertal gilts treated with eCG followed by an injection of the GnRH analogue had induced ovulation in all gilts; however, GnRH alone induced a few ovulations (Brüssow et al., 2007). Studies have shown that the route of administration is important in determining the effectiveness of the dosage. The intravaginal administration of triptorelin, a GnRH agonist, induced an LH surge within 4–16 h. The highest dose of GnRH agonist through intravaginal administration induced more sows to ovulate within 48 h after treatment while also showing a better fertility rate during synchronized follicle development and induced ovulation in sows compared to gilts (Martinat-Botté et al., 2010).

GnRH agonists-antagonists in male livestock

Both the GnRH agonists and antagonists are commercially available and used widely in reproductive research. GnRH antagonist treatment has been used in animals for the conventional pharmacological inhibition of endogenous GnRH action. Furthermore, in such gonadotropin-deprived animals, classical restoration experiments have been performed with the supplementation of LH and/or FSH (either alone or in combination) to examine the selective or synergistic functions of gonadotropins in fertility regulation (Hazum and Conn, 1988).

On the other hand, due to the GnRH agonist's higher affinity toward GnRH-R with a relatively longer half-life in circulation compared to that of the natural GnRH sequence, it can be administered in animals at a lower dose to manipulate farm productivity/fertility (Karten and Rivier, 1986a, b). There are two different modes of GnRH agonist treatment practiced in cattle with differential outcomes: (i) in both bulls and heifers, an acute treatment, which is usually restricted to a few weeks, shows an instant increase in circulatory LH and/or FSH, followed by a sharp decline reaching its basal levels (Melson et al., 1986; Gong et al., 1995, 1996; D'Occhio et al., 1996) and (ii) in both bulls and heifers, when the agonist treatment is continued beyond 3–4 months, the rise in the plasma LH is blocked, despite maintenance of its basal level (Melson et al., 1986; D'Occhio et al., 1996; D'Occhio and Aspden, 1996; Jimenez-Severiano et al., 1999; Evans and Rawlings, 1994; Gong et al., 1995). Persistent exposure of the GnRH agonist desensitizes GnRH-R, resulting in the selective inhibition of pulsatile LH (not FSH) secretion (D'Occhio and Aspden, 1996; Melson et al., 1986; Hazum and Conn, 1988). Furthermore, the administration of potent GnRH analogues such as nafarelin or deslorelin to bulls showed a significant decrease in GnRH-R, LH-β, and FSHH-β transcripts in the pituitary gonadotrophic cells (Melson et al., 1986; Aspden et al., 1996, 1997, 1998). However, elevated levels of LH were maintained in some of these long-term treatment groups. Bulls treated with deslorelin showed increased plasma testosterone (T) concentrations (D'Occhio and Aspden, 1996). Similarly, deslorelin induced a consistent increase in T secretion in red deer (Lincoln, 1987). Increased levels of LH are considered to induce upregulation of either the steroid acute regulatory (StAR) protein or other steroidogenic enzymes, which correlates to the rise of T concentration. However, in male Zebu calves, deslorelin treatment led to testicular hypertrophy (D'Occhio and Aspden, 1996). In younger bulls, testicular growth was retarded in short-term studies (Melson et al., 1986; Ronayne et al., 1993; Chandolia et al., 1997). However, more studies in bulls are necessary to understand the potential implications of GnRH agonist-induced testicular growth in the future.

Rodents and GnRH analogues

GnRH analogues in rodent reproduction are of utmost significance. Even though rodents do not come under the category of livestock, they are the most available and suitable model system used in biomedical research.

In addition, rodent species that are laboratory created and genetically modified such as transgenic, mutants, and knockout models lack equal potential to reproduce and expand the colony. Different reproductive traits such as litter size, sperm production, and hormone-induced ovulation rate are characteristic of a strain that might affect its reproductive performance (Sztein et al., 2000; Spearow and Barkley, 1999). IVF has become the most popular technique to sort out the availability of those model animals.

Female rodents and GnRH analogues

Exogenous gonadotropin treatment is widely used to induce superovulation in mice, where hCG is administered to mice for 46–48 h after treatment with pregnant mare serum gonatotropin (PMSG). Studies on improving the IVF fertility rate in rodents suggested prolonging the time interval between PMSG and hCG as well as additional treatment with buserelin (GnRH agonist) 24 h after PMSG. hCG treatment significantly improves the fertility rate in mice (Vasudevan and Sztein, 2012).

Generally, antagonist treatment is given first, followed by an agonist treatment to trigger a supraphysiological surge of LH (Olivennes et al., 1996; Fauser et al., 2002). A mechanistic study suggested that treatment with the GnRH-antagonist in primed rats reduced the amount of GnRH-R. However, they remain active and responsive to pharmacologic doses of the GnRH-agonist. Interestingly, the GnRH-antagonist is unable to block the synthesis and accumulation of gonadotropin in the pituitary. Treatment with the GnRH-agonist stimulates GnRH-R and causes the release of large pools of gonadotropins (Taylor et al., 2010).

In comparison to superovulated rats treated with eCG-hCG, GnRH agonist treatment is known to be more effective to increase the matured oocyte number and ovulation rate in rats (Kanter et al., 2004; Coskun and Kanter, 2005). A study showed a significant increase in the number of oocytes and offspring in mice after a combination treatment of inhibin antiserum (IAS) and eCG (Takeo and Nakagata, 2015). However, in the ART where ovarian hyperstimulation syndrome (OHSS) is a concern, during ovulation stimulation the combined strategy of the GnRH antagonist and agonist is also useful. A group of researchers suggested that in the OHSS mice model, the GnRH agonist showed a positive impact (Moradi et al., 2018). This suggests that the GnRH agonist is able to improve the quality and quantity of oocytes, the IVF rate, and the periimplantation development in comparison to control and superovulated mice (Moradi et al., 2018).

The positive effect of the GnRH agonist is not only limited to oocyte maturation, as it also provides significant improvements in periimplantation events such as implantation, decidualization, and term pregnancy when compared to historical hCG treatment. A study showed that buserelin, a potent agonist of GnRH, did not alter ovarian steroidogenesis with no alteration in progesterone/estradiol synthesis and secretion (Ezoe et al., 2019). It was demonstrated that in comparison to hCG, buserelin does not adversely affect progesterone receptor expression and its downstream signaling. Rather, it successfully inhibited estrogenic activity in the luminal epithelium during preimplantation and showed no adverse effect on endometrial proliferation, differentiation switching, etc. This addresses its harmless ability in blastocyst implantation in mice. A combination treatment of low-dose buserelin and hCG shows improved blastocyst implantation and a successful pregnancy outcome in mice (Ezoe et al., 2019). GnRH agonists also improve implantation through altering uterine receptivity in the adenomyosis mice model. The GnRH agonist caused significant upregulation of Hoxa10, Hoxa11, Lif, and integrin b3 expression in glandular epithelial cells and endometrial luminal epithelial cells. These genes are responsible for endometrial receptivity for the implantation of the embryo. Therefore, by upregulating its expression, GnRH agonists improve the success rate of ART in the adenomyosis condition (Guo et al., 2018).

Fertility preservation is an important aspect of ART. Studies have shown a persistent decrease in ovarian follicular numbers as a result of chemotherapy in various malignant and nonmalignant diseases in young or adult animals. To protect follicles from the deleterious effect of chemotherapeutic agents, GnRH analogues have been used. GnRH agonists (and antagonists) could protect ovarian gonad toxicity if applied before chemotherapy in rodents (Matsuo et al., 2007; Ozcelik et al., 2009; Tan et al., 2010; Kishk and Ali, 2013; Lemos et al., 2010). A study showed that triptorelin (GnRH agonist) is capable of preserving primordial and primary follicles in a dose-dependent manner (Tan et al., 2010). Unlike the agonist, the GnRH antagonist promotes an immediate effect on the gonad by the instant inhibition of gonadotropic stimulus on the ovaries.

Male rodents and GnRH analogues

The GnRH antagonist cetrorelix inhibits the pituitary-gonadal axis in peripubertal male rats and prevents sexual maturation, and thus could be effective in treating central precocious puberty in males (Roth et al.,

2000). GnRH agonists if treated chronically suppress testicular function, such as the production of testosterone, and impair sperm production (Swerdloff et al., 1983).

Genotoxic agents mainly affect proliferating spermatogonial cells. These agents not only kill most of the spermatogonial stem cells, but also block their differentiation in a testosterone-dependent manner (Zhang et al., 2007). Extreme gonad toxicity causes apoptosis of the spermatogonial stem cells and damages sertoli cells, which leads to complete sterility. To restore fertility in rodents, the suppression of gonadotropins and testosterone with GnRH analogues during or before toxic treatments promotes protection to spermatogenesis (Meistrich and Kangasniemi, 1997; Meistrich et al., 2001; Udagawa et al., 2001; Meistrich and Shetty, 2003). GnRH-analogues can be administered after chemotoxicity to revive spermatogenesis (Wang et al., 2010; Shetty et al., 2013). After testicular damage, treating the rat with GnRH-analogues can also stimulate the proliferation and differentiation of spermatogonia, independent of stem cell factors (Ohmura et al., 2003). Long-term treatment with the GnRH antagonist orntide, released from a depot formulation, helps to return fertility in chemically castrated male rats (Kostanski et al., 2001).

Conclusion and future directions

The commercial uses of GnRH analogues show significant improvement in the fertility and/or productivity of laboratory rodents and farm mammals. Rodents show remarkable success in conception with the supplementation of buserelin (a potential GnRH agonist) after PMSG-hCG treatment during experimental IVF programs. Both GnRH antagonists and/or agonists show an increase in GnRH-R expression and activate the pituitary further to release gonadotropins in rodents. Treatment with the GnRH agonist shows a promising rise in the number of maturing oocytes, thereby enhancing the ovulation rate in rats. In male rodents, GnRH antagonists have been shown to induce precocious puberty. On the other hand, a critical commercial importance of GnRH analogues has been noticed in cattle for successful ART. Recent studies have shown an increase in the conception rate by synchronization of ovulation using OvSynch prior to the AI schedule. The GnRH agonist has also been used to induce ovarian follicles. However, in male bulls, GnRH agonists induce testicular hypertrophy with elevated levels of plasma LH and T. In summary, despite some adverse effects such as metabolic disturbances and osteoporosis, such agonists and antagonists of GnRH collectively show a potential plethora of actions in supporting the fertility and productivity of laboratory rodents and cattle, respectively. However, in order to face the emerging market demand of the livestock industry, future research is necessary to design and develop more effective GnRH analogues for better farm productivity.

Funding

I.B. is partially supported by the UGC Start Up Grant: UNDER THE MHRD/UGC–Empowered Committee's Basic Science Research Program F.30-104/2015(BSR), and the Early Career Award by the Science and Engineering Research Board, Department of Science and Technology, Government of India, reference ECR/2018/000868. A.B. is partially supported by SERB EMR/2017/004290. M.D. is a national postdoctoral fellow from the Science and Engineering Research Board, New Delhi, India (File PDF/2017/000989). J.F. acknowledges the financial support from ICMR.

References

Aspden, W.J., Rao, A., Scott, P.T., Clarke, I.J., Trigg, T.E., Walsh, J., D'Occhio, M.J., 1996. Direct actions of the luteinizing hormone-releasing hormone agonist, deslorelin, on anterior pituitary contents of luteinizing hormone LH and follicle-stimulating hormone FSH, LH and FSH subunit messenger ribonucleic acid, and plasma concentrations of LH and FSH in castrated male cattle. Biol. Reprod. 55, 386–392.

Aspden, W.J., Rao, A., Rose, K., Scott, P.T., Clarke, I.J., Trigg, T.E., Walsh, J., D'Occhio, M.J., 1997. Differential responses in anterior pituitary luteinizing hormone LH. Content and LH b- and a-subunit mRNA, and plasma concentrations of LH and testosterone, in bulls treated with the LH-releasing hormone agonist deslorelin. Domest. Anim. Endocrinol. 14, 429–437.

Aspden, W.J., Rodgers, R.J., Stocco, D.M., Scott, P.T., Wreford, N.G., Trigg, T.E., Walsh, J., D'Occhio, M.J., 1998. Changes in testicular steroidogenic acute regulatory StAR protein, steroidogenic enzymes and testicular morphology associated with increased testosterone secretion in bulls receiving the luteinizing hormone releasing hormone agonist deslorelin. Domest. Anim. Endocrinol. 15, 227–238.

Baruselli, P.S., Marques, M.O., Carvalho, N.A.T., Valentim, R., Berber, R.C.A., Carvalho Filho, A.F., Madureira, E.H., Costa Neto, W.P., 2000. Ovsynch protocol with fixed-time embryo transfer increasing pregnancy rates in bovine recipients. Arq. Fac. Vet. UFRGS 28, 205.

Brüssow, K.P., Schneider, F., Tuchscherer, A., Rátky, J., Kraeling, R.R., Kanitz, W., 2007. Luteinizing hormone release after administration of the gonadotropin-releasing hormone agonist Fertilan (goserelin) for synchronization of ovulation in pigs. J. Anim. Sci. 85 (1), 129–137.

Chandolia, R.K., Evans, A.C.O., Rawlings, N.C., 1997. Effect of inhibition of increased gonadotrophin secretion before 20 weeks of age in bull calves on testicular development. J. Reprod. Fertil. 109, 65–71.

Coccia, M.E., Comparetto, C., Bracco, G.L., Scarselli, G., 2004. GnRH antagonists. Eur. J. Obstet. Gynecol. Reprod. Biol. 115, S44–S56.

Coskun, O., Kanter, M., 2005. GnRH agonist stimulates oocyte number and maturation in mice superovulated with eCG and hCG. Ann. N. Y. Acad. Sci. 2, 1–4.

D'Occhio, M.J., Aspden, W.J., 1996. Characteristics of luteinizing hormone LH. and testosterone secretion, pituitary responses to LH-releasing hormone LHRH and reproductive function in young bulls receiving the LHRH agonist deslorelin: effect of castration on LH responses to LHRH. Biol. Reprod. 54, 45–52.

D'Occhio, M.J., Aspden, W.J., Whyte, T.R., 1996. Controlled, reversible suppression of oestrous cycles in beef heifers and cows using agonists of luteinizing hormone-releasing hormone. J. Anim. Sci. 74, 218–225.

Denniston, D.J., Crabb, J.J., Bruemmer, J.E., Squires, E.L., 2006. Effects of a gonadotropin-releasing hormone antagonist and altrenogest on luteinizing hormone concentration and ovulation in the mare. J. Equine Vet. 26 (3), 95–101.

Derar, R.I., Maeda, Y., Tsunoda, N., Hoque, M.S., Osawa, T., Miyake, Y.I., 2002. The peripheral levels of luteinizing hormone (LH), follicle-stimulating hormone (FSH), Immunoreactive (ir-) inhibin, progesterone (P) and estradiol-17β (E 2) at the time of control of ovulation with gonadotropin releasing hormone (GnRH) agonist (Deslorelin) in pony mares. J. Equine Sci. 13 (3), 83–87.

Dippert, K.D., Hofferer, S., Palmer, E., Jasko, D.J., Squires, E.L., 1992. Initiation of superovulation in mares 5 or 12 days after ovulation using equine pituitary extract with or without GnRH analogue. Theriogenology 38 (4), 695–710.

Evans, A.C.O., Rawlings, N.C., 1994. Effects of a long-acting gonadotrophin-releasing hormone agonist leuprolide on ovarian follicular development in prepubertal heifer calves. Can. J. Anim. Sci. 74, 649–656.

Ezoe, K., Murata, N., Yabuuchi, A., et al., 2019. Evaluation of uterine receptivity after gonadotropin releasing hormone agonist administration as an oocyte maturation trigger: a rodent model. Sci. Rep. 9, 12519.

Fauser, B.C., De Jong, D., Olivennes, F., Wramsby, H., Tay, C., Itskovitz-Eldor, J., et al., 2002. Endocrine profiles after triggering of final oocyte maturation with GnRH agonist after cotreatment with the GnRH antagonist ganirelix during varian hyperstimulation for in vitro fertilization. J. Clin. Endocrinol. Metab. 87, 709–715.

Gaievski, F.R., Lamb, G.C., Weiss, R.R., Bertol, M.A.F., Segui, M.S., Abreu, A.C.M.R., 2015. Gonadotropin releasing hormone (GnRH) and equine chorionic gonadotropin (ECG) improve the pregnancy rate on protocols for timed-artificial insemination in beef cattle. Veterinária E Zootecnia 22 (3), 471–480.

Gong, J.G., Bramley, T.A., Gutierrez, C.G., Peters, A.R., Webb, R., 1995. Effects of chronic treatment with a gonadotrophin-releasing hormone agonist on peripheral concentrations of FSH and LH, and ovarian function in heifers. J. Reprod. Fertil. 105, 263–270.

Gong, J.G., Campbell, B.K., Bramley, T.A., Gutierrez, C.G., Peters, A.R., Webb, R., 1996. Suppression in the secretion of follicle-stimulating hormone and luteinizing hormone, and ovarian follicle development in heifers continuously infused with a gonadotropin-releasing hormone agonist. Biol. Reprod. 55, 68–74.

Guo, S., Li, Z., Yan, L., Sun, Y., Feng, Y., 2018. GnRH agonist improves pregnancy outcome in mice with induced adenomyosis by restoring endometrial receptivity. Drug Des. Devel. Ther. 12, 1621–1631.

Hazum, E., Conn, P.M., 1988. Molecular mechanism of gonadotropin releasing hormone GnRH action: I. The GnRH receptor. Endocr. Rev. 9, 379–386.

Herbison, A.E., 2016. Control of puberty onset and fertility by gonadotropin-releasing hormone neurons. Nat. Rev. Endocrinol. 12, 452.

Jimenez-Severiano, H., Mussard, M., Ehnis, L., Koch, J., Zanella, E., Lindsey, B., Enright, W., D'Occhio, M., Kinder, J., 1999. Secretion of LH in bull calves treated with analogs of GnRH. J. Reprod. Fertil. 2000 (Suppl), 54.

Kanter, M., Yildiz, C., Meral, I., Koc, A., Tasal, I., 2004. Effects of a GnRH agonist on oocyte number and maturation in mice superovulated with eCG and hCG. Theriogenology 61, 393–398.
Karten, M.J., Rivier, J.E., 1986a. Gonadotropin-releasing hormone analog design. Structure-function studies toward the development of agonists and antagonists: rationale and perspective. Endocr. Rev. 7 (1), 44–66.
Karten, M.J., Rivier, J.E., 1986b. Gonadotropin-releasing hormone analog design. Structure-function studies towards the development of agonists and antagonists: rationale and perspectives. Endocr. Rev. 7, 44–66.
Kishk, E.A.F., Ali, M.H.M., 2013. Effect of a gonadotropin-releasing hormone analogue on cyclophosphamide-induced ovarian toxicity in adult mice. Arch. Gynecol. Obstet. 287, 1023–1029.
Kostanski, J.W., Jiang, G., Dani, B.A., Murty, S.B., Qiu, W., Schrier, B., Thanoo, B.C., DeLuca, P.P., 2001. Return to fertility after extended chemical castration with a GnRH antagonist. BMC Cancer 1, 18.
Kumar, P., Sharma, A., 2014. Gonadotropin-releasing hormone analogs: understanding advantages and limitations. J. Human Reprod. Sci. 7 (3), 170–174.
Lemos, C.N., Reis, F.M., Pena, G.N., Silveira, L.C., Camargos, A.F., 2010. Assessment of fertility protection and ovarian reserve with GnRH antagonist in rats undergoing chemotherapy with cyclophosphamide. Reprod. Biol. Endocrinol. 8, 51.
Limonta, P., Manea, M., 2013. Gonadotropin-releasing hormone receptors as molecular therapeutic targets in prostate cancer: current options and emerging strategies. Cancer Treat. Rev. 39, 647–663.
Limonta, P., Marelli, M.M., Mai, S., Motta, M., Martini, L., Moretti, R.M., 2012. GnRH receptors in cancer: from cell biology to novel targeted therapeutic strategies. Endocr. Rev. 33, 784–811.
Lincoln, G.A., 1987. Long-term stimulatory effects of a continuous infusion of LHRH agonist on testicular function in male red deer Cerĺus elaphus. J. Reprod. Fertil. 80, 257–261.
Macmillan, K.L., Thatcher, W.W., 1991. Effects of an agonist of gonadotropin-releasing hormone on ovarian follicles in cattle. Biol. Reprod. 45 (6), 883–889.
Martinat-Botté, F., Venturi, E., Guillouet, P., Driancourt, M.A., Terqui, M., 2010. Induction and synchronization of ovulations of nulliparous and multiparous sows with an injection of gonadotropin-releasing hormone agonist (Receptal). Theriogenology 73 (3), 332–342.
Matsuo, G., Ushijima, K., Shinagawa, A., et al., 2007. GnRH agonist acts as ovarian protection in chemotherapy induced gonadotoxicity: an experiment using a rat model. Kurume Med. J. 54, 25–29.
Mayor, J.C., Tribulo, H.E., Bó, G.A., 2008. Pregnancy rates following fixed-time embryo transfer in Bos indicus recipients synchronized with progestin devices and estradiol or GnRH and treated with eCG. Reprod. Domest. Anim. 43, 180.
McCue, P.M., Farquhar, V.J., Carnevale, E.M., Squires, E.L., 2002. Removal of deslorelin (Ovuplant) implant 48h after administration results in normal interovulatory intervals in mares. Theriogenology 58, 865–870.
Meistrich, M.L., Kangasniemi, M., 1997. Hormone treatment after irradiation stimulates recovery of rat spermatogenesis from surviving spermatogonia. J. Androl. 18, 80–87.
Meistrich, M.L., Shetty, G., 2003. Inhibition of spermatogonial differentiation by testosterone. J. Androl. 24, 135–148.
Meistrich, M.L., Wilson, G., Shuttlesworth, G., Huhtaniemi, I., Reissmann, T., 2001. GnRH agonists and antagonists stimulate recovery of fertility in irradiated LBNF1 rats. J. Androl. 22, 809–817.

Melson, B.E., Brown, J.L., Schoenemann, H.M., Tarnavsky, G.K., Reeves, J.J., 1986. Elevation of serum testosterone during chronic LHRH agonist treatment in the bull. J. Anim. Sci. 62, 199–207.

Minakata, H., 2010. Oxytocin/vasopressin and gonadotropin-releasing hormone from cephalopods to vertebrates. Ann. N. Y. Acad. Sci. 1200 (1), 33–42.

Miyamoto, K., Hasegawa, Y., Nomura, M., Igarashi, M., Kangawa, K., Matsuo, H., 1984. Identification of the second gonadotropin-releasing hormone in chicken hypothalamus: evidence that gonadotropin secretion is probably controlled by two distinct gonadotropin-releasing hormones in avian species. Proc. Natl. Acad. Sci. 81 (12), 3874–3878.

Montaner, A.D., Somoza, G.M., King, J.A., Bianchini, J.J., Bolis, C.G., Affanni, J.M., 1998. Chromatographic and immunological identification of GnRH (gonadotropin-releasing hormone) variants. Occurrence of mammalian and a salmon-like GnRH in the forebrain of an eutherian mammal: Hydrochaeris hydrochaeris (Mammalia, Rodentia). Regul. Pept. 73 (3), 197–204.

Moradi, Y., Hosseini, S., Mohammadi-Yeganeh, S., Salehpour, S., Salehi, M., 2018. Ovarian hyperstimulation syndrome model in mice: effect of GnRH agonist. Biomed. Res. 29 (15), 3037–3042.

Morehead, J.A., Blanchard, T.L., 2000. Clinical experience with deslorelin (OvuplantTM) in a Kentucky thoroughbred broodmare practice (1999). J. Equine Vet. 20 (6), 358–402.

Nogueira, M.F.G., Barros, B.J., Teixeira, A.B., Trinca, L.A., Michael, J.D., Barros, C.M., 2002. Embryo recovery and pregnancy rates after the delay of ovulation and fixed time insemination in superstimulated beef cows. Theriogenology 57 (6), 1625–1634.

Ohmura, M., Ogawa, T., Ono, M., Dezawa, M., Hosaka, M., Kubota, Y., Sawada, H., 2003. Increment of murine spermatogonial cell number by gonadotropin-releasing hormone analogue is independent of stem cell factor c-kit signal. Biol. Reprod. 68 (6), 2304–2313.

Olivennes, F., Fanchin, R., Bouchard, P., Taieb, J., Frydman, R., 1996. Triggering of ovulation by a gonadotropin-releasing hormone (GnRH) agonist in patients pretreated with a GnRH antagonist. Fertil. Steril. 66, 151–153.

Ozcelik, B., Turkyilmaz, C., Ozgun, M.T., Serin, I.S., Batukan, C., Ozdamar, S., Ozturk, A., 2009. Prevention of paclitaxel and cisplatin induced ovarian damage in rats by a gonadotropin-releasing hormone agonist. Fertil. Steril. 30, 30–36.

Perkins, N.R., Grimmett, J.B., 2001. Pregnancy and twinning rates in thoroughbred mares following the administration of human chorionic gonadotropin (hCG). N. Z. Vet. J. 49 (3), 94–100.

Perry, E.J., 1960. The artificial insemination of farm animals, third rev. ed.

Plant, T.M., 2015. The hypothalamo-pituitary-gonadal axis. J. Endocrinol. 226, T41–T54.

Pursley, J.R., Mee, M.O., Wiltbank, M.C., 1995. Synchronization of ovulation in dairy cows using PGF2α and GnRH. Theriogenology 44 (7), 915–923.

Ronayne, E., Enright, W.J., Roche, J.F., 1993. Effects of continuous administration of gonadotropin-releasing hormone _GnRH or a potent GnRH analogue on blood luteinizing hormone and testosterone concentrations in prepubertal bulls. Domest. Anim. Endocrinol. 10, 179–189.

Roth, C., Leonhardt, S., Seidel, C., Lakomek, M., Wuttke, W., Jarry, H., 2000. GnRH antagonist cetrorelix prevents sexual maturation of peripubertal male rats. Exp. Clin. Endocrinol. Diabetes 108, 358–363.

Sealfon, S.C., Weinstein, H, Millar, R.P., 1997. Molecular mechanisms of ligand interaction with the gonadotropin-releasing hormone receptor. Endocr. Rev. 18 (2), 180–205.

Shetty, G., Uthamanthil, R.K., Zhou, W., Shao, S.H., Weng, C.C., Tailor, R.C., et al., 2013. Hormone suppression with GnRH antagonist promotes spermatogenic recovery from transplanted spermatogonial stem cells in irradiated cynomolgus monkeys. Andrology 1, 886–898.

Spearow, J.L., Barkley, M., 1999. Genetic control of hormone induced ovulation rate in mice. Biol. Reprod. 61, 851–856.

Suszka-Świtek, A., Ryszka, F., Dolińska, B., Dec, R., Danch, A., Filipczyk, Ł., Wiaderkiewicz, R., 2017. Pharmacokinetics and bioavailability of the GnRH analogs in the form of solution and Zn2+-suspension after single subcutaneous injection in female rats. Eur. J. Drug Metab. Pharmacokinet. 42 (2), 251–259.

Swerdloff, R.S., Heber, D., Bhasin, S., Rajfer, J., 1983. Effect of GnRH superactive analogs (alone and combined with androgen) on testicular function in man and experimental animals. J. Steroid Biochem. 19 (1B), 491–497.

Sztein, J.M., Farley, J.S., Mobraaten, L.E., 2000. In vitro fertilization with cryopreserved inbred mouse sperm. Biol. Reprod. 63, 1774–1780.

Takeo, T., Nakagata, N., 2015. Superovulation using the combined administration of inhibin antiserum and equine chorionic gonadotropin increases the number of ovulated oocytes in C57BL/6 female mice. PLoS One 10 (5), e0128330.

Tan, S.J., Yeh, Y., Shang, W., Wua, G., Liu, J., Chen, C., 2010. Protective effect of a gonadotropin-releasing hormone analogue on chemotherapeutic agent-induced ovarian gonadotoxicity: a mouse model. Eur. J. Obstet. Gynecol. Reprod. Biol., 12–16.

Taylor, J.E., Miller, B.T., Gray, K.D., Scott Jr., R.T., Catherino, W.H., Segars, J.H., 2010. The mechanism responsible for the supraphysiologic gonadotropin surge in females treated with gonadotropin-releasing hormone (GnRH) agonist and primed with GnRH antagonist. Fertil. Steril. 93 (5), 1668–1675.

Tukun, F.L., Olberg, D.E., Riss, P.J., Haraldsen, I., Kaass, A., Klaveness, J., 2017. Recent development of non-peptide GnRH antagonists. Molecules 22 (12), 2188.

Udagawa, K., Ogawa, T., Watanabe, T., Yumura, Y., Takeda, M., Hosaka, M., 2001. GnRH analog, leuprorelin acetate, promotes regeneration of rat spermatogenesis after severe chemical damage. Int. J. Urol. 8, 615–622.

Vasudevan, K., Sztein, J.M., 2012. In vitro fertility rate of 129 strain is improved by Buserelin (GnRH) administration prior to superovulation. Lab Anim. 46 (4), 299–303.

Wang, G., Shao, S.H., Weng, C.C., Wei, C., Meistrich, M.L., 2010. Hormonal suppression restores fertility in irradiated mice from both endogenous and donor-derived stem spermatogonia. Toxicol. Sci. 117, 225–237.

Yamada, K., 2005. Development of ovulation synchronization and fixed time artificial insemination in dairy cows. J. Reprod. Dev. 51 (2), 177–186.

Zhang, Z., Marvin, S., Meistrich, L., 2007. The radiation-induced block in spermatogonial differentiation is due to damage to the somatic environment, not the germ cells. J. Cell. Physiol. 211 (1), 149–158.

CHAPTER TEN

Oxidative stress in modulation of immune function in livestock

K.M. Kiran Kumar[a], Rashmi Nagesh[b], M. Naveen Kumar[b], S.J. Prashanth[c], and R.L. Babu[d]

[a]Yen Onco Centre, Yenepoya Medical College, Yenepoya Deemed to be University, Mangaluru, Karnataka, India
[b]Department of Biotechnology and Genetics, M.S. Ramaiah College of Arts, Science and Commerce, Bengaluru, India
[c]Department of Food Processing and Nutrition, Karnataka State Akkamahadevi Women's University, Vijayapura, India
[d]Department of Bioinformatics and Biotechnology, Karnataka State Akkamahadevi Women's University, Jnanashakthi Campus, Vijayapura, Karnataka, India

Introduction

Oxidative stress is a biological status of the cellular system instigated as a result of an imbalance between the production and accumulation of free radicals (Pizzino et al., 2017). Additionally, the same biological system exhibits the ability to detoxify the accumulated free radicals to maintain cellular homeostasis (Kurutas, 2016). Further, the inequity between the generation and level of detoxification of oxidants by antioxidants could lead to cell damage and tissue injury (Kiran Kumar et al., 2016, 2017; Nagesh et al., 2017). Oxidants (highly reactive free radicals) are substances that are able to reduce themselves and act as oxidizing agents, the characteristic of atoms or molecules that have in their outer shell one or more unpaired electrons (Davies and Parrott, 1978; McDonagh and Sheehan, 2008; Phaniendra et al., 2015). Reactive oxygen species (ROS) and reactive nitrogen species (RNS) (Galon et al., 2002) are the major forms of free radicals usually coupled with an oxygen atom (O) that are highly reactive with other molecules as compared to molecular O_2. Mostly, ROS/RNS are generated as byproducts of the cellular metabolism and ionizing radiation, indicating the following species of free radicals: superoxide anion (O_2^-), hydrogen peroxide (H_2O_2), hydroxyl radical (OH), and singlet oxygen ($^1O_2$). The other biologically important free radicals are lipid hydroperoxide (ROOH), lipid peroxyl radical (ROO), and lipid alkoxyl radical (RO) that are associated with membrane lipids; nitric oxide (NO), nitrogen dioxide (NO_2), and peroxy

Emerging Issues in Climate Smart Livestock Production
https://doi.org/10.1016/B978-0-12-822265-2.00008-9

nitrite ($ONOO^-$) are RNS; and thiol radical (RS), which has an unpaired electron on the sulfur atom (Kurutas et al., 2005; Valko et al., 2007).

Apart from the cellular metabolism, internal or external stressors are also responsible for ROS liberation in the mitochondria through the electron transport chain reaction. They are known to play a crucial role in mediating the cellular responses to endogenous/exogenous physical, chemical, and biological stress through signal transduction and inflammatory/immune/toxic responses (Iorio, 2017). ROS are potent metabolites that can increase the oxygenation of biological macromolecules involved in regulating important cellular functions such as proliferation and differentiation (Halliwell and Gutteridge, 2007). They are highly reactive and cause oxidative damage as a consequence of the peroxidation of membrane lipids, the modification of DNA, the loss of sulfhydryls, and carbonylation in proteins (Goyns, 2002; Sohal et al., 1990; Sohal and Weindruch, 1996). Although ROS have numerous beneficial effects on immune and inflammatory responses, damage to host cells can occur with excessive accumulation of these highly reactive molecules. While small fluctuations in the steady-state concentrations of ROS are necessary for optimal immune and inflammatory responses, dramatic imbalances can result in tissue damage and loss of normal cell function (Sordillo and Aitken, 2009; Valko et al., 2007; Victor et al., 2004). To mitigate the negative effects caused by ROS, the improved approach of antioxidant barriers by both exogenous (diet-derived antioxidants such as vitamin E) and endogenous antioxidants (uric acid or antioxidant enzymes: superoxide dismutase and peroxidases) supports converting the harmful level of reactive species to less reactive molecules as well as maintaining cellular homeostasis (Fig. 1).

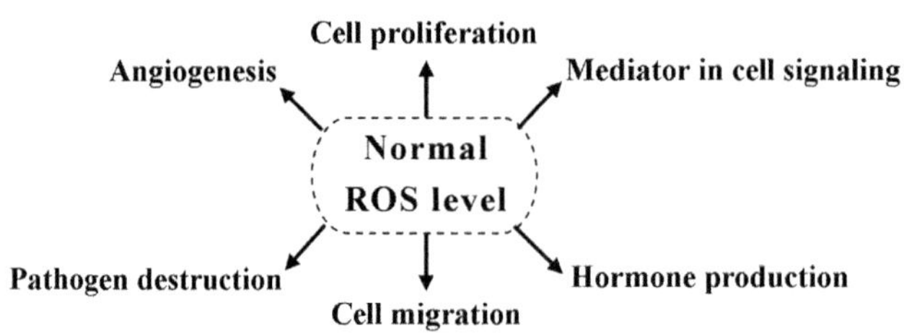

Fig. 1 Beneficial effects of normal reactive oxygen species levels in a cellular system.

Role of free radicals in evoking immune responses

The oxidative stress pronounces the initiation of numerous deleterious progressions consequential of an imbalance between excessive liberation of ROS and/or reduced antioxidant defenses (Sies, 1986). Several endogenous antioxidant defense mechanisms are present to tightly regulate ROS accumulation within tissues (Valko et al., 2007). Antioxidant defenses are capable of slowing or preventing the oxidation of other molecules and can be characterized as either radical scavengers or detoxifying enzyme systems (Halliwell, 2007). Altered ratios of pro and antioxidants can result in substantial damage to nearby tissues by oxidizing biological molecules and can stimulate the immune response as a consequence of stress. ROS are involved in signal transduction pathways leading to the expression of cytokines, eicosanoids, and other immunoregulatory factors essential to host defense during infection (Kvietys and Granger, 2012; Patil et al., 2016). Additionally, the redox status can alter the immune response at the site of infection. When vertebrates are attacked by pathogens, the host system first mounts an innate nonspecific immune response during which fluids, molecules, and immune cells such as phagocytes are delivered into the infected tissue (Sorci and Faivre, 2009). Phagocytosis is an essential component of the cellular innate immune response, involving the generation of toxic ROS necessary for the oxygen-dependent destruction of invading pathogens. The ROS act as effective antimicrobial agents, but they do not discriminate between pathogens and host cells and can therefore generate stress in the host tissues (Sorci and Faivre, 2009). White blood cells such as neutrophils are the first line of defense against invading pathogens, and are the key effectors of the innate immune response. Activated neutrophils and macrophages are another important source of. O_2^- that are formed enzymatically by NADPH oxidase during inflammatory reactions (Sorg, 2004). Neutrophils trigger phagocytes that release highly reactive species to kill pathogens (Swindle and Metcalfe, 2007). Neutrophils can also exert a regulatory role on the adaptive immune response (Christoffersson and Phillipson, 2018) (Fig. 2).

ROS and RNS are also known to exert beneficial effects at low concentrations and involve physiological cellular signaling systems. H_2O_2 is produced in response to cytokines and growth factors and is involved in regulating immune cell activation and vascular remodeling in mammals (Veal et al., 2007). It can also interact with several targets involved in cell signaling pathways. NO• is generated in vivo by specific NO synthases

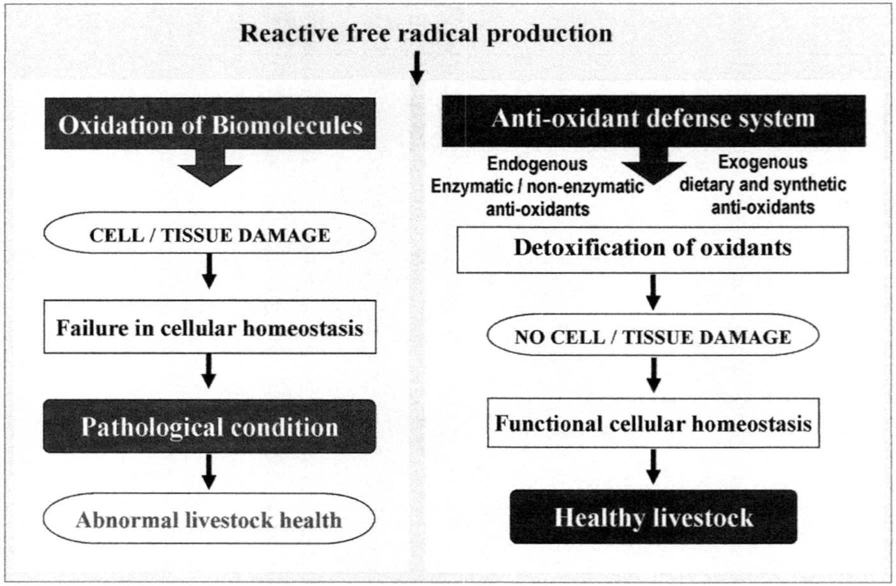

Fig. 2 Modulation of oxidative stress status against free radical production.

(NOS) and the nitrate-nitrite-NO pathway, and is a critical regulator of vascular homeostasis, neurotransmission, and host defense (Lundberg et al., 2008). Excessive production of NO•, under pathological environments, leads to detrimental effects of this molecule on tissues, which can be attributed to its reaction with the superoxide anion (O2•−) to form ONOO−. ONOO− is 1000 times more potent as an oxidizing compound than H_2O_2 (Van Dyke, 1997). Hence, the extracellular accumulation of ROS and RNS can affect the mounting of the immune response and contribute to the generation of a systemic inflammatory status, characterized by oxidative stress (Belikov et al., 2015). Oxidative stress can cause senescence in cells and it is therefore hypothesized to be an important modulator of life-history trade-offs in vertebrates (Costantini, 2008; Nussey et al., 2009). It has consequently been regarded as the main cause of aging in the literature in past decades (Harman, 1956). Thus, oxidative stress plays a crucial role in maintaining cellular homeostasis. In case of any disturbance, it can lead to numerous pathophysiological states such as early aging, autoimmune diseases, cardiovascular diseases, and malignancies.

Oxidative stress occurs as a consequence of increased cellular metabolism and external stressors that alters the balance of pro- and antioxidants which disturbs the cellular homeostasis state that lead to severe health issues and their act in evoking immune response against the infectious diseases caused as a result of increased redox status in livestock population which are reared for various commercial purpose such as for milk, meat and even for domestic benefits for humans. This chapter emphasizes the importance of oxidative stress in immune response and its deleterious effects on animal populations that could affect its productivity. Further, it helps us to understand the merits and demerits of the redox state, which can be addressed by employing numerous available antioxidants to boost the immune response against pathogenic attack, especially when animals are at risk due to the excessive accumulation of free radicals. Therefore, it is necessary to maintain the balance between the generation of oxidants and the levels of antioxidant defense to counteract the damaging effect of reactive species in livestock to improve the quality of the product (Fig. 3).

Implication of oxidative stress on immune responses in dairy cattle

Recent evidence has suggested the association of oxidative stress on the cellular metabolism as well as on the initiation of immune response in

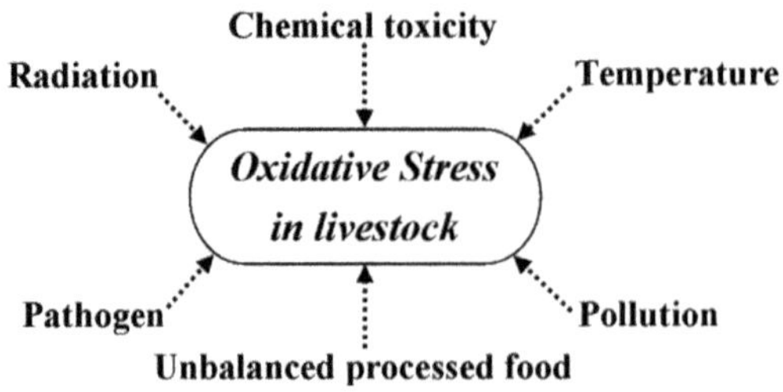

Fig. 3 Factors responsible for oxidative stress in livestock.

dairy cattle (Colitti et al., 2019; Mavangira and Sordillo, 2018; Sordillo and Aitken, 2009). Oxidative stress, an active area of research in veterinary medicine, has been implicated in various disease processes such as sepsis, mastitis, enteritis, pneumonia, and respiratory and joint diseases (Lykkesfeldt and Svendsen, 2007). Dairy cows undergo considerable metabolic and physiological adaptations during the transition from pregnancy to lactation. Studies have indicated that adult dairy cows experience oxidative stress around the time of calving (Sordillo, 2005). Coordinated shifts in nutrient partitioning must occur in order to meet the increased demand for energy and other nutrients necessary for fetal growth and lactation (Drackley, 1999). The increased demand for energy and other nutrients and the shuttle of nutrients are of prime importance to necessitate fetal growth and development (Sordillo, 2005). Oxygen demand increased during enhanced metabolic processes leads to the enhanced production of ROS with increased oxidative tissue damage accompanied by lipid peroxidation (Nagesh et al., 2017; Salahshoor et al., 2020). Lipid peroxidation is a consequence of oxidative stress, wherein lipids are oxidized by oxidants and malondialdehyde, a biomarker of lipid degradation and oxidative damage. Polyunsaturated fatty acids (PUFAs), with more than one double bond, are easily oxidized by free radical-mediated chain oxidation and are the major substrates of lipid peroxidation (Dervishi et al., 2016; Niki, 2009, 2014). ROS production is followed by oxidative stress, which may be one of the reasons for the higher incidence of diseases after parturition with the rapid differentiation of secretory parenchyma, intense mammary gland growth, and the onset of milk synthesis (Miller et al., 1993).

Milk cattle farming is constantly growing and is considered an important activity with economic interest in the world. However, health maintenance and the nutritional strategy of dairy cattle are of prime importance to reduce losses in milk production. Previous studies have reported the role of oxidative stress in the etiology of dairy cattle disorders, which are known to ameliorate the severity of various metabolic and infectious diseases with a lack of antioxidant defense (Miller et al., 1993). Recent studies support the concept that oxidative stress is a significant underlying factor for the dysfunction of host immune and inflammatory responses that can increase the susceptibility of dairy cattle to a range of health ailments, particularly at the period of transition (Allison and Laven, 2000; Bernabucci et al., 2005; Castillo et al., 2005; Sordillo, 2005). Pre and postcalving are considered critical periods for lactating dairy cows and a keen inspection on development of nutritional management after the birth is to be monitored because it is related to low intake of dry matter and demands high energy for milk production, which may lead to an imbalance in energy in dairy cattle (Esposito et al., 2014). Numerous genetic, physiological, and environmental factors can compromise the cow's immunological defenses and increase the incidence of disease during the perinatal period (Sordillo, 2005). Physiological changes as well as losses in overall antioxidant potential were recognized in both humans and cows (Castillo et al., 2005; Gitto et al., 2002; Stefanon et al., 2005). The possibility that oxidative stress during the transition period may be a major underlying cause of inflammatory and immune dysfunction in dairy cattle is supported by both in vivo and in vitro studies. The altered balance in energy may result in physiological variations, characterized by high free radical formation in the bloodstream that affects the normal metabolism (Sordillo, 2016). Further, an increase in lipid peroxidation occurs during postcalving due to a greater mobilization of body fat as a result of the lower intake of dry matter and the greater nutritional requirement for increased milk production (Wathes et al., 2007). Throughout the transition period, there is a normal response to free radical formation in the body through antioxidant enzymes that are regulated by minerals. The minerals play a key role in the immune and reproductive systems and in the growth of dairy cows (Shankar and Prasad, 1998).

Supplementing dairy cows with adequate levels of vitamin E and selenium (Se) have been shown to increase phagocytosis, bacterial killing, and the oxidative metabolism of peripheral blood and mammary gland neutrophils when compared to cows that were otherwise deficient in this micronutrient (Grasso et al., 1990; Hogan et al., 1990). Macrophage functions also could be enhanced by vitamin E supplementation. Cattle supplemented with vitamin E have shown an increased production of macrophage-derived interleukin 1 and MHC class II expression (Politis et al., 1995). Studies have reported that the in vitro supplementation of α-tocopherol to bovine peripheral blood mononuclear cells resulted in enhanced pokeweed mitogen-induced production of IgM (Stabel et al., 1992). More recent studies suggest that leukocyte–endothelial cell interactions could be affected by improved Se status. Oxidative stress can enhance the adhesion of activated neutrophils to mammary endothelial cells, an effect that can be reversed with Se supplementation (Maddox et al., 1999). Selenium also has a significant impact on restoring vascular homeostasis during prooxidant challenge. Several studies showed that Se supplementation of bovine endothelial cell populations can decrease ROS-induced apoptosis, adhesion molecule expression, platelet activating factor synthesis, and proinflammatory eicosanoid production (Cao et al., 2001). Thus, Se may affect the magnitude and speed of leukocyte migration into tissues and influence the efficient resolution of the inflammatory response through its impact on the vascular endothelium. However, micronutrients are highly potent to counteract the deleterious effects of intracellular ROS accumulation and evoke the immune response in this regard.

Immunological reflex in poultry birds on redox burden

Poultry are among fastest growing animal industries, and they have a substantial contribution to food security and nutrition. Poultry meat and eggs are among the most common animal sources of food consumed at the global level. Globally, industries such as poultry and egg production are among the largest agricultural commodities. Massive technological improvements have been made in genetics, feed conversion ratio, fat reduction, and increased breast size of broiler chickens; also, substantial improvement has been made in egg production and egg quality in laying hens (Reyer et al., 2015; Richards and Proszkowiec-Weglarz, 2007). In poultry farming, stress has become an inevitable experience for domestic birds. Commercial poultry production is facing various stress factors that compromise the productive and reproductive performance of growing chickens, parent birds, and laying hens. The feed is a major component that influences growth and egg laying in birds (Ulluwishewa et al., 2011). It is fact that the stress response stands as a built-in security system, alerting animals on possible dangers; successive acute or chronic stress on a long run could be detrimental.

Poultry feed is often contaminated with a wide range of environmental toxicants as well as bacterial and fungal toxins, and it is known to affect gut health. The intestinal luminal epithelial cells and tight junction proteins between two adjacent epithelial cells form a barrier, thus preventing the paracellular absorption of toxins. Oxidative stress predominantly alters not only the cellular processes but also the intestinal barrier function. Feed exposes the birds to a wide variety of factors through the gastrointestinal tract (GIT) (Gitto et al., 2002), and can affect poultry health and production. The GIT provides the biological environment for nutrient digestion and absorption as well as protection from pathogens and toxins. Physiologically, ROS and RNS are generated by GIT epithelial cells, either from the oxygen metabolism or by enteric commensal bacteria regulate gut health. The GIT is a highly complex and dynamic organ that plays a critical role in nutrient absorption and immune response (Lan et al., 2007). The intestinal mucosa is a site for nutrient absorption while the intestinal epithelia are exposed continuously to a wide variety of potentially harmful substances. They act as a selective barrier between the tissues and luminal environment of the GIT. Several stressors/infections lead to inflammation in the GIT, which is mediated through ROS generation and interrupted redox balance. Many of the stressors such as feed toxin; infectious agents which induce the generation of cellular free radicals further results in redox imbalance. Stress can negatively affect the productivity of poultry products which caused by delicate balance among the components of the chicken GIT, which in turn, affect the health status of poultry. Mycotoxins (feed contaminant) are metabolites produced in a strain-specific way by a wide range of fungi, particularly molds. The common mycotoxins are aflatoxin, zearalenone, deoxynivalenol, trichothecenes, fumonisin, T-2 toxin, and ochratoxin. Once these toxins come in contact with epithelial cells or during absorption, the GIT is greatly impacted by the induction of oxidative stress. It has been shown that chronic long-term exposure to even low levels of mycotoxins may impact the immune system and intestinal integrity as well as compromise the blood phagocytic activity in chickens (Wu et al., 2010). These mycotoxins generate ROS, which induces lipid peroxidation while altering cellular redox signaling, the antioxidant status, and the membrane integrity of the cells with increased intestinal epithelial permeability in GIT. Together, mycotoxins increase cellular apoptosis and affect poultry health and production.

Arsenic contamination of water and food is a potent environmental threat. It is highly toxic and causes adverse effects on the digestion and absorption of nutrients, resulting in potential losses to poultry growth. Heavy or chronic arsenic exposure induces lipid peroxidation, decreases antioxidants, and eventually triggers apoptosis in several body tissues of poultry (Zhao et al., 2017, b). Copper, arsenic, and their combination induce the inflammation and destruction of intestinal mucosa (Wang et al., 2018). Ammonia is a primary source of air contamination in a poorly ventilated poultry house. High levels of ammonia decrease the growth rate and body weight while altering feed efficiency (Miles et al., 2004). Long time exposure of ammonia clues to several health issues and confrontation the welfare of broilers. The absorption capacity of the intestine depends on the number and size of villi. Poultry or chickens exposed to a high concentration of ammonia had a problem in segments of small intestine showed lower villus height and crypt depth. Ammonia also exerts negative impacts on the immune organ development of chickens, which may cause enormous damage to nutrient absorption and the immune system (Wei Feng-Xian et al., 2012). It has also been reported that ammonia exposure increases the activity of creatine kinase and decreases the activity of serum T-SOD, producing oxidative stress and apoptosis of the mucosal structure (Zhang et al., 2015a).

High temperature is one of the most challenging environmental stressors associated with poultry production (Altan et al., 2003). Heat stress is a major source of systemic oxidative stress, which thoroughly alters the redox imbalance between the pro- and antioxidants in favor of prooxidants. Heat stress has been shown to alter the feed intake while also causing poor growth performance, immunosuppression, hypoxia, and high mortality (Mujahid et al., 2007; Yahav et al., 1995). Heat stress also deteriorates the meat quality of chicken (Hashizawa et al., 2013). Birds under cyclic heat stress display less crypt depth, mucous area, and villus height of small intestine (Marchini et al., 2011), leading to negative impact on nutrient absorption. In addition, heat stress causes intestinal epithelial cell injury as well as apoptosis, which countersigns to intestinal hyper permeability conduits influx of bacterial products from the intestinal lumen into the circulatory system and affects organ systems. ROS components such as superoxide anions, hydrogen peroxide, and hydroxyl radicals are produced in the mitochondria and act as signaling intermediates (Kurutas, 2016). Under physiological conditions, antioxidant enzymes scavenge the liberated ROS. Several studies link oxidative stress with heat stress or lipopolysaccharide (LPS) and suggest the synergistic augmenting of cell death and increased ROS generation in specific cells (Quinteiro-Filho et al., 2010). Hypothalamus of chicken, pituitary adrenal axis, resulting in elevated serum corticosterone were activated by heat stress which in turn declines food intake, body weight gain, relative immune organ weight, and innate immunity, this neuro-immune dysfunction alters intestinal-immune barrier, permitting pathogenic bacteria to migrate through the intestinal mucosa and generating an inflammatory infiltrate. Inflammation of the intestine also decreases nutrition absorption and consequently decreases weight gain in poultry (Zhao et al., 2017).

The dietary inclusion of antioxidant compounds reduces intestinal free radicals, and also helps in maintaining the intestinal mucosa. Several studies showed that oxidative stress has an adverse effect that predisposes the birds to various pathological and welfare situations. Therefore, it is essential to formulate a cost-effective strategy to mitigate oxidative stress. Supplementations of vitamin C and E improve antioxidant ability and immune performance (Min et al., 2018). Alpha lipoic acid, which possesses fat and is water soluble, is a potent antioxidant that is protective against oxidative damage in the poultry intestine (El-Senousey et al., 2018). The inclusion of polyphenol compounds also exhibits potent antioxidant activity (Gerasopoulos et al., 2015). Dietary galacto-oligosaccharides, as a prebiotic, stabilize the intestinal integrity and prevent oxidative damage (Varasteh et al., 2015). Dietary supplementation of L-glutamine also prevents necrotic enteritis in antibiotic-free diets (Xue et al., 2018). Dietary glutamate and *N*-acetyl-cysteine induce several antioxidant genes and inflammatory biomarkers in the intestinal mucosa that alleviate LPS-induced intestinal inflammation (Hou et al., 2013); they can potentially be used in the chicken diet. Therefore, based on the stress, an individual ingredient or a combination can be used to mitigate oxidative stress in the intestinal tissue of poultry to improve the quality of meat and egg production.

Impact of stress on the goat immune system

Commercial businesses and recreational farming associated with domestic goats is developing substantially, and a variety of quality goat-derived products are being consumed around the world (Wilkinson and Stark, 1987). Due to this upswing in the importance of goats, they must be raised healthily and be observed carefully to maintain individual goats as well as the herd (Jaudas and Mobini, 2006). Additionally, the decline in reproductive function and aging testes are associated with loss of muscle mass and a concomitant increase in fat mass (Katznelson et al., 1996), decreased muscle strength , and decreased bone mineral density (Riggs et al., 1982), all of which can adversely affect the well-being of goats. However, the relationship between oxidative stress and testicular aging has not been investigated. Aging is a natural process associated with the time-dependent gradual deterioration of the physiological functions necessary for survival and reproduction due to several endogenous and exogenous factors. However, the precise molecular mechanisms underlying aging are not well established. Nevertheless, the excessive liberation of reactive free radicals and the lifespan have a strong correlation, indicating increased ROS accumulation to be a primary mechanism of aging (Sergiev et al., 2015). Morphological and molecular alterations in the reproductive organs with a decline in reproductive capacity are signs of aging, thereby affecting the reproductive system (Handelsman and Staraj, 1985). Gonads, especially the testis, are of particular interest in aging research due to the decline of reproductive capacity. Furthermore, testis aging has direct implications for longevity, as decreased testicular function is associated with decreased longevity in several species (Partridge et al., 2005).

Age-dependent elevation in mitochondrial oxidative stress can contribute toa decrease in testicular function. Studies have reported that aging increased markers of redox biomarkers such as ROS and nitrite with decreased mitochondrial complex activity, diminished antioxidant enzymes, and elevated levels of free radicals in goat testis (Majrashi et al., 2020). Sexual functions of goats have been shown to decline during aging (Smith et al., 2006). Mitochondria is the primary source of ROS often subjected to oxidative damage at a level that displaces the protective capacity of the antioxidant response. Lethal effects of ROS on cellular components lead to the accumulation of oxidative damage, which causes cellular dysfunction with age (Di Meo et al., 2016). Earlier studies reported that the increased accumulation of reactive ROS can cause oxidative stress with a significant decline in the levels of glutathione antioxidants with increased susceptibility in the elderly. Hence, organs with high rates of oxygen consumption and limited respiration levels such as the brain and heart are highly susceptible to this phenomenon. This partially explains the high prevalence of cardiovascular diseases and neurological disorders in the elderly (Corbi et al., 2008). A decline in antioxidant system in the testis could account for the decline in sexual and reproductive capacities. Redox status and dysregulation of the immune system were observed in aging with an exaggerated systemic inflammatory response with the activation of inflammatory mediators. Hence, chronic inflammation in aging may serve as a pathophysiologic association that converts normal functional changes to age-related degenerative diseases (Viola and Soehnlein, 2015).

Increased ROS in juvenile and adult goats may be associated with the innate immune system. The respiratory burst associated with innate immunity is a process that involves enzymes and produces several forms of ROS.

Further, ROS essentially impact cellular processes by affecting cell signaling, the transcription factor, and the translation regulator under controlled physiological conditions that in turn affect cell growth. However, the excessive production and prolonged exposure of ROS can induce lipid peroxides and protein carbonyls in both young and adult goat testis (Hauck and Bernlohr, 2016). Lipid peroxidation has been shown to affect sperm survival and fertility (Guthrie and Welch, 2012). Decreased sperm motility is due to ROS-induced lesions in ATP utilization or in the contractile apparatus of the flagellum, and can result in testicular hypoplasia and infertility (Majrashi et al., 2020). The relationship between oxidative stress, inflammation, and aging, the oxidation-inflammatory theory of aging has been proposed: aging is a loss of homeostasis due to a chronic oxidative stress. The consequent activation of the immune system induces an inflammatory state that creates a vicious circle in which chronic oxidative stress and inflammation feed on each other; consequently, this increases age-related morbidity and mortality (De la Monica and Jaime, 2009).

Additionally, the quality of dairy products depends on the raw milk, which is, in turn, primarily affected by animal health, milking intervals, milk storage, and time until processing (Leitner et al., 2008b; Roupas, 2001). A reduction in goat milk quality will have a negative influence on its industrial value, as most of the milk is processed into fermented products and cheese. The quality and quantity of milk in dairy animals, including goats, were found to be mainly affected by subclinical intramammary infections (Leitner et al., 2008; Silanikove et al., 2010), mainly due to various coagulase-negative staphylococci (Djabri et al., 2002; Leitner et al., 2004; Taponen et al., 2006). In dairy cows, subclinical mastitis has only a moderate projection on bulk milk because of the low increase in somatic cell count in the infected glands (Leitner et al., 2008b; Pitkälä et al., 2004) and the dilution effect of the four quarters. However, the influence of the infection reflected by the somatic cell count on the animal level is relatively high in sheep and goats compared with cows because of the strong immune response to the infection and the existence of only two mammary glands (Silanikove et al., 2011).

Sequential effects of oxido-reductive stress in piglets

Increased ROS accumulation triggers proinflammatory substances that further contribute to the induction of oxidants (Hernandez et al., 1987), thereby causing tissue injury. This is implicated in the pathogenesis of many diseases. In pigs, oxidative stress has been linked to several diseases and clinical conditions such as experimental endotoxemia or septic shock (Basu and Eriksson, 2000), respiratory infections (Basu and Eriksson, 2000), porcine reproductive and respiratory syndrome (Stukelj et al., 2013), and skin disease (Dimri et al., 2010). Built-in levels of ROS induced damage in the intestinal tissue, which can facilitate bacterial translocation and compromise the intestinal barrier integrity in pigs (Hernandez et al., 1987). In addition to diseases, changes in oxidative stress biomarkers in pigs can occur at different situations during their productive period, which are associated with variations in the oxidant balance of the animals. Studies have recorded increased ROS generation during transportation as a consequence of oxidative stress (Onmaz et al., 2011; Young et al., 2005; Zhang et al., 2015b). Assays such as Trolox equivalent antioxidant capacity (TEAC), total antioxidant capacity (TAC) assay, cupric reducing antioxidant capacity (CUPRAC), ferric reducing ability of plasma (FRAP), advanced oxidation protein products (AOPP), individual antioxidant (uric acid) assays to evaluate oxidant concentrations and hydrogen peroxide (H_2O_2) was measured in serum of pigs to check the elevated levels of changed dietary supplement. The collection of blood from pigs has an adverse effect that causes stress to the animals. It also implies the immobilization of the animal, and how exactly the stress can affect the different oxidant biomarkers at different levels. Total antioxidant capacity measurement in the plasma of piglets undergoing weaning-induced stress with increased ROS levels (Sauerwein et al., 2005) which often leads to growth depression and are more susceptible to diseases known as post-weaning stress syndrome (Campbell et al., 2013). In addition, the serum levels of ROS and thiobarbituric acid reactive substances (TBARS) are higher during gestation and lactation than in early gestation (Tan et al., 2015). It is known that during the perinatal period, particularly the delivery, and early lactation, there is an increased demand for high energy and oxygen, which attained by rapid differentiation of secretory parenchyma, intense mammary gland growth, and the onset of copious milk synthesis and secretion with elevated metabolic rate and tissue oxygen (Gitto et al., 2002). This physiological situation can lead to an oxidative stress state due to the overproduction of oxidants (Agarwal et al., 2005).

Pigs are exposed to various types of stress during their lives such as dietary, social, and environmental, including metabolic stress through high performance. Some stages are considered to be especially critical and can result in oxidative stress. Even a small change in the energy metabolism or immune response can disturb the delicate balance between oxidant and antioxidant status. Oxidative stress is directly linked with inflammation because oxidants are activators of the nuclear transcription factor NF-κB, the key regulator of inflammation (Pantano et al., 2006). NF-κB regulates proteins such as cytokines that stimulate the production of oxidants by activated neutrophils and promote oxidative stress that finally ends in a vicious cycle. Oxidative stress decreases immunity and causes muscle degeneration, loss of appetite, diarrhea, liver damage, and finally cell death. Oxidative stress and related inflammation are suggested to be involved in a number of disorders in pigs, such as mulberry heart disease, leaky gut syndrome, pneumonia, and the MMA complex in sows. Piglets could be stressed when separated from their mother and shifted to other barns (Kannan and Jain, 2000). Dietary stress is typically caused by poor-quality feed while dietary stress results due to replacement of mainly plant-based solid feed rather easily digestible sow milk. Hence, in animal husbandry, stable diet formulations are necessary to improve disease complications caused by oxidative stress through exogenous agents. The advancements in today's scientific technology in terms of food and animal husbandry increase opportunities to develop a sustainable livestock population. However, maintaining normal biological status in livestock with limited immune complexity by catering to proper health and nutritional supplements is of prime importance in view of the global economic interest, commercial benefits, and conservation of wild livestock breeds (Table 1).

Table 1 Consequences associated with oxidative stress through endogenous and exogenous risk factors.

Livestock	Risk factors for stress	Consequence
Poultry	Heat, toxins, polluted water, decreased food metabolism, neuroimmune disorder, enzyme dysfunction	Poor growth, decreased production of meat, ineffective egg laying and hatching
Goats	Mitochondrial dysfunction, reduced glutathione antioxidant enzyme, ATP utilization	Reproductive incapability, increase in fat mass, decreased muscle mass and bone density
Dairy cows	Dietary stress due to unbalanced processed food, feed contaminants, xenobiotics	Affects milk storage, loss in milk production, abnormal fetal development, reduced fertility, sepsis, bovine mastitis, and enteritis
Pigs	Dietary metabolic stress, temperature variation, pathogen susceptibility	Early lactation, endotoxemia, septic shock, respiratory infection, and skin disease

References

Agarwal, A., Gupta, S., Sharma, R.K., 2005. Role of oxidative stress in female reproduction. Reprod. Biol. Endocrinol. 3, 28.

Allison, R.D., Laven, R.A., 2000. Effect of vitamin E supplementation on the health and fertility of dairy cows: a review. Vet. Rec. 147, 703–708.

Altan, Ö., Pabuçcuoğlu, A., Altan, A., Konyalioğlu, S., Bayraktar, H., 2003. Effect of heat stress on oxidative stress, lipid peroxidation and some stress parameters in broilers. Br. Poult. Sci. 44, 545–550.

Basu, S., Eriksson, M., 2000. Vitamin E in relation to lipid peroxidation in experimental septic shock. Prostaglandins Leukot. Essent. Fat. Acids 62, 195–199.

Belikov, A.V., Schraven, B., Simeoni, L., 2015. T cells and reactive oxygen species. J. Biomed. Sci. 22, 85.

Bernabucci, U., Ronchi, B., Lacetera, N., Nardone, A., 2005. Influence of body condition score on relationships between metabolic status and oxidative stress in periparturient dairy cows. J. Dairy Sci. 88, 2017–2026.

Campbell, J.M., Crenshaw, J.D., Polo, J., 2013. The biological stress of early weaned piglets. J. Anim. Sci. Biotechnol. 4, 19.

Cao, Y.Z., Cohen, Z.S., Weaver, J.A., Sordillo, L.M., 2001. Selenium modulates 1-O-alkyl-2-acetyl-sn-glycero-3-phosphocholine (PAF) biosynthesis in bovine aortic endothelial cells. Antioxid. Redox Signal. 3, 1147–1152.

Castillo, C., Hernandez, J., Bravo, A., Lopez-Alonso, M., Pereira, V., Benedito, J.L., 2005. Oxidative status during late pregnancy and early lactation in dairy cows. Vet. J. 169, 286–292.

Christoffersson, G., Phillipson, M., 2018. The neutrophil: one cell on many missions or many cells with different agendas? Cell Tissue Res. 371, 415–423.

Colitti, M., Stefanon, B., Gabai, G., Gelain, M.E., Bonsembiante, F., 2019. Oxidative stress and nutraceuticals in the modulation of the immune function: current knowledge in animals of veterinary interest. Antioxidants (Basel) 8, 28.

Corbi, G., Acanfora, D., Iannuzzi, G.L., Longobardi, G., Cacciatore, F., Furgi, G., Filippelli, A., Rengo, G., Leosco, D., Ferrara, N., 2008. Hypermagnesemia predicts mortality in elderly with congestive heart disease: relationship with laxative and antacid use. Rejuvenation Res. 11, 129–138.

Costantini, D., 2008. Oxidative stress in ecology and evolution: lessons from avian studies. Ecol. Lett. 11, 1238–1251.

Davies, D.I., Parrott, M.J., 1978. Free Radicals in Organic Synthesis. Springer-Verlag Berlin, New York.

De la Monica, F., Jaime, M., 2009. An update of the oxidation-inflammation theory of aging: the involvement of the immune system in oxi-inflammaging. Curr. Pharm. Des. 15, 3003–3026.

Dervishi, E., Zhang, G., Hailemariam, D., Goldansaz, S.A., Deng, Q., Dunn, S.M., Ametaj, B.N., 2016. Alterations in innate immunity reactants and carbohydrate and lipid metabolism precede occurrence of metritis in transition dairy cows. Res. Vet. Sci. 104, 30–39.

Di Meo, S., Reed, T.T., Venditti, P., Victor, V.M., 2016. Role of ROS and RNS sources in physiological and pathological conditions. Oxidative Med. Cell. Longev., 1–44.

Dimri, U., Sharma, M.C., Yamdagni, A., Ranjan, R., Zama, M.M., 2010. Psoroptic mange infestation increases oxidative stress and decreases antioxidant status in sheep. Vet. Parasitol. 168, 318–322.

Djabri, B., Bareille, N., Beaudeau, F., Seegers, H., 2002. Quarter milk somatic cell count in infected dairy cows: a meta-analysis. Vet. Res. 33, 335–357.

Drackley, J.K., 1999. ADSA Foundation Scholar Award. Biology of dairy cows during the transition period: the final frontier? J. Dairy Sci. 82, 2259–2273.

El-Senousey, H.K., Chen, B., Wang, J.Y., Atta, A.M., Mohamed, F.R., Nie, Q.H., 2018. Effects of dietary vitamin C, vitamin E, and alpha-lipoic acid supplementation on the antioxidant defense system and immune-related gene expression in broilers exposed to oxidative stress by dexamethasone. Poult. Sci. 97, 30–38.

Esposito, G., Irons, P.C., Webb, E.C., Chapwanya, A., 2014. Interactions between negative energy balance, metabolic diseases, uterine health and immune response in transition dairy cows. Anim. Reprod. Sci. 144, 60–71.

Galon, J., Franchimont, D., Hiroi, N., Frey, G., Boettner, A., Ehrhart-Bornstein, M., O'Shea, J.J., Chrousos, G.P., Bornstein, S.R., 2002. Gene profiling reveals unknown enhancing and suppressive actions of glucocorticoids on immune cells. FASEB J. 16, 61–71.

Gerasopoulos, K., Stagos, D., Kokkas, S., Petrotos, K., Kantas, D., Goulas, P., Kouretas, D., 2015. Feed supplemented with byproducts from olive oil mill wastewater processing increases antioxidant capacity in broiler chickens. Food Chem. Toxicol. 82, 42–49.

Gitto, E., Reiter, R.J., Karbownik, M., Tan, D.X., Gitto, P., Barberi, S., Barberi, I., 2002. Causes of oxidative stress in the pre- and perinatal period. Biol. Neonate 81, 146–157.

Goyns, M.H., 2002. Genes, telomeres and mammalian ageing. Mech. Ageing Dev. 123, 791–799.

Grasso, P.J., Scholz, R.W., Erskine, R.J., Eberhart, R.J., 1990. Phagocytosis, bactericidal activity, and oxidative metabolism of milk neutrophils from dairy cows fed selenium-supplemented and selenium-deficient diets. Am. J. Vet. Res. 51, 269–274.

Guthrie, H.D., Welch, G.R., 2012. Effects of reactive oxygen species on sperm function. Theriogenology 78, 1700–1708.

Halliwell, B., 2007. Biochemistry of oxidative stress. Biochem. Soc. Trans. 35, 1147–1150.

Halliwell, B., Gutteridge, J.M.C., 2007. Free Radicals in Biology and Medicine. Oxford University Press.

Handelsman, D.J., Staraj, S., 1985. Testicular size: the effects of aging, malnutrition, and illness. J. Androl. 6, 144–151.

Harman, D., 1956. Aging: a theory based on free radical and radiation chemistry. J. Gerontol. 11, 298–300.

Hashizawa, Y., Kubota, M., Kadowaki, M., Fujimura, S., 2013. Effect of dietary vitamin E on broiler meat qualities, color, water-holding capacity and shear force value, under heat stress conditions. Anim. Sci. J. 84, 732–736.

Hauck, A.K., Bernlohr, D.A., 2016. Oxidative stress and lipotoxicity. J. Lipid Res. 57, 1976–1986.

Hernandez, L.A., Grisham, M.B., Twohig, B., Arfors, K.E., Harlan, J.M., Granger, D.N., 1987. Role of neutrophils in ischemia-reperfusion-induced microvascular injury. Am. J. Physiol. 253, H699–H703.

Hogan, J.S., Smith, K.L., Weiss, W.P., Todhunter, D.A., Schockey, W.L., 1990. Relationships among vitamin E, selenium, and bovine blood neutrophils1. J. Dairy Sci. 73, 2372–2378.

Hou, Y., Wang, L., Yi, D., Ding, B., Yang, Z., Li, J., Chen, X., Qiu, Y., Wu, G., 2013. N-acetylcysteine reduces inflammation in the small intestine by regulating redox, EGF and TLR4 signaling. Amino Acids 45, 513–522.

Iorio, E.L., 2017. Oxidative stress and immune system: a review from the literature. Immunome Res.

Jaudas, U., Mobini, S., 2006. The Goat Handbook, first ed. Barron's, New York.

Kannan, K., Jain, S.K., 2000. Oxidative stress and apoptosis. Pathophysiology 7, 153–163.

Katznelson, L., Finkelstein, J.S., Schoenfeld, D.A., Rosenthal, D.I., Anderson, E.J., Klibanski, A., 1996. Increase in bone density and lean body mass during testosterone administration in men with acquired hypogonadism. J. Clin. Endocrinol. Metab. 81, 4358–4365.

Kiran Kumar, K., Naveen Kumar, M., Patil, R.H., Nagesh, R., Hegde, S.M., Kavya, K., Babu, R., Ramesh, G.T., Sharma, S.C., 2016. Cadmium induces oxidative stress and apoptosis in lung epithelial cells. Toxicol. Mech. Methods 26, 658–666.

Kiran Kumar, K., Naveen Kumar, M., Nagesh, R., Patil, R.H., Babu, R., Sharma, S.C., Manjunath, K., 2017. Arsenic induces oxidative stress and apoptosis in human lung cancer cells. Int. J. Life Sci. 5, 551–560.

Kurutas, E.B., 2016. The importance of antioxidants which play the role in cellular response against oxidative/nitrosative stress: current state. Nutr. J. 15, 71.

Kurutas, E.B., Ciragil, P., Gul, M., Kilinc, M., 2005. The effects of oxidative stress in urinary tract infection. Mediat. Inflamm. 2005, 242–244.

Kvietys, P.R., Granger, D.N., 2012. Role of reactive oxygen and nitrogen species in the vascular responses to inflammation. Free Radic. Biol. Med. 52, 556–592.

Lan, Y., Verstegen, M.W.A., Tamminga, S., Williams, B.A., 2007. The role of the commensal gut microbial community in broiler chickens. Worlds Poult. Sci. J. 61, 95–104.

Leitner, G., Merin, U., Silanikove, N., Ezra, E., Chaffer, M., Gollop, N., Winkler, M., Glickman, A., Saran, A., 2004. Effect of subclinical intramammary infection on somatic cell counts, NAGase activity and gross composition of goats' milk. J. Dairy Res. 71, 311–315.

Leitner, G., Silanikove, N., Merin, U., 2008. Estimate of milk and curd yield loss of sheep and goats with intrammamary infection and its relation to somatic cell count. Small Rumin. Res. 74, 221–225.

Leitner, G., Krifucks, O., Jacoby, S., Lavi, Y., Silanikove, N., 2008b. Concentrations of ganglioside type M1 and immunoglobulin G in colostrum are inversely related to bacterial infection at early lactation in cows. J. Dairy Sci. 91, 3337–3342.

Lundberg, J.O., Weitzberg, E., Gladwin, M.T., 2008. The nitrate-nitrite-nitric oxide pathway in physiology and therapeutics. Nat. Rev. Drug Discov. 7, 156–167.

Lykkesfeldt, J., Svendsen, O., 2007. Oxidants and antioxidants in disease: oxidative stress in farm animals. Vet. J. 173, 502–511.

Maddox, J.F., Aherne, K.M., Reddy, C.C., Sordillo, L.M., 1999. Increased neutrophil adherence and adhesion molecule mRNA expression in endothelial cells during selenium deficiency. J. Leukoc. Biol. 65, 658–664.

Majrashi, M., Fujihashi, A., Almaghrabi, M., Fadan, M., Fahoury, E., Ramesh, S., Govindarajulu, M., 2020. Augmented oxidative stress and reduced mitochondrial function in ageing goat testis. Vet. Med. Sci. 6 (4), 766–774.

Marchini, C.F.P., Silva, P., Nascimento, M.R.B.M., Beletti, M.E., Silva, N.M., Guimaraes, E.C., 2011. Body weight, intestinal morphometry and cell proliferation of broiler chickens submitted to cyclic heat stress. Int. J. Poult. Sci. 10, 455–460.

Mavangira, V., Sordillo, L.M., 2018. Role of lipid mediators in the regulation of oxidative stress and inflammatory responses in dairy cattle. Res. Vet. Sci. 116, 4–14.

McDonagh, B., Sheehan, D., 2008. Effects of oxidative stress on protein thiols and disulphides in Mytilus edulis revealed by proteomics: actin and protein disulphide isomerase are redox targets. Mar. Environ. Res. 66, 193–195.

Miles, D.M., Branton, S.L., Lott, B.D., 2004. Atmospheric ammonia is detrimental to the performance of modern commercial broilers. Poult. Sci. 83, 1650–1654.

Miller, J.K., Brzezinska-Slebodzinska, E., Madsen, F.C., 1993. Oxidative stress, antioxidants, and animal function. J. Dairy Sci. 76, 2812–2823.

Min, Y.N., Niu, Z.Y., Sun, T.T., Wang, Z.P., Jiao, P.X., Zi, B.B., Chen, P.P., Tian, D.L., Liu, F.Z., 2018. Vitamin E and vitamin C supplementation improves antioxidant status and immune function in oxidative-stressed breeder roosters by up-regulating expression of GSH-Px gene. Poult. Sci. 97, 1238–1244.

Mujahid, A., Akiba, Y., Toyomizu, M., 2007. Acute heat stress induces oxidative stress and decreases adaptation in young white leghorn cockerels by downregulation of avian uncoupling protein. Poult. Sci. 86, 364–371.

Nagesh, R., Kumar, K.K., Kumar, M.N., Patil, R.H., Kavya, K., Babu, R., Ramesh, G.T., Sharma, S.C., 2017. Aqueous areca nut extract induces oxidative stress in human lung epithelial A549 cells: probable role of p21 in inducing cell death. Gene Rep. 6, 103–111.

Niki, E., 2009. Lipid peroxidation: physiological levels and dual biological effects. Free Radic. Biol. Med. 47, 469–484.

Niki, E., 2014. Biomarkers of lipid peroxidation in clinical material. Biochim. Biophys. Acta Gen. Subj. 1840, 809–817.

Nussey, D.H., Pemberton, J.M., Pilkington, J.G., Blount, J.D., 2009. Life history correlates of oxidative damage in a free-living mammal population. Funct. Ecol. 23, 809–817.

Onmaz, A.C., Hoven, R., Güne, V., Cinar, M., Kucuk, O., 2011. Oxidative stress in horses after a 12-hours transport period. Revue Méd. Vét. 162, 213–217.

Pantano, C., Reynaert, N.L., van der Vliet, A., Janssen-Heininger, Y.M., 2006. Redox-sensitive kinases of the nuclear factor-kappaB signaling pathway. Antioxid. Redox Signal. 8, 1791–1806.

Partridge, L., Gems, D., Withers, D.J., 2005. Sex and death: what is the connection? Cell 120, 461–472.

Patil, R.H., Babu, R., Kumar, M.N., Kumar, K.K., Hegde, S.M., Nagesh, R., Ramesh, G. T., Sharma, S.C., 2016. Anti-inflammatory effect of apigenin on LPS-induced pro-inflammatory mediators and AP-1 factors in human lung epithelial cells. Inflammation 39, 138–147.

Phaniendra, A., Jestadi, D.B., Periyasamy, L., 2015. Free radicals: properties, sources, targets, and their implication in various diseases. Indian J. Clin. Biochem. 30, 11–26.

Pitkälä, A., Haveri, M., Pyörälä, S., Myllys, V., Honkanen-Buzalski, T., 2004. Bovine mastitis in Finland 2001—prevalence, distribution of bacteria, and antimicrobial resistance. J. Dairy Sci. 87, 2433–2441.

Pizzino, G., Irrera, N., Cucinotta, M., Pallio, G., Mannino, F., Arcoraci, V., Squadrito, F., Altavilla, D., Bitto, A., 2017. Oxidative stress: harms and benefits for human health. Oxidative Med. Cell. Longev. 2017, 8416763.

Politis, I., Hidiroglou, M., Batra, T.R., Gilmore, J.A., Gorewit, R.C., Scherf, H., 1995. Effects of vitamin E on immune function of dairy cows. Am. J. Vet. Res. 56, 179–184.

Quinteiro-Filho, W.M., Ribeiro, A., Ferraz-de-Paula, V., Pinheiro, M.L., Sakai, M., Sá, L. R., Ferreira, A.J., Palermo-Neto, J., 2010. Heat stress impairs performance parameters, induces intestinal injury, and decreases macrophage activity in broiler chickens. Poult. Sci. 89, 1905–1914.

Reyer, H., Hawken, R., Murani, E., Ponsuksili, S., Wimmers, K., 2015. The genetics of feed conversion efficiency traits in a commercial broiler line. Sci. Rep. 5, 16387.

Richards, M.P., Proszkowiec-Weglarz, M., 2007. Mechanisms regulating feed intake, energy expenditure, and body weight in poultry. Poult. Sci. 86, 1478–1490.

Riggs, B.L., Wahner, H.W., Seeman, E., Offord, K.P., Dunn, W.L., Mazess, R.B., Johnson, K.A., Melton III, L.J., 1982. Changes in bone mineral density of the proximal femur and spine with aging: differences between the postmenopausal and senile osteoporosis syndromes syndromes. J. Clin. Invest. 70, 716–723.

Roupas, P., 2001. On-farm practices and post farmgate processing parameters affecting composition of milk for cheesemaking. Aust. J. Dairy Technol. 56, 219.

Salahshoor, M.R., Abdolmaleki, A., Faramarzi, A., Ziapour, A., Roshankhah, S., 2020. Thymus vulgaris attenuates myleran-induced reproductive damage by decreasing oxidative stress and lipid peroxidation in male rats. J. Hum. Reprod. Sci. 13, 38–45.

Sauerwein, H., Schmitz, S., Hiss, S., 2005. The acute phase protein haptoglobin and its relation to oxidative status in piglets undergoing weaning-induced stress. Redox Rep. 10, 295–302.

Sergiev, P.V., Dontsova, O.A., Berezkin, G.V., 2015. Theories of aging: an ever-evolving field. Acta Naturae 7, 9–18.

Shankar, A.H., Prasad, A.S., 1998. Zinc and immune function: the biological basis of altered resistance to infection. Am. J. Clin. Nutr. 68, 447s–463s.
Sies, H., 1986. Biochemistry of oxidative stress. Angew. Chem. Int. Ed. Engl. 25, 1058–1071.
Silanikove, N., Leitner, G., Merin, U., Prosser, C.G., 2010. Recent advances in exploiting goat's milk: quality, safety and production aspects. Small Rumin. Res. 89, 110–124.
Silanikove, N., Rauch-Cohen, A., Shapiro, F., Blum, S., Arieli, A., Leitner, G., 2011. Lipopolysaccharide challenge of the mammary gland in bovine induced a transient glandular shift to anaerobic metabolism. J. Dairy Sci. 94, 4468–4475.
Smith, K.C., Brown, P., Parkinson, T.J., 2006. Hypospadias in rams. Vet. Rec. 158, 789–795.
Sohal, R.S., Weindruch, R., 1996. Oxidative stress, caloric restriction, and aging. Science (New York, N.Y.) 273, 59–63.
Sohal, R.S., Svensson, I., Brunk, U.T., 1990. Hydrogen peroxide production by liver mitochondria in different species. Mech. Ageing Dev. 53, 209–215.
Sorci, G., Faivre, B., 2009. Inflammation and oxidative stress in vertebrate host-parasite systems. Philos. Trans. R. Soc. Lond. Ser. B Biol. Sci. 364, 71–83.
Sordillo, L.M., 2005. Factors affecting mammary gland immunity and mastitis susceptibility. Livest. Prod. Sci. 98, 89–99.
Sordillo, L.M., 2016. Nutritional strategies to optimize dairy cattle immunity. J. Dairy Sci. 99, 4967–4982.
Sordillo, L.M., Aitken, S.L., 2009. Impact of oxidative stress on the health and immune function of dairy cattle. Vet. Immunol. Immunopathol. 128, 104–109.
Sorg, O., 2004. Oxidative stress: a theoretical model or a biological reality? C. R. Biol. 327, 649–662.
Stabel, J.R., Reinhardt, T.A., Stevens, M.A., Kehrli Jr., M.E., Nonnecke, B.J., 1992. Vitamin E effects on in vitro immunoglobulin M and interleukin-1 beta production and transcription in dairy cattle. J. Dairy Sci. 75, 2190–2198.
Stefanon, B., Sgorlon, S., Gabai, G., 2005. Usefulness of nutraceutics in controlling oxidative stress in dairy cows around parturition. Vet. Res. Commun. 29 (Suppl. 2), 387–390.
Stukelj, M., Toplak, I., Svete, A.N., 2013. Blood antioxidant enzymes (SOD, GPX), biochemical and haematological parameters in pigs naturally infected with porcine reproductive and respiratory syndrome virus. Pol. J. Vet. Sci. 16, 369–376.
Swindle, E.J., Metcalfe, D.D., 2007. The role of reactive oxygen species and nitric oxide in mast cell-dependent inflammatory processes. Immunol. Rev. 217, 186–205.
Tan, C., Wei, H., Sun, H., Ao, J., Long, G., Jiang, S., Peng, J., 2015. Effects of dietary supplementation of oregano essential oil to sows on oxidative stress status, lactation feed intake of sows, and piglet performance. Biomed. Res. Int. 2015, 525218.
Taponen, S., Simojoki, H., Haveri, M., Larsen, H.D., Pyorala, S., 2006. Clinical characteristics and persistence of bovine mastitis caused by different species of coagulase-negative staphylococci identified with API or AFLP. Vet. Microbiol. 115, 199–207.
Ulluwishewa, D., Anderson, R.C., McNabb, W.C., Moughan, P.J., Wells, J.M., Roy, N. C., 2011. Regulation of tight junction permeability by intestinal bacteria and dietary components. J. Nutr. 141, 769–776.
Valko, M., Leibfritz, D., Moncol, J., Cronin, M.T., Mazur, M., Telser, J., 2007. Free radicals and antioxidants in normal physiological functions and human disease. Int. J. Biochem. Cell Biol. 39, 44–84.
Van Dyke, K., 1997. The possible role of peroxynitrite in Alzheimer's disease: a simple hypothesis that could be tested more thoroughly. Med. Hypotheses 48, 375–380.
Varasteh, S., Braber, S., Akbari, P., Garssen, J., Fink-Gremmels, J., 2015. Differences in susceptibility to heat stress along the chicken intestine and the protective effects of galacto-oligosaccharides. PLoS One 10, e0138975.

Veal, E.A., Day, A.M., Morgan, B.A., 2007. Hydrogen peroxide sensing and signaling. Mol. Cell 26, 1–14.

Victor, V.M., Rocha, M., De la Fuente, M., 2004. Immune cells: free radicals and antioxidants in sepsis. Int. Immunopharmacol. 4, 327–347.

Viola, J., Soehnlein, O., 2015. Atherosclerosis—a matter of unresolved inflammation. Semin. Immunol. 27, 184–193.

Wang, Y., Zhao, H., Liu, J., Shao, Y., Li, J., Luo, L., Xing, M., 2018. Copper and arsenic-induced oxidative stress and immune imbalance are associated with activation of heat shock proteins in chicken intestines. Int. Immunopharmacol. 60, 64–75.

Wathes, D.C., Fenwick, M., Cheng, Z., Bourne, N., Llewellyn, S., Morris, D.G., Kenny, D., Murphy, J., Fitzpatrick, R., 2007. Influence of negative energy balance on cyclicity and fertility in the high producing dairy cow. Theriogenology 68 (Suppl. 1), S232–S241.

Wei Feng-Xian, X.B., Hu, X.-F., Shao-Yu, L., Fu-Zhu, L., Quan-You, S., Yu-Ping, J., Wang, L.-Y., 2012. The effect of ammonia and humidity in poultry houses on intestinal morphology and function of broilers. J. Anim. Vet. Adv. 11, 3641–3646.

Wilkinson, J.M., Stark, B., 1987. Commercial Goat Production.

Wu, Q., Dohnal, V., Huang, L., Kuca, K., Yuan, Z., 2010. Metabolic pathways of trichothecenes. Drug Metab. Rev. 42, 250–267.

Xue, G.D., Barekatain, R., Wu, S.B., Choct, M., Swick, R.A., 2018. Dietary L-glutamine supplementation improves growth performance, gut morphology, and serum biochemical indices of broiler chickens during necrotic enteritis challenge. Poult. Sci. 97, 1334–1341.

Yahav, S., Goldfeld, S., Plavnik, I., Hurwitz, S., 1995. Physiological responses of chickens and turkeys to relative humidity during exposure to high ambient temperature. J. Therm. Biol. 20, 245–253.

Young, J.F., Rosenvold, K., Stagsted, J., Nielsen, J.H., Andersen, H.J., 2005. Significance of vitamin E supplementation, dietary content of polyunsaturated fatty acids, and pre-slaughter stress on oxidative status in pig as reflected in cell integrity and antioxidative enzyme activities in porcine muscle. J. Agric. Food Chem. 53, 745–749.

Zhang, J., Li, C., Tang, X., Lu, Q., Sa, R., Zhang, H., 2015a. Proteome changes in the small intestinal mucosa of broilers (Gallus gallus) induced by high concentrations of atmospheric ammonia. Proteome Sci. 13, 9.

Zhang, T., Zhou, Y.F., Zou, Y., Hu, X.M., Zheng, L.F., Wei, H.K., Giannenas, I., Jin, L. Z., Peng, J., Jiang, S.W., 2015b. Effects of dietary oregano essential oil supplementation on the stress response, antioxidative capacity, and HSPs mRNA expression of transported pigs. Livest. Sci. 180, 143–149.

Zhao, H., He, Y., Li, S., Sun, X., Wang, Y., Shao, Y., Hou, Z., Xing, M., 2017. Subchronic arsenism-induced oxidative stress and inflammation contribute to apoptosis through mitochondrial and death receptor dependent pathways in chicken immune organs. Oncotarget 8, 40327–40344.

CHAPTER ELEVEN

Cow urine: Potential resource for sustainable agriculture

M. Devasena[a] and V. Sangeetha[b]
[a]Department of Civil Engineering, PSG Institute of Technology and Applied Sciences, Coimbatore, India
[b]Department of Civil Engineering, Coimbatore Institute of Technology, Coimbatore, India

1 Introduction

Of late, there has been significant sensitization at both the national and international levels to promote sustainable agriculture and encourage safe consumption of food by the use of organic fertilizers. While organic fertilizers involve the use of manure and compost, the decomposed products of organic materials, inorganic or chemical fertilizers are based on synthesized formulations that aid in the quick absorption of nutrients. Chemical fertilizers used copiously are unsustainable for a plethora of reasons, including high application inefficiency in terms of plant tissue drying out, tremendous energy and water inputs for production, and potential negative environmental implications. The achievement and sustainability of food security are momentous global challenges demanding modified agricultural practices to efficiently combat the consequences of climate change, population increase, and arable land loss. Organic fertilizers potentially enhance crop growth and increase yield without overfertilizing, unlike commercial chemical fertilizers. It is evident that organic fertilizer-induced plant nutrition vividly improves crop defense against pathogenic diseases while also enhancing crop production.

On a global scale, widespread chemical fertilizer use has increased over the last 4 decades; on the contrary, nutrient use efficiency by plants has decreased. Organic fertilizers maintain a nutrient-rich soil, improving the standard of crops produced (Yin et al., 2017). Also, a harmonious balance among various complex ecosystems is achieved with organic fertilizers sourced from livestock. Cow urine, the main constituent of organic fertilizer, and its benefits are exploited by direct application or in a novel way such as nanoparticles. They form significant alternatives to expensive synthetic

Emerging Issues in Climate Smart Livestock Production
https://doi.org/10.1016/B978-0-12-822265-2.00007-7

fertilizers that pose potential risk to farmers, consumers, marketers, and the environment in the long run.

A healthy plant essentially requires nitrogen (N), phosphorus (P), and potassium (K), as they form three major macronutrients. Nitrogen is an important component of proteins essential for stem, leaf, and root growth while phosphorus aids in developing strong roots and potassium is vital for tuber and fruit growth. The integration of crops and livestock facilitates nutrient cycling for sustainable agriculture. It has been reported that only 20% of the nitrogenous material consumed by cows is absorbed while 52% of the dietary nitrogen intake is expelled in cow urine and 28% in cow dung. Apart from urea, which constitutes a major portion of the total nitrogenous material excreted (69%), cow urine also contains 5.8% hippuric acid, 7.3% allantoin, 3.7% creatinine, 1.3% uric acid, 2.5% creatine, 0.5% hypoxanthine, 2.8% ammonia, and 1.3% free amino acid N. A healthy cow generates 6 L of urine a day. A farmer rearing two cows at home will benefit from 4380 L of urine annually, which equals 65 kg of nitrogen or 136 kg of urea. In return, the crop residues are utilized as animal feed to maintain the nutrient cycle and promote resource recovery and sustainable agriculture. Unassimilated nutrients from the body are expelled via urine and dung. Livestock urine is an imperative facilitator of the nutrient cycle in agricultural systems. This chapter focuses on the potential aspects of cow urine to promote sustainable agriculture in terms of direct foliar spray or by indirect application methods.

2 Composition of cow urine

Cow urine possesses versatile applications in agriculture. The practice dates back to the Vedic period. It is a natural nutrient-rich resource that improves soil fertility and crop yield while promoting sustainability in agriculture through direct or indirect usage. Cow dung as manure has been established in a plethora of formulations and applications, but cow urine is generally drained or wasted from farms and gaushalas. It is a micro- and macronutrient source that revives soil fertility and increases the protein and chlorophyll content of crops. Cow urine is an aqueous solution of nitrogenous compounds and trace minerals; it has low level carbon (C) concentrations with a pH of 7.5–8.5. It contains 95% water with urea (2.5%) and trace mineral salts (2.5%), including plant enzymes and hormones. It contains significant levels of major nutrients, including nitrogen, phosphorous, and potassium. The relative ratios are typically around 11 parts nitrogen (80%) to one part phosphorus (55%) to 2.5 parts potassium (80%–90%), making it suitable as a biofertilizer.

Apart from nitrogenous compounds, cow urine constitutes sulfur, which is directly proportional to dietary intake. The sulfur concentration varies between 350 and 810 mg/L and is a mixture of organic and inorganic forms such as SO_4^{2-} (Miah et al., 2017). The potassium concentration is up to 10 g/L, and around 33%–81% is expelled in cattle urine as the primary urinary cation content. Other cations such as magnesium (Mg^{2+}) and calcium (Ca^{2+}) are very low (<1 g/L) with an expulsion of 12%–22% Mg^{2+} and 3% of the Ca^{2+} consumed. Anions such as chloride and bicarbonate vary between 20% and 50% and 40%–70%, respectively, with trace concentrations of boron (B^-) and iodide (I^-). A phosphorus content of less than 0.1 g/L is predominantly excreted in the dung with other trace unabsorbed elements by the digestive tract; these elements include zinc (Zn), copper (Cu), manganese (Mn), cadmium (Cd), iron (Fe), and lead (Pb). Deficient P content gives rise to competition between the soil microbial community and plants. However, it is rich in organic molecules, signifying it to be a rich fertilizer (Prabhu and Mutnuri, 2014).

Cow urine possesses pesticidal and therapeutic properties. The composition and properties of cow urine vary with respect to cattle diet, physiological factors, cattle breed, and physiochemical changes of cow urine upon storage. It has been reported by Apte and Bhalchandra (2002) that cow urine contains traces of silver and gold apart from estrogens, corticosteroids, and keto steroids. Cow urine, popularly known as "gomutra," is well known for its germicidal, antibiotic, antimicrobial, and medicinal properties that are evident from ancient Ayurveda (Kekuda et al., 2010). It can inhibit *Salmonella typhi*, *Escherichia coli*, *Pseudomonas aeruginosa*, *Staphylococcus epidermidis*, *Streptococcus pyogenes*, and many more. Cow urine added to many medicines is considered the best low-cost bioenhancer (Randhawa, 2010). The antibiotic and antifungal properties of cow urine have gained importance and showcased tremendous potential in the field of medicine (Harshad et al., 2017); it is also used as a disinfectant (Dhama et al., 2005). The presence of creatinine, urea, phenols, calcium, and manganese contributes to the germicidal property of cow urine. Table 1 shows the properties of cow urine. Due to the considerable mineral and nutrient composition, emerging technologies for cow urine utilization have been under research for the past few decades, except for small-scale practical applications.

Table 1 Properties of cow urine.

S. no.	Constituents	Concentration
1	pH	7.4–8.4[a]
2	Specific gravity	1.025–1.045[a]
3	EC	>23.7 mS/cm[b]
4	Urea nitrogen	23,000–28,000 mg/L per day[a]
5	Total nitrogen	6800–21,600 mg/L per day[a]
6	Ammonia nitrogen	1000–1700 mg/L per day[a]
7	Allantoin	770–3400 mg/L per day[c]
8	Calcium	100–140 mg/L per day[a]
9	Chloride	10–110 mg/L per day[a]
10	Creatinine	15–20 mg/L per day[a]
11	Phenols	4.7580 mg/100 mL[a]
12	Urea	440 mg/L[c]
13	Hippuric acid	5.96–8.93 mg/L[c]
14	Total phosphorous	305 mg/L[d]
15	Amylase	90.236 units[a]
16	Magnesium	3.7 mg/L per day[a]
17	Potassium	14.4–27 mg/L per day[a]
18	Sodium	4300–6100 mg/L per day[a]
19	Ammonia	20 mg/L[c]
20	Sulfate	3–5 mg/L per day[a]
21	Uric acid	150–530 mg/L per day[c]
22	Leucocyte	<15 μL[a]

[a]Data from Harshad et al. (2017).
[b]Data from Marickar (2009).
[c]Data from Bristow et al. (1992).
[d]Data from Prabhu and Mutnuri (2014).

The total natural N content is 6.8–21.6 g/L in cow urine, consisting of 7.3% allantoin, 69% urea, 5.8% hippuric acid, 1.3% uric acid, 3.7% creatinine, 0.5% hypoxanthine, 2.5% creatine, 2.8% ammonia, and 1.3% free amino acid N. It has been reported that cow urine accelerated rye grass yield by increasing the soil nitrogen content and led to a 10% nitrogen fixation decrement in clover. Urine also increased the potassium concentration of both rye grass and clover. As per the report by Veeresha and Gopakkali (2014), the application of 34,300 L of cow urine per hectare increased the available nitrogen to 272.4 kg/ha, phosphorus to 23.5 kg/ha, and potassium to 199.9 kg/ha.

3 Diet effects on composition of cow urine

Urea, the dominant form of nitrogen, is influenced directly by the livestock diet. Urine volume and urine nitrogen concentrations are congruent with the diet consumed by livestock. Among cows, sheep, and goats, the urea nitrogen in cow urine was around 65%–90% compared to 56%–78% in goat urine and 25% in sheep urine, according to Bristow et al. (1992). The difference in the urea nitrogen level is attributed to variations in the diet of extremely rich fertilized grass fed to cattle compared to the low protein diet fed to sheep. Studies by Kebreab et al. (2010), Huhtanen et al. (2008), and Castillo et al. (2000) reported a direct relationship between dietary intake and urinary nitrogen output.

Both total nitrogen excretion and urinary nitrogen excretion are dependent on protein diet degradability. A high protein content diet leads to high concentrations of urea urine and therefore urea-nitrogen. A report by Cantalapiedra-Hijar et al. (2014) depicted a 16.5%–12% protein content decrease for Jersey cows, which decreased urinary N excretion by more than 50% and increased the overall N efficiency from 26.4% to 31.3%. However, both the milk and milk protein yield decreased. Hay, straw, grass, or other roughage fed to livestock produced an alkaline urine low in phosphate content, whereas livestock fed with concentrated diets such as maize and barley produced acidic urine with a high phosphate content. However, the influence of diet on the difference in phosphorus content is not well established. A greater possibility of phosphorus reabsorption by the kidneys could be a contributing factor for the phosphorus content decrease in cattle urine (Scott, 1972).

4 Applications of cow urine

The effects of the direct or indirect application of cow urine on crops and plants such as mustard, lettuce, maize, mango, green gram, chilies, methi, and brinjal have been well established (Choudhary et al., 2017). Cow urine also finds application as a pest inhibitor or biopesticide in association with plant and seed extracts (Jandaik et al., 2015). The indigenous soil microbial population and soil enzymatic activities also increased (Veeresha and Gopakkali, 2014). The impact of cow urine application on crops is determined by physical parameters such as the dry and fresh weights of roots, shoots, and leaves; the number of leaves; the leaf length and width; the root and shoot length; the number of flowers; the number of fruits; and biochemical parameters such as chlorophyll content, protein content, and mineral content such as N, K, P, and Mg. Chemical fertilizers are not crop specific and show fixed impact on plants irrespective of their species as they desynchronize with the understanding of individual plant nutrition and rhizosphere processes.

Synthetic fertilizers affect soil properties in the long run and render the soil ineffective (Pradhan et al., 2018). Through traditional knowledge, generic fertilizer use for most crops and soils can be curtailed. The application of cow urine is classified based on the mode of application as direct application or indirect application, as shown in Fig. 1.

5 Direct applications

Direct application includes fresh or fermented cow urine as a foliar spray and application to the soil. A 30%–55% foliar cattle urine application on *Mangifera indica* L. (Alphonso) at various spraying frequencies improved the fruit weight, fruit number, and yield (kg/plant and ton/ha) (Damodhar and Shinde, 2010). Foliar application was more effective than soil application, as the former had a positive impact on the physical parameters and commercial yield (de Oliveira et al., 2009). Apart from its solitary use, cow urine in association with various indigenous plant and seed extracts is beneficial for plant growth and the control of pests and insects. A 3% mixture of neem kernel extract and cow urine foliar application has shown significant inhibition of *Lipaphis erysimi* (mustard aphid) on mustard plants (Gupta, 2005). In vivo cow urine-based plant extract application significantly curbed the mycelial growth of *Rhizoctonia solani*, the causative agent of sheath blight in the rice plant (Tiwari and Das, 2011). Jandaik et al. (2015) determined the antifungal property of cow urine through a growth suppression study by a poison food technique on three strains of fungus isolated from infected *Abelmoschus esculentus* (Bhindi) and *Trigonella foenum-graecum* (Methi) plants.

A cow urine application of 15% concentration brought a growth suppression of 78.57% of *Fusarium oxysporum*, 78.37% of *Rhizoctonia solani*, and 73.84% of *Sclerotium rolfsii*. Also, the study of cow urine influence on *A. esculentus* and *T. foenum-graecum* revealed that a 5% cow urine spray significantly increased the protein and carbohydrate content and enhanced the

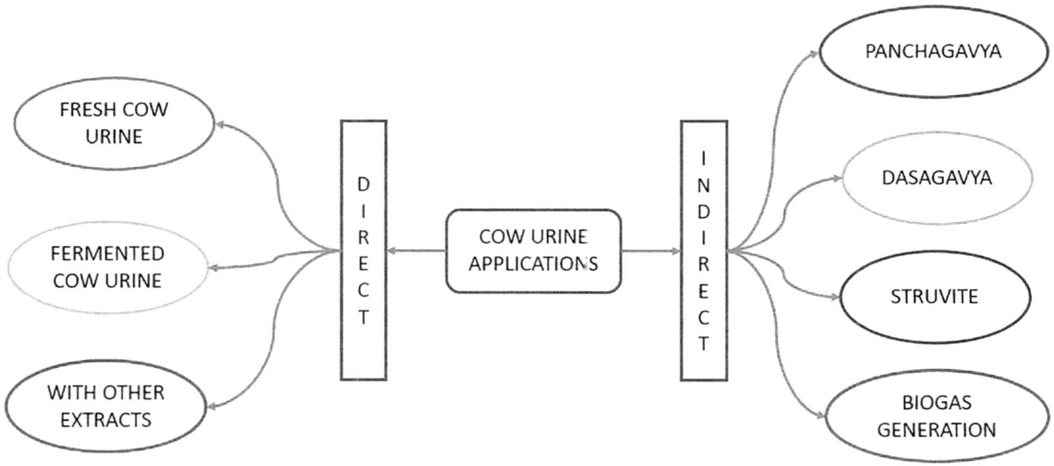

Fig. 1 Applications of cow urine for plant growth enhancement.

phenotype characteristics. Direct cow urine application to soil increased the microbial concentration, thereby enhancing the physiochemical and biological properties as well as the crop yield percentage (Choudhary et al., 2017).

The effectiveness of cow urine as a plant growth enhancer was investigated by Gottimukkala et al. (2019), where the growth of plants such as wheat, grass, maize, and methi was monitored using different cow urine concentrations ranging from 5% to 25% for 30 days continuously. It was found that a 5% cow urine concentration exhibited a positive effect on plant growth. Out of 16 compounds identified in cow urine, six were antifungal, three were antibacterial, 4-(p-methoxyphenyl)- 3-buten-2-one (National Center for Biotechnology Information, 2020a) and Triacontane (National Center for Biotechnology Information, 2020b) were identified as plant growth enhancers. Though cow urine is proven to be the best fertilizer, it is likely that the excess direct application of cow urine may reduce soil nitrogen fixing ability and cause eutrophication of water bodies. Upon addition to soil, urea nitrogen is first converted to ammonia hydrogen carbonate, which is governed by soil pH; after this, it is converted to ammonia nitrogen by urease enzymes (Whitehead and Bristow, 1990). However, scientific methods to minimize the loss intensify resource recovery and closer nutrient recycling. Around 85% of nonvolatilized ammonia undergoes nitrification and is made available to plants and crops. Cow urine investigated independently and in combination with insecticides inhibits pests infesting crops such as chili, soybean, pigeon pea, sesame, linseed, mustard, cabbage, beans, and blackgram. Snail attacks on crops are inhibited by mixing natural oils and herbal root powders with cow urine and solar drying for 15 days (Korati and Dengale, 2011).

6 Indirect applications

Indirect application is using organic products with cow urine as one of the components or products derived from the cow urine-based substrate. The indirect application of cow urine to crops includes panchagavya, dasagavya, biogas production, and struvite production (magnesium ammonium phosphate) salt.

6.1 Panchagavya

Panchagavya is an organic mixture of five ("*panch*" meaning "five") cow products, namely milk, ghee, curd, urine, and dung. Other ingredients include banana, jaggery, tender coconut water, and water (Fig. 2). The

Fig. 2 Constituents of panchagavya.

composition and ratio of ingredients for panchagavya preparation are summarized in Table 2. Panchagavya application to crops at various dosages significantly enhances crop and plant growth, which is determined by physical parameters such as stem height, number of branches, number of fruits, root length, wet and dry weights of root and shoot, leaf length and width, leaf area index, number of pods, pod length, and number of seeds per pod. A 3% panchagavya application on *Capsicum frutescens*, *Vigna radiata,* and *Brassica campestris* have increased the yield over the control by 300.65%, 208.30%, and 288.88%, respectively (Ali et al., 2011). Combined application of panchagavya as foliar spray and organic manure potentially influenced the growth attributes of rice crops. Moreover, organic seed treatment with panchagavya improved the germination and vigor of the seeds. Similarly, the grain yield showed an increase of 429 kg/ha and straw yield of 544 kg/ha.

A trial to evaluate panchagavya application outcome on the growth, yield, quality, and biochemical components of maize revealed a 3.9%–10.5% increase in cob length and a 4.9%–18.6% increase in the number of grains per cob while the overall harvest index increased from 1.5% to 3.6%. Akhter et al. (1970) observed conidial germination inhibition of *Bipolaris sorokiniana*, the causative agent of leaf blight disease in wheat crops, with a mixture of cow urine and extract (10:1 mL) of *Calotropis procera* (91% inhibition); leaf extracts of *A. indica* (85%), *L. camara* (87%), *O. sanctum* (80%), *V. rosea* (83%), and the flower extract of *O. sanctum* (59%) at a 2.5% concentration.

Table 2 Composition of ingredients for panchagavya preparation.

Constituents	Quantity[a]	Quantity[b]
Cow dung	5 kg	7 kg
Cow ghee	500 g	1 kg
Cow urine	3 L	10 L
Water or sugarcane juice	3 L	10 L
Cow milk	2 L	3 L
Cow curd	2 L	2 L
Tender coconut water	3 L	3 L
Jaggery	500 g	3 kg
Well-ripened banana	12 nos.	12 nos.

[a]Data from Kumar et al. (2016).
[b]Data from TNAU Agritech Portal (2020a).

6.2 Dasagavya

Dasagavya, also known as enriched panchagavya, constitutes of ingredients of panchagavya along with five (in temperate regions) or seven (in tropical regions) other plant extracts: *Leucas aspera*, *Lantana camara*, *Datura metal*, *Artemisia nilgirica*, and *Phytolacca octandra* (temperate regions) or *Adathoda vasica*, *Calotrophis*, *Tephrosia purpurea*, *Vitex negundo*, *Jatropha curcas*, *Azadirachta indica*, and *Pongamia pinnata* (tropical areas) (TNAU Agritech Portal, 2020b). One liter of plant leaves soaked in cow urine (1:1) for 10 days is added to 5 L of panchagavya. The mixture is kept for 25 days, filtered and used as a 3% foliar spray to control aphids, mites, pests, and the causative agents of leaf blight, leaf spot, and mildew diseases (Kumar et al., 2016). A 3% dasagavya treatment for seeds and roots of seedlings before plantation for 30 min maximized seed germination and root development. Dasagavya-sprayed plants consistently yielded bigger leaves forming a denser canopy; the formed robust root system enabled higher nutrient intake and enhanced taste and shelf life.

With the estimated dasagavya production cost being 0.54 USD/L, it can be sold for up to 1.22 USD/L, which is more economical than commercial fertilizers. Dasagavya production has the potential to evolve as a commercial venture ably managed by small-scale farmers and women self-help groups. Compost tea is also prepared with fresh cow urine and dung taken in equal proportion, to which 10 times water is added and left for 12 h. For 10 L of tea, 100 g of jaggery is added and used after dilution with water as fertigation or drip irrigation method. 100 L of compost tea would be sufficient for five acres of farmland. In addition, compost tea increases the growth of indigenous bacteria, fungi, and actinomycetes. For every gram of soil, 47×10^5 CFU bacterial colonies, 34.67×10^4 fungi colonies, and 40×10^3 actinomycete colonies were grown. Cow urine must be utilized when the pH lies between 6.5 and 7.5, which generally is the pH of fresh cow urine. The pH changes upon hydrolysis of urine and may not be effective when used for panchagavya or dasagavya application.

6.3 Struvite

A large portion of the dormant phosphorus in the soil resulting from overdosing has prompted phosphorus ore depletion and the occurrence of peak phosphorus. Unassimilated phosphorus progresses to water bodies and soil in the form of excreta or urine (Rahman et al., 2014). Livestock wastes including cow urine and swine wastewater, domestic and industrial inputs such as municipal wastewater and brine are few potential sources to recover soluble phosphorus (Schroder et al., 2011). Phosphorus is recovered from cow urine as struvite crystals, a crystalline mineral comprised of magnesium, ammonium, and phosphate in an equimolar ratio of 1:1:1. The parameters related to struvite derived from cow urine collected from the available literature are summarized in Table 3. The optimum struvite precipitation includes pH maintenance (8.5–9.5) in the presence of magnesium and supersaturation conditions (Prabhu and Mutnuri, 2014). With its low solubility, struvite remains in soil, minimizing eutrophication rates. It is also available for plant intake during later growth stages. In addition to phosphorus recovery, struvite production derives economic benefits and enables effective waste management (Talboys et al., 2016).

Nitrogen enhances plant morphology and photosynthesis while phosphorus is essential for early plant growth; cell division and reproduction; plant metabolism; and harvest, storage, and energy use (Razaq et al., 2017). A study by Prabhu and Mutnuri (2014) revealed the efficacy of the cow urine and brine mixture (pH –9.0) struvite precipitation and its functionality as an organic fertilizer. The optimum struvite application level to *Vigna radiata* was 2 g/kg of soil. Plants fertilized with struvite showed a more significant increase in total nitrogen and total phosphorus in the root and stem than diammonium phosphate (DAP) and organic fertilizer. Tarragó et al. (2016) showed the advantages of struvite over commercial fertilizers as they release nutrients slowly; are suitable for magnesium demanding sugar beets (*Beta vulgaris*); are suitable for annual applications of fertilizer; and cause no considerable damage with higher doses. A research gap exists in the optimum recovery of nutrients from cow urine and its impact post-application to crops, as only a few studies have been conducted on this.

Cow urine constitutes recoverable soluble phosphate and ammonia that are retrieved by the crystallization method. The crystal size influences its

Table 3 Parameters related to struvite derived from cow urine collected from the available literature.

Parameters	Quantity
Specific gravity	1.6[a]
Solubility	0.2 g/L[a]
Composition of magnesium, ammonium, and phosphate ions	3.16:0.56:5.85[b]
Quantity of struvite from 1 L of cow urine	40 g/L[b]

[a]Data from Barrak and Stafford (2006).
[b]Data from Prabhu and Mutnuri (2014).

storage, transportation, and application to crops. Struvite, the organic phosphorus fertilizer, proliferates in plant crop production by increasing the yield and soil fertility by enhancing soil microbes. This phosphorus recovery technique from cattle urine can be made primary in rural regions with dairy farming as the predominant occupation. In addition, cow urine used for nanoparticle synthesis to control bacterial and fungal pathogens serves as a green alternative to hazardous agents. Domestic struvite production augments food crop production with a view of attaining self-sufficiency, scaling down fertilizer import, and generating additional revenue for rural people.

6.4 Biogas production and biogas spent slurry

Biogas is a mixture of 55%–65% methane and 35%–45% carbon dioxide produced by the anaerobic breakdown of organic matter from wastes, including livestock waste such as cattle dung. Biogas production and process stability are based mainly on the waste composition. Cow urine addition has been investigated by Fardous et al. (2020). Cow urine in addition to cow dung is proven to enhance biogas production and produce enriched fertilizer. Also, biogas slurry reflected the high nitrogen content of cow urine when added, and exchangeable K also increased. Cow urine enhanced C/N ratio of manure thus promoting the agronomical use of compost. Besides cow urine addition, cattle urine-soaked materials such as straw and leaves also stimulate and increase the biogas yield. Though the C/N ratio of urine is lower than cow dung, nitrogen-rich urine improves gas production in the anaerobic process.

Biogas spent slurry from digested cow dung and urine possesses more nitrogen, accelerating the composting process in a relatively shorter span of time. This improves the C/N ratio in spent slurry and aids in nutrient availability to crops and soil biota. Combined biogas slurry and conventional fertilizer application improved the yield of crops such as cotton, maize, wheat, and rice by 6.5%, 15.2%, 8.9%, and 15.9%, respectively. Biogas slurry has extensive potential to satiate both macro- and micronutrients for crops. It has been reported that 1 m^3 of biogas spent slurry contains 0.16–1.05 kg N, which is equivalent to 0.35–2.5 kg urea. At the same time, the expense of commercial fertilizers can be minimized by 15%–20% (Kumar et al., 2015).

7 Conclusion

Unprecedented world population growth has posed a severe threat toward food security and sustainability in recent years. The increased use of mineral fertilizers and intensified cultivation methods are detrimental toward human health and the environment. Exorbitant synthetic fertilizer usage to tackle the increasing food demands has led to serious health risks and decreases in soil productivity. With the advent of high-yielding crops and cropping intensity, mineral fertilizer usage has gained a considerable rise. Chemical fertilizer use could not be overruled, considering the need for a high crop production rate. Biogas spent slurry consisting of cattle urine with 0.9%–1.2% of initial N and 0.5%–1% of initial K shows 30%–40% greater NPK value when compared to 10%–15% under normal composting. Cow urine-based fertilizers augment soil fertility, inhibit pests, and manage diseases effectively, thus forming a major component in ecofriendly agricultural systems. They are used in a plethora of agricultural operations as a biofertilizer and biopesticide. Urine addition increases the soil microbial community and fungal biomass. The plant uptake of soil nutrients is higher when cow urine is applied directly or indirectly. Hence, cow urine application must be promoted as a robust farming practice to farmers worldwide.

References

Akhter, N., Begum, M.F., Alam, S., Alam, M.S., 1970. Inhibitory effect of different plant extracts, cow dung and cow urine on conidial germination of *Bipolaris sorokiniana*. J. Biosci. 14, 87. https://doi.org/10.3329/jbs.v14i0.449.

Ali, M.N., Ghatak, S., Ragul, T., 2011. Biochemical analysis of Panchagavya and Sanjibani and their effect in crop yield and soil health. J. Crop Weed. 7 (2), 84–86.

Apte, Bhalchandra, 2002. Analysis of Go-mutra and its possible therapeutic implications. In: 5th International Congress on Traditional Asian Medicine. Halle (Saale), pp. 18–24.

Barrak, P., Stafford, A., 2006. Struvite: a recovered and recycled phosphorus fertilizer. In: Proc. of the 2006 Wisconsin Fertilizer, Aglime & Pest Management Conference. 45, pp. 199–204.

Bristow, A.W., Whitehead, D.C., Cockburns, J.E., 1992. Nitrogenous constituents in the urine of cattle, sheep and goats. J. Sci. Food Agric. 59, 387. https://doi.org/10.1002/jsfa.2740590316.

Cantalapiedra-Hijar, G., Peyraud, J.L., Lemosquet, S., Molina-Alcaide, E., Boudra, H., 2014. Dietary carbohydrate composition modifies the milk N efficiency in late lactation cows fed low crude protein diets. Animal 8 (2), 275–285.

Castillo, A.R., Kebreab, E., Beever, D.E., France, J., 2000. A review of efficiency of nitrogen utilisation in lactating dairy cows and its relationship with environmental pollution. J. Anim. Feed Sci. 9, 1. https://doi.org/10.22358/jafs/68025/2000.

Choudhary, S., Kushwaha, M., Seema Singh, P., Sodani, R., Kumar, S., 2017. Cow urine: a boon for sustainable agriculture. Int. J. Curr. Microbiol. Appl. Sci. 6, 1824. https://doi.org/10.20546/ijcmas.2017.602.205.

Damodhar, V.P., Shinde, V.V., 2010. Effect of cattle urine sprays on yield and quality of mango (*Mangifera Indica*). Asian J. Horticult. 5 (2), 307–308.

de Oliveira, N.L.C., Puiatti, M., Santos, R.H.S., Cecon, P.R., Rodrigues, P.H.R., 2009. Soil and leaf fertilization of lettuce crop with cow urine. Hortic. Bras. 27, 431. https://doi.org/10.1590/S0102-05362009000400006.

Dhama, K., Chauhan, R.S., Singhal, L.K., 2005. Anti-cancer activity of cow urine: current status and future directions. Int. J. Cow Sci. 1 (2), 1–25.

Fardous, M.R., Nasrin, M.S., Shakil, M.E., Islam, M.S., Hoque, M.A., Islam, M.N., 2020. Cow urine for enhancement of biogas production and fertilizer quality of biogas slurry. Aust. J. Sci. Res. 12, 135. https://doi.org/10.3329/jsr.v12i1.43031.

Gottimukkala, K.S.V., Mishra, B., Joshi, S., Reddy, M.K., 2019. Cow urine: plant growth enhancer and antimicrobial agent. J. Horticult. Plant Res. 8, 30. https://doi.org/10.18052/www.scipress.com/JHPR.8.30.

Gupta, M., 2005. Efficacy of neem in combination with cow urine against mustard aphid and its effect on Coccinellid predators. Indian J. Nat. Prod. Resour. 4 (2), 102–106.

Harshad, G., Amit, N., Nilesh, M., Sunanda, B., Amrut, S., 2017. Gomutra (cow urine): a multidimensional drug review article. Int. J. Res. Ayurveda Pharm. 8, 1. https://doi.org/10.7897/2277-4343.085231.

Huhtanen, P., Nousiainen, J.I., Rinne, M., Kytölä, K., Khalili, H., 2008. Utilization and partition of dietary N in dairy cows fed grass silage-based diets. J. Dairy Sci. 91, 3589. https://doi.org/10.3168/jds.2008-1181.

Jandaik, S., Thakur, P., Kumar, V., 2015. Efficacy of cow urine as plant growth enhancer and antifungal agent. Adv. Agricult. 2015, 1. https://doi.org/10.1155/2015/620368.

Kebreab, E., Strathe, A.B., Dijkstra, J., Mills, J.A.N., Reynolds, C.K., Crompton, L.A., Yan, T., France, J., 2010. Energy and protein interactions and their effect on nitrogen excretion in dairy cows. In: Crovetto, G.M. (Ed.), Parma, Italy. 3rd EAAP Inaternational Symposium on Energy and Protein Metabolism and Nutrition. 6–10 September 2010. Wageningen Academic Publishers, Wageningen, pp. 417–425.

Kekuda, T.R.P., Onkarappa, R., Gautham, S.A., Sunita, C., Raghavendra, H.L.M., 2010. Antimicrobial, antioxidant and cytotoxic activity of Streptomyces species from Western Ghat soils of Karnataka, India. Sci. Technol. Arts Res. J. 4, 164. https://doi.org/10.4314/star.v4i2.20.

Korati, D.M., Dengale, K.R., 2011. Cow urine—a natural animal by product for insect pest suppression. Green Farm. 2 (4), 502–504.

Kumar, S., Malav, L.C., Malav, M.K., Khan, S.A., 2015. Biogas slurry: source of nutrients for eco-friendly agriculture. Int. J. Ext. Res. 2, 42–46.

Kumar, T.K., Rao, B.B., Dasu, C.S., Rao, P.G., Ramana, M.V., 2016. Application of indigenous technical knowledge in organic fruit production. Progr. Res. Int. J. 11 (7), 4761–4765.

Marickar, F., 2009. Electrical conductivity and total dissolved solids in urine. Urol. Res. 38 (4), 233–235.

Miah, M.N.A., Miah, M.R.U., Alam, M.Z., 2017. Determining chemical composition of cattle urine and indigenous plant extracts. Int. Ann. Sci. 3, 23. https://doi.org/10.21467/ias.3.1.23-26.

National Center for Biotechnology Information, 2020a PubChem Compound Summary for CID 9603450. https://pubchem.ncbi.nlm.nih.gov/compound/9603450. (accessed 01.11.2020).

National Center for Biotechnology Information, 2020b. PubChem Compound Summary for CID 12535, Triacontane. https://pubchem.ncbi.nlm.nih.gov/compound/Triacontane. (accessed 01.11.2020).

Prabhu, M., Mutnuri, S., 2014. Cow urine as a potential source for struvite production. Int. J. Recycl. Org. Waste Agricult. 3, 1. https://doi.org/10.1007/s40093-014-0049-z.

Pradhan, S.S., Verma, S., Kumari, S., Singh, Y., 2018. Bio-efficacy of cow urine on crop production: a review. Int. J. Chem. 6 (3), 298–301.

Rahman, M.M., Salleh, M.A.M., Rashid, U., Ahsan, A., Hossain, M.M., Ra, C.S., 2014. Production of slow release crystal fertilizer from wastewaters through struvite crystallization—a review. Arab. J. Chem. 7, 139. https://doi.org/10.1016/j.arabjc.2013.10.007.

Randhawa, G.K., 2010. Cow urine distillate as bioenhancer. J. Ayurveda Integr. Med. 1, 240. https://doi.org/10.4103/0975-9476.74089.

Razaq, M., Zhang, P., Shen, H., Salahuddin, 2017. Influence of nitrogen and phosphorous on the growth and root morphology of Acer mono. PLoS One 12. https://doi.org/10.1371/journal.pone.0171321, e0171321.

Schroder, J.J., Smit, A.L., Cordell, D., Rosemarin, A., 2011. Improved phosphorus use efficiency in agriculture: a key requirement for its sustainable use. Chemosphere 84, 822. https://doi.org/10.1016/j.chemosphere.2011.01.065.

Scott, D., 1972. Excretion of phosphorus and acid in the urine of sheep and calves fed either roughage or concentrate diets. Q. J. Exp. Physiol. 57, 379. https://doi.org/10.1113/expphysiol.1972.sp002174.

Talboys, P.J., Heppell, J., Roose, T., Healey, J.R., Jones, D.L., Withers, P.J.A., 2016. Struvite: a slow-release fertiliser for sustainable phosphorus management. Plant and Soil 401, 109. https://doi.org/10.1007/s11104-015-2747-3.

Tarragó, E., Puig, S., Ruscalleda, M., Balaguer, M.D., Colprim, J., 2016. Controlling struvite particles' size using the up-flow velocity. Chem. Eng. J. 302, 819. https://doi.org/10.1016/j.cej.2016.06.036.

Tiwari, R.K.S., Das, K., 2011. Inhibitory effect of cow urine based plant extracts against *Rhizoctonia solani* causing sheath blight of Rice. Indian Phytopath. 64 (3), 265–268.

TNAU Agritech Portal, 2020a. Panchagavya. http://agritech.tnau.ac.in/org_farm/orgfarm_panchakavya.html (accessed 21.10.2020).

TNAU Agritech Portal, 2020b. Dasakavya. http://agritech.tnau.ac.in/org_farm/orgfarm_dasakavya.html (accessed 21.10.2020).

Veeresha, S., Gopakkali, P., 2014. Effect of organic production practices on yield and soil health of irrigated maize (*Zea mays* L.) as influenced by various levels of FYM and cattle urine application. Environ. Ecol. 32 (2A), 627–630.

Whitehead, D.C., Bristow, A.W., 1990. Transformations of nitrogen following the application of 15N-labelled cattle urine to an established grass sward. J. Appl. Ecol. 27 (2), 667–678.

Yin, J., Wang, Y., Gilbertson, L.M., 2017. Opportunities to advance sustainable design of nano-enabled agriculture identified through a literature review. Environ. Sci. Nano 2018 (5), 11–26. https://doi.org/10.1039/C7EN00766C.

CHAPTER TWELVE

Effect of climate change on vector-borne disease

B.K. Biswas
West Bengal University of Animal and Fishery Sciences, Kolkata, West Bengal, India

Introduction

In evolution, animals are thriving against enemies such as natural calamities, want of food, and different types of diseases. They are also struggling for existence between intra- or interspecies. In the advancement of civilization, humans are facing the same problems. Rather, humans are also suffering from some diseases of zoonotic importance. In the present era, the effect of animal diseases is measured by the terms of economic loss. But this cannot be done for humans. A great part of human nutrition comes from animal sources, which include milk, meat eggs, and their byproducts. The livelihoods of billions of people directly or indirectly depend on animal farming as well animal husbandry.

To meet the growing demand for animal-originated food, the livestock sector, such as the farming processing industry, as well as marketing are growing. The organized long market chains employ at least 1.3 billion people globally. The livelihoods of 600 million poor smallholder farmers are getting direct support from the animal husbandry sector in the developing world (Thornton et al., 2006).

There are great roles of diseases in profitable animal farming systems. Diseases also hinder the production of healthy and quality consumable animal products. Disease has negative effects on the livestock production system, such as low production, reduced income, and substandard livelihoods. These effects also include the cost of treatment, market disturbances, and the costs of prevention or disease control (McInerney, 1996).

Among various mammalian infectious diseases, vector-borne diseases (BVD) play a major role in terms of health concerns and as well production. Of 593 known mammalian viral diseases, about 29% are exclusively vector-borne (Johnson et al., 2015). These vector-borne viruses have three times the host range compared to nonvector-borne pathogens (Olival et al., 2017).

Emerging Issues in Climate Smart Livestock Production
https://doi.org/10.1016/B978-0-12-822265-2.00006-5

This means that particular vector-borne pathogens can infect multiple animal species that may also act as hosts or reservoirs of the same pathogen. In addition, particular vector-borne viruses can be transmitted by multiple related vector species. For these reasons, it is a great challenge to control VBDs. Of all recorded parasitic and infectious diseases, about 17% represent BVDs only (Townson et al., 2005). The VBD epidemiology is being influenced by the climate. The VBDs can serve as "the canary in the coal mine" because they are too sensitive to climate; this results in the first alert to disease cycle changes due to climate (Randolph, 2009).

These diseases cause socioeconomic loss worldwide. Ticks and tick-borne diseases (TTBDs) have a major effect on profitable livestock production and productivity. Hemoprotozoan diseases, especially babesiosis, anaplasmosis, theileriosis, and trypanosomosis, are considered major obstacles for better health and productive performance of cattle (Rajput et al., 2006). Vectors affect the economy of livestock farming in three ways: (i) direct losses from their bite and irritation, (ii) diseases they spread, and (iii) expenses incurred for their control. Not only mosquitoes and ticks, but flies such as *Culicoides* spp. and *Musca* spp. also play significant roles in spreading disease in addition to their direct effects (Narladkar, 2018). As per the record of the Food and Agriculture Organization (FAO) (2004), 80% of the worldwide cattle population has been exposed to tick infestation, causing losses of about 7.3 US $/head/year. Heavily tick-infested cattle had a reduction of 65% of their body weight due to loss of appetite (Seebeck et al., 1971). The biting midge Culicoides transmits bovine ephemeral fever and bluetongue, which cause an 18.97% loss of milk production (Narladkar and Shivpuje, 2012), resulting in $6 million in losses (Metacalf et al., 1980). Bath estimated the global impact of bluetongue disease was $3 billion (Bath, 1989). A study on 121 Holstein Friesian female cattle that formed eight cohorts (1981–1988) of a dairy farm in Salta, Argentina, was undertaken to evaluate direct financial losses caused by babesiosis. Six cohorts out of eight that suffered from clinical babesiosis cases amounted to $1624.6 (prices in October 1990). Of the total loss, 62% were the result of physical losses and 38% to costs of control (Guglielmone et al., 1992).

Resistance against infectious diseases depends upon the individual health status. Good health does not depend on only nutrition, but also on the environmental conditions where the individual lives. Among many factors, climate has a potential role in altering disease states and probably exerts an overwhelming negative effect on human and animal health (Rabinowitz and Conti, 2013). Increased temperatures among the different components of climate have direct effects on animal and human health. This may primarily be due to changes in the frequency and intensity of heat waves (Gaughan et al., 2010). The animal's feed intake is reduced during the hotter part of the day but it increases when the temperature cools down to the ambient. These types of feeding can cause acidosis, which is considered a major cause of laminitis. The development of immunity against any invasive pathogenic organism is a complex of mechanisms. Chronic exposure to heat stress impaired the immune response in avian species (Regnier and Kelley, 1981) and severe heat stress reduced colostral immunoglobulins (IgG and IgA) in dairy cows with negative consequences on immunization, which affects the survival of newborn calves (Nardone et al., 1997). Impressive depression in lymphocyte functions are observed in severely heat-stressed periparturient dairy cows, and it increases vulnerability to pathogens while also reducing the efficacy of vaccinations (Lacetera et al., 2005).

Weather and climate change can plausibly affect the biology and distribution of vectors, and consequently vector-borne infections. The changes of temperatures, global wind and precipitation patterns, and relative humidity in temperate climates increase the reproduction of insects and, consequently, their population density. Thus, the vector-borne tropical diseases may move from their natural area to other countries, which means newer virgin areas.

Bluetongue disease primarily affects domestic and wild ruminants and is prevailing across the world. It is a VBD transmitted by the arthropod culicoides midges. This virus has been spreading significantly since 1990 due to changing climatic and environmental conditions that provide necessary support to the *Culicoides* vectors for survival as well as reproduction. It is expected that the global mean temperature will rise 2°C during the next 100 years due to global warming resulting the range of *Culicoides imicola* in Europe could be extended to a northward shift of approximately 200 km (Wittmann et al., 2001).

Infestations of gastrointestinal nematodes are also increasing extensively due to increases in environmental temperature, precipitation, and moisture. Because a significant part of the life cycle of these parasites is completed outside the host, their survival and development are very suitable to climate change (Rose et al., 2015).

This chapter deals with the relation between climate change and VBDs. The life cycles of different arthropods are briefly described. The most common and important VBDs are discussed.

Vector-borne disease (VBD)

VBDs are those diseases transmitted by arthropods or small animal vectors. They cause enormous suffering in humans and animals. Vectors act as the main mode of transmission of the pathogen from one host to another, where the essential stage of the transmission cycle took place in the vector.

Types of VBD transmission

- Anthroponotic infections or human-vector-human transmissions, where humans are the only reservoir of the disease.
- Zoonotic infections or animal-vector-human transmission, where animals are the main reservoir of the disease and humans are considered secondary hosts. Humans do not generally take part in the disease transmission cycle as their levels of circulating pathogen are often too low to help maintain transmission.

Many vectors are escalating into new types of diseases and hosts. Chikungunya, dengue, and West Nile have recently caused significant human epidemics. Arthropod-borne animal diseases such as bluetongue, African horse sickness, and Rift Valley fever are substantial threats to livestock economies around the world. The life cycle and behavioral pattern of arthropod vectors depend upon the components of the climate. Climate change impacts VBD epidemiology and pathogen distribution as well as changes the ability of arthropods to transmit pathogens. Climate change has an effect on the way pathogens interact with both the arthropod vector and the human or animal host. Predicting and mitigating the effects of future changes in the environment such as climate change on the complex arthropod-pathogen-host epidemiological cycle must understand the variety of complex mechanisms from the population to the molecular level. Although there has been substantial progress on many fronts, the challenges to effectively understand and face the impact of potential changes in the environment on vector-borne pathogens are frightening and at an early stage of development. The challenges will be explored using several arthropod-borne pathogen systems as illustrations, and potential avenues to meet the challenges will be presented.

Common arthropod-borne diseases.

Mosquito-borne diseases

- Chikungunya
- Dengue fever
- Eastern equine encephalitis
- Japanese encephalitis
- La Crosse encephalitis
- Malaria
- St. Louis encephalitis
- West Nile virus
- Zika virus

Tick-borne diseases

- Babesiosis
- Ehrlichiosis
- Lyme disease
- Rocky Mountain spotted fever
- Tularemia
- Bluetongue

Climate change and vector

The average temperature of the Earth's climate system is increasing day by day. This warming has been demonstrated by direct temperature measurements and also observations of numerous various effects of warming (IPCC, 2013). The effect of global warming on the climate is a rise in global surface temperatures (IPCC, 2014), and also includes these effects, such as changes in rainfall. Global warming is not a new phenomenon but is a prehistoric occurrence (IPCC, 2013). However, the changes observed since the mid-20th century are alarming and unprecedented in rate and scale (IPCC, 2013).

The Intergovernmental Panel on Climate Change (IPCC) concluded that "human influence on climate has been the dominant cause of observed warming since the mid-20th century" (IPCC, 2018a, b). The national science academies of major nations have recognized these observations and no scientific body of national or international standing has raised any question on the statement. More than 90% of the impact on climate change by carbon dioxide and methane (Gleick, 2017) is due to the emission of these man-made greenhouse gases (Olivier and Peters, 2019). Other major sources of these gases are from fossil fuel burning, with agricultural emissions and deforestation also playing significant roles. Visible effects on climate by these gases are loss of snow cover, increased water vapor, and melting permafrost (Met Office, 2016).

The land surfaces are used to be hot faster than the ocean surface (NASA, 2020). The land surface becomes hotter than ocean surface in day time. These phenomenon's causes heat waves, wildfires, and also the expansion of deserts (IPCC, 2014). With the increasing temperature as atmospheric energy, the evaporation rates are also increased, which causes more intense storms and extreme weather, damaging infrastructure and agriculture (Campbella et al., 2016). Surface temperature has increased in temperate and arctic zones at a greater rate, which cause melting of glaciers, permafrost, and sea ice (National Geographic, 2019). The impacts of these environmental changes include relocation or the extinction of many species as their ecosystems change, primarily in coral reefs, mountains, and the arctic. Getting the ambient temperature, moisture, stagnated water for breeding the arthropods, insects and other parasites are making habitation in new horizons (EPA, 2017). Surface temperatures could be decreased through reduced emissions, but other impacts such as rising sea levels from melting ice sheets, rising ocean temperatures, and ocean acidification from elevated levels of carbon dioxide will continue for centuries (IPCC, 2018a; Kehse, 2020). Pronounced increases in night and winter temperatures, decreasing frost days, and changes in moisture conditions cause the emergence of the blue tongue viral disease in Europe (Purse et al., 2005). As the tick Culicoides is sensitive to environmental temperatures for its developmental stages, the competence of Culicoides vectors is directly enhanced by warm temperatures due to global warming (Paweska et al., 2002). The tick-borne encephalitis (TBE) of human is increasing day by day. The reasons are; availability of favorable conditions for natural transmission cycles of tics which results higher densities of ticks, movement of animals toward tick infested areas and increased tendency to consume raw milk. These first two factors are climate-dependent; TBE is enlisted in the vector-borne pathogens those are anticipated to become greater threats to humans in predict warmer areas (Lindgren, 1998; Martens et al., 1999).

Culex mosquito

The mosquitoes under the *Culex* genus serve as vectors of one or more important diseases of humans, birds, and other animals. The diseases they transmit include arbovirus infections such as Japanese encephalitis or St. Louis encephalitis and West Nile virus, but also filariasis and avian malaria. They exist throughout the world, except the extreme northern parts of the temperate zone. They are the most common form of mosquito encountered in Asian and African countries as well as the United States.

Description

The adult *Culex* mosquito may measure from 4 to 10 mm (0.2–0.4 in.) but it varies species to species. The adult has a distinguished head, thorax, and abdomen and it has two forewings held horizontally over the abdomen when at rest. Proper identification of the mosquito is important to take control measures. Experience is needed to separate this mosquito from others (Mike Service, 2008). The larvae of anopheline can be differentiated based on behavioral character. The anopheline larvae tend to float horizontal at the surface of the water when not in motion, whereas culicine larvae float with head low and only the siphon at the tail held at the surface.

Lifecycle

It takes about 2 weeks for the developmental cycle of most species in warm weather. This mosquito's developmental cycle is holometabolism, which includes four life stages: egg, larva, pupa, and imago or adult. The female lays eggs in rafts of as many as 300 on the water's surface. Small bodies of standing freshwater are suitable habitats for egg-laying, including puddles, pools, tin cans, buckets, bottles, ditches, and water storage tanks (tree boles are suitable for only a few species). The eggs are tiny, cigar-shaped, and dark brown in color. They adhere to each other through adhesion forces without any kind of cement, and can be easily separated. Eggs can't hatch without water, and the larvae are aquatic and linear in form. Maintaining their position, they mostly move in a vertical attitude in water on their bristly mouthparts. To swim, they lash their bodies back and forth through the water (Mosquito Ecology and Surveillance Laboratory, 2014; Mike Service, 2008). The insect lives submerged in water during the larval stage. They used to live on microscopic organisms or plant material, particles of organic matter. After several instars, it develops into a pupa. The pupa becomes comma-shaped. Though it does not take feed, it can swim in rapid jerking motions to get rid of potential predators. It must keep in regular contact with the water's surface to breathe, but it must not become dried out. After 24–48 h, the pupa ruptures and the adult emerges from the exoskeleton.

Diseases from culex

Several species of *Culex* mosquitoes are persistent vectors of some major diseases or epidemics while some species rarely and only incidentally transmit a few diseases. Arbovirus infections including West Nile virus, St. Louis encephalitis, Japanese encephalitis, and western and eastern equine encephalitis are transmitted by various species of *Culex*. Zika viruses are also suspected to be transmitted by Culex species, as reported by Brazilian scientists. Nematodes such as filaria may be borne by *Culex* species as well as by other mosquitoes and bloodsucking flies. The inhabitation areas and target animals vary with the bioform of the *Culex species*, that is, the *Culex pipiens pipiens* bioform usually inhabits areas above ground and targets avian species. The *C. pipiens molestus* bioform, in contrast, usually inhabits underground cavities and targets human hosts. If these two bioforms hybridize, the progeny becomes depolarized in both habitat and host preference. Consequently, these hybridized species play a convenient role in disease transmission.

Aedes mosquito

Aedes aegypti is a mosquito that can spread chikungunya, Zika fever, dengue fever, mayaro, and yellow fever viruses as well as other disease agents that can cause great loss to mankind. White markings on its legs and a marking in the form of a lair on the upper surface of its thorax are the cardinal signs of this mosquito. Nowadays, mosquitoes are found in tropical, subtropical, and temperate regions (Eisen and Moore, 2013) as well as throughout the world (Womack, 1993).

Spread of disease and prevention methods

A. aegypti is a vector for transmitting several tropical fevers such as chikungunya, dengue, yellow fever, etc. Only the female mosquitoes bite for blood, which she needs to mature her eggs. The mosquitoes are attracted to chemical compounds emitted by mammal hosts including ammonia, carbon dioxide, lactic acid, and octenol molecules (O'Brien, 2010).

Common feature of the environment for breeding mosquitoes

A. aegypti mosquitoes feed at dusk and dawn, indoors, in shady areas, or in cloudy weather. They bite and spread infection all year long and at any time of day (Womack, 1993; O'Brien, 2010). The preferable areas for breeding are stagnated water, such as uncovered barrels, buckets, flower vases, and discarded tires. The most dangerous areas are wet shower floors and toilet tanks, as they allow the mosquitoes to breed in the residence. Certain chemicals emanating from bacteria in water containers stimulate female mosquitoes to lay their eggs. The water with the chemicals associated with the microbial stew is far more stimulating to discerning female mosquitoes than the plain or filtered water in which the bacteria once lived (Banfield et al., 1965). It is better to control mosquitoes to prevent diseases. A mosquito net is very useful to keep away these diseases. To control the mosquitoes, initiatives must be taken to reduce the sources. This also includes "fogging" for adult control, the use of mosquito traps such as the lethal ovitrap, and pesticide spraying for larval control. The lifespan of an adult *Ae. aegypti* is 2–4 weeks depending on environmental conditions (Travelers' Health Outbreak Notice, 2010). The eggs can remain viable for more than a year in a dry state, which allows the mosquito to reemerge after a dry spell or cold winter. The endosymbiotic bacterium adopts mosquitoes to be resistant to certain arboviruses such as dengue fever and Zika virus strains (Travelers' Health Outbreak Notice, 2010). Invasion of these bacteria in the mosquitoes may be a biocontrol measure to reduce the diseases.

Anopheles mosquito

J.W. Meigen first described and named *Anopheles* as a genus of the mosquito in 1818 (Meigen, 1818). Among 460 species under the genus *Anopheles*, more than 100 species can transmit human malaria, and only 30–40 common species transmit parasites of the genus *Plasmodium*, which causes malaria in humans in endemic areas. *Anopheles gambiae* transmit the most dangerous malaria parasite species (to humans), that is, *Plasmodium falciparum*.

Life stages

Like all other mosquitoes, anophelines also grow through four stages in their life cycles: egg, larva, pupa, and imago or adult. The developments of the first three stages take place in aquatic environments. Together, this takes 5–14 days, depending on the species and the ambient temperature. In the adult stage, the female *Anopheles* mosquito sucks human blood and acts as a malaria vector. The life span of adult females may be up to a month (or more in captivity), but it is limited to 2 weeks in nature (Anopheles Mosquitoes, 2015).

Eggs

Generally, adult females lay 50–200 eggs per oviposition; these eggs are quite small (about 0.5 × 0.2 mm). Females lay eggs singly and directly on water. It is unique that the eggs float on either side. Eggs cannot resist drying and they hatch in water within 2–3 days; depending on the environmental temperature, it may take up to 2–3 weeks in colder climates (Anopheles Mosquitoes, 2015).

Larvae

In the developmental stage after hatching, larvae get a well-developed head with mouth brushes that are used for feeding, a large thorax, and an abdomen having nine segments. Larvae have no legs as well as no respiratory siphon. It positions itself parallel to the surface of the water (Anopheles Mosquitoes, 2015). In contrast, the feeding larva of a nonanopheline mosquito species attaches itself to the water surface with its posterior siphon, with its body pointing downward.

Larvae must come to the surface frequently to breathe through spiracles on the eighth abdominal segment. The larvae live on algae, bacteria, and other microorganisms in the surface microlayer. The larvae swim either by jerky movements of the entire body or through propulsion with the mouth brushes and dive below the surface only when disturbed (Anopheles Mosquitoes, 2015). Larvae go through four stages, or instars, after which they metamorphose into pupae. At the end of each instar, the larvae molt, shedding their exoskeletons, or skin, to allow for further growth (Anopheles Mosquitoes, 2015). The length of first-stage larvae is about 1 mm; a fourth-stage larva is normally about 5–8 mm. The length of development from egg-laying to adult emergence depends on the environmental temperature. It takes a minimum of 7 days. *Anopheles* mosquito larvae generally prefer clean, unpolluted water for their development. The larvae crop up in a wide range of habitats such as rice fields, grassy ditches, the edges of streams and rivers, small temporary rain pools, freshwater or saltwater marshes, mangrove swamps, etc.; they also occur even in shaded breeding sites in forests, tree holes, or the leaf axils of some plants (Anopheles Mosquitoes, 2015).

Pupa

In this stage, the head and thorax are merged into a cephalothorax with the abdomen curving around underneath. They look like a comma-shaped body when viewed from the side. The pupa is also known as a tumbler due to its tumbling movement. As with the larvae, pupae must come to the surface frequently to breathe, which they do through a pair of respiratory trumpets on their cephalothoraces (Anopheles Mosquitoes, 2015). The pupal stage lasts around 2–3 days in temperate areas.

Adults

After about 2–3 days as a pupa, the adult mosquito emerges through splitting the dorsal surface of the cephalothorax. The time taken to develop an adult mosquito from an egg varies considerably among species, and it is strongly influenced by ambient temperature. In tropical conditions, mosquitoes can develop from an egg to an adult in 5–14 days. Mating of adult mosquitoes usually starts within a few days after emerging from the pupal stage. In maximum species, at evening the females fly into the large swarms of males to mate (Anopheles Mosquitoes, 2015). Males live on nectar and other sources of sugar and live for about a week. Males cannot feed on blood, as it produces toxic effects and kills them within a few days ("ScienceDaily", sciencedaily.com). Females can live on sugar sources as n energy source, but usually a blood meal is required to develop eggs. After a full blood meal, the female will rest for a few days; at that time, the blood is digested and eggs are developed. The total time needed for this process depends on the temperature, but usually it takes 2–3 days in tropical conditions. The female lays the fully developed eggs and then again continues host-seeking (Anopheles Mosquitoes, 2015). This cycle repeats itself until the female dies. The lifespan of female mosquitoes depends on temperature, humidity, and their ability to obtain a blood meal while avoiding host defenses (Anopheles Mosquitoes, 2015). Female mosquitoes carrying malaria parasites are significantly more attracted to human breath and odors than uninfected mosquitoes.

Habitat

Malaria was limited to tropical areas because Anopheles mosquitoes lived there, but nowadays many species of the mosquitoes are found in colder regions such as Canada (Wylie, 1983). *Plasmodium* parasite (not the *Anopheles* mosquito) has been eliminated from First World countries. Still, these areas are at risk of reintroduction of the disease.

The hemolytic C-type lectin CEL-III from *Cucumaria echinata* (sea cucumber) found in the Bay of Bengal impairs the development of the malaria parasite when produced by transgenic *A. stephensi*. This could potentially be used to control malaria by spreading genetically modified mosquitoes (Yoshida et al., 2007).

Susceptibility to become a vector of disease

All the species are not vectors of malaria; some are poor vectors as the parasites do not develop well (or at all) within them. There is also a variation within species. Some strains have an immune response that encapsulates and kills the malaria parasites after they have invaded the mosquito's stomach wall.

Culicoides midges

The biting midge genus *Culicoides* is under the family Ceratopogonidae. More than 1000 species are identified in the genus (Meiswinkel et al., 2004; Connelly, 2013). The genus is divided into many subgenera. Some species are known to be vectors of different diseases and parasites that can affect animals (Szadziewski et al., 2019).

Description

The adult insects are dark and small, about 1–3 mm long. The 15 segmented antennae are long, densely haired in males and less hairy in females. The thorax is banded and carries a pair of broad spotted wings. The first two longitudinal veins are only distinct. Midges and mosquitoes are morphologically distinct from each other. Midges lack a proboscis, limiting their ability to bite through clothing. Biting midges are sometimes incorrectly referred to as sand flies. Sand flies are insects that belong to a different biological group and should not be confused with biting midges.

Habitation and behavior

Many biting midge species live in substantial salt marsh areas. Some species, such as the bluetongue virus vector *Culicoides sonorensis*, live in areas that include highly organic soil that is wet but not underwater, such as those found with high manure loads in swine, sheep, and cattle farms. These insects do not establish inside homes, apartments, or inside humans or other animals. Both males and females nourish on nectar. But only the females take a blood meal, which is essential for the maturation of fertilized eggs (Connelly, 2013). Females characteristically bite at dusk or dawn, often in dense swarms, and usually in the surrounding areas of water, marshes, or rotting vegetation.

Life cycle

The life cycle of *Culicoides*: Biting midges are holometabolous, progressing from egg to larva to pupa and finally to the adult stage. The complete cycle can occur in 2–6 weeks, but is dependent on the species and environmental conditions. Females lay their eggs *en masse* in habitats that include water vegetation, slow running streams, damp soil, or manure heaps. The number of eggs produced varies among species and with the size of the blood meal. For example, *Culicoides furens* (Poey) can lay 50–110 eggs per bloodmeal, and *C. mississippiensis* (Hoffman) can lay 25–50 eggs per blood meal. The eggs hatch into tiny smooth white larvae with four pairs of anal gills. Larvae are not strictly aquatic or terrestrial; they require water, air, and food for survival. They need moisture for their development. They feed on small organisms. Larvae of most species cannot survive more than a few inches below the air-water interface. In the tropics, the larvae of many species live in rotting fruit, bromeliads, and other water-holding plants. They can even survive and develop in debris at the edges of ponds, lakes, springs, mud, sand, tree holes, and slime-covered bark. The required time for the development of larva depends on the species, temperatures, and geographic area. It may last from 2 weeks to a year. Larvae molt to pupae. The bodies of pupa consist of a fused cephalothorax with slender respiratory trumpets and a segmented abdomen. From pupae, adults emerge through a straight slit within 3–7 days. During biting, *Culicoides* prick the animals sharply; this is often followed by irritating swelling that may last for a few hours or a few days.

Disease transmit

Different *Culicoides* species have acted as vectors for the following organisms: *Leucocytozoon*, *Plasmodium agamae*, *Mansonella* spp., *Onchocerca gibsoni*, *O. cervicalis*, bluetongue virus, Schmallenberg virus, African horse sickness, bovine ephemeral fever (*C. osystoma* and *C. nipponesis*), Akabane virus, Queensland itch, and epizootic hemorrhagic disease (Szadziewski et al., 2019; Bessell et al., 2016).

Ixodes tick

Ixodes ricinus is a hard-bodied tick known as the castor bean tick. They are mainly found in Europe. Lyme disease, a bacterial disease, and tick-borne encephalitis, a viral disease, are transmitted by these ticks. The length of the adult male is 2.4–2.8 mm and the female is 3.0–3.6 mm. The length of an unfed nymph is 1.3–1.5 mm, which become 11 mm (0.43 in.) long when engorged with blood (Ruedisueli and Manship, 2010). Ixodes have no eyes and are not complete. They have no festoons (wrinkles along the posterior margin). It has an anal groove above the anus. The palpi are longer than the width (Walker, 2018). It has a hard dorsal shield that covers the entire opisthosoma (abdomen), but only part of it is in females and nymphs (Samour, 2000).

Distribution

I. ricinus is traced throughout Europe, even in Iceland and the far east of Russia. The ticks are found in neighboring parts of North Africa and also in the Middle East (Ruedisueli and Manship, 2010). Tick habitats seem to be determined by environmental factors, including temperature. The ticks have extended their habitation northward, that is, the Netherlands, Sweden, and Denmark due to milder winters (Lindgren et al., 2000). The ticks prefer the areas where their hosts are easily available. They also prefer hot and humid conditions and search for woodlands, heaths, and forest for habitation (Ruedisueli and Manship, 2010).

Lifecycle

Ixodes spp. has a three-host life cycle that takes about 2–3 years to complete. The length of the complete life cycle may be extended up to 6 years depending upon the environment. Adults or active ticks at their parasitic stage feed on the blood of large mammals such as sheep, cattle, dogs, deer, humans, and horses for 6–13 days. Mating also takes place there. Then the engorged female drops to oviposit. The female lays several thousand eggs in some sheltered microenvironment and subsequently dies (Ruedisueli and Manship, 2010). The eggs hatch to larvae and find rodents, rabbits, birds, reptiles, or bats as hosts. They also feed on insectivores (orders Erinaceomorpha and Soricomorpha) (Ruedisueli and Manship, 2010). The larvae feed for 3–5 days and are dropped off for molting to nymph. The nymphs then climb grasses or twigs and seek their next host. If it is late to get a suitable host, the nymph returns to the moist microclimate at the soil surface if they become dehydrated (Capinera, 2008). The nymphs feed on small- to medium-sized mammals and being engorged, they drop off and molt to adult and again search for a host.

Disease transmission

Ixodes spp. can transmit a number of diseases caused by bacteria, virus, protozoa, etc., in different mammals (Ruedisueli and Manship, 2010). They transmit Lyme disease (borreliosis), caused by the spirochaete bacteria *Borrelia burgdorferi*, *B. afzelii*, and *B. garinii* to dogs. They cause redwater fever (caused by the protozoa *Babesia divergens*, *B. bovis*, and *B. ovis*), tick-borne fever (*Anaplasma phagocytophila*), Q fever (*Coxiella burnetii*), Boutonneuse fever (*Rickettsia conorii*), and the bacterium *Anaplasma marginale* and Lyme disease (from *B. burgdorferi*) in cattle and tick pyemia (*Staphylococcus aureus*) in sheep. Even horses may be infected with Lyme disease, *Anaplasma phagocytophila*, and the viral infection louping ill, which are all transmitted by the ticks (Ruedisueli and Manship, 2010).

Malaria

Malaria is a noncontagious mosquito-borne infectious disease that affects humans and other animals. It is caused by single-celled microorganisms of the *Plasmodium* group (WHO, 2014). There are about 30 parasitic *Plasmodium* species that naturally infect nonhuman primates (Baird, 2009) and nearly 200 species that have been identified that infect birds, reptiles, and other mammals (Rich and Ayala, 2006). The common symptoms of the disease include fever, tiredness, vomiting, and headaches. It can be fatal and cause yellow skin, seizures, and coma (Caraballo, 2014). Manifestation of symptoms usually starts 10–15 days after a bite by an infected mosquito. The disease may recur months later if not properly treated (WHO, 2014). Patients who survive achieve partial resistance against the disease, but this resistance disappears over months to years if the person has no continuing exposure to malaria. If the patient is reinfected by a malaria parasite, it usually causes milder symptoms (Caraballo, 2014).

Generally, infected female *Anopheles* mosquitos spread the disease. The mosquito introduces the parasites from the mosquito's saliva into a person's blood through a bite (WHO, 2014). The introduced parasites pass through to the liver where they mature and reproduction takes place. Then, disease manifestation starts. Most fatalities are caused by *P. falciparum* while others such as *P. vivax*, *P. ovale*, and *P. malariae* generally cause a milder form of malaria (WHO, 2014; Caraballo, 2014).

The threat of disease can be reduced by controlling mosquito breeding as well as keeping mosquitoes away. Physical methods such as using mosquito nets or chemicals for killing mosquitoes are effective measures for this disease control.

The disease widely prevails in hot and humid areas, particularly in the tropical and subtropical regions around the equator (Caraballo, 2014).

Signs and symptoms

The onset of malaria symptoms starts 8–25 days following the bite from the infected mosquito (Fairhurst and Wellems, 2010), but may arise later in those who have taken antimalarial medications as prevention (Nadjm and Behrens, 2012). Initial symptoms for all malaria species manifested are similar to flu-like symptoms (Bartoloni and Zammarchi, 2012). They can also resemble other conditions such as sepsis, gastroenteritis, and viral diseases (Nadjm and Behrens, 2012). The common symptoms include headache, fever, shivering, joint pain, vomiting, hemolytic anemia, jaundice, hemoglobin in the urine, retinal damage, and convulsions (Beare et al., 2006).

Cyclical occurrence of the symptoms such as sudden coldness followed by shivering and then fever and sweating are the classic symptoms of malaria. Every 2 days (tertian fever) in case of *P. vivax* and *P. ovale* infections, and every 3 days (quartan fever) in case of *P. malariae* infection, the symptoms recur. But the *P. falciparum* infection can cause recurrent fever every 36–48 h, or a less evident and almost continuous fever (Ferri, 2009).

P. falciparum causes severe malaria where manifestations of the disease arise 9–30 days after infection (Bartoloni and Zammarchi, 2012). It may result in cerebral malaria and victims frequently exhibit neurological symptoms, including abnormal posturing, nystagmus, conjugate gaze palsy (failure of the eyes to turn together in the same direction), opisthotonus, seizures, or coma (Bartoloni and Zammarchi, 2012).

Cause

Malaria parasites belong to the genus Plasmodium (phylum Apicomplexa). The common species that cause malaria in humans are *P. falciparum*, *P. malariae*, *P. ovale*, *P. vivax*, and *P. knowlesi* (Mueller et al., 2007; Collins, 2012). Among those species, *P. falciparum* is the most common species identified (~75%), followed by *P. vivax* (~20%) (Caraballo, 2014). *P. falciparum* traditionally accounts for the majority of deaths (Sarkar et al., 2009). Recent evidence suggests that *P. vivax* malaria is also fatal, like the *P. falciparum* infection (Baird, 2013).

Life cycle

Only female mosquitoes feed on blood because they need blood meals for maturation of their eggs. Male mosquitoes feed on plant nectar and cannot transmit the disease. Female mosquitoes of the genus Anopheles prefer to feed at night. They usually start searching for a meal at dusk, and continue through the night until they succeed (Arrow et al., 2004). Female mosquitoes receive gametocytes, a form of Plasmodium through blood by biting an infected person. The gametocytes mature in the mosquito gut. The male and female gametocytes fuse and form a fertilized, motile zygote called an ookinete. Ookinetes develop into a motile infective form (called the sporozoite) that migrates to the insect's salivary glands and becomes ready to infect a new vertebrate host. The sporozoites are injected into the skin of the secondary host in the saliva, when the mosquito takes a successive blood meal (Cowman et al., 2012).

The sporozoites then travel to liver cells (hepatocytes) through blood vessels, where they reproduce asexually (tissue schizogony), producing thousands of merozoites. These infect new red blood cells and initiate a series of asexual multiplication cycles (blood schizogony) that produce 8–24 new infective merozoites, at which point the cells burst and the infective cycle begins anew (Schlagenhauf-Lawlor, 2008). Other merozoites develop into immature gametocytes, which are the precursors of male and female gametes.

Recurrent malaria

Symptoms of malaria may come back after symptom-free periods, which vary depending on the types of causative agents. These recurrences can be classified as recrudescence, relapse, and reinfection. Recrudescence is when parasites survive in the blood as a result of inadequate or ineffective treatment and return into disease condition (WHO, 2010). Relapse is when the parasites have been eliminated from the blood but persist as dormant hypnozoites in liver cells and subsequently symptoms reappear (Markus, 2011). In a relapse, the recurrence of symptoms commonly occurs between 8 and 24 weeks and is often seen in *P. vivax* and *P. ovale* infections (Caraballo, 2014). Reinfection may be confused with recrudescence and relapse. All the parasites that caused the past infection are eliminated from the body through proper treatment, but a new parasite was introduced within a short period and the cardinal symptoms arise. It seems that this is typically attributed to treatment failure (WHO, 2010).

Global climate change such as greater change in rainfall affects malaria transmission, but the degree of effect and the areas affected are uncertain (WHO, 2016a, b). More rainfall is associated with increased mosquito numbers (WHO, 2018).

Diagnosis

Malaria is usually confirmed by a microscopic examination observing the presence of the parasite in the blood films. The blood film is prepared using blood from peripheral circulation, that is, the fingertip, during the febrile condition of the patient. It is also confirmed by antigen-based rapid diagnostic tests (RDT) (Abba et al., 2011; Kattenberg et al., 2011). RDTs can distinguish infection by *Plasmodium falciparum* with other species of parasites, as this is because needed for treatment strategies (Abba et al., 2014). Polymerase chain reaction-based tests can also give confirmatory results, but these tests are not widely used due to their complexity (Fairhurst and Wellems, 2010). They need well-equipped laboratories to perform the tests, and the accuracy of the results depends on both the skill of the person examining the blood film and the levels of parasite in the blood (Wilson, 2012).

Prevention

The disease can be prevented through adopting some measures. Vector control is an imperative method to decrease parasite transmission to individuals. Keeping mosquitoes away from individuals by any means, that is, the use of mosquito nets with or without insecticide treatment, will reduce the disease. Even the use of mosquito repellents will provide protection to the individual. Vaccination against malaria was also approved by European regulators in 2015 (Walsh, 2015).

Epidemiology

As per the estimation published by the World Health Organization (WHO), about 228 million new cases of malaria were reported in 2018, resulting in 405,000 deaths (WHO, 2019). Malaria is endemic in a broad band area around the equator. These areas are in the Americas, many parts of Asia, and much of Africa. About 85%–90% of malaria fatalities occur in sub-Saharan Africa (Layne, 2006). Tropical and subtropical regions are prevalent with malaria. These areas are suitable for breeding mosquitoes because of rainfall, consistent high temperatures, and high humidity. There are plenty of stagnant water pools where mosquito larvae readily mature, providing them with the environment they need for continuous breeding (Jamieson et al., 2006). Outbreaks of malaria in drier areas can be forecasted with reasonable accuracy by mapping rainfall (Abeku, 2007).

Dengue

Dengue fever is a mosquito-borne disease prevailing in tropical zones caused by the dengue virus (WHO, 2015). Symptoms are observed 3–14 days after infection (Kularatne, 2015). Common symptoms are high fever, headache, vomiting, muscle and joint pains, and a characteristic skin rash (WHO, 2015; Kularatne, 2015). Most patients generally recover within 2–7 days (WHO, 2015). In a very small number of cases, the disease develops into a severe form known as dengue hemorrhagic fever. In these cases, the blood platelet number is reduced abruptly, resulting in bleeding and blood plasma leakage. Another form is dengue shock syndrome with dangerously low blood pressure (WHO, 2015; Kularatne, 2015).

Dengue viruses are spread by several species of female mosquitoes under the genus *Aedes*, principally *A. aegypti* (WHO, 2015; Kularatne, 2015). The dengue virus has five types (Normile, 2013; Mustafa et al., 2015). Infection with one type of dengue virus usually gives lifelong immunity against the same type. But infection with multiple types of the dengue virus increases the risk of severe complications (WHO, 2015).

Since World War II, dengue has become a global problem and is common in more than 120 countries, mainly in Asia and South America (WHO,

2019, 2020; Gubler, 1998). About 390 million people are infected each year and approximately 40,000 die globally (WHO, 2020; Roth et al., 2018).

Symptoms

About 80% of people infected with dengue virus remain asymptomatic, and some others show only mild symptoms such as an uncomplicated fever (Whitehorn and Farrar, 2010; WHO, 2009a,b; Reiter, 2010). A few (5%) have more severe illness, and a very few have life-threatening conditions (Rodenhuis-Zybert et al., 2010; Wilder-Smith et al., 2009). Generally, the incubation period of the disease ranges from 3 to 14 days, but most often it is 4–7 days (Gubler, 2010). If travelers returning from endemic areas don't show any symptoms, such as high fever, headache, vomiting, muscle and joint pains, or a characteristic skin rash, within 14 days, they may be considered free from dengue (Gubler, 1998). Children often experience symptoms such as the common cold and gastroenteritis (vomiting and diarrhea) (Varatharaj, 2010). Though initial symptoms of the children are generally mild but they suffer from high fever (Gubler, 1998; Simmons et al., 2012). The children have a greater risk of severe complications (Simmons et al., 2012).

Clinical course

The characteristic symptoms of dengue are sudden-onset fever, headache (typically located behind the eyes), muscle and joint pains, and a rash. As one of the cardinal symptoms of the disease is muscle and joint pains, the alternative name for dengue is "breakbone fever" (Whitehorn and Farrar, 2010; Chen and Wilson, 2010).

The course of infection is divided into three phases: febrile, critical, and recovery (WHO, 2009a, b).

The febrile phase includes high fever, potentially over 104 °F, and is associated with generalized pain and a headache. This febrile stage usually lasts for 2–7 days (WHO, 2009a, b; Chen and Wilson, 2010). Nausea and vomiting are also associated (Simmons et al., 2012). Among these symptoms, 50%–80% of cases will include a rash (Chen and Wilson, 2010; Wolff and Johnson, 2009). The fever itself is classically biphasic or saddleback in nature (Gould and Solomon, 2008). as it breaks and then returns for 1–2 days (Knoop et al., 2010; Gould and Solomon, 2008).

In some cases, disease prognosis depends on the fever in the critical phase (Simmons et al., 2012). At that time, plasma leaks from the blood vessels to the chest and abdominal cavity while fluid is depleted from circulation and the blood supply is decreased to vital organs. Typically, this condition lasts for 1–2 days (WHO, 2009a, b). As a result, organ dysfunction and severe bleeding, typically from the gastrointestinal tract, start (Gubler, 1998; WHO, 2009a, b). These conditions occur in less than 5% of all cases (Gubler, 1998). People who have previously been infected with other serotypes of dengue virus (secondary infection) are at an increased risk (Gubler, 1998; Rodenhuis-Zybert et al., 2010). Though the phase is rare, it occurs relatively more commonly in children and young adults (Simmons et al., 2012).

Maximum cases are recovered with restoration of the leaked fluid into the bloodstream (WHO, 2009a, b). This phase usually lasts for 2–3 days (Gubler, 1998). A feeling of fatigue may last for weeks in adults (Simmons et al., 2012).

Cause

Dengue fever viruses are transmitted by arthropods (mosquitos or ticks), which is why they are also referred to as arboviruses (arthropod-*bo*rne viruses) (Gould and Solomon, 2008). It is an RNA virus of the family *Flaviviridae*, genus *Flavivirus*. These viruses are distinguished into five strains, also called serotypes (Normile, 2013). These strains are differentiated based on their antigenicity (Solomonides, 2010).

Transmission

Dengue viruses are generally transmitted by *Aedes* mosquitos, particularly *A. aegypti* (WHO, 2009a, b). The virus may be transmitted through infected blood products, organ donation (Wilder-Smith et al., 2009; Stramer et al., 2009), vertical transmission (from mother to child) (Wiwanitkit, 2009), and by direct contact with infected people, that is, sexual transmission (Chen and Wilson, 2010). The habitats of these Aedes mosquitoes are between the latitudes of 35 degrees north and 35 degrees south and below an elevation of 1000 m (3300 ft.) (WHO, 2009a, b). The mosquito may bite around the clock (Centers for Disease Control and Prevention, 2010), but it typically bites during the early morning and in the evening (WHO, 2009a; WHO, 2009b; WHO, 2021). Humans are the primary host of the virus (WHO, 2009a, b; Gould and Solomon, 2008). but nonhuman primates also suffer from infection (WHO, 2011). A dingle bite is enough to acquire an infection (Center for Disease Control and Prevention, 2010). The female mosquito bites an infected person with dengue fever during febrile condition to take a blood meal then it becomes itself contaminated. Then the viruses enter cells lining of its gut (St. Georgiev, 2009). After 8–10 days, the virus reaches other tissues, including the mosquito's salivary glands. Subsequently, the viruses are released into its saliva and the hosts receive these viruses through bites. The mosquitoes are not detrimentally affected by the viruses, but the infection remains lifelong. These *A. aegypti* mosquitoes prefers to lay eggs in artificial water containers, even in clean water at human habitats and nearby areas. The mosquito likes to feed on people rather than other vertebrates (Gubler, 2010).

Predisposition

Healthy well-nourished young children are more susceptible to this severe disease in contrary to many other infections (Gubler, 1998). Other factors that make a person more vulnerable to the disease are the female gender, a high body mass index (Simmons et al., 2012), and the viral load (Martina et al., 2009). Each serotype can cause the full spectrum of disease (Gould and Solomon, 2008), so the virus strain is not a risk factor (Simmons et al., 2012). Infection with one serotype may produce lifelong immunity to that type, but may provide only short-term protections against the other three (WHO, 2009a, b; Chen and Wilson, 2010).

Diagnosis

The diagnosis of dengue is done based on clinical symptoms and a physical examination, particularly in endemic areas (Whitehorn and Farrar, 2010). It is difficult to differentiate from other viral infections at the early stages of the disease because the symptoms are similar to other flus (Gubler, 1998). The cardinal symptoms are high fever along with nausea and vomiting, rash, generalized pains, low white blood cell count, and a positive tourniquet test (WHO, 2009a, b). Differential diagnosis should be made from chikungunya (Chen and Wilson, 2010) malaria, leptospirosis, viral hemorrhagic fever, typhoid fever, meningococcal disease, measles, and influenza (Gubler, 1998; WHO, 2009a, b) and Zika fever (Musso et al., 2014).

Tab tests

The diagnosis of dengue fever may be confirmed by virus isolation in cell cultures, nucleic acid detection by PCR, viral antigen detection (such as for NS1), or specific antibodies (serology) (Guzman et al., 2010; WHO,

2009a, b). Isolation of the virus and the detection of nucleic acid are more accurate than antigen detection, but these tests are not widely available due to their greater cost and time-consuming nature (WHO, 2009a, b). Serological tests for dengue virus-specific antibodies, types IgG and IgM, can be useful in confirming a diagnosis in the later stages of the infection, that is, 14–21 days after infection (Gubler, 2010, Chen and Wilson, 2010; Guzman et al., 2010).

Prevention

Prevention of the mosquito-borne disease can be achieved only by controlling mosquitoes that transmit the viruses and protection from the mosquito bites (WHO, 2009a, b). Breeding as well as development of *A. aegypti* can be controlled by eliminating its habitats. Open sources of water can be treated with insecticides such as an organophosphate or pyrethroid. Biological control agents also can be used in free water bodies (WHO, 2009a, b). Individuals can protect themselves from mosquito bites by wearing clothing that fully covers the skin. The use of a mosquito net while resting, and/or the application of insect repellent (DEET being the most effective) can also be effective (Center for Disease Control and Prevention, 2010). The range of the disease also appears to be expanding, possibly due urbanization and to climate change (Normile, 2013).

Vaccine

Vaccines are used in some countries, but they can't give full protection against dengue fever (East, 2016; Archive, 2016). They have great limitations for efficacy while cost is also a concern.

Epidemiology

Almost 120 countries have dengue (WHO, 2020). It is one of the most common VBDs worldwide (Yacoub and Wills, 2014). Dengue virus is retained in nature, like most other arboviruses. The virus circulates from vertebrate to vertebrate hosts through blood-sucking vectors, which are the female Aedes mosquito (Gubler, 2010). In urban areas, the virus is primarily transmitted by the highly domesticated *A. aegypti*, but in rural areas, the virus is transmitted to humans by *A. aegypti* along with other species of *Aedes* such as *A. albopictus* (Gubler, 2010).

From 1960 to 2010, dengue increased 30-fold (WHO, 2009a, b). This might be due to urbanization, which increases mosquito habitats.

Population growth increased international travel and also global warming may increase dengue (Whitehorn and Farrar, 2010). Dengue is counted as one of 17 neglected tropical diseases by the WHO (2014).

Japanese encephalitis

Japanese encephalitis (JE) is a disease of the brain caused by infection from Japanese encephalitis virus (JEV) (CDC, 2015a, b, c). Among infected persons few show cardinal signs of the disease rest with little or no symptoms. In very few cases occasional inflammation of the brain occurs Cardinal symptoms of the disease are headaches, vomiting, fever, confusion, and seizures. These symptoms come about 5–15 days after infection (CDC, 2015a, b, c). *Culex* mosquitoes generally spread JEV. The disease mostly occurs in the countryside or rural unhygienic areas with hot and humid weather. Pigs and some wild birds are reservoirs for the virus (WHO, 2015).

Sign and symptom

The incubation period of JEV is about of 2–26 days (Moloney et al., 2014). Only one out of 250 infections develops into encephalitis, as the rest remain asymptomatic (Simon and Kruse, 2018). At the onset of this disease, the patient shows severe rigors. Common symptoms are fever, headache, nausea, vomiting, and malaise; these may last for 1–6 days. In the acute encephalitic stage, the signs are neck rigidity, cachexia, hemiparesis, convulsions, and a raised body temperature between 38°C and 41°C (100.4–105.8 °F). Mental retardation is also developed (Ghoshal et al., 2007). Children face a higher mortality rate compare to others. Vertical transmissions of the viruses are noted. The survived patients have lifelong neurological defects such as deafness, emotional liability, and hemiparesis among the patient with central nervous system involvement.

Cause

The disease is caused by Japanese encephalitis virus (JEV) transmitted by mosquito (Solomon, 2006). This virus comes under the family Flaviviridae. There are nine genetically and antigenically related viruses, some which are particularly severe in horses while four are known to infect humans, including West Nile virus (Lobigs and Diamond, 2012). The virus consists of a single-stranded RNA genome that is packaged in the capsid, which is made up of the capsid protein. The outer envelope is formed by an envelope protein that acts as the protective antigen, which assists in entry of the virus into the cell (Su et al., 2002).

Diagnosis

JEV can be diagnosed through the identification of JE virus-specific IgM antibodies in serum and/or cerebrospinal fluid, for example by IgM capture ELISA (Shrivastva et al., 2008). Commercial companies are now producing kits for the test. JE virus-specific IgM antibodies are generally detectable 3–8 days after the onset of illness and persist for 30–90 days, possibly longer. The presence of JE virus-specific IgM antibodies may also reflect a past infection or vaccination. Sometimes serum collected within 10 days of illness onset may not have detectable IgM, which is why repeated tests should be done and the sample collected in the convalescent stage. For confirmation, neutralizing antibody testing should be performed. Autopsy samples can be used for the isolation and identification of the virus through culture, and also identification can be done through nucleic acid amplification. Viral antigens can be shown in tissues by the indirect fluorescent antibody staining technique.

Prevention

Japanese encephalitis infection confers lifelong immunity to the individual. Currently, three different types of vaccines are available. All are based on the genotype III virus. A formalin-inactivated mouse brain-derived vaccine was first produced in Japan in the 1930s and was validated for use in different Asian countries. This disease mainly prevails in rural village areas. Urbanization reduces the habitat and breeding areas of mosquitoes that led to minimize the population of vectors, i.e., mosquito which indirectly control the disease. The widespread use of vaccines also helps control the disease in some countries (Solomon, 2006). The effective immunity as well as neutralizing antibodies persist in the circulation for at least 2–3 years and perhaps longer (Gambel et al., 1995; Kurane and Takashi, 2000). Vaccine boosters are recommended every 3 years for people who remain at risk.

Epidemiology

Japanese encephalitis (JE) is the major cause of encephalitis in Asia, with about 70,000 cases reported annually (Campbell et al., 2011). Casualty rates may be from 0.3% to 60% and the rates vary with population and age. People in rural areas in endemic locations are at greatest risk; generally, people in urban areas do not suffer from the disease. The major countries with epidemics such as Japan, Taiwan, China, South Korea, and Thailand have controlled the disease primarily by vaccination. Other countries such as India, Vietnam, Cambodia, Myanmar, Nepal, and Malaysia are still considered to have periodic epidemics. Human, cattle, and horses are considered dead-end hosts, as the disease manifests as fatal encephalitis. Pigs act as reservoirs where viral amplifying takes place, and they have a very significant role in the epidemiology of the disease. Infection in swine is generally asymptomatic, except in pregnant sows, when abortion and fetal abnormalities are common. Birds are natural hosts of JEV, not humans, and many believe that the disease will never be completely eradicated (Ghosh et al., 2009).

Blue tongue

Bluetongue disease is a noncontagious, insect-borne, viral disease of ruminants, mainly sheep and less frequently cattle (BBC, 2008), yaks (Mauroy et al., 2008), goats, buffalo, deer, dromedaries, and antelopes (Roy, 2008). It is caused by the orbivirus, which is also known as buetongue virus (BTV). The virus is spread by bites from black gnats and it is very seasonal, only occurring during summer and fall. The virus is transmitted by the midges *Culicoides imicola* and *Culicoides variipennis* as well as other culicoids. It is an acute disease that causes high morbidity and mortality in sheep. BTV also infects goats, cattle, and other domestic animals as well as wild ruminants (blesbuck, white-tailed deer, elk, and pronghorn antelope) (Roy, 2008).

Signs and symptoms

The cardinal symptoms of the disease are high fever, swelling of the face and tongue, cyanosis of the tongue, and excessive salivation. Typical blue appearance of tongue occurs due to swelling and cyanosis. Nasal discharge with stertorous respiration is also a prominent sign. Some animals may develop foot lesions, beginning with coronitis, followed by lameness. In case of sheep, the symptom like knee-walking may develop. The sheep will walk keeping its weight on knees. In cattle, the constant changing of the position of the feet gives this disease the nickname the dancing disease (McGrath, 2007). In severely affected animals, torsion of the neck (opisthotonos or torticollis) is observed. Not all the infected animals develop the clinical disease. Once the animals develop signs, they lose condition rapidly, and the sickest die within a week. If the affected animals survive, a slow recovery takes several months. The incubation period of the disease is 5–20 days, and within a month all signs usually develop. The mortality rate is not normally high, but it is high in susceptible breeds of sheep. There is no mortality in local breeds of sheep in Africa, but in imported breeds, it may reach up to 90% (Robertson, 1976). Only red deer suffer from the disease as acutely as sheep. Other ruminants such as cattle, goats, and wild ruminants are usually asymptomatic despite high virus levels in the blood (Jensen and Swift, 1982).

Cause

Bluetongue disease is caused by *Bluetongue virus* (*BTV*) (Roy, 2008), under the genus *Orbivirus* of the Reoviridae family. Twenty-six serotypes are recognized for this virus (Maan et al., 2011). The BT virus particle consists of 10 strands of double-stranded RNA surrounded by two protein shells, but has no lipid envelope.

Epidemiology

Culicoides are killed by cold winter temperatures such as frost temperature, and BTV infections typically do not last for more than 60 days. Nowadays, bluetongue disease is common in Australia, the United States, Africa, the Middle East, Asia, and Europe. In India, the disease outbreak was in the epidemic form. The disease depends upon the survival of the vector, that is, the midge Culicoides spp. It is believed that the virus somehow survives in overwintering midges or animals. The occurrence of the disease is seasonal in temperate zones, but it may exist year-round in tropical weather (Baylis et al., 2017). It is not a threat to humans, but common domestic ruminants are most vulnerable.

Livestock management and insect control

Practical midge control measures through simple husbandry practice changes may help break the livestock infection cycle. During maximum biting time from dusk to dawn, livestock may be housed. Fine mesh netting or coarser material impregnated with insecticide may be used in livestock shelters, which will reduce contact with midges. The virus-infested *Culicoides* midges usually breed on animal dung and moist soils, either bare or covered in short grass. Proper management or treatment of these breeding grounds to break the breeding cycle will significantly reduce the local midge population as well as bite midges (Gairdner, 2007; Abel, 2008).

Vaccines

European countries are using vaccines against different strains (Anderson et al., 2013). In January 2015, Indian researchers also launched a vaccine named Raksha Blu that is designed to protect livestock against five strains of the bluetongue virus prevalent in the country (Staff Reporter, 2015).

Babesiosis

Babesia is thought to be the second-most common blood parasite of mammals after trypanosomes. They also have major adverse effects on the health of domestic animals in worm or moderate worm areas. Babesiosis is similar to the malaria-like parasitic disease caused by infection with the eukaryotic parasite *Babesia*, an alveolate under the phylum Apicomplexa (Berger and Marr, 2006a, b). Babesiosis transmission occurs via tick bite. The disease is most common in the Northeastern and Midwestern United States and parts of Europe, and is sporadic throughout the rest of the world (CDC, 2015a, b, c). Humans or animals can get infected with babesia parasites through the bite of infected ticks. The ticks are infected by getting a blood transfusion from an infected donor of blood products. Babesiosis also has congenital transmission (an infected mother to her baby) (CDC, 2018). In cattle, the disease is known as Texas cattle fever, redwater, or piroplasmosis (Hunfeld et al., 2008).

Signs and symptoms

Most children and a good number of previously healthy adults with *Babesia* infection are asymptomatic. The most common symptoms are fever and hemolytic anemia, with fevers up to 40.5 °C (105 °F) and shaking chills similar to those of malaria (Vannier and Krause, 2012). Respiratory distress syndrome following organ failure may be seen in adults. Sepsis in people who have had a splenectomy can occur rapidly, consistent with overwhelming postsplenectomy infection. Very young, very old, and other persons with immunodeficiency suffer severely. The symptoms are generally noticed 1–4 weeks after the bite, or 1–9 weeks after transfusion of contaminated blood products if the receiver becomes ill. The infected person with babesiosis gradually develops malaise and fatigue, followed by a fever. Red blood cells are destroyed and removed from the blood and become anemic. Other common symptoms are chills, sweats, thrombocytopenia, enlargement of the liver and spleen, jaundice, anorexia, nausea, etc. These symptoms may last from several days to several months. Babesiosis was diagnosed in humans in the 2000s. People with immunodeficiency when coming in contact with the vector ticks, they become more susceptible to the disease (Hunfeld et al., 2008). It causes hemolytic anemia, which is quite similar to malaria. The species that cause disease in nonhuman animals are *Babesia canis rossi*, *Babesia bigemina*, and *Babesia bovis*. They cause severe hemolytic anemia, hemoglobinuria "red-water, "disseminated intravascular coagulation, and cerebral babesiosis. Babesia causes hemolytic anemia in bovine species. Initially, the mucous membrane becomes pale. In progressive cases, the red blood cells are lyses and bilirubin is produced as a byproduct. Also, the visible mucous membranes become yellow in color, which is called icterus. It happens due to the failure of the liver to metabolize the excess bilirubin while hemoglobinuria is seen due to the excretion of red-blood-cell lysis byproducts through urine. Body temperature becomes high, about 40.5 °C (105 °F) due to the release of inflammatory byproducts (Noskoviak and Broome, 2008).

Life cycle

Babesia spp. needs a biological stage in a rodent or deer host to complete the life cycle. The tick Ixodes transmits the parasite between hosts. The tick becomes infected with gametocytes during a blood meal from rodents. It also inject the *Babesia* into the intermediate host (e.g., cattle) when taking a blood meal. *Babesia* enters the animal's red blood cells (erythrocytes); they are called sporozoites. In the developmental process in the red blood cell, the protozoa become cyclical and change into a trophozoite ring. Molting of trophozoites takes place in RBC and develops merozoites, which have a tetrad structure called the Maltese cross form (Herwaldt et al., 2003). These trophozoites and merozoites rupture the host erythrocyte and release vermicules, the infectious parasitic bodies. These protozoa rapidly spread throughout the blood (Noskoviak and Broome, 2008). Some of the merozoites produce gametocytes, which are fertilized in the tick gut, develop into sporozoites, and enter the salivary glands. During biting or taking a blood meal, the infected tick introduces the sporozoites into the intermediate host (Despommier, 1995).

Diagnosis

Clinical signs such as hemoglobinuria and anemia with high fever in endemic zones are considered babesiosis in animals. It can be confirmed by the observation of merozoites on a thin film blood smear examination. Blood film is stained by methylene blue and eosin and is examined under maximum magnification using oil. Clinical signs like persistent fever with hemolytic anemia are to be correlated with medical history in human. Whether the person lives in endemic zone or visits to endemic zone within nine weeks of onset of clinical signs or history of blood transfusion are to be considered for diagnosis of the disease. It can be confirmed by the identification of parasites on a Giemsa-stained thin film blood smear (Wormser et al., 2006). Differential diagnosis can be done from malaria by observing the so-called Maltese cross formations on the blood film. These Maltese cross formations are absent in malaria (Despommier, 1995). It needs a very specialized laboratory and skilled technicians to identify this formation. It may need multiple smears and careful examination because *Babesia* infects less than 1% of circulating red blood cells, thus it can be easily overlooked (Krause, 2003). Serological tests are more sensitive for confirmatory diagnosis. Detecting antibodies against *Babesia* (both IgG and IgM) can be done through ELISA (Krause et al., 1994). *Babesia* from the peripheral blood can be identified through polymerase chain reaction (PCR) (Persing et al., 1992).

Transmission

Babesia spp. is generally transmitted by several spp. of ticks. *B. microti* is transmitted by *Ixodes scapularis* ticks (Vannier and Krause, 2012). *B. divergens* is also the cause of infectious babesiosis and is transmitted by *I. ricinus* (Hunfeld et al., 2008). Ticks of domestic animals such as *Rhipicephalus (Boophilus) microplus*, *R. (B.) decoloratus*, and *Rhipicephalus microplus* transmit several species of *Babesia* to livestock, causing considerable economic losses to farmers in tropical and subtropical regions (Archived, 2012). The organism can also be transmitted through a blood transfusion (Herwaldt et al., 2011; Leiby, 2011).

Kyasanur forest disease (KFD)

Kyasanur forest disease (KFD) is a hemorrhagic fever endemic to the southwestern part of India. The cause of the disease is a virus belonging to the family *Flaviviridae*. The KFD virus is transmitted to humans through the bite of infected hard ticks (*Haemaphysalis spinigera*), which act as a reservoir (Gould and Solomon, 2008). The fatality rate of the disease is 3%–10%, and it affects 400–500 people annually (Holbrook, 2012; Mourya et al., 2012).

Signs and symptoms

The cardinal symptoms of the disease are high fever, including frontal headaches, chills, and rigorous muscle pain. There are vomiting and gastrointestinal such as stomach aches and crimping. Bleeding problems may also occur 3–4 days after the beginning of the initial symptoms. The blood pressure may fall abnormally; the blood cell count, that is, the platelets, red blood cells, and even white blood cells become lower. Some patients recover without complication after 1–2 weeks of the initiation of symptoms. In some patients, the disease is biphasic. For them, a second wave of symptoms appears at the beginning of the third week. The second wave starts with the symptoms of first phase, fever and headaches, and new complications include signs of neurological manifestations, such as severe mental disturbances, tremors, and vision deficits (Gerhard, 2010; Dobler, 2010; Mourya et al., 2013). The recovery period is very long. It may take several months. In this period, muscle aches and weakness continue and the affected person is fit for any physical activities.

Transmission

Rats, squirrels, mice, shrews, and porcupines are thought to be reservoir hosts for the kyasanur forest disease virus (Gerhard, 2010). Primates such as monkeys, Langur-presbytis entellus and the Bonnet monkey–*Macaca radiata* are very susceptible and also sufferers from the KFD virus. Monkeys are the main amplifying hosts for the KFD virus. During the tremendous viremic stage, they infect the ticks. *H. spinigera*, a forest tick, acts as a vector for disease transmission (Holbrook, 2012). Humans are infected by the bite of tick nymphs. This disease is transmitted to man only from these primates. There is no man-to-man transmission because the ticks cannot sustain in a domestic environment.

Diagnosis

Initially, to detect the virus, the serum from a suspected case was inoculated into suckling mice (Swiss Albino mice), and the subsequent death of the mice was confirmed as a KFD positive case. Nowadays, other methods of diagnosis include hemagglutination inhibition (HI), complement fixation, and neutralization tests (Upadhyaya et al., 1975). More efficient molecular based methods, that is, RT-PCR, nested RT-PCR, and TaqMan-based real-time RT-PCR, are used to diagnose KFDV. To detect immunoglobin M antibodies and immunoglobin G, ELISA tests are done.

For PCR, test samples are to be collected 8–10 days from the onset of symptoms. The anti-KFDV antibodies in patients are produced from the fifth day of onset of symptoms and remain up to 3 months; they can be detected by ELISA-based methods (Mourya et al., 2012).

Prevention and treatment

Prevention includes vaccination and population control of ticks. Tick bites can be reduced by using protective clothing as well by killing the tick reservoir. Formalin-inactivated KFDV vaccines are used, but the effectiveness rate is 62.4% for those who receive two doses. The effectiveness increases to 82.9% for an additional dose (Kasabi et al., 2013). No specific treatment is available.

Risk factor and risk group

The disease is endemic in the Kyasanur Forest of Karnataka in India. The people who are directly involved with or travel in the forest are the most vulnerable to infection. The forest is infested with the ticks *H. spinigera*, which are the virus reservoir. Villagers in nearby forests or at the crossroads of the animal-human interaction, especially villages adjoining forest areas and interstate borders, are at risk. Villagers who frequently visit the forest for their livelihood get the infection through tick bites. The people in adjacent states such as Kerala, Maharastra, and Tamilnadu also suffer from the infection. A seroprevalence study showed that the people of the Kutch and Saurashtra regions of the Gujarat state, Kingaon, and Parbatpur of West Bengal state (Pattnaik, 2006) and Andaman and Nicobar islands (Padbidri et al., 2002) are positive to the specific infection.

Yellow fever disease

Yellow fever is a mosquito-borne viral disease of typically short duration (WHO, 2013). The mortality rate is almost 50% for those who get the severe disease, although the overall fatality rate is 3%–7.5% (WHO, 2013; CDC, 2015a, b, c). Common symptoms are fever, chills, loss of appetite, nausea, and headaches. There may be moderate to severe muscle pains, particularly in the back. These symptoms normally recede within 5 days (WHO, 2013) but about 15% of patient with in short period of improving the fever they come back with typical problem such as abdominal pain and liver damage, which begins causing yellow skin (WHO, 2013; Scully, 2014). Those patients are at risk of bleeding and kidney problems (WHO, 2013).

Human and other primates receive the yellow fever virus through the bite of an infected *A. aegypti* female mosquito. These mosquitoes are found throughout the tropics and subtropics (WHO, 2013). The virus is an RNA virus of the genus *Flavivirus* (Lindenbach et al., 2007). It is difficult to differentiate the disease from other illnesses, especially in the early stages (WHO, 2013). Polymerase chain reaction is required to confirm a suspected case through testing of blood samples.

The disease is common in tropical and subtropical areas of the continents of South America and Africa, but not in Asia (WHO, 2013; CDC, 2012). Nearly 90% occur in African nations (Tolle, 2009). The incidence of the disease increasing day by day may be due to people moving frequently, and changing climate increasing the habitat for mosquitoes. Some countries already use a vaccine against yellow fever for travelers. Mass vaccinations including mosquito-controlling measures are to be undertaken to control the disease (WHO, 2013).

Signs and symptoms

The incubation period of yellow fever is 3–6 days (CDC, 2010). Overall, 80%–85% cases are not very serious, showing common symptoms such as chills, fever, headaches, loss of appetite, back pain, fatigue, muscle pain, nausea, and vomiting (WHO, 2009a, b). In these cases, the symptoms last only 3–4 days. In severe cases, 15% of the infected people enter a second, toxic phase of the disease with recurring fever. At the same time, symptoms of abdominal pain as well as jaundice appear due to liver damage (David and Heymann, 2015). Bleeding from mouth, nose, and eyes and vomiting containing blood starts (Chastel, 2003). There may also be kidney failure, hiccups, and delirium (Sherman, 2017; Franklin and Sutherland, 1984). Surviving people achieve lifelong immunity, and normally no permanent organ damage results (Rogers et al., 2006).

Cause

Yellow fever is caused by an enveloped RNA virus 40–50 nm in width, under the family *Flaviviridae* and named the yellow fever virus (Lindenbach et al., 2007). This disease was the first known illness transmitted by mosquitoes. The causative agents are transmissible by filtered human serum (Staples and Monath, 2008). Yellow fever belongs to the group of hemorrhagic fevers. The viruses infect monocytes, macrophages, Schwann cells, and dendritic cells.

Transmission

Yellow fever viruses are transmitted through the bite of Aedes spp. mosquitoes mainly by *A. aegypti*, but other mosquitoes such as the tiger mosquito (*Aedes albopictus*) can also serve as a vector for this virus. Like other mosquito-borne infections, that is, arboviruses, the *yellow fever virus* is also taken up by a female mosquito during taking the blood of an infected human or another primate. After ingestion, viruses reach the stomach of the mosquito, infect epithelial cells, and replicate there. After replication, the host epithelial cells burst, releasing viruses that reach the blood system of mosquitoes. From blood circulation, the viruses lodge in salivary glands. When this virus-infested mosquito sucks blood from a person, it injects its saliva into the dermis layer of skin, so the virus reaches the bloodstream. The eggs of the virus-infested mosquito also carry the viruses during their life cycle; the adult one also carries the viruses. They need not suck blood from any diseased person. This is called transovarial and transstadial transmission (Fontenille et al., 1997).

There are three different epidemiologically infectious cycles (Barrett and Higgs, 2007) in which the virus is transmitted from mosquitoes to humans or other primates (CDC, 2016a, b). In the "urban cycle," only *A. aegypti* transmits the virus from an infected person to a new one. These mosquitoes are well adopted in urban areas. In the sylvatic cycle (forest or jungle cycle), Aedesafricanus (in Africa) or mosquitoes of the genus Haemagogus and Sabethes (in South America) act as vectors. In the cycle, the mosquitoes infect mainly nonhuman primates of the jungle. The disease is mostly asymptomatic in African primates. This cycle is a way to carry the infection from jungle to urban areas. In urban areas, only *A. aegypti* acts as a vector. It is not possible to eradicate the disease unless oyu eradicate the vector mosquitoes (Barrett and Higgs, 2007). In the "savannah cycle" or intermediate cycle, different mosquitoes of the genus *Aedes* are responsible for carrying the infection from jungle to urban areas (WHO, 2006).

Pathogenesis

After transmissions from mosquito to the host body, the viruses replicate in the lymph nodes and then infect particularly dendritic cells. Getting release from dendritic cells the viruses reach the liver and infect hepatocytes, which cause to eosinophilic degradation of hepatocytes and lead to release of cytokines. Effects of the phenomenon is emerging of apoptotic masses known as Councilman Bodies in the cytoplasm of hepatocytes (Ryan and Ray, 2004; Quaresma et al., 2006). Fatality may occur due to multiple organ failure following cytokine storm and shock (Monath, 2008).

Diagnosis

Diagnosis of yellow fever is done on the basis of clinical symptoms and travel history. Confirmation of the disease can be done virologically. The virus cannot be confirmed before 6–10 days of the illness. Detection of a viral antigen can be obtained by reverse transcription polymerase chain reaction. Another approach is the isolation of the virus and its growth in a cell culture using blood plasma. Confirmation of infection also can be done through a serological test, that is, the enzyme-linked immunosorbent assay.

Prevention

The disease can be prevented by two ways: mass vaccination programs or avoiding mosquito bites by any means. Vaccination is recommended for tourists before traveling to affected or endemic zones because nonnative people tend to develop more severe illness when they are infected. About 95% of people become immune by the 10th day after vaccine administration (Barrett and Teuwen, 2009) and protective immunity lasts for at least 10 years. A single dose of vaccine is sufficient to confer lifelong immunity against yellow fever disease (WHO, 2013). This live attenuated vaccine was developed in 1937 by Max Theiler (Barrett and Teuwen, 2009). It is most important to control *A. aegypti*, the yellow fever mosquito, especially because the same mosquito also transmits dengue fever and chikungunya disease. Two main approaches are employed to reduce *A. aegypti* populations. One tactic is to kill the developing larvae and reduce the water-accumulated areas where mosquitoes breed and subsequently larvae develop. Chemical larvicides are used to kill larvae. Biologically, larvae can be controlled using larvae-eating fish and copepods in the water-accumulating areas (Tolle, 2009). The second approach is to reduce populations of the adult mosquito using lesser amounts of pesticide because this targets the pest directly. To keep away the mosquitoes, insecticide-treated mosquito nets can be used (Tolle, 2009).

Zika virus disease

The Zika virus disease is an infectious disease caused by an RNA virus named the Zika virus (WHO, 2016a, b). Common symptoms are fever, red eyes, joint pain, headaches, and a maculopapular rash that may be confused with dengue fever (WHO, 2016a, b; Chen and Hamer, 2016; Musso et al., 2014, b). In most cases, patients don't show any symptoms, and in mild cases, symptoms generally last less than 7 days (Chen and Hamer, 2016). The virus was first identified in the Zika forest near Entebbe, Uganda, and the name was given after the forest name (Sikka et al., 2016). The virus is transmitted through mosquitoes, sex, blood transfusions, and even vertically (Chen and Hamer, 2016; Rasmussen et al., 2016; CDC, 2016a, b; WHO, 2016a, b).

Signs and symptoms

Generally, the infected people are asymptomatic (Rasmussen et al., 2016). In the manifestation of the disease, the most common signs and symptoms of Zika fever are fever, rash, conjunctivitis (red eyes), muscle and joint pain, and headaches, which are similar to other viral diseases such as dengue and chikungunya fever (Heang et al., 2012). The incubation period, that is, the time from a mosquito bite to developing symptoms may be a few days to a week (Centers for Disease Control and Prevention, 2014). The manifestation of the symptoms lasts for a maximum of 1 week and generally it does not need any treatment (WHO, 2016a, b; Sikka et al., 2016). In pregnant women, the virus is transmitted from mother to child in the womb and can cause multiple problems such as abnormalities of the fetus (Brasil et al., 2016). The common abnormalities observed in brain and eye development are microcephaly, chorioretinal scarring (de Paula et al., 2016), hydrops fetalis, and abnormal accumulation of fluid in the fetus (Sarno et al., 2016; Li et al., 2016).

Cause

The disease is caused by a mosquito-borne flavivirus that is closely related to the dengue and yellow fever viruses. Mosquitoes act as the vector and West African monkeys and rodents may be the main reservoirs (Hayes, 2009; Brack, 2012).

Transmission

A. aegypti mosquitoes are primarily responsible for transmission of the Zika virus in tropical regions. But other species under the genus Aedes, such as *Ae. africanus*, *Ae. apicoargenteus*, *Ae. luteocephalus* (Walter Reed Biosystematics Unit, 2016), *Ae. Albopictus* (Grard et al., 2014; Wong et al., 2013). *Ae. vittatus*, and *Ae. furcifer* (Hayes, 2009) are also reservoirs of Zika. In a 2007 outbreak on Yap Island in the South Pacific, *Aedes hensilli* was the vector, and *Aedes polynesiensis* spread the virus in French Polynesia in 2013 (Archive, 2015). The Zika virus is found 10,000 times more in semen than blood or urine after 2 weeks since infection (Mansuy et al., 2016). It is a sexually transmitted disease. The virus can also spread from infected men to their partners (Oster et al., 2016; Centers for Disease Control and Prevention, 2016; Hills et al., 2016). There is no data of sexual transmissions of the virus from women to their sexual partners (Hills et al., 2016). Vertical perinatal transmissions also have been reported (Gatherer and Kohl, 2015). A blood transfusion is also a source of virus transmission and several affected countries have developed strategies to screen blood donors (Sikka et al., 2016; Franchini and Velati, 2016).

Diagnosis

Clinical signs and symptoms are not enough to diagnose the disease because these signs and symptoms overlap with other diseases caused by arboviruses at similar areas (Sikka, et al., 2016; Fauci and Morens, 2016). Other diseases such as leptospirosis, malaria, rickettsia, group A streptococcus, rubella, measles, parvovirus, enterovirus, adenovirus, and alphavirus infections (e.g., chikungunya, Mayaro, Ross River, Barmah Forest, O′nyong'nyong, and Sindbis viruses) also show similar signs and symptoms (Centers for Disease Control and Prevention, 2015a, b). The disease can be diagnosed through identification of the Zika virus from the patient by reverse transcriptase PCR (RT-PCR). The time of sample collection is vital to get the virus in the sample. For RT-PCR testing, the serum is to be collected within 1–3 days of symptom onset or on saliva samples collected during the first 3–5 days of disease inception (Archive, 2015). Urine samples can be collected and tested up to 14 days after the onset of symptoms (MMWR, 2016). The disease can be diagnosed through the detection of specific IgM and IgG antibodies against Zika virus through ELISA. IgM antibodies can be detectable within 3 days of the onset of illness (Hayes, 2009).

Prevention

The virus is spread by mosquito bites in most cases. Keeping away from mosquito bites is how to avoid the viral infection. This may be done by two ways: personal protection from mosquito bites and eliminating or reducing the mosquito population by any means. Personal protection may be done with these methods:

- Permethrin-treated long-sleeved shirts and long pants covering exposed skin should be worn (Archive, 2016).
- Insect repellent containing picaridin, oil of lemon eucalyptus, ethyl butylacetylaminopropionate, or DEET may be used during exposure to mosquito habitat areas (Louis, 2016).
- Air-conditioned rooms should be covered with screens.
- Bed and window nets should be mandatory in mosquito-prone areas (Centers for Disease Control and Prevention, 2015a, b, 2016).

The mosquito population may be controlled following these methods:

- Mosquito breeding areas are to be reduced through eliminating standing water in household areas, farming areas, etc.
- Larvicidal chemicals may be used in stagnant water, that is, breeding places.
- Insecticide may be sprayed to kill flying mosquitoes (WHO, 2016a, b).
- Genetically modified mosquitoes may be released for breeding to prevent transmitting viruses.
- Mosquitoes may be infected with *Wolbachia* bacterium to inhibit the spread of viruses (Sikka et al., 2016; Gale, 2016).

Zika virus also is transmitted sexually. Precautionary measures should be taken for men who have gone to an area where Zika fever is occurring (Sikka et al., 2016; Oster et al., 2016; Petersen et al., 2016).

African horse sickness (AHS)

African horse sickness (AHS) is a noncontagious deadly disease of horses, mules, donkeys, and zebras. The most susceptible host is the horse and the mortality rate is about 90% (The Merck Veterinary Manual, 2015) among the affected animals, followed by mules (50%) and donkeys (10%). African donkeys and zebras are generally asymptomatic and are considered the natural reservoir of the virus. It is a highly infectious VBD caused by the *African horse sickness virus*. The virus is the genus *Orbivirus* under the family *Reoviridae*. There are nine serotypes of this virus and any of the nine serotypes can cause the disease (ICTV, 2019).

Transmission

This disease is transmitted through biting by infected arthropods. Mostly, the disease is spread by the biological vector of the virus, that is, the *Culicoides* (midges) species. Even mosquitoes such as *Culex*, *Anopheles*, and *Aedes*, and ticks such as *Hyalomma* and *Rhipicephalus* can spread the viruses (Du Toit, 1944; Bouayoune et al., 1998; Meiswinkel, 1998). Generally, dogs become infected by eating infected horse meat. But there are recent reports that, without any history of horse meat ingestion, dogs were infected (van Sittert Sybrand et al., 2013).

Clinical signs

In the per-acute form, the respiratory system is affected and the disease occurs. Common symptoms are high fever, depression, and respiratory troubles. The affected animals suffer from trouble breathing, coughing, and frothy fluid from the nostril and mouth. Within 4 days, signs of pulmonary edema arise. The animal dies within 24 h of onset of clinical symptoms due to serious lung congestion, which causes respiratory failure (Erasmus, 1972). In the subacute form of the disease, generally the cardiac system is affected. The incubation period of this form is longer than that of the pulmonary form. The symptoms of the disease are observed at day 7–12 after infection. Common symptoms are high fever along with conjunctivitis, abdominal pain, and progressive dyspnea. There may be fluid accumulation under the skin of the head and neck, even swelling of the supraorbital fossae, intermandibular space, and palpebral conjunctiva (Coetzer and Erasmus, 1994). Zebras and African donkeys suffer from mild to subclinical forms of the disease. Clinical diagnosis of the infected animals may show a low-grade fever and congested mucous membrane. Mortality does not occur in the mild form of AHS (Erasmus, 1972). In some cases, the affected horse may express both the pulmonary and cardiac signs of AHS. This mixed form of AHS can be diagnosed through necropsy (Erasmus, 1972).

Diagnosis

Diagnosis of the disease is assumed by reading characteristic clinical signs, observing the presence of competent vectors, and identifying specific *post-mortem* lesions. The disease can be confirmed by isolation of the virus. Viral RNA can be detected with techniques such as quantitative PCR, antigen capture (ELISA), and also immunofluorescence of infected tissues (Crafford et al., 2003; Hamblin et al., 1991).

Epidemiology

The disease was first reported in the southern part of the Sahara Desert in the mid-1600s. It became endemic to the equatorial, eastern, and southern regions of Africa with the introduction of horses in those areas (CIDRAP, 2010; Howell, 1963; Mellor and Boorman, 1995). AHS has spread to Morocco, the Middle East, India, and Pakistan from endemic sub-Saharan Africa. Recently, outbreaks of the disease have been reported in the Iberian Peninsula and Thailand. The occurrence and spread of the disease depend on host-vector interaction and the presence of high numbers of competent vectors. The significant vector for transmitting the virus in endemic areas is the biting midge *Culicoides imicola* (Du Toit, 1944). These vectors prefer warm, humid conditions for their survival as well as for breeding.

Control of the disease

The disease can be controlled through quarantine, vector control, and vaccination (Animal Research Info, 2020). The affected horses are to be slaughtered, and the uninfected horses are vaccinated against the virus (Iowa State University, 2020). Different types of vaccines, including polyvalent vaccine, monovalent vaccine, and monovalent inactivated vaccine, are now available (Du Toit, 1944). This disease can also be prevented by destroying the insect vector habitats and by using insecticides.

Conclusion

Vector-borne pathogens have been the scourge of man and animals since the beginning of time. They cause huge suffering to humans and animals. Many diseases are expanding their range into new areas. Because of that, the vectors are changing their habitat due to climate change. Climate changes impacts the VBD epidemiology. Life cycles and life histories of the vectors are changing with the change of climate. These changes influence arthropod vectors to invade new agro-climatic zones. It results in changing both vector and pathogen distribution, changes in the ability of arthropods to transmit pathogens, and pathogen-vector-host interaction. Dengue, malaria, West Nile, Chikungunya, etc., have recently caused substantial human epidemics. Arthropod-borne animal diseases such as bluetongue, Rift Valley fever, babesiosis, KFD, Lyme disease, African horse sickness, etc., pose considerable threats to livestock economies around the world.

These diseases question the profitability of animal farming systems. Diseases hinder the production of healthy and quality consumable animal products. Diseases have a negative influence on the livestock production system, such as low production, reduced income, and substandard livelihood. To predict and mitigate these grave situations, we need to understand the changes in the environment such as climate change on the complex arthropod–pathogen–host epidemiological cycle from the molecular to the population level. It is noted that there has been sizeable progress on many fronts to effectively understand and mitigate the impact of potential changes in the environment on vector-borne pathogens. Strong measures need to be taken to combat global warming and restrict arthropod propagation and invasion in new or virgin areas. The challenge to be faced is to control the diseases through controlling the vectors and developing vaccines against these diseases.

References

Abba, K., Deeks, J.J., Olliaro, P., Naing, C.M., Jackson, S.M., Takwoingi, Y., Donegan, S., Garner, P., 2011. Rapid diagnostic tests for diagnosing uncomplicated P. falciparum malaria in endemic countries. Cochrane Database Syst. Rev. Abba, K. (Ed.), (7): CD008122. doi: 10.1002/14651858.CD008122.pub2. PMC 6532563. PMID 21735422.

Abba, K., Kirkham, A.J., Olliaro, P.L., Deeks, J.J., Donegan, S., Garner, P., Takwoingi, Y., 2014. Rapid diagnostic tests for diagnosing uncomplicated non-falciparum or Plasmodium vivax malaria in endemic countries. Cochrane Database Syst. Rev. 12 (12). https://doi.org/10.1002/14651858.cd011431, CD011431. PMC 4453861. PMID 25519857.

Abeku, T.A., 2007. Response to malaria epidemics in Africa. Emerg. Infect. Dis. 13 (5), 681–686. https://doi.org/10.3201/eid1305.061333. PMC 2738452. PMID 17553244.

Abel, C., 2008. Bluetongue Vaccine BTV8 Questions Answered. Archived 2008-06-11 at the Wayback Machine, Farmers Weekly 2 April 2008.

Anderson, J., Hägglund, S., Bréard, E., Comtet, L., Lövgren Bengtsson, K., Pringle, J., Zientara, S., Valarcher, J.F., 2013. Evaluation of the immunogenicity of an experimental DIVA subunit vaccine against Bluetongue virus serotype 8 in cattle. Clin. Vacc. Immunol. 20 (8), 1115–1122. https://doi.org/10.1128/CVI.00229-13. PMC 3754508. PMID 23720365.

Animal Research Info, 2020. African Horse Sickness. Animal Research Info. Retrieved 29 June 2020.

Anon., 2015. Anopheles Mosquitoes. Centers for Disease Control and Prevention. October 21, Retrieved December 21, 2016. This article incorporates text from this source, which is in the public domain.

Archive, 2015. Zika Virus. Archived from the original on 25 December 2015. Retrieved 24 December 2015.

Archive, 2016. Zika Virus—Protect Yourself From Mosquito Bites. 5 November 2014. Archived from the original on 1 October 2016. Retrieved 1 October 2016. Permethrin-treated clothing will protect you after multiple washings.

Archived, 2012. Tick Fever (bovine babesiosis). Queensland Government, Department of Primary Industries and Fisheries. Archived from the original on 5 April 2011. Retrieved 21 February 2012.

Arrow, K.J., Panosian, C., Gelband, H., 2004. Saving Lives, Buying Time: Economics of Malaria Drugs in an Age of Resistance. National Academies Press, p. 141. ISBN 978-0-309-09218-0. Archived from the original on 2016-05-15.

Baird, J.K., 2009. Malaria zoonoses. Travel Med. Infect. Dis. 7 (5), 269–277. https://doi.org/10.1016/j.tmaid.2009.06.004. PMID 19747661.

Baird, J.K., 2013. Evidence and implications of mortality associated with acute Plasmodium vivax malaria. Clin. Microbiol. Rev. 26 (1), 36–57. https://doi.org/10.1128/CMR.00074-12. PMC 3553673. PMID 23297258.

Banfield, W.G., Woke, P.A., MacKay, C.M., Cooper, H.L., 1965. Mosquito transmission of a reticulum cell sarcoma of hamsters. Science 148 (3674), 1239–1240. https://doi.org/10.1126/science.148.3674.1239. Bibcode: 965Sci...148.1239B. PMID 14280009 (web archive link).

Barrett, A.D., Higgs, S., 2007. Yellow fever: a disease that has yet to be conquered. Annu. Rev. Entomol. 52, 209–229. https://doi.org/10.1146/annurev.ento.52.110405.091454. PMID 16913829.

Barrett, A.D., Teuwen, D.E., 2009. Yellow fever vaccine—how does it work and why do rare cases of serious adverse events take place? Curr. Opin. Immunol. 21 (3), 308–313. https://doi.org/10.1016/j.coi.2009.05.018. PMID 19520559.

Bartoloni, A., Zammarchi, L., 2012. Clinical aspects of uncomplicated and severe malaria. Mediterr. J. Hematol. Infect. Dis. 4 (1). https://doi.org/10.4084/MJHID.2012.026, e2012026. PMC 3375727. PMID 22708041.

Bath, G.O., 1989. Bluetongue. In: 2nd International Congress for Sheep Veterinarians. Massey University, Wellington, New Zealand, pp. 349–357.

Baylis, M., Caminade, C., Turner, J., Jones, A.E., 2017. The role of climate change in a developing threat: the case of bluetongue in Europe: -EN- -FR- L'influence du changement climatique dans les menaces sanitaires évolutives: l'exemple de la fièvre catarrhale ovine en Europe -ES- La influencia del cambio climático en una amenaza creciente: el caso de la lengua azul en Europa. In: Revue Scientifique et Technique—Office International des Épizooties. 36, pp. 467–478, https://doi.org/10.20506/rst.36.2.2667. (2). ISSN 0253-1933. PMID 30152470.

BBC, 2008. Q&A: Bluetongue Disease. BBC. 2008-09-17. Retrieved 2010-01-01.

Beare, N.A., Taylor, T.E., Harding, S.P., Lewallen, S., Molyneux, M.E., 2006. Malarial retinopathy: a newly established diagnostic sign in severe malaria. Am. J. Trop. Med. Hyg. 75 (5), 790–797. https://doi.org/10.4269/ajtmh.2006.75.790. PMC 2367432. PMID 17123967.

Berger, S.A., Marr, J.S., 2006a. Human Parasitic Diseases Sourcebook. books.google.com.

Berger, S.A., Marr, J.S., 2006b. Human Parasitic Diseases Sourcebook. Jones & Bartlett Learning. ISBN 978-0-7637-2962-2.

Bessell, P.R., Searle, K.R., Auty, H.K., Handel, I.G., Purse, B.V., de Bronsvoort, B.M.C., 2016. Assessing the potential for Bluetongue virus 8 to spread and vaccination strategies in Scotland. Sci. Rep. 6, 8940. https://doi.org/10.1038/srep38940. ISSN 2045-2322. PMC 5154200. PMID 27958339.

Bouayoune, H., Touti, J., El Hasnaoui, H., Baylis, M., Mellor, P.S., 1998. The Culicoides vectors of African horse sickness virus in Morocco: distribution and epidemiological implications. Arch. Virol. Suppl. 14, 113–125.

Brack, M., 2012. Agents Transmissible From Simians to Man. Springer Science & Business Media. ISBN 978-3-642-71911-0. Archived from the original on 17 June 2016.

Brasil, P., Pereira, J.P., Moreira, M.E., Nogueira, R.M.R., Damasceno, L., Wakimoto, M., Rabello, R.S., Valderramos, S.G., Halai, U.-A., 2016. Zika virus infection in pregnant women in Rio de Janeiro. N. Engl. J. Med. 375 (24), 2321–2334. https://doi.org/10.1056/nejmoa1602412. PMC 5323261. PMID 26943629 (web archive link).

Campbell, G.L., Hills, S.L., Fischer, M., Jacobson, J.A., Hoke, C.H., Hombach, J.M., Marfin, A.A., Solomon, T., Tsai, T.F., Tsu, V.D., Ginsburg, A.S., 2011. Estimated global incidence of Japanese encephalitis: a systematic review. Bull. World Health Organ. 89 (10), 766–774. https://doi.org/10.2471/BLT.10.085233. PMC 3209971 PMID22084515.

Campbell, B.M., Vermeulen, S.J., Aggarwal, P.K., Corner-Dolloff, C., Girvetz, E., Loboguerrero, A.M., Ramirez-Villegas, J., Rosenstock, T., Sebastian, L., Thornton, P.K., Wollenberg, E., 2016. Reducing risks to food security from climate change. Glob. Food Sec. 11, 34–43.

Capinera, J.L., 2008. Ticks (Acari: Ixodida). In: Encyclopedia of Entomology, second ed. vol. 3. Springer, pp. 3733–3802. ISBN 978-1-4020-6242-1.

Caraballo, H., 2014. Emergency department management of mosquito-borne illness: Malaria, dengue, and west nile virus. Emerg. Med. Pract. 16 (5), 1–23. quiz 23–4. PMID 25207355. Archived from the original on 2016-08-01.

CDC, 2010. CDC: Yellow Fever—Symptoms and Treatment. Archived from the original on 2012-03-21. Retrieved 2010-11-10.

CDC, 2012. CDC Yellow Fever. Archived from the original on 2012-12-21. Retrieved 2012-12-12.

CDC, 2015a. Centers for Disease Control and. "CDC—Babesiosis". www.cdc.gov. Retrieved 2015-10-25.

CDC, 2015b. Japanese Encephalitis. Archived from the original on 24 May 2017. Retrieved 29 October 2017.

CDC, 2015c. Frequently Asked Questions About Yellow Fever. Archived from the original on 23 March 2016. Retrieved 18 March 2016.

CDC, 2016a. CDC Concludes Zika Causes Microcephaly and Other Birth Defects. Archived from the original on 14 April 2016. Retrieved 14 April 2016.

CDC, 2016b. Infectious diseases related to travel. In: Yellow Book. Centers for Disease Control and Prevention. Archived from the original on 20 March 2016. Retrieved 20 March 2016.

CDC, 2018. Centers for Disease Control and "CDC2—Babesiosis". www.cdc.gov. (PDF). Retrieved 2018-09-20.

Center for Disease Control and Prevention, 2010. Chapter 5 Dengue fever (DF) and dengue hemorrhagic fever (DHF). In: 2010 Yellow Book. Archived from the original on 29 December 2010. Retrieved 23 December 2010.

Centers for Disease Control and Prevention, 2010. Travelers' Health Outbreak Notice. Archived from the original on 26 August 2010. Retrieved 27 August 2010.

Centers for Disease Control and Prevention, 2014. Signs and Symptoms. In: Zika Virus Home. Archived from the original on 29 January 2016. Retrieved 30 January 2016.

Centers for Disease Control and Prevention, 2015a. Clinical evaluation & disease. In: For Health Care Providers. Archived from the original on 25 December 2015. Retrieved 24 December 2015.

Centers for Disease Control and Prevention, 2015b. Help control mosquitoes that spread dengue, chikungunya, and Zika viruses. In: Chikungunya Virus Home: Fact Sheets and Posters. Archived (PDF) from the original on 1 February 2016.

Centers for Disease Control and Prevention, 2016. CDC encourages following guidance to prevent sexual transmission of Zika virus. In: CDC Newsroom Releases. Archived from the original on 27 February 2016.

Chastel, C., 2003. Centenary of the discovery of yellow fever virus and its transmission by a mosquito (Cuba 1900–1901). Bull. Soc. Pathol. Exot. 96 (3), 250–256. (in French). PMID 14582304.

Chen, L.H., Hamer, D.H., 2016. Ideas and opinions. Zika virus: rapid spread in the Western hemisphere. Ann. Intern. Med. 164 (9), 613–615. https://doi.org/10.7326/M16-0150. ISSN 0003-4819. PMID 26832396.
Chen, L.H., Wilson, M.E., 2010. Dengue and chikungunya infections in travelers. Curr. Opin. Infect. Dis. 23 (5), 438–444. https://doi.org/10.1097/QCO.0b013e32833c1d16. PMID 20581669.
CIDRAP, 2010. African Horse Sickness. Archived July 12, 2010, at the Wayback Machine.
Coetzer, J.A.W., Erasmus, B.J., 1994. African horse sickness. In: Coetzer, J.A.W., Thomson, G.R., Tustin, R.C. (Eds.), Infectious Diseases of Livestock With Special Reference to Southern Africa. vol. 1. Oxford University Press, Cape Town, pp. 460–475.
Collins, W.E., 2012. Plasmodium knowlesi: a malaria parasite of monkeys and humans. Annu. Rev. Entomol. 57, 107–121. https://doi.org/10.1146/annurev-ento-121510-133540. PMID 22149265.
Connelly, C.R., 2013. Biting Midges: Culicoides spp. Featured Creatures, Entomology and Nematology Department, University of Florida IFAS. August 2013 Edition.
Cowman, A.F., Berry, D., Baum, J., 2012. The cellular and molecular basis for malaria parasite invasion of the human red blood cell. J. Cell Biol. 198 (6), 961–971. https://doi.org/10.1083/jcb.201206112. PMC 3444787. PMID 22986493.
Crafford, J.E., Guthrie, A.J., van Vuuren, M., Mertens, P.P.C., Burroughs, J.N., Howell, P. G., Hamblin, C., 2003. A group-specific, indirect sandwich ELISA for the detection of equine encephalosis virus antigen. J. Virol. Methods 112 (2003), 129–135.
David, L., Heymann, M.D., 2015. Control of Communicable Diseases Manual, twentieth ed. American Public Health Association. ISBN 978-0-87553-018-5.
de Paula, F.B., de Oliveira, D.J., Prazeres, J., et al., 2016. Ocular findings in infants with microcephaly associated with presumed zika virus congenital infection in Salvador, Brazil. JAMA Ophthalmol. 134 (5), 529–535. https://doi.org/10.1001/jamaophthalmol.2016.0267. ISSN 2168-6165. PMC5444996. PMID 26865554.
Despommier, D.D., 1995. Parasitic Diseases, third ed. Springer-Verlag, pp. 224–226. ISBN 978-0-387-94223-0.
Dobler, G., 2010. Zoonotic tick-borne flaviviruses. Vet. Microbiol. 140 (3/4), 221–228. https://doi.org/10.1016/j.vetmic. 2009.08.024. PMID 19765917.
Du Toit, R.M., 1944. The transmission of bluetongueand horse sickness by Culicoides, Onderstepoort. J. Vet. Sci. Anim. Ind. 19 (1944), 7–16.
East, S., 2016. World's First Dengue Fever Vaccine Launched in the Philippines. CNN. Archived from the original on 18 October 2016. Retrieved 17 October 2016.
Eisen, L., Moore, C.G., 2013. Aedes (Stegomyia) aegypti in the continental United States: a vector at the cool margin of its geographic range. J. Med. Entomol. 50 (3), 467–478. https://doi.org/10.1603/ME12245. PMID 23802440.
EPA, 2017. Climate Impacts on Ecosystems. Archived from the original on 27 January 2018. Retrieved 5 February 2019.
Erasmus, B.J., 1972. The pathogenesis of African horse sickness. In: Bryans, J.T., Gerber, H. (Eds.), Equine Infectious Diseases III, Proc. 3rd Int. Conf. Equine Inf. Dis., Paris, France, 1972. Karger, Basel, pp. 1–11. 1973.
Fairhurst, R.M., Wellems, T.E., 2010. Chapter 275. Plasmodium species (malaria). In: Mandell, G.L., Bennett, J.E., Dolin, R. (Eds.), Mandell, Douglas, and Bennett's Principles and Practice of Infectious Diseases, seventh ed. vol. 2. Churchill Livingstone/Elsevier, Philadelphia, pp. 3437–3462. ISBN 978-0-443-06839-3.
FAO, 2004. Ticks: acaricide resistance, diagnosis, management and prevention. In: Resistance Management and Integrated Parasite Control in Ruminants: Guidelines Module 1. Animal Production and Health Division, Food and Agriculture Organization of the United Nations, Rome, pp. 25–77.

Fauci, A.S., Morens, D.M., 2016. Zika virus in the Americas—yet another arbovirus threat. N. Engl. J. Med. 374 (7), 601–604. https://doi.org/10.1056/NEJMp1600297. PMID 26761185.

Ferri, F.F., 2009. Chapter 332. Protozoal infections. In: Ferri's Color Atlas and Text of Clinical Medicine. Elsevier Health Sciences, p. 1159. ISBN 978-1-4160-4919-7. Archived from the original on 2016-06-03.

Fontenille, D., Diallo, M., Mondo, M., Ndiaye, M., Thonnon, J., 1997. First evidence of natural vertical transmission of yellow fever virus in *Aedes aegypti*, its epidemic vector. Trans. R. Soc. Trop. Med. Hyg. 91 (5), 533–535. https://doi.org/10.1016/S0035-9203(97)90013-4. PMID 9463659.

Franchini, M., Velati, C., 2016. Blood safety and zoonotic emerging pathogens: now it's the turn of Zika virus! Blood Transfus. 14 (2), 93–94. https://doi.org/10.2450/2015.0187-15. PMC 4781775. PMID 26674809.

Franklin, J., Sutherland, J., 1984. Guinea Pig Doctors: The Drama of Medical Research Through Self-Experimentation. by Jon Franklin (Author), John Sutherland (Author), William Morrow & Co. ISBN 0-688-02666-4.

Gairdner, J., 2007. Bluetongue (Blue Tongue) Outbreak in the UK—FWi's Special Report. Archived 2007-10-12 at the Wayback Machine, Farmers Weekly, 24 September 2007.

Gale, J., 2016. The Best Weapon for Fighting Zika? More Mosquitoes. Bloomberg. Archived from the original on 6 April 2017.

Gambel, J.M., DeFraites, R., Hoke, C., et al., 1995. Japanese encephalitis vaccine: persistence of antibody up to 3 years after a three-dose primary series (letter). J. Infect. Dis. 171 (4), 1074. https://doi.org/10.1093/infdis/171.4.1074. PMID 7706798.

Gatherer, D., Kohl, A., 2015. Zika virus: a previously slow pandemic spreads rapidly through the Americas. J. Gen. Virol. 97 (2), 269–273. https://doi.org/10.1099/jgv.0.000381. PMID 26684466.

Gaughan, J.B., Bonner, S., Loxton, I., 2010. Effect of shade on body temperature and performance of feedlot steers. J. Anim. Sci. academic.oup.com.

Ghosh, D., Basu, A., Brooker, S., 2009. Japanese encephalitis—a pathological and clinical perspective. PLoS Negl. Trop. Dis. 3 (9). https://doi.org/10.1371/journal.pntd.0000437, e437. (Ed.) PMC 2745699. PMID 19787040.

Ghoshal, A., Das, S., Ghosh, S., Mishra, M.K., Sharma, V., Koli, P., Sen, E., Basu, A., 2007. Proinflammatory mediators released by activated microglia induces neuronal death in Japanese encephalitis. Glia 55 (5), 483–496. https://doi.org/10.1002/glia.20474. PMID 17203475.

Gleick, 2017. Scientific consensus: earth's climate is warming. In: Climate Change: Vital Signs of the Planet. NASA JPL. Archived from the original on 28 March 2020. Retrieved 29 March 2020.

Gould, E.A., Solomon, T., 2008. Pathogenic flaviviruses. Lancet 371 (961), 500–509. https://doi.org/10.1016/S0140-6736(08)60238-X. ISSN 0140-6736. PMID 18262042.

Grard, G., Caron, M., Mombo, I.M., Nkoghe, D., Ondo, S.M., Jiolle, D., Fontenille, D., Paupy, C., Leroy, E.M., 2014. Zika virus in Gabon (Central Africa)—2007: a new threat from *Aedes albopictus*? PLoS Negl. Trop. Dis. 8 (2). https://doi.org/10.1371/journal.pntd.0002681, e2681. ISSN 1935-2735. PMC 3916288. PMID 24516683.

Gubler, D.J., 1998. Dengue and dengue hemorrhagic fever. Clin. Microbiol. Rev. 11 (3), 480–496. https://doi.org/10.1128/cmr.11.3.480. PMC 88892. PMID 9665979.

Gubler, D.J., 2010. Dengue viruses. In: Mahy, B.W.J., Van Regenmortel, M.H.V. (Eds.), Desk Encyclopedia of Human and Medical Virology. Academic Press, Boston, pp. 372–382. ISBN 978-0-12-375147-8. Archived from the original on 9 June 2016.

Guglielmone, A.A., Aguirre, D.H., Späth, E.J.A., Gaido, A.B., Mangold, A.J., de Ríos, L.G., 1992. Long-term study of incidence and financial loss due to cattle babesiosis in an Argentinian dairy farm. Prev. Vet. Med. 12 (3–4), 307–312. Elsevier.

Guzman, M.G., Halstead, S.B., Artsob, H., Buchy, P., Farrar, J., Gubler, D.J., Hunsperger, E., Kroeger, A., Margolis, H.S., Martínez, E., Nathan, M.B., Pelegrino, J.L., Simmons, C., Yoksan, S., Peeling, R.W., 2010. "dengue: a continuing global threat". Nature reviews. Microbiology 8 (12 Suppl), S7–16. https://doi.org/10.1038/nrmicro2460. PMC 4333201. PMID 21079655.

Hamblin, C., Mertens, P.P.C., Mellor, P.S., Burroughs, J.N., Crowther, J.R., 1991. A serogroup specific enzyme-linked immunosorbent assay for the detection and identification of African horse sickness viruses. J. Virol. Methods 31, 285–292.

Hayes, E.B., 2009. Zika virus outside Africa. Emerg. Infect. Dis. 15 (9), 1347–1350. https://doi.org/10.3201/eid1509.090442. PMC 2819875. PMID 19788800.

Heang, V., Yasuda, C.Y., Sovann, L., et al., 2012. Zika virus infection, Cambodia, 2010. Emerg. Infect. Dis. 18 (2), 349–351. https://doi.org/10.3201/eid1802.111224. ISSN 1080-6040. PMC 3310457. PMID 22305269.

Herwaldt, B.L., Cacciò, S., Gherlinzoni, F., et al., 2003. Molecular characterization of a non-Babesia divergens organism causing zoonotic babesiosis in Europe. Emerg. Infect. Dis. 9 (8), 942–948. https://doi.org/10.3201/eid0908.020748. PMC 3020600. PMID 12967491.

Herwaldt, B.L., Linden, J.V., Bosserman, E., Young, C., Olkowska, D., Wilson, M., 2011. Transfusion-associated babesiosis in the United States: a description of cases. Ann. Intern. Med. 155 (8), 509–519. https://doi.org/10.7326/0003-4819-155-8-201110180-00362. PMID 21893613.

Hills, S.L., Russell, K., Hennessey, M., et al., 2016. Transmission of Zika virus through sexual contact with travelers to areas of ongoing transmission—continental United States, 2016. Morb. Mortal. Wkly. Rep. 65 (8). https://doi.org/10.15585/mmwr.mm6508e2er. ISSN 0149-2195. Archived from the original on 15 March 2016.

Holbrook, M., 2012. Kyasanur forest disease. Antiviral Res. 96 (3), 353–362. https://doi.org/10.1016/j.antiviral.2012.10.005. PMC 3513490. PMID 23110991.

Howell, P.G., 1963. African horse sickness. In: Emerging Diseases of Animals. FAO, Rome FAO Agricultural Studies, pp. 71–108. No. 61.

Hunfeld, K.P., Hildebrandt, A., Gray, J.S., 2008. Babesiosis: recent insights into an ancient disease. Int. J. Parasitol. 38 (11), 1219–1237. https://doi.org/10.1016/j.ijpara.2008.03.001. PMID 18440005.

ICTV, 2019. ICTV 9th Report (2011) Reoviridae. International Committee on Taxonomy of Viruses (ICTV). (html). Retrieved 17 January 2019.

Iowa State University, 2020. "African Horse Sickness" (PDF). Iowa State University. Retrieved 29 June 2020.

IPCC, 2013. AR5 WG1 Ch5, 5: Information From Paleoclimate Archives: The PETM [Around 55.5–55.3 Million Years Ago] was Marked by .. Global Warming of 4 °C to 7 °C Deglacial Global Warming Occurred in Two Main Steps From 17.5 to 14.5 ka [Thousand Years Ago] and 13.0 to 10.0 ka., p. 389. 399–400.

IPCC, 2014. AR5 SYR Glossary, Global Warming Refers to the Gradual Increase, Observed or Projected, in Global Surface Temperature, as One of the Consequences of Radiative Forcing Caused by Anthropogenic Emissions., p. 124. IPCC SR15 Ch1 2018, p. 51: Global warming is defined in this report as an increase in combined surface air and sea surface temperatures averaged over the globe and over a 30-year period. Unless otherwise specified, warming is expressed relative to the period 1850–1900, used as an approximation of pre-industrial temperatures in AR5.

IPCC SR15 Ch1 2018a, ssp. 53.

IPCC, 2018b. SR15 Ch1, Sustained Net Zero Anthropogenic Emissions of CO2 and Declining Net Anthropogenic Non-CO2 Radiative Forcing Over a Multi-Decade Period Would Halt Anthropogenic Global Warming Over That Period, Although It Would Not Halt Sea Level Rise or Many Other Aspects of Climate System Adjustment., p. 64.

Jamieson, A., Toovey, S., Maurel, M., 2006. Malaria: A Traveller's Guide. Struik, p. 30. ISBN 978-1-77007-353-1. Archived from the original on 2016-05-11.

Jensen, R., Swift, B.L., 1982. Diseases of Sheep. Lea and Febiger, Philadelphia. ISBN 0-8121-0836-1.

Johnson, C.K., Hitchens, P.L., Smiley Evans, T., Goldstein, T., Thomas, K., Clements, A., Joly, D.O., Wolfe, N.D., Daszak, P., Karesh, W.B., Mazet, J.K., 2015. Spillover and pandemic properties of zoonotic viruses with high host plasticity. Sci. Rep. 5, 14830. [PMC free article] PubMed.

Kasabi, G., Murhekar, M., Sandhya, V., Raghunandan, R., Kiran, S., Channabasappa, G., Mehendale, S., 2013. Coverage and effectiveness of kyasanur forest disease (KFD) vaccine in Karnataka, South India, 2005–10. PLoS Negl. Trop. Dis. 7 (1). https://doi.org/10.1371/journal.pntd.0002025, e2025. PMC 3554520. PMID 23359421.

Kattenberg, J.H., Ochodo, E.A., Boer, K.R., Schallig, H.D., Mens, P.F., Leeflang, M.M., 2011. Systematic review and meta-analysis: rapid diagnostic tests versus placental histology, microscopy and PCR for malaria in pregnant women. Malaria J. 10, 321. https://doi.org/10.1186/1475-2875-10-321. PMC 3228868. PMID 22035448.

Kehse, U., 2020. Global Warming Doesn't Stop When the Emissions Stop Excess Energy is Mainly Absorbed by the Oceans. Water has a High Heat Capacity and Consequently a Long Reaction Time. As a Result, at Present the Oceans are Cooling the Air. Over time, However, the Deep Ocean Layers Will Also Warm Up, Until Finally, After Thousands of Years, the Climate Will Settle Down at a Higher Average Ocean and Atmospheric Temperature. phys.org. Retrieved 19 April.

Knoop, K.J., Stack, L.B., Storrow, A., Thurman, R.J. (Eds.), 2010. Tropical medicine. In: Atlas of Emergency Medicine, third ed. McGraw-Hill Professional, New York, pp. 658–659. ISBN 978-0-07-149618-6.

Krause, P.J., 2003. Babesiosis diagnosis and treatment. Vector Borne Zoonotic Dis. 3 (1), 45–51. https://doi.org/10.1089/153036603765627451. PMID 12804380.

Krause, P.J., Telford, S.R., Ryan, R., et al., 1994. Diagnosis of babesiosis: evaluation of a serologic test for the detection of Babesia microti antibody. J. Infect. Dis. 169 (4), 923–926. https://doi.org/10.1093/infdis/169.4.923. PMID 8133112.

Kularatne, S.A., 2015. Dengue fever. BMJ 351, h4661. https://doi.org/10.1136/bmj.h4661. PMID 26374064.

Kurane, I., Takashi, T., 2000. Immunogenicity and protective efficacy of the current inactivated Japanese encephalitis vaccine against different Japanese encephalitis virus strains. Vaccine 18 (Suppl), 33–35. https://doi.org/10.1016/S0264-410X(00)000414. PMID 10821971.

Lacetera, N., Bernabucci, U., Scalia, D., Ronchi, B., Kuzminsky, G., Nardone, A., 2005. Lymphocyte functions in dairy cows in hot environment. Int. J. Biometeorol. 50, 105–110.

Layne, S.P., 2006. Principles of Infectious Disease Epidemiology. UCLA Department of Epidemiology. (PDF). EPI 220. Archived from the original (PDF) on 2006-02-20. Retrieved 2007-06-15.

Leiby, D.A., 2011. Transfusion-associated babesiosis: shouldn't we be ticked off? Ann. Intern. Med. 155 (8), 556–557. https://doi.org/10.7326/0003-4819-155-8-201110180-00363. PMID 21893616.

Li, H., Saucedo-Cuevas, L., Shresta, S., Gleeson, J.G., 2016. The neurobiology of Zika virus. Neuron 92 (5), 949–958. https://doi.org/10.1016/j.neuron.2016.11.031. PMID 27930910.

Lindenbach, B.D., et al., 2007. Flaviviridae: the viruses and their replication. In: Knipe, D. M., Howley, P.M. (Eds.), Fields Virology, fifth ed. Lippincott Williams & Wilkins, Philadelphia, PA, p. 1101. ISBN 978-0-7817-6060-7.

Lindgren, E., 1998. Climate change, tick-borne encephalitis and vaccination needs in Sweden—a prediction model. Ecol. Model. 110, 55–63.

Lindgren, E., Tälleklint, L., Polfeldt, T., 2000. Impact of climatic change on the northern latitude limit and population density of the disease-transmitting European tick *Ixodes ricinus*. Environ. Health Perspect. 108 (2), 119–123. https://doi.org/10.2307/3454509. JSTOR 3454509. PMC 1637900. PMID 10656851.

Lobigs, M., Diamond, M.S., 2012. Feasibility of cross-protective vaccination against flaviviruses of the Japanese encephalitis serocomplex. Expert. Rev. Vaccines 11 (2), 177–187. https://doi.org/10.1586/erv.11.180. PMC 3337329. PMID 22309667.

Louis, C.S., 2016. DEET Seen as Safe for Pregnant Women to Avoid Zika Despite Few Studies. The New York Times. Archived from the original on 17 September 2016. Retrieved 3 March 2017. DEET Seen as Safe for Pregnant Women to Avoid Zika Despite Few Studies.

Maan, S., Maan, N.S., Nomikou, K., Veronesi, E., Bachanek-Bankowska, K., Belaganahalli, M.N., Attoui, H., Mertens, P.P., 2011. Complete genome characterisation of a novel 26th bluetongue virus serotype from Kuwait. PLoS One 6 (10), e26147. Bibcode: 2011PLoSO...626147M https://doi.org/10.1371/journal.pone.0026147. PMC 3198726. PMID 22031822.

Mansuy, J.M., Dutertre, M., Mengelle, C., et al., 2016. Zika virus: high infectious viral load in semen, a new sexually transmitted pathogen? Lancet Infect. Dis. 16 (4), 405. https://doi.org/10.1016/S1473-3099(16)00138-9. ISSN 1473-3099. PMID 26949027.

Markus, M.B., 2011. Malaria: origin of the term "Hypnozoite". J. Hist. Biol. 44 (4), 781–786. https://doi.org/10.1007/s10739-010-9239-3. PMID 20665090.

Martens, P., Kovats, R.S., Nijhof, S., de Vries, P., Livermore, M.T.J., Bradley, D.J., Cox, J., McMichael, A.J., 1999. Climate change and future populations at risk of malaria. Glob. Environ. Chang. 9, S89–S107.

Martina, B.E., Koraka, P., Osterhaus, A.D., 2009. Dengue virus pathogenesis: an integrated view. Clin. Microbiol. Rev. 22 (4), 564–581. https://doi.org/10.1128/CMR.00035-09. PMC 2772360. PMID 19822889.

Mauroy, A., Guyot, H., De Clercq, K., Cassart, D., Thiry, E., Saegerman, C., 2008. Bluetongue in captive yaks. Emerg. Infect. Dis. 14 (4), 675–676.

McGrath, M., 2007. 'Dancing' Disease Set for Long Run. BBC News. Retrieved 2008-10-24.

McInerney, J., 1996. Old economics for new problems—livestock disease. Presid. Addr. 47 (3), 295–314.

Meigen, J.W., 1818. Systematische Beschreibung der Bekannten Europäischen Zweiflügeligen Insekten. vol. 1 Forstmann, Aachen. 332 pp.

Meiswinkel, R., 1998. The 1996 outbreak of African horse sickness in South Africa—the entomological perspective. Arch. Virol. Suppl. 14 (1998), 69–83.

Meiswinkel, R., et al., 2004. The taxonomy of Culicoides vector complexes–unfinished business. Vet. Ital. 40 (3), 151–159.

Mellor, P.S., Boorman, J., 1995. The transmission and geographical spread of African horse sickness and bluetongue viruses. Ann. Trop. Med. Parasitol. 89 (1995), 1–15.

Met Office, 2016. Met Office State of the UK Climate report shows 2016 the 13th warmest year. Press Office; 11:00 (UTC+1) on Thursday, 27 July 2017.

Metacalf, H.E., Lomm, J., Beal Jr., V.C., 1980. Estimate of incidence and direct economic losses due to bluetongue in Mississippi cattle during 1979. Proc. Ann. Mtng. U.S. Anim. Hlth. Assoc. 84, 186–202.

Mike Service, 2008. Medical Entomology for Students. Cambridge University, pp. 53–54. ISBN 978-0-521-70928-6.
MMWR, 2016. Interim guidance for Zika virus testing of urine—United States, 2016. Morb. Mortal. Wkly. Rep. 65 (18), 474. 13 May 2016 10.15585/mmwr.mm6518e1. PMID 27171368.
Moloney, R.M., Kmush, B., Rudolph, K.E., Cummings, D.A.T., Lessler, J., 2014. Incubation periods of mosquito-borne viral infections: a systematic review. Am. J. Trop. Med. Hyg. 90 (5), 882–891. https://doi.org/10.4269/ajtmh.130403. PMC 4015582. PMID 24639305.
Monath, T.P., 2008. Treatment of yellow fever. Antiviral Res. 78 (1), 116–124. https://doi.org/10.1016/j.antiviral.2007.10.009. PMID 18061688.
"Mosquito Ecology and Surveillance Laboratory". 2014. Retrieved 7 August.
Mourya, D., Yadav, P., Mehla, R., Barde, P., Yergolkar, P., Kumar, S., Thakare, J., Mishra, A., 2012. Diagnosis of Kyasanur forest disease by nested RT-PCR, real-time RT-PCR and IgM capture ELISA. J. Virol. Methods 186 (1/2), 49–54. https://doi.org/10.1016/j.jviromet.2012.07.019. PMID 22874757.
Mourya, D., Yadav, P., Sandhya, V., Reddy, S., 2013. Spread of Kyasanur forest disease, Bandipur tiger reserve, India, 2012–2013. Emerg. Infect. Dis. 19 (9), 1540–1541. https://doi.org/10.3201/eid1909.121884. PMC 3810911. PMID 23977946.
Mueller, I., Zimmerman, P.A., Reeder, J.C., 2007. Plasmodium malariae and Plasmodium ovale—the "bashful" malaria parasites. Trends Parasitol. 23 (6), 278–283. https://doi.org/10.1016/j.pt.2007.04.009. PMC 3728836. PMID 17459775.
Musso, D., Nilles, E.J., Cao-Lormeau, V.M., 2014. Rapid spread of emerging Zika virus in the Pacific area. Clin. Microbiol. Infect. 20 (10), O595–O596. https://doi.org/10.1111/1469-0691.12707. PMID 24909208.
Mustafa, M.S., Rasotgi, V., Jain, S., Gupta, V., 2015. Discovery of fifth serotype of dengue virus (DENV-5): a new public health dilemma in dengue control. Med. J. Armed Forces India 71 (1), 67–70. https://doi.org/10.1016/j.mjafi.2014.09.011. PMC 4297835. PMID 25609867.
Nadjm, B., Behrens, R.H., 2012. Malaria: an update for physicians. Infect. Dis. Clin. N. Am. 26 (2), 243–259. https://doi.org/10.1016/j.idc.2012.03.010. PMID 22632637.
Nardone, A., Lacetera, N., Bernabucci, U., Ronchi, B., 1997. Composition of colostrum from dairy heifers exposed to high air temperatures during late pregnancy and the early postpartum period. J. Dairy Sci. 80, 838–844.
Narladkar, B.W., 2018. Projected economic losses due to vector and vector-borne parasitic diseases in livestock of India and its significance in implementing the concept of integrated practices for vector management. Vet. World 11, 151–160.
Narladkar, B.W. and Shivpuje, P.R. (2012). Research Project Report of Development of Integrated Pest Management Module Against Important Dipteran Pests of Livestock with Reference to Culicoides Midges (Diptera: Ceratopogonidae) Financed and Accepted by Department of Biotechnology, Government of India, New Delhi. p 1–167.
NASA, 2020. Global Annual Mean Surface Air Temperature Change. Retrieved 23 February 2020.
National Geographic, 2019. NPR 2010.
Normile, D., 2013. Tropical medicine. Surprising new dengue virus throws a spanner in disease control efforts. Science 342 (6157), 415. https://doi.org/10.1126/science.342.6157.415. PMID 24159024.
Noskoviak, K., Broome, E., 2008. Images in clinical medicine. Babesiosis. N. Engl. J. Med. 358 (17). https://doi.org/10.1056/NEJMicm070903, e19. PMID 18434647.
O'Brien, D., 2010. ARS Study Provides a Better Understanding of How Mosquitoes Find a Host. U.S. Department of Agriculture. Archived from the original on 8 October 2010. Retrieved 2010-08-27.

Olival, K.J., Hosseini, P.R., Zambrana-Torrelio, C., Bogich, T.L., Daszak, P., 2017. Host and viral traits predict zoonotic spillover from mammals. Nature 546 (7660), 646–650. Public Health Emergency Collection. In Review [PMC free article] PubMed.

Olivier, J.G.J., Peters, J.A.H.W., 2019. Trends in Global CO_2 and Total Greenhouse Gas Emission. Report May 2020. Revised version 26 May 2020.

Oster, A.M., Brooks, J.T., Stryker, J.E., et al., 2016. Interim guidelines for prevention of sexual transmission of Zika virus—United States, 2016. Morb. Mortal. Wkly. Rep. 65 (5), 120–121. https://doi.org/10.15585/mmwr.mm6505e1. ISSN 0149-2195. PMID 26866485.

Padbidri, V.S., Wairagkar, N.S., Joshi, G.D., Umarani, U.B., Risbud, A.R., Gaikwad, D.L., Bedekar, S.S., Divekar, A.D., Rodrigues, F.M., 2002. A serological survey of arboviral diseases among the human population of the Andaman and Nicobar Islands, India. Southeast Asian J. Trop. Med. Public Health 33 (4), 794–800. ISSN 0125-1562. PMID 12757228.

Pattnaik, P., 2006. Kyasanur forest disease: an epidemiological view in India. Rev. Med. Virol. 16 (3), 151–165. https://doi.org/10.1002/rmv.495. ISSN 1052-9276. PMID 16710839.

Paweska, J.T., Venter, G.J., Mellor, P.S., 2002. Vector competence of south African Culicoides species for bluetingue virus serotype 1 (BTV-1) with special reference to the effect of temperature on the rate of virus replication in C. imicola and C. bolitinos. Med. Vet. Entomol. 16, 10–21.

Persing, D.H., Mathiesen, D., Marshall, W.F., et al., 1992. Detection of Babesia microti by polymerase chain reaction. J. Clin. Microbiol. 30 (8), 2097–2103. PMC 265450. PMID 1500517.

Petersen, E.E., Polen, K.N.D., Meaney-Delman, D., et al., 2016. Update: interim guidance for health care providers caring for women of reproductive age with possible zika virus exposure—United States, 2016. Morb. Mortal. Wkly. Rep. 65 (12), 315–322. https://doi.org/10.15585/mmwr.mm6512e2. ISSN 0149-2195. PMID 27031943. Archived from the original on 28 March 2016.

Purse, B.V., Mellor, P.S., Rogers, D.J., Samuel, A.R., Mertens, P.P.C., Baylis, M., 2005. Climate change and the recent emergence of bluetongue in Europe. Nat. Rev. Microbiol. 3, 171–181.

Quaresma, J.A., Barros, V.L., Pagliari, C., Fernandes, E.R., Guedes, F., Takakura, C.F., Andrade, H.F., Vasconcelos, P.F., Duarte, M.I., 2006. Revisiting the liver in human yellow fever: virus-induced apoptosis in hepatocytes associated with TGF-beta, TNF-alpha and NK cells activity. Virology 345 (1), 22–30. https://doi.org/10.1016/j.virol.2005.09.058. PMID 16278000.

Rabinowitz, P., Conti, L., 2013. Links among human health, animal health, and ecosystem health. Annu. Rev. Public Health 34, 189–204.

Rajput, Z.I., Hu, S., Chen, W., Arijo, A.G., Xiao, C., 2006. Importance of ticks and their chemical and immunological control in livestock. J. Zhejiang Univ. Sci. B 7, 912–921.

Randolph, S.E., 2009. Perspectives on climate change impacts on infectious diseases. Ecology 90, 927–931.

Rasmussen, S.A., Jamieson, D.J., Honein, M.A., Petersen, L.R., 2016. Zika virus and birth defects—reviewing the evidence for causality. N. Engl. J. Med. 374 (20), 1981–1987. https://doi.org/10.1056/NEJMsr1604338. ISSN 0028-4793. PMID 27074377.

Regnier, J.A., Kelley, K.W., 1981. The influence of hot and cold ambient temperatures on cellular immune responses of chickens was assessed, utilizing dinitrofluorobenzene-induced contact sensitivity, phytohemagglutinin (PHA) skin test, and PHA-stimulated, whole blood lymphocyte. Am. J. Vet. Res. europepmc.org.

Reiter, P., 2010. Yellow fever and dengue: a threat to Europe? Euro Surveill. 15 (10), 19509. https://doi.org/10.2807/ese.15.10.19509-en. (inactive 4 March 2020). PMID 20403310. Archived from the original on 7 July 2011.

Rich, S.M., Ayala, F.J., 2006. Evolutionary origins of human malaria parasites. In: Dronamraju, K.R., Arese, P. (Eds.), Malaria: Genetic and Evolutionary Aspects. Springer, New York, pp. 125–146. ISBN 978-0-387-28294-7.

Robertson, A., 1976. Handbook on Animal Diseases in the Tropics. Centre for Tropical Veterinary Medicine. [Corporate Author] University of Edinburgh [Corporate Author] British Veterinary Association [Corporate Author].

Rodenhuis-Zybert, I.A., Wilschut, J., Smit, J.M., 2010. Dengue virus life cycle: viral and host factors modulating infectivity. Cell. Mol. Life Sci. 67 (16), 2773–2786. https://doi.org/10.1007/s00018-010-0357-z. PMID 20372965.

Rogers, D.J., Wilson, A.J., Hay, S.I., Graham, A.J., 2006. The global distribution of yellow fever and dengue. Adv. Parasitol. 62, 181–220. https://doi.org/10.1016/S0065-308X(05)62006-4. ISBN 978-0-12-031762-2. PMC 3164798. PMID 16647971.

Rose, H., Wang, T., van Dijk, J., Morgan, E.R., 2015. GLOWORM-FL: A Simulation Model of the Effects of Climate and Climate 2 Change on the Free-Living Stages of Gastro-Intestinal Nematode Parasites of 3 Ruminants. https://livrepository.liverpool.ac.uk/3050743/1/GLOWORM-FL.

Roth, G.A., Abate, D., Abate, K.H., Abay, S.M., Abbafati, C., 2018. Global, regional, and national age-sex-specific mortality for 282 causes of death in 195 countries and territories, 1980–2017: a systematic analysis for the Global Burden of Disease Study 2017. Lancet 392 (10159), 1736–1788. https://doi.org/10.1016/S0140-6736(18)32203-7. PMC 6227606. PMID 30496103.

Roy, P., 2008. Molecular Dissection of Bluetongue Virus. Animal Viruses: Molecular Biology, Caister Academic Press, pp. 305–354. ISBN 978-1-904455-22-6.

Ruedisueli, F.L., Manship, B., 2010. Background Information: Ixodes ricinus. University of Lincoln. Retrieved July 22, 2010.

Ryan, K.J., Ray, C.G. (Eds.), 2004. Sherris Medical Microbiology, fourth ed. McGraw Hill. ISBN 978-0-8385-8529-0.

Samour, J., 2000. Ticks. In: Avian Medicine. Elsevier Health Sciences, pp. 223–224. ISBN 978-0-7234-2960-9.

Sarkar, P.K., Ahluwalia, G., Vijayan, V.K., Talwar, A., 2009. Critical care aspects of malaria. J. Intensive Care Med. 25 (2), 93–103. https://doi.org/10.1177/0885066609356052. PMID 20018606.

Sarno, M., Sacramento, G.A., Khouri, R., do Rosário, M.S., Costa, F., Archanjo, G., Santos, L.A., Nery Jr., N., Vasilakis, N., 2016. Zika virus infection and stillbirths: a case of hydrops fetalis, hydranencephaly and fetal demise. PLoS Negl. Trop. Dis. 10 (2). https://doi.org/10.1371/journal.pntd.0004517, e0004517. ISSN 1935-2735. PMC 4767410. PMID 26914330.

Schlagenhauf-Lawlor, P., 2008. Travelers' Malaria. PMPH-USA. ISBN 978-1-55009-336-0.

Scully, C., 2014. Scully's Medical Problems in Dentistry. Elsevier Health Sciences, p. 572. ISBN 978-0-7020-5963-6.

Seebeck, R.M., Springell, P.H., O'Kelly, J.C., 1971. Alterations in the host metabolism by the specific and anorectic effects of the cattle tick (*Boophilus microplus*) in food intake and body weight growth. Aust. J. Biol. Sci. 24, 373–380.

Sherman, I., 2017. Twelve Diseases that Changed Our World. ASM Press, p. 144. 2007. ISBN 978-1-55581-466-3. OCLC 141178241.

Shrivastva, A., Tripathi, N.K., Parida, M., Dash, P.K., Jana, A.M., Lakshmana Rao, P.V., 2008. Comparison of a dipstick enzyme-linked immunosorbent assay with commercial assays for detection of Japanese encephalitis virus-specific IgM antibodies. J. Postgrad. Med. 54 (3), 181–185. https://doi.org/10.4103/0022-3859.40959. PMID 18626163.

Sikka, V., Chattu, V.K., Popli, R.K., Galwankar, S.C., Kelkar, D., Sawicki, S.G., Stawicki, S.P., Papadimos, T.J., 2016. The emergence of zika virus as a global health security threat: a review and a consensus statement of the INDUSEM Joint working Group

(JWG). J. Glob. Infect. Dis. 8 (1), 3–15. https://doi.org/10.4103/0974-777X.176140. ISSN 0974-8245. PMC 4785754. PMID 27013839.
Simmons, C.P., Farrar, J.J., Nguyen, V.V., Wills, B., 2012. Dengue. N. Engl. J. Med. 366 (15), 1423–1432. https://doi.org/10.1056/NEJMra1110265.hdl. 11343/191104. PMID 22494122. (PDF).
Simon, L.V., Kruse, B., 2018. Encephalitis, Japanese. StatPearls. PMID 29262148.
Solomon, T., 2006. Control of Japanese encephalitis—within our grasp? N. Engl. J. Med. 355 (9), 869–871. https://doi.org/10.1056/NEJMp058263. PMID 16943399.
Solomonides, T., 2010. Healthgrid applications and core technologies. In: Proceedings of HealthGrid 2010. IOS Press, Amsterdam, p. 235. [Online-Ausg.]. ed. ISBN 978-1-60750-582-2. Archived from the original on 1 May 2016.
St. Georgiev, V., 2009. National Institute of Allergy and Infectious Diseases, NIH, first ed. Humana, Totowa, NJ, p. 268. ISBN 978-1-60327-297-1. Archived from the original on 1 May 2016.
Staff Reporter, 2015. Vaccine for Bluetongue Disease Launched. The Hindu.
Staples, J.E., Monath, T.P., 2008. Yellow fever: 100 years of discovery. JAMA 300 (8), 960–962. https://doi.org/10.1001/jama.300.8.960. PMID 18728272.
Stramer, S.L., Hollinger, F.B., Katz, L.M., Kleinman, S., Metzel, P.S., Gregory, K.R., Dodd, R.Y., 2009. Emerging infectious disease agents and their potential threat to transfusion safety. Transfusion 49 (Suppl. 2), 1S–29S. https://doi.org/10.1111/j.1537-2995.2009.02279.x. PMID 19686562.
Su, H.L., Liao, C.L., Lin, Y.L., 2002. Japanese encephalitis virus infection initiates endoplasmic reticulum stress and an unfolded protein response. J. Virol. 76 (9), 4162–4171. https://doi.org/10.1128/JVI.76.9.4162-4171. PMC 155064. PMID 11932381.
Szadziewski, R., Dominiak, P., Sontag, E., Krzemiński, W., Wang, B., Szwedo, J., 2019. Haematophagous biting midges of the extant genus Culicoides Latreille (Diptera: Ceratopogonidae) evolved during the mid-Cretaceous. Zootaxa 4688 (4), 535–548. https://doi.org/10.11646/zootaxa.4688.4.5. ISSN 1175-5334.
The Merck Veterinary Manual, 2015. African Horse Sickness—Clinical Findings and Lesions.
Thornton, P.K., Jones, P.G., Owiyo, T.M., Kruska, R.L., Herrero, M., Kristjanson, P., Notenbaert, A., Bekele, N., Orindi, V., Otiende, B., Ochieng, A., Bhadwal, S., Anantram, K., Nair, S., Kumar, V., Kulkar, U., 2006. Mapping Climate Vulnerability and Poverty in Africa. ILRI, Nairobi (Kenya). See http://www.dfid.gov. uk/research/mapping-climate.pdf [Google Scholar]. 200 p.
Tolle, M.A., 2009. Mosquito-borne diseases. Curr. Probl. Pediatr. Adolesc. Health Care 39 (4), 97–140. https://doi.org/10.1016/j.cppeds.2009.01.001. PMID 19327647.
Townson, H., Nathan, M.B., Zaim, M., Guillet, P., Manga, L., Bos, R., Kindhauser, M., 2005. Exploiting the potential of vector control for disease prevention. Bull. World Health Organ. 83 (12), 942–947. Dec, Epub 2006 Jan 30. PMID: 16462987 Free PMC article.
Anon., 2010. Travelers' Health Outbreak Notice. Centers for Disease Control and Prevention. June 2, Archived from the original on 26 August 2010. Retrieved 2010-08-27.
Upadhyaya, S., Narasimha Murthy, D.P., Yashodhara Murthy, B.K., 1975. Viraemia studies on the Kyasanur Forest Disease human cases of 1966. Indian J. Med. Res. 63 (7), 950–953. ISSN 0971-5916. PMID 175006.
van Sittert Sybrand, J., Drew, T.M., Kotze, J.L., Strydom, T., Weyer, C.T., Guthrie, A.J., 2013. Occurrence of African horse sickness in a domestic dog without apparent ingestion of horse meat. J. S. Afr. Vet. Assoc. 84 (1), 5. https://doi.org/10.4102/jsava.v84i1.948.
Vannier, E., Krause, P.J., 2012. Human Babesiosis. N. Engl. J. Med. 366 (25), 2397–2407. https://doi.org/10.1056/NEJMra1202018. PMID 22716978.

Varatharaj, A., 2010. Encephalitis in the clinical spectrum of dengue infection. Neurol. India 58 (4), 585–591. https://doi.org/10.4103/0028-3886.68655. PMID 20739797.
Walker, M.D., 2018. The biology and ecology of the sheep Tick Ixodes ricinus. Antenna 42 (2), 61–65. Royal Entomological Society. (PDF).
Walsh, F., 2015. Malaria Vaccine Gets 'Green Light'. BBC News Online. Archived from the original on 21 December 2016.
Walter Reed Biosystematics Unit, 2016. Aedes luteocephala. In: Medically Important Mosquitoes. Archived from the originalon 28 January 2016. Retrieved 1 February 2016.
Whitehorn, J., Farrar, J., 2010. Dengue. Br. Med. Bull. 95, 161–173. https://doi.org/10.1093/bmb/ldq019. PMID 20616106.
WHO, 2006. Yellow Fever Fact Sheet. WHO—Yellow Fever. Archived from the original on 2006-04-18. Retrieved 2006-04-18.
WHO, 2009a. Yellow Fever. Archived from the original on 2009-08-17. Retrieved 2009-08-13.
WHO. (2009b).
WHO, 2010. World Malaria Report. WHO.
WHO. (2011).
WHO, 2013. Yellow Fever Vaccination Booster Not Needed. Archived 2013-06-09 at the Wayback Machine. Who.int (2013-05-17). Retrieved on 2014-05-12.
WHO, 2014. The 17 neglected tropical diseases. In: Neglected Tropical Diseases. World Health Organization. Archived from the original on 22 February 2014. Retrieved 10 April 2013.
WHO, 2015. Dengue and Severe Dengue Fact Sheet N 117. Archived from the original on 2 September 2016. Retrieved 3 February 2016.
WHO, 2016a. Climate change and infectious diseases. In: Climate Change and Human Health—Risk and Responses. World Health Organization. (PDF). Archived (PDF) from the original on 2016-03-04.
WHO, 2016b. Zika Virus. Archived from the original on 3 February 2016. Retrieved 3 February 2016.
WHO, 2018. World Malaria Report 2019. World Health Organization, Switzerland, pp. xii–xiii. 4–10. ISBN 978-92-4-156572-1.
WHO, 2019. Dengue and Severe Dengue.
WHO, 2020. Dengue and Severe Dengue.
WHO, 2021. Dengue and Severe Dengue. WHO (19 May, 2021).
Wilder-Smith, A., Chen, L.H., Massad, E., Wilson, M.E., 2009. Threat of dengue to blood safety in dengue-endemic countries. Emerg. Infect. Dis. 15 (1), 8–11. https://doi.org/10.3201/eid1501.071097. PMC 2660677. PMID 19116042.
Wilson, M.L., 2012. Malaria rapid diagnostic tests. Clin. Infect. Dis. 54 (11), 1637–1641. https://doi.org/10.1093/cid/cis228. PMID 22550113.
Wittmann, E.J., Mellor, P.S., Baylis, M., 2001. Using climate data to map the potential distribution of Culicoides imicola (Diptera;Ceratopogonidae) in Europe. Rev. Sci. Tech. 20, 731–740.
Wiwanitkit, V., 2009. Unusual mode of transmission of dengue. J. Infect. Dev. Ctries. 4 (1), 51–54. https://doi.org/10.3855/jidc.145. PMID 20130380.
Wolff, K., Johnson, R.A. (Eds.), 2009. Viral infections of skin and mucosa. In: Fitzpatrick's Color Atlas and Synopsis of Clinical Dermatology, sixth ed. McGraw-Hill Medical, New York, pp. 810–812. ISBN 978-0-07-159975-7.
Womack, M., 1993. The yellow fever mosquito, Aedes aegypti. Wing Beats 5 (4), 4.
Wong, P.J., Li, M.I., Chong, C., Ng, L., Tan, C., 2013. Aedes (Stegomyia) albopictus (Skuse): a potential vector of Zika virus in Singapore. PLoS Negl. Trop. Dis. 7 (8). https://doi.org/10.1371/journal.pntd.0002348, e2348. ISSN 1935-2735. PMC 3731215. PMID 23936579.

Wormser, G.P., Dattwyler, R.J., Shapiro, E.D., et al., 2006. "The clinical assessment, treatment, and prevention of Lyme disease, human granulocytic anaplasmosis, and babesiosis: clinical practice guidelines by the Infectious Diseases Society of America" (PDF). Clin. Infect. Dis. 43 (9), 1089–1134. https://doi.org/10.1086/508667. PMID 17029130.

Wylie, W.N.T., 1983. Poverty, distress, and disease: labour and the construction of the Rideau Canal, 1826–32. Labour/Le Travail 11, 7–29. https://doi.org/10.2307/25140199. JSTOR 25140199.

Yacoub, S., Wills, B., 2014. Predicting outcome from dengue. BMC Med. 12 (1), 147. https://doi.org/10.1186/s12916-014-0147-9. PMC 4154521. PMID 25259615.

Yoshida, S., Shimada, Y., Kondoh, D., Kouzuma, Y., Ghosh, A.K., Jacobs-Lorena, M., Sinden, R.E., 2007. Hemolytic C-type lectin CEL-III from sea cucumber expressed in transgenic mosquitoes impairs malaria parasite development. PLoS Pathog. 3 (12). https://doi.org/10.1371/jouArnal.ppat.0030192, e192. PMC 2151087. PMID 18159942.

CHAPTER THIRTEEN

Role of biotechnology on animal breeding and genetic improvement

D.N. Das[a], D. Paul[a], and Sukanta Mondal[b]
[a]Genetics Lab, ICAR-National Dairy Research Institute, Southern Regional Station, Bengaluru, Karnataka, India
[b]Animal Physiology Division, ICAR-National Institute of Animal Nutrition and Physiology, Adugodi, Bengaluru, Karnataka, India

Introduction

Genetic improvement of farm animals has been a major concern over the years among livestock farmers across the countries. Since the beginning of human history, biotechnology has been practiced in animal husbandry. In 1919, Karl Ereky, a Hungarian engineer, first coined the term "biotechnology" and described it as the process by which products could be synthesized from raw materials with the aid of living organisms (Food and Agriculture Organization (FAO), 2010). In general, biotechnology is referred to as the application of scientific and engineering principles that use living organisms or substances from these organisms tos make or alter products, improve plant or animal productivity, or develop microorganisms for specific uses that are beneficial to humans (Steinberg and Raso, 1998).

As there are some built-in problems in animals such as breed composition and its genetic architecture, long generation interval, performance opportunities, and of course the environment where the animals are nurtured, several reproductive technologies are employed from time to time to attain the required goals. Assisted reproductive technologies (ART) such as artificial insemination, super ovulation, *in vitro* fertilization (IVF), embryo transfer technology, and ovum pick-up technique have been introduced to overcome reproductive shortcomings, increase the number of offspring from selected females, and reduce the generation interval in farm animals. With the development of various ARTs, tremendous progress has been made over the past few decades. Reproductive capacity and efficiency have been improved tremendously since the introduction of artificial insemination.

Emerging Issues in Climate Smart Livestock Production
https://doi.org/10.1016/B978-0-12-822265-2.00015-6

The development of animal cloning using various cells from the animal body has opened up a new scientific arena. These technologies were introduced to conserve and propagate indigenous livestock breeds. Also, these reproductive techniques have been shown to be effective in preventing the transmission of vertically transmitted diseases as well as manipulating reproductive processes. Some successful reproductive techniques, such as artificial insemination and embryo transfer, should be implemented in field/farms on a broad scale. Emerging biotechnologies such as MOET, IVF, and cloning provide livestock populations a powerful tool for remarkable genetic change. In essence, the most important effect of reproductive technologies is to enhance the reproductive potential in breeding populations. This means that fewer parents are required to produce a given number of offspring compared with natural service. Generally, this results in increased intensity of selection, which in turn can result in an increase in the average genetic merit of offspring. Hence, it is established that if a reproductive technology is used on a recurrent basis in a closed population, the rate of genetic improvement in the population can be increased. This is the vital genetic advantage offered by reproductive technologies. A brief overview of the reproductive technologies is described in this chapter.

Animal breeding and biotechnologies

Various types of biotechnologies can impact animal breeding, which lead to genetic improvement by many ways:

a. Biotechnologies (such as artificial insemination, sperm sexing, embryo transfer, cloning, etc.) can influence the efficiency of reproduction and selection programs (Van Vleck, 1981; Ruane and Thompson, 1991).

b. Biotechnologies can also improve the identification of animal genetic values (such as identification of genetic markers, candidate genes, etc.) (Georges et al., 1995; Renaville et al., 1997). The use of molecular markers has provided a number of benefits over conventional breeding methods. The identification and use of genetic markers for carcass quality and fertility traits provide faster and preferred growth in bovines, whereas genetic markers for disease resistance, heat tolerance, and milk quality will ensure better health and productivity (Singh et al., 2014). The use of various genetic markers for assessing biodiversity, such as microsatellites, will aid in the conservation of the indigenous germplasm.

Table 1 Biotechnologies used in animal genetics, breeding, and reproduction.

Sl. no.	Biotechnologies used in animal genetics, breeding, and reproduction
1.	Artificial insemination (AI)
2.	Semen and embryo sexing
3.	Estrus synchronization
4.	Embryo transfer (ET)
5.	*In vitro* fertilization (IVF)
6.	Cryopreservation
7.	Cloning
8.	Recombinant DNA technology and transgenesis
9.	Molecular markers; Marker-assisted and genomic selection
10.	Genetic engineering/genome editing

Adapted from Food and Agriculture Organization (FAO). (2010). Current status and options for livestock biotechnologies in developing countries. FAO International Technical Conference, Agricultural Biotechnologies in Developing Countries: Options and Opportunities in Crops, Forestry, Livestock, Fisheries and Agro-Industry to Face the Challenges of Food Insecurity and Climate Change. Guadalajara, Mexico, March 1–4 2010. Report ABDC-10/5. 57 p. http://www.fao.org/fileadmin/user_upload/abdc/documents/livestock.pdf.

c. Biotechnologies (such as genetic engineering, gene transfer, etc.) can artificially transform the genome at the DNA level (Solter, 1981) to achieve favorable function.

As listed in Table 1, some biotechnologies, especially reproductive technologies, have played a key role in advances in animal genetics and breeding that are also associated with genetic improvement (Food and Agriculture Organization (FAO), 2010).

Factors governing genetic gain (ΔG)

The widespread use of biotechnologies in animal production systems has resulted in a significant effect on genetic progress, as stated by Van Vleck (1981). Various factors involving genetic progress are influenced by various types of biotechnological tools. These factors include selection accuracy, generation interval, genetic variance, and intensity of selection. Breeding programs using a combination of advanced ART and genomic techniques are aimed at manipulating component(s) of the breeder's equation to increase the rate of genetic gain (Van Eenennaam, 2016). Therefore, these technologies should impact one or more components of the breeder's equation in order to influence the rate of genetic change (Falconer, 1989).

$$\Delta G = i\,\sigma_p\,h^2 = i\,r_{G\hat{G}}\,\sigma_G$$

where $r_{G\hat{G}}$ is the correlation between the predicted and actual additive genetic value, also called the accuracy of selection; i is the selection intensity (also expressed as the standard normal selection differential); h^2 is the heritability; σ_p is the phenotypic standard deviation; and σ_G is the genetic standard deviation.

The equation of genetic gain, $\Delta G = i r_{G\hat{G}} \sigma_G$, can be modified to reflect genetic gain by time period (year). Therefore, genetic gain *per annum* (ΔG/year $= \Delta g$) is equal to

$$\Delta g = \left(i r_{G\hat{G}} \sigma_G\right)/L$$

where L is the generation interval in years.

The above-mentioned formula is only theoretical based on a hypothesis of equal generation intervals, selection accuracies, and selection intensities in different selection paths. But in practice, factors such as selection intensity and accuracy of selection vary among male and female animals in small herds. Thus, there is a need to account for the unequal contribution of male and female parents to genetic gain (ΔG). Rendel and Robertson (1950) used four paths of gene transfer from parents to progeny to predict genetic gain. Here, accuracy of selection and selection intensity vary along the four paths of gene transfer, that is, sire to produce future sires (SS), sire to produce future dams (SD), dam to produce future sires (DS), and dam to produce future dams (DD).

The genetic gain *per annum* from the four paths is as follows:

$$\Delta G \text{ per year} = \frac{(\Delta G_{SS} + \Delta G_{SD} + \Delta G_{DS} + \Delta G_{DD})}{(L_{SS} + L_{SD} + L_{DS} + L_{DD})}$$

where ΔG_{SS}, ΔG_{SD}, ΔG_{DS}, and ΔG_{DD} are ΔG per generation along the sire to sire path, sire to dam path, dam to sire path, and dam to dam path, respectively. L_{SS} is the average age of sires when their male offspring (bulls) were born and used as breeding bulls; L_{SD} is the average age of sires when their female offspring were born and entered the herd; L_{DB} is the average age of dams when their male offspring (bulls) were born and used as breeding bulls; and L_{DD} is the average age of dams when their female offspring were born and entered the herd.

Biotechnologies can primarily act by allowing for increased selection intensity as there is a pronounced chance of improving reproductive efficiency due to the use of high-throughput reproductive tools, and by increasing the accuracy of selection with additional information or an increase in the number of progeny. However, the widespread use of some biotechnologies can also influence genetic variance (e.g., reducing genetic diversity or increasing the inbreeding trend). Some biotechnologies could be able to improve genetic gain by reducing generation intervals.

The effect of biotechnology is limited to additive genetic effects. Inbreeding and nonadditive genetic effects such as dominance effects, however, influence the expected progeny difference. Some of these parameters are highly influenced by different reproductive tools (VanRaden and Hoeschele, 1991). Moreover, inbreeding is a significant issue in some cases because certain biotechnologies may result in less-active breeding animals or even new types of animals with a higher or high inbreeding coefficient. Further, offspring produced from biotechnological interventions such as cloning and embryo transfer are expected to carry a higher inbreeding coefficient due to the presence of common genes.

There are two main characteristics of traditional animal farming. To begin with, it makes use of existing genetic diversity rather than creating it artificially. Cross-breeding with different populations and/or the selection of animals in existing populations was also used to introduce new alleles or characteristics. Secondly, superior animals are identified using phenotype and advanced statistical techniques (best linear unbiased prediction, BLUP), which enable genetic and environmental causes to be separated.

Artificial insemination (AI)

AI is the most useful and practical method in the genetic improvement of animals. AI is the introduction of male sperm from one superior male into the female reproductive tract by means of an instrument without mating; it is one of the earliest established reproductive technologies. AI is now an integral part of dairy herd management. This technology is mostly used in cattle production systems, especially in the dairy industry. The introduction of artificial insemination in the 1930s represented the start of a revolution in traditional animal breeding. Every year, more than 100 million cattle, 40 million pigs, 3.3 million sheep, and 500,000 goats are inseminated around the world (Boa-Amponsem and Minozzi, 2006). In 2007, 34 million inseminations were performed in India alone (DAH, 2008). In breeding programs, AI is more effective and versatile due to preservation techniques and cooled or cryopreserved semen from genetically top-quality male animals. Breeding schemes that make use of AI and embryo transfer help to enhance genetic progress. This has been practiced on a worldwide scale for more than

50 years. It is still the major technology available for the improvement of a farm animal's reproductive efficiency. The opportunity for genetic improvement through progeny testing and breeding programs would be extremely limited without AI for the widespread dissemination of semen from superior sires at a reasonable cost. AI offers many potential advantages over natural service. AI has been the main vehicle for the improvement of genetic quality in. herds around the world. In addition, various technologies have evolved that led to the efficient use of AI in developing desired livestock, including freezing semen or cryopreservation and sperm sexing.

Advantages of artificial insemination

(a) AI allows for a significant increase in selection intensity (Van Vleck, 1981).

(b) The number of bulls required to breed the existing cows has been drastically decreased. Simultaneously, the extensive use of tested older bulls as well as progeny testing of younger bulls allows for highly accurate breeding value estimations.

(c) AI can disseminate favorable alleles in a larger population, not just through countries but also across the world.

(d) AI can contribute to the development of improved breeds, in so far as the breeding animals used for insemination have a genetic merit obvious to the breeder.

(e) AI can control the transmission of venereal disease and libido problems, reduce the need for farms to maintain breeding males, produce healthy calves, allow more precise pedigree recording, and reduce the cost of introducing improved genetics.

(f) The current trend of using molecular markers in combination with reproduction technologies such as AI and *in vitro* embryo production is likely to hasten more genetic change to obtain animals with desirable characteristics.

(g) The sex-sorted sperm can subsequently be used for AI to obtain offspring of the desired sex.

Disadvantage of artificial insemination

(a) Wider and rapid use of a single outstanding bull may accelerate genetic gain, but it also has the potential to decrease genetic variability.

(b) Increase in the inbreeding coefficient affecting the maternal traits can be constraints of AI, due to the use of fewer male animals.

(c) Dissemination of genetic defects or uncontrolled (or unknown) diseases.
(d) AI requires well-trained technicians and special equipment for inseminating the semen.

Semen sexing

Presently, semen sexing is considered one of the best methods to enhance either the male or female elite germplasm as per our desire for conservation and genetic improvement (Das and Mohan, 2020). Here, flow cytometry is mostly used for determining the sex of future progeny by analyzing the DNA content of individual sperm cells currently in use. It determines whether they have the larger X chromosome (resulting in female offspring) or the smaller Y chromosome (resulting in male offspring). Depending on the species, X chromosome-bearing sperm contain 2%–5% more DNA than sperm bearing the Y-chromosome (as the X chromosome is larger than the Y chromosome). When sperm are stained with a fluorescent dye and exposed to light, they emit distinct emission patterns. A flow cytometry will separate the sperm because of this difference. To obtain offspring of the desired sex, the sex-sorted sperm can subsequently be used for AI. When ultraviolet light is exposed during flow cytometry, X spermatozoa fluoresce brighter than Y spermatozoa.

Other methods to differentiate between X and Y spermatozoa include the Percoll method (X spermatozoa is larger while Y spermatozoa is relatively smaller), the swim up method (X spermatozoa swims slower while Y spermatozoa swims faster), free flow electrophoresis (X spermatozoa migrates to cathode fast while Y spermatozoa migrates to cathode slow), immunological sexing methods (absence of the HY antigen in X spermatozoa and the presence of the HY antigen in Y spermatozoa) (Prasad et al., 2010).

Advantages of semen sexing

(a) The sexed semen bearing the X chromosome could be used in elite cows to produce superior high-yielding cows at a faster rate than the conventional unsexed semen (Kumar et al., 2016).
(b) By introducing sexed male sperm, superior bulls could be produced from the limited number of elite cows available.

(c) Sexed semen technology ensures the required number of progeny per bull under a progeny testing program, thereby increasing the accuracy of bull testing.
(d) The sexed female sperms could be used in test mating to ensure the production of the required number of daughters in a short time, thus increasing the genetic gain.
(e) Sexed semen is an excellent way to expand the dairy herd without spending large amounts on replacement; virtually, there is no need for a breeder to buy new heifers.

Disadvantages of semen sexing

(a) A major limitation with the implementation of sexed semen technology is the lower fertility of sorted semen (Seidel Jr et al., 1999).
(b) High cost of machine.
(c) Lack of skilled manpower.
(d) Conception rate is 10%–20% lower in sorted sexed semen compared to conventional semen (Norman et al., 2010).
(e) Standardization is needed in Indian conditions (Kumar et al., 2016).
(f) The low number of elite bulls will limit the options for semen sexing in genetically superior bulls.

Embryo sexing

The beef industry mostly prefers bulls while the dairy industry prefers heifers. In general, the pig industry prefers females due to the higher quality and lower cost of production. As a method of controlling the sex of offspring, embryo sexing has become possible with the introduction of embryo transfer techniques. Five different methods have been developed for embryo sexing: assay of X-linked enzymes, cytogenetic analysis, detection of male-specific antigens, analysis of differential development rates, and the use of Y-specific DNA probes (Bondioli, 1992). Of these, Y-specific DNA probes, specifically those that involve DNA amplification methods, have shown the greatest commercial potential (Bondioli, 1992). A bovine Y-chromosome probe is used to establish embryo sexing methods in cattle. Generally, technical experts remove a few cells from the embryo and assess the DNA in these cells for the presence of a Y-chromosome. A male embryo is identified by the presence of a Y-chromosome. Embryo sexing has not been widely used in traditional embryo transfer for economic and technical reasons, but these processes, particularly Y-chromosome-specific DNA amplification methods, are very suitable to embryo cloning and will be critical if cloning becomes economically feasible (Bondioli, 1992). Standardization of techniques for laboratory production of embryos both by normal and the nuclear transfer method would lead to the establishment of embryo banks for the conservation of indigenous animal genetic resources and endangered species.

Estrus synchronization

Estrus synchronization is the process of manipulating the estrus cycle that results in standing estrus (heat) in the majority of animals in a short time. It is very effective to increase the number of animals bred at a particular breeding season. Estrus synchronization allows the more efficient use of artificial insemination with the use of elite sires having superior genetic makeup and embryo transfer procedures. It reduces problems related to estrus detection, such as estrus detection errors and labor costs. This technique should be used such a way so that permissible inbreeding limit (<5%) should be taken care (Das and Mohan, 2020).

Animal cloning

The recent assisted breeding tool in livestock industry has been included is cloning. This is a very reliable methodology for production high genetic merit animals. In this process, the genetic architecture of the animal remains unchanged with the original DNA structure of the parent. This technique allows breeders to create an exact genetic copy of an existing animal, with a target of producing identical twin. A cloned animal carries the same DNA as its parent, much like identical twins share the same DNA. They are genetically identical to each other to their common ancestor. In 1996, Scottish scientists successfully created Dolly, the first mammal ever cloned from an adult somatic cell of a sheep using the nuclear transfer process (Wilmut et al., 2000). Cloning in plants and microbes has also received a lot of attention. However, the techniques used in plants are not applicable for animals. At present, two types of techniques for embryo cloning–nuclear transplantation (transfer) and embryonic stem cell cloning–are being developed. Though outstanding cloned animals are produced by proven reproductive tool, the performances done by the same animal in successive parities may not show same kind of performances due lower repeatability (Van Vleck, 1999). Hence, multiplying phenotypically outstanding animal does not warrant to result into outstanding clones in next generation (Van Vleck, 1999).

In the dairy industry, cloning is occasionally used to reproduce animals of high genetic merit, including bulls that are repeatedly ranked high on the basis of production traits but cannot produce enough semen to meet the demand. Hence, cloning is an effective way of conservation of breeds and increasing population size. However, studies revealed that cloned animals tend to have comparatively more problems during gestation and birth, resulting in higher rates of miscarriage and death among host mothers. According to the literature, there were severe physical deformities in cloned animals, including oversized navels, oddly shaped heads, immune deficiencies, heart and lung damage, kidney failure, brain irregularities, and malformed arteries. Some reports revealed that the success rate of animal cloning was limited to 10%. Further, as only a few cows are cloned toward future use in the entire industry, the genetic variability could be reduced leading to inbreeding depression that will have a detrimental impact on environmental adaptation.

The world's first cloned buffalo calf was born in India at the National Dairy Research Institute (NDRI), Karnal, in February 2009, using the somatic cell of a new-born calf. The animal from which the donor cells were derived was only 12 months old when the cloned calf was born (Report on The Golden Quinquennium 2007–2012, NDRI, 2012). The world's second cloned buffalo calf (Garima) was produced by the hand-guided cloning method on June 2009, using the somatic cell of a fetus obtained from a slaughterhouse. Genotyping Garima was established by microsatellite markers. In August 2010, the buffalo calf Shresth was born from a frozen–thawed cloned embryo (Report on The Golden Quinquennium 2007–2012, NDRI, 2012). Another male cloned buffalo calf named Swarn produced through the new and advanced hand-guided cloning method was born on March 18, 2013. This cloned buffalo calf was unique and different from the earlier clones because, in this case, the donor somatic cell used was isolated from the seminal plasma of a bull currently being used.

Advantages of animal cloning

An advantage of cloning is that it is possible to integrate dominance effects into models, as each clone of the same family has the same dominance effect and reproduces them.

Disadvantages of animal cloning

Multiplying a phenotypically outstanding individual is not a guarantee that the resulting clones will be outstanding. Again, phenotypically outstanding animals are also very sensitive to preferential treatment.

Similarly, only animals with accurately calculated genetic merits, particularly sires, may be interesting candidates for cloning, and there are questions about why we should clone an outstanding sire who could readily reproduce through AI or other technique. Cloning needs a precise evaluation of each clone family, and thus, a lot of clone testing with resulting costs (Gengler and Druet, 2001). Though cloning may increase selection intensity in the first generation, it would result in little or substantially decreased genetic progress in subsequent generations because we would be duplicating the same animals, which would lead to a more critical issue. By definition, cloning is the process of conserving an existing germplasm. This means that, except for mutations and aging of DNA, at least the nuclear DNA will not change. At every mating, however, genetic improvement demands the production of novel, eventually advantageous, combinations of the parental germplasm (Gengler and Druet, 2001).

It is well understood that rigorously selected animals produce optimally only in specific conditions. under new environmental conditions of new reproductive techniques, the animals would exhibit poor adaptability to new conditions and there would be no way back. Hence, cloning resists any change in selection path because animals are fixed, not adaptable to new needs, new economic conditions, hence, they will not resist to new disease.

Transgenesis

The creation of transgenic animals using genetic engineering (GE) technologies is a prominent field of current biotechnology research. The mechanism of introducing an exogenous gene known as a "transgene" (introduced DNA) into a living organism so that the organism will exhibit a new property and transmit that property to its offspring is called transgenesis. DNA microinjection, retrovirus-mediated gene transfer, and embryonic stem cell-mediated gene transfer are three of the most common methods for creating transgenic animals. Transgenic animals are created by introducing a selected DNA fragment into an embryo, enabling the animal to express a desired trait. The gene that is intended to express favorable

Table 2 Examples of transgenic animals for agricultural applications.

Species	Transgene	Origin	Effect on
Cattle	Lysozyme	Human	Milk composition
	Lysostaphin	Bacterial	Mastitis resistance
	PrP	Knockout	Animal health
	α-,κ-Casein	Bovine	Milk composition
	Omega-3	Nematode	Milk composition
Goat	Monosaturated fatty acid	Rat-Bovine	Mastitis resistance
	Lysozyme	Human-Bovine	Animal health
	Lactoferrin	Human	Prophylactic treatment
	Human beta-defensin 3	Human	Milk composition
Sheep	IGF-1	Ovine	Wool growth
	CsK	Bacterial	Wool growth
	Visna resistance	Viral	Disease resistance
	PrP	Knockout	Animal health
Pig	Phytase	*E. coli*-Mouse	Feed uptake
	Growth hormone	Human-Porcine	Growth rate
	cSKI	Chicken	Muscle development
	Lysozyme	Human	Piglet survival
	Unsaturated fatty acid	Spinach	Meat composition
	Omega-3	Nematode	Meat composition
	α-lactalbumin	Bovine	Piglet survival
	Mx1	Murine	Influenza resistance

qualities during the recipient animal's growth and development is constructed mostly using recombinant DNA technology. Some examples of transgenic animals for agricultural applications involving genetic modification are shown in Table 2 (adapted from VanRaden, 2007; Lievens et al., 2015).

Multiple ovulation and embryo transfer (MOET)

Multiple ovulation and embryo transfer (MOET) is a biotechnology that is similar to AI, but affects the reproduction ability of females (Seidel and Seidel, 1981). MOET may be used to allow for extra genetic gain by producing embryos obtained from selected females and males. The most significant feature of MOET is increasing selection intensity in the female selection paths (Ruane and Thompson, 1991), especially in the selection of bull dams with fewer but better dams, thereby improving the bull dam pathway. The number of full-sibs families increases by using MOET techniques (VanRaden et al., 1992). In order to obtain more accurate breeding values, it is more important to have dominance effects (VanRaden et al., 1992). Animals produced by MOET are frequently used as top reproducers, as the gain of accuracy of breeding values (BV) through the inclusion of dominance effects is quite important. Using the standardized protocol for superovulation and embryo transfer, it has been possible to produce more than 10 claves from a single cow in one calendar year (Das and Mohan, 2020). As the genetic merit of male animals is judged depending on the performance of their sisters and half-sisters, called the sibling test, and not on the performances of their progeny, the generation interval is considerably reduced (Das and Mohan, 2020). Mass MOET combined with the open nucleus breeding system (ONBS) will allow for more accuracy in selection and faster genetic improvement in the field. Embryo transfer technology offers the opportunity for efficient exploitation of the female gamete pool and shortening of the generation interval (Hansen and Block, 2004). In addition, it reduces the transmission of infection from one herd to another. However, the use of this technology as a tool for improving reproductive efficiency is limited due to the high cost of producing embryos, the low embryo yield, etc.

Two modern enhancements to MOET are *in vitro* fertilization (IVF) and ovum pick-up (OPU). OPU can greatly reduce the generation interval because it provides a way to use very young animals in selection schemes. OPU is a procedure that allows the aspiration of the follicles, and thus the collection of the oocytes, from live animals after the endoscopic or ultrasound visualization of the ovaries. The OPU/*in vitro* embryo production (IVP) technology involves harvesting oocytes (immature) from genetically superior living animals by ultrasound-guided transvaginal follicular aspiration, followed by *in vitro* maturation (IVM), fertilization, and culture until the embryos enter the stage of morula or blastocyst and can be transferred nonsurgically into recipients or frozen and used for research purposes at any stage of development. Moreover, the OPU technology helps to select elite animals for embryo production. OPU/IVP offers more advantages over AI and MOET in terms of the number of offspring that can be produced per cow per time span; on average, one calf per year can be achieved by AI and 20–25 calves per year can be achieved by MOET, whereas 80–100 calves per year can be achieved by OPU/IVP, as stated by Van Wagtendonk-de Leeuw (2006). There are some disadvantages of OPU such as the cost of equipment, need for highly skilled technicians and a well-equipped laboratory for IVP production, etc.

Marker-assisted selection (MAS) and gene mapping

Over the last 2 decades, the genes responsible for various traits in domestic animals have been mapped to their specific loci using information obtained on conserved genes through comparative gene mapping and identifying microsatellite markers linked with certain phenotypes. Information recently published for cattle includes quantitative traits, including growth rate, milk yield, milk fat content, back fat, ovulation rate, twinning rate, and other reproductive and health traits, all of which may help in marker-assisted selection (MAS) in the future. MAS is expected to be particularly useful in improving low-heritability traits including reproduction, longevity, carcass traits, and disease resistance. It may be used at an early age, reducing the generation interval and improving the rate of genetic progress by increasing selection accuracy.

Genome editing biotechnologies transforming the genome at the DNA level

The above-mentioned biotechnologies were based on naturally existing DNA, whereas genetic engineering exists and can be used artificially in animals. Genome editing technologies may allow animal breeders to improve animal welfare, performance, and efficiency, paving the way to a more sustainable future for livestock production systems. Genome editing is referred to as a technique that allows site-specific nucleases to precisely edit or change the genetic code in the targeted genome. Genome editing can be used to accurately introduce desired alleles (e.g., disease resistance, heat tolerance) and haplotypes into native, locally adapted cattle breeds, thereby helping to improve their productivity (Van Eenennaam, 2019). These technologies enable researchers to insert, delete, or replace DNA at a specific site with the aid of programmed nucleases. Recently, several genome editing technologies have been developed to cut the genome in a specific targeted position using engineered nucleases, including zinc finger nucleases (ZFNs), transcription activator-like effector nucleases (TALENs), and clustered regularly interspaced short palindromic repeats/CRISPR-associated endonuclease 9 (CRISPR/Cas9). Targeted genome editing with the CRISPR/Cas9 tool has recently emerged as a revolutionary approach to modify endogenous genes in various cell types. CRISPR/Cas9 genome editing comprises two components, the Cas9 endonuclease for cutting DNA at the desired

Table 3 Examples of successful gene edited applications in livestock species.

Species	Target	Effect/Goal	References
Cattle	Polled	No horns	Tan et al. (2013)
	Myostatin	Increased muscle growth	Proudfoot et al. (2015)
	Insertion of TB-resistant gene *NRAMP-1* into bovine fetal cell	Suppression of tuberculosis	Gao et al. (2017)
Goat	Myostatin	Increased muscle growth	Ni et al. (2014)
	Prion protein	Elimination of prion protein	
	Beta-lactoglobulin	Elimination of milk allergen	
Pig	CD163	PRRSV resistance	Whitworth et al. (2015) and Sanchez-Torres et al. (2003)
	RELA	African swine fever resistance	Lillico et al. (2016)
	NANOS2	Germline ablation	Park et al. (2017)
Sheep	Myostatin	Increased muscle growth	Proudfoot et al. (2015)

location in the target genome, and single-guide RNA for recognizing the sequence of the DNA to be edited. The CRISPR/Cas9 system is a naturally occurring prokaryotic immune defense mechanism that has been recently adapted for eukaryotic genome engineering, and it offers a wide variety of biomedical, therapeutic, agricultural, and research applications (Unniyampurath et al., 2016).

Table 3 represents some examples of successful gene edited applications in livestock species.

Genome editing technology vs traditional animal breeding

Traditional animal breeding is limited by genetic linkage and the amount of genetic variation accessible within a breed. However, genome editing enables animal breeders to overcome these biological obstacles and insert polymorphisms artificially that are not present in the gene pool of an elite brood stock, or even produce novel modifications that are expected to result in enhanced genetic gain (Laible et al., 2015). These advanced technologies allow animal breeders to alter an animal's DNA in a precise and efficient manner. In addition, precise genetic engineering allows introducing favorable traits such as disease resistance, heat tolerance, and hornlessness or suppressing deleterious mutations such as recessive lethal genetic variants without affecting other genetic features of the herd (Davis et al., 2017; Mueller et al., 2019; Proudfoot et al., 2019). A study revealed that combining genome editing with genomic selection could improve the response to selection fourfold after 20 generations (Jenko et al., 2015).

Gene editing allows researchers to translate the thousands of SNP markers discovered by livestock sequencing projects, information from numerous genome-wide association studies, and the discovery of causative SNPs (quantitative trait nucleotides, QTNs) into useful genetic variations for animal breeding programs (Hickey, 2013). It may also be used to change a less-desirable gene (called an allele) to a more-desirable allele without attempting to introgress (i.e., repeatedly backcross) or outcross with an animal that has the same allele. As a result, genome editing is almost similar to precision breeding, in which breeders may introduce specific sequences that they choose to use for genome editing. Genomic selection and selective breeding have had a huge influence on animal productivity. Now, transgenic and genome editing technologies are opening new and exciting possibilities for producing healthier, fitter, disease-resistant, and more productive animals. With a more sophisticated regulatory framework, these genome editing technologies may help to accelerate livestock genetic progress and help to achieve a sustainable future for livestock. Hence, genome editing can be integrated into genomic selection practices to alter the genetic variation toward the desired direction and reduce the generation interval to reach higher genetic progress (Das et al., 2019).

Major constraints on applying biotechnologies in developing countries

A variety of factors affect the application of novel molecular biotechnologies and breeding strategies to livestock breeds used in smallholder production systems in developing countries. Due to limitations in infrastructure resources, large number of cattle breeds; the application of biotechnology in developing countries is slower. But in India biotechnological tools like use of sexed semen, cloning, embryo transfer are being applied in targeted indigenous dairy breeds. In developing countries, the limitations and constraints for biotechnology in livestock production are influenced by a number of factors such as poverty, malnutrition, disease, poor hygiene, and unemployment among the human population in these countries (Madan, 2003). Some major constraints on applying biotechnology in developing countries are specifically listed below:

(a) Lack of accurate and complete digitization as well as a real-time database on livestock and animal owners in most of the developing world.
(b) Large biodiversity present within species and breeds in available agro-ecological systems.
(c) Models of biotechnological intervention differ significantly between developed and developing economies.
(d) Many animal species and breeds are unique to the developing world, having their own developmental, adaptation, production, disease resistance, and nutrient utilization characteristics.
(e) Biotechnologies evolved in developed countries are not suitable for developing countries due different situation.
(f) Lack of trained scientists, technicians, and field workers to develop and apply the technologies.
(g) Absence of a mechanism among industries, universities, and institutions for technology transfer.
(h) Expensive technology to be purchased from the developed world.
(i) High cost of technological inputs such as materials, biologicals, and equipment.
(j) Failure to address issues of biosafety and to conduct risk analyses of new biologicals, gene products, transgenics, modified food items, etc.
(k) Negligible investment in animal biotechnology.
(l) Lack of clear policy and commitment from the government.

Conclusion

The use of reproductive biotechnologies such as artificial insemination, semen and embryo sexing, cloning, and embryo transfer is essential in animal breeding programs to achieve increased genetic improvement. Improved selection intensities, the faster distribution of a superior germplasm, and a reduction of the generation interval are three main areas where these ARTs play important roles. However, there are some negative consequences, such as increased inbreeding and decreased genetic diversity due to the use of some biotechnologies. Secondly, the ultimate goal of quantitative or economical trait loci (QTL/ETL) is to enhance accuracy toward the determination of genetic merit. Despite extensive research, these methods would almost certainly never be able to explain all genetic variation. The evaluation and introduction of molecular knowledge into genetic evaluation as well as its use in selection schemes with the advantage of advanced computerized mating schemes will enable breeders to bring genetic progress within a short generation interval. Recently, genome editing has enabled animal breeders to overcome the current biological obstacles and insert nucleotides that are not present in the gene pool of an elite breed stock, or even produce novel modifications that are expected to result in enhanced genetic gain. But some technical and ethical issues related to genome editing technologies are not yet solved. Genomic selection and selective breeding have a huge influence on animal productivity. Recently, transgenic and genome editing technologies have opened new and exciting possibilities for producing healthier, fitter, disease-tolerant, and more efficient and productive animals. With a more sophisticated regulatory framework, these genome editing technologies may help to accelerate livestock genetic progress and help to achieve sustainable progress for the livestock industry. Therefore, it is concluded that biotechnology plays a significant role on animal breeding and genetic improvement.

References

Boa-Amponsem, K., Minozzi, G., 2006. The state of development of biotechnologies as they relate to the management of animal genetic resources and their potential application in developing countries. In: Background Study Paper, p. 33.

Bondioli, K.R., 1992. Embryo sexing: a review of current techniques and their potential for commercial amdication in livestock production. J. Anim. Sci. 70 (suppl_2), 19–29.

DAH, 2008. Animal Husbandry. Annual Report 2007–2008. Department of Animal Husbandry, Dairying and Fisheries. New Delhi, Ministry of Agriculture, Government of India.

Das, D.N., Mohan, M., 2020. Chapter 5: Conservation of animal genetic resources: problems and prospects. In: Innovative Approaches for Conservation and Improvement of Indigenous Bovine Genetic Resources. International Books & Periodical Supply Service, Delhi, pp. 45–53.

Das, D.N., Shanmugapriya, G., Sivaram, M., Shah, U.N., Mondal, S., Malakar, D., 2019. Application of CRISPR-CAS9 tools for livestock improvement. In: Winter School on Conservation and Improvement of Indigenous Bovine Genetic Resources through Integrated Omic Approaches and Optimal Resource Utilization. ICAR NDRI, SRS, Bengaluru, pp. 6–26. November 2019.

Davis, S.R., Spelman, R.J., Littlejohn, M.D., 2017. Breeding and genetics symposium: breeding heat tolerant dairy cattle: the case for introgression of the "slick" prolactin receptor variant into dairy breeds. J. Anim. Sci. 95 (4), 1788–1800.

Falconer, D.S., 1989. Introduction to Quantitative Genetics, third ed. Longman, New York.

Food and Agriculture Organization (FAO), 2010. Current status and options for livestock biotechnologies in developing countries. In: FAO International Technical Conference,

Agricultural Biotechnologies in Developing Countries: Options and Opportunities in Crops, Forestry, Livestock, Fisheries and Agro-Industry to Face the Challenges of Food Insecurity and Climate Change. Guadalajara, Mexico, March 1–4 2010. Report ABDC-10/5. 57 p. http://www.fao.org/fileadmin/user_upload/abdc/documents/livestock.pdf.

Gao, Y., Wu, H., Wang, Y., Liu, X., Chen, L., Li, Q., Zhang, Y., 2017. Single Cas9 nickase induced generation of NRAMP1 knock in cattle with reduced off-target effects. Genome Biol. 18 (1), 1–15.

Gengler, N., Druet, T., 2001. Impact of biotechnology on animal breeding and genetic progress. In: Biotechnology in Animal Husbandry. Springer, Dordrecht, pp. 33–45.

Georges, M., Nielsen, D., Mackinnon, M., Mishra, A., Okimoto, R., Pasquino, A.T., Sargeant, L.S., Sorensen, A., Steele, M.R., Zhao, X., 1995. Mapping quantitative trait loci controlling milk production in dairy cattle by exploiting progeny testing. Genetics 139 (2), 907–920.

Hansen, P.J., Block, J., 2004. Towards an embryocentric world: the current and potential uses of embryo technologies in dairy production. Reprod. Fertil. Dev. 16 (2), 1–14.

Hickey, J.M., 2013. Sequencing millions of animals for genomic selection 2.0. J. Anim. Breed. Genet. 130 (5), 331–332.

Jenko, J., Gorjanc, G., Cleveland, M.A., Varshney, R.K., Whitelaw, C.B.A., Woolliams, J.A., Hickey, J.M., 2015. Potential of promotion of alleles by genome editing to improve quantitative traits in livestock breeding programs. Genet. Sel. Evol. 47 (1), 1–14.

Kumar, A., Vineeth, M.R., Sinha, R., Singh, R.K., Thakur, A., Gupta, S.K., 2016. Current status, scope and constraints of sexed semen—an Indian perspective. Agric. Rev. 37 (3).

Laible, G., Wei, J., Wagner, S., 2015. Improving livestock for agriculture—technological progress from random transgenesis to precision genome editing heralds a new era. Biotechnol. J. 10 (1), 109–120.

Lievens, A., Petrillo, M., Querci, M., Patak, A., 2015. Genetically modified animals: options and issues for traceability and enforcement. Trends Food Sci. Technol. 44 (2), 159–176.

Lillico, S.G., Proudfoot, C., King, T.J., Tan, W., Zhang, L., Mardjuki, R., Paschon, D.E., Rebar, E.J., Urnov, F.D., Mileham, A.J., McLaren, D.G., Whitelaw, C.B.A., 2016. Mammalian interspecies substitution of immune modulatory alleles by genome editing. Sci. Rep. 6 (1), 1–5.

Madan, M.L., 2003. Opportunities and constraints for using gene-based technologies in animal agriculture in developing countries and possible role of international donor agencies in promoting R and D in this field (IAEA-CN-110). In: FAO/IAEA International Symposium on Applications of Gene-Based Technologies for Improving Animal Production and Health in Developing Countries, Vienna, Austria. Food and Agriculture Organisation/International Atomic Energy Agency, Vienna, pp. 103–104. 6–10 October 2003.

Mueller, M.L., Cole, J.B., Sonstegard, T.S., Van Eenennaam, A.L., 2019. Comparison of gene editing versus conventional breeding to introgress the POLLED allele into the US dairy cattle population. J. Dairy Sci. 102 (5), 4215–4226.

NDRI, 2012. Report on The Golden Quinquennium 2007–2012. vol. 76 National Dairy Research Institute, Karnal (India), pp. 3–10.

Ni, W., Qiao, J., Hu, S., Zhao, X., Regouski, M., Yang, M., Polejaeva, I.A., Chen, C., 2014. Efficient gene knockout in goats using CRISPR/Cas9 system. PLoS One 9 (9), e106718.

Norman, H.D., Hutchison, J.L., Miller, R.H., 2010. Use of sexed semen and its effect on conception rate, calf sex, dystocia, and stillbirth of Holsteins in the United States. J. Dairy Sci. 93 (8), 3880–3890. https://doi.org/10.3168/jds.2009-2781.

Park, K.E., Kaucher, A.V., Powell, A., Waqas, M.S., Sandmaier, S.E., Oatley, M.J., Park, C.H., Tibary, A., Donovan, D.M., Blomberg, L.A., Lillico, S.G., Whitelaw, C.B., Mileham, A.,

Telugu, B.P., Oatley, J.M., 2017. Generation of germline ablated male pigs by CRISPR/Cas9 editing of the NANOS2 gene. Sci. Rep. 7, 40176.
Prasad, S., Rangasamy, S., Satheshkumar, S., 2010. Sex preselection in domestic animals—current status and future prospects. Vet. World 3 (7), 346–348.
Proudfoot, C., Carlson, D.F., Huddart, R., Long, C.R., Pryor, J.H., King, T.J., Lillico, S.G., Mileham, A.J., McLaren, D.G., Whitelaw, C.B., Fahrenkrug, S.C., 2015. Genome edited sheep and cattle. Transgenic Res. 24 (1), 147–153.
Proudfoot, C., Lillico, S., Tait-Burkard, C., 2019. Genome editing for disease resistance in pigs and chickens. Anim. Front. 9 (3), 6–12.
Renaville, R., Gengler, N., Vrech, E., Prandi, A., Massart, S., Corradini, C., Bertozzi, C., Mortiaux, F., Burny, A., Portetelle, D., 1997. Pit-1 gene polymorphism, milk yield, and conformation traits for Italian Holstein-Friesian bulls. J. Dairy Sci. 80 (12), 3431–3438.
Rendel, J.M., Robertson, A., 1950. Estimation of genetic gain in milk yield by selection in a closed herd of dairy cattle. J. Genet. 50 (1), 1–8.
Ruane, J., Thompson, R., 1991. Comparison of simulated and theoretical results in adult MOET nucleus schemes for dairy cattle. Livest. Prod. Sci. 28 (1), 1–20.
Said, S., Agung, P.P., Putra, W.P.B., Kaiin, E.M., 2020. The role of biotechnology in animal production. In: IOP Conference Series: Earth and Environmental Science. vol. 492. IOP Publishing, p. 012035. No. 1.
Sanchez-Torres, C., Gomez-Puertas, P., Gomez-del-Moral, M., Alonso, F., Escribano, J.M., Ezquerra, A., Dominguez, J., 2003. Expression of porcine CD163 on monocytes/macrophages correlates with permissiveness to African swine fever infection. Arch. Virol. 148 (12), 2307–2323.
Seidel Jr., G.E., Schenk, J.L., Herickhoff, L.A., Doyle, S.P., Brink, Z., Green, R.D., Cran, D.G., 1999. Insemination of heifers with sexed sperm. Theriogenology 52 (8), 1407–1420.
Seidel, G.E., Seidel, S.M., 1981. The embryo transfer industry. In: Brackett, B.G., Seidel, G. E., Seidel, S.M. (Eds.), New Technologies in Animal Breeding. Academic Press, London, pp. 41–80.
Singh, U., Deb, R., Alyethodi, R.R., Alex, R., Kumar, S., Chakraborty, S., Dhama, K., Sharma, A., 2014. Molecular markers and their applications in cattle genetic research: a review. Biomark. Genom. Med. 6 (2), 49–58.
Solter, D., 1981. Gene transfer in mammalian cells. In: Brackett, B.G., Seidel, G.E., Seidel, S. M. (Eds.), New Technologies in Animal Breeding. Academic Press, London, pp. 201–220.
Steinberg, F.M., Raso, J., 1998. Biotech pharmaceuticals and biotherapy: an overview. J. Pharm. Pharm. Sci. 1 (2), 48–59.
Tan, W., Carlson, D.F., Lancto, C.A., Garbe, J.R., Webster, D.A., Hackett, P.B., Fahrenkrug, S.C., 2013. Efficient nonmeiotic allele introgression in livestock using custom endonucleases. Proc. Natl. Acad. Sci. 110 (41), 16526–16531.
Unniyampurath, U., Pilankatta, R., Krishnan, M., 2016. RNA interference in the age of CRISPR: will CRISPR interfere with RNAi? Int. J. Mol. Sci. 17 (3), 291.
Van Eenennaam, A.L., 2016. The current and future uses of biotechnology in animal agriculture. Ceiba 54 (1), 72–81.
Van Eenennaam, A.L., 2019. Application of genome editing in farm animals: cattle. In: Transgenic Research. vol. 28. Springer International Publishing, pp. 93–100. No. 2.
Van Vleck, L.D., 1981. Potential genetic impact of artificial insemination, sex selection, embryo transfer, cloning, and selfing in dairy cattle. In: New Technologies in Animal Breeding. Academic Press, New York, pp. 221–242.
Van Vleck, L.D., 1999. Implications of cloning for breed improvement strategies: are traditional methods of animal improvement obsolete? J. Anim. Sci. 77 (suppl_2), 111–121.

Van Wagtendonk-de Leeuw, A.M., 2006. Ovum pick up and in vitro production in the bovine after use in several generations: a 2005 status. Theriogenology 65 (5), 914–925.

VanRaden, P.M., 2007. Improving animals each generation by selecting from the best gene sources. In: Duke Animal Law Conference. 4 p.

VanRaden, P.M., Hoeschele, I., 1991. Rapid inversion of additive by additive relationship matrices by including sire-dam combination effects. J. Dairy Sci. 74 (2), 570–579.

VanRaden, P.M., Lawlor, T.J., Short, T.H., Hoeschele, I., 1992. Use of reproductive technology to estimate variances and predict effects of gene interactions. J. Dairy Sci. 75 (10), 2892–2901.

Whitworth, K.M., Rowland, R.R., Ewen, C.L., Trible, B.R., Kerrigan, M.A., Cino-Ozuna, A.G., Samuel, M.S., Lightner, J.E., McLaren, D.G., Mileham, A.J., Wells, K. D., Prather, R.S., 2015. Gene-edited pigs are protected from porcine reproductive and respiratory syndrome virus. Nat. Biotechnol. 34 (1), 20–22.

Wilmut, I., Young, L., DeSousa, P., King, T., 2000. New opportunities in animal breeding and production—an introductory remark. Anim. Reprod. Sci. 60, 5–14.

CHAPTER FOURTEEN

Biotechnology in animal nutrition and feed utilization

Bharti Gupta[a], S. Lokeswara Balakrishna[a], Kshitij R.B. Singh[b], Parikipandla Sridevi[a], and Ravindra Pratap Singh[a]
[a]Department of Biotechnology, Indira Gandhi National Tribal University, Amarkantak, MP, India
[b]Department of Chemistry, Govt. V. Y. T. PG. Autonomous College, Durg, Chhattisgarh, India

Introduction

Biotechnology has an important role in the overall improvement of life and human standards, and there has been tremendous progress in the biotechnology field, including animal science, medical science, agricultural science, food science, marine science, and many others. Biotechnology in livestock production has emerged as a tool for improving nutrition, animal health, genetics, reproduction, and production to fulfill the increasing demand for food from livestock products due to population and income growth (Alemneh, 2019). Biotechnology plays a crucial role in livestock advancement by using three techniques—chemical, physical, and biological—that influence animal nutrition, breeding, health, and reproduction (Kaasschieter et al., 1992). The use of feed improvement, feed additives, and the antibiotic genetic manipulation of feed and gut microbes, is widely accepted in developed as well as developing countries such as South Africa, India, Brazil, etc. However, the application of these techniques in some developing continents such as Africa and Asia lacks the use of feed improvement due to various economic factors such as technical, educational, and poor infrastructural facilities (Bigliardi et al., 2005). Due to a growing world population, urbanization, and income hikes in developing countries, the demand for livestock, their feed improvement, and animal products is also increasing (Tona, 2018; Asmare, 2014).

Biotechnological developments have led to the use of harmless fungi to enhance fibrous feed such as silage or poor-quality roughage; particularly, white-rot fungi are widely utilized because of their ability to degrade whole wood components such as lignin, cellulose, and hemicellulose into simpler forms. Apart from antibiotics, an extensive range of feed additives can be

Emerging Issues in Climate Smart Livestock Production
https://doi.org/10.1016/B978-0-12-822265-2.00003-X

used for rumen fermentation modification (McSweeney et al., 1999). These feed additives contain specific substrate components such as antibiotics, microbes, and oligosaccharides, which reduce methanogenesis and protein degradation, increase propionic acid production, and improve microbial protein synthesis, which further inhibits protozoan growth (Sallam et al., 2020). Furthermore, the production of effective enzymes in great quantities can be relatively inexpensive and beneficial. Therefore, dietary supplements as a source for refining nutritional value are gaining much attention. The use of biotechnology for improving animal nutrition is the ultimate goal, which will refine the nutritious value by collective nutrients from the feed and by wastage reduction from feed (Asmare and Mekuriaw, 2015; Asmare, 2014).

Further, modern biotechnology potentialities will open new avenues for enhanced livestock productivity, nutrition, and food security; thus, it will also reduce poverty and promote the sustainable utilization of naturally occurring resources (Cheah and Looi, 2008). For animal nutrition and feed improvement, biotechnology will improve nutritional value by protecting protein and amino acids, increasing the use of crude proteins, and enzymes; which will increase the digestibility of feed; use of prebiotics and probiotics to improve gut health, also inhibit the harmful microbiota of gut (Tona, 2018). The most promising biotechnology applications in livestock systems improve animal health and production in assisted reproduction, increased disease resistance, nano-based refined diagnostic techniques, and improved vaccines with effective delivery systems. The application of recombinant DNA technology to improve animal health also may contribute significantly to improved animal disease control, thereby stimulating both food production and livestock trade (Zaman, 2013).

The current need is to increase animal productivity in developing countries to fulfill society's needs. It is necessary to safeguard and utilize scant resources to generate income from rising agricultural demands. The traditional methods used previously to improve livestock effectively increase livestock production, but animal feed nutritional value and feeding practices utilizing biotechnology are more effective than conventional methods. These improvements are largely due to the dilution of maintenance, and other improvements result from increased digestibility/nutrient availability/genetic selection for feed efficiency (Mertz, 1987). Recently, livestock has become increasingly important to boost the economies of developing countries, and the biotechnological applications are widely applied for profitable considerations and to fulfill socioeconomic demands. Biotechnological intervention in livestock improvement and feed utilization will provide new opportunities by enhancing the productivity of animal growth, the nutritional value of feed and fodder, and the quality of animal health and safety while reducing the environmental waste generated by livestock (van Deynze et al., 2004). Hence, this chapter will elaborate on the impact of biotechnological applications on livestock nutrition and feed utilization.

Composition and nutrient requirements of livestock

Nutrients are essential for all living organisms for their development, reproduction, and survival. Essential nutrients and a proper nutritious diet must be fed to animals as they are an important aspect to keep livestock healthy and help them produce (Asmare, 2014). Ruminant livestock, which includes cattle, goats, sheep, giraffes, deer, and gazelles, and all domesticated and wild bovines are capable of obtaining their nutrients from plant-derived foods. Ruminants possess a distinctive digestive system, allowing them to efficiently digest and absorb nutrients, and use energy from fibrous plant material more than other herbivores (Alderman, 1989). The ruminant stomach is multichambered and divided into four consecutive compartments—the rumen, reticulum, omasum, and abomasum—designed to ferment feedstocks and gain energy for livestock to use. The rumen contains nearly 50 gal of fluid and ingested forage; most of the initial digestion process happens in the stomach. The rumen also harbors a diverse population of microbes, mainly bacteria, protozoa, and fungi, that help break down complex fibrous materials into simpler ones (Bonneau and Laarveld, 1999). The nutritional requirements of livestock have four major components: maintenance, growth, reproduction, and lactation. Furthermore, carbohydrates, proteins, fats, water, vitamins, and minerals are the six major food groups that are primarily concerned with maintaining livestock health and product viability; these are explained below in detail (Spears and Weiss, 2014; Guyer and Owen, 1979).

Carbohydrates

Carbohydrates are formed from carbon, hydrogen, and oxygen molecules, and they provide most of the energy needed by animals. Carbohydrates make up 65%–75% of the dry weight of crop residues, plants, forages, and grains, along with some other plant-based materials such as starch, sugars, lignin, cellulose, and hemicellulose. Crude fiber and nitrogen-free extract (NFE) are the two major carbohydrate fractions used to assess livestock feed carbohydrates (Perry, 1995). The rumen microbiont utilizes cellulose and hemicellulose as energy sources (Halmemies-Beauchet-Filleau et al., 2018).

Proteins

They are complex compounds made up of amino acids, and they play an important role in growth and milk production in animals. They are also needed daily by animal bodies as the animal body constantly repairs itself and replaces dead cells and tissues. Proteins have an average nitrogen content of 16% (FAO, 2012). Plant proteins are a major protein source in livestock diets; mature cattle and heavy stockers also use urea nonprotein nitrogen (NPN) as a protein source. Rumen microbiota helps in NPN formation, which is later utilized. For the evolution of protein content in feedstuff, the crude protein (CP) values are calculated. First, the nitrogen is measured and later converted into protein by multiplying to 6.25 (100/16; proteins contain 16% nitrogen on average). However, CP does not provide overall details on the true proteinogenic or nonproteinogenic content present in a feed. The structuring of hair, skin, muscles, and internal organs is one of the only food sources of nitrogen. CP is considered the total protein content of a feed and nourishment. An average 16% nitrogen constituted by the proteins and the estimation of total nitrogen will determine the amount of protein in the feed and ration. The absorption and proper digestion of amino acids and nitrogen also vary in different livestock species (Belay and Negesse, 2018).

Fats

They contain more calories than carbohydrates and sugars per unit weight. They also increase the energy density in feed production, which can be analyzed from the end product of either extract. Fats are an important energy source; adding up to 5% fat in a livestock diet supplies a great energy source (Morris, 2001). Energy is required to do physical activity and it also helps in growth, reproduction, and other physiological functions such as digestion. It is the total caloric value of a feed, and other mechanical, chemical, and mathematical methods also determines the feed energy values of a ration. The important energies are net energy, digestible energy, metabolizable energy, and total digestible nutrients (Mahima et al., 2012).

Water

It is a basic need of all livestock animals to survive and thrive; it also causes variations in feed value because high water content dilutes the nutrients in feedstuffs. For better feed quality, water removal is necessary because the nutrients are calculated per unit of dry matter of feed. However, high moisture-containing feeds are considered valuable in minimizing the dustiness in some rations. It allows the majority of the physiological functions in the body. Therefore, it is one of the most critical components in livestock nutrients as water is usually provided free of choice, making it a difficult nutrient to determine the actual requirement for many livestock. Minimizing the water content in animal feed results in reducing dry matter intake and restricting the number of nutrients in the ration (National Dairy Board, 2012).

Vitamins

For performing various body processes, livestock need vitamins, as they increase body processes and functions; they also make animals resistant to diseases. Vitamins D and A play crucial roles, as D is required for animals that get little sunlight, and A is a major component of feedstuffs. About 14 different vitamins are known to have a specific role in livestock metabolism, nutrition, and feed utilization (Bryan, 1957). Vitamins are similar to minerals as they both take part in many physiological functions, such as coenzymes, which help prevent damage to cells by their metabolic and antioxidant functions. Vitamins are divided into two major categories: water-soluble and fat soluble. Many of the essential vitamins for feedstuff-eating animals are obtained from sunlight, synthesized by gut microbes or stored in the liver. Vitamins that are not made by animals should be provided in the forage (Morris, 2001).

Minerals

Minerals aid in maintaining growth, reproduction, and health in animals. Calcium (Ca), phosphorus (P), sulfur (S), and potassium (K) are the most important minerals in the feedstuffs. Calcium is important for strengthening bones, teeth, and eggshells in farm animals; phosphorus is the second most abundant mineral in the animal diet, as it helps in gestation and lactation; sulfur is required for the normal growth and reproduction of bacteria in the guts of animals; thus, these all aid in better digestion and nutrient absorption. Potassium plays a key role in maintaining the salt balance between cells and body fluid and regulating cardiovascular, renal, and nerve functions in animals (Wanapat et al., 1996).

The nutrient requirement varies with environmental conditions, breed, species, body weight, age, sex, milk production level, stages of gestation, and other factors such as disease and genetic potential. The requirement of livestock feed can be estimated by the above components and food groups, as it gives a detailed insight into the essential factors required and can be helpful for producers to make adjustments accordingly (FAO, 2012; Ajila et al., 2012).

Feed of ruminants

Ruminants digest plant-based food by processing it in chambers in their digestive system. More than 150 ruminant species are identified, including cattle, goats, sheep, giraffes, bison, moose, elk, yaks, water buffalo, deer, camels, alpacas, llamas, and antelopes; they have a four-chambered stomach that differs from the monogastric (Halas et al., 2018). The rumen is connected with the reticulum and works in concert, so it is called the "reticulo-rumen" (Wanapat et al., 2013). An illustration of the ruminant digestive system is given in Fig. 1. Ruminants can meet all their energy needs for growth, reproduction, maintenance, and health with a feed that consists of 100% good-quality fibrous fodder. However, some additional supplementation is also required during certain periods such as late gestation and lactation. Along with roughage, some feed such as whole corn, barley, wheat, oats, or other high-carbohydrate feeds are also required in feedstuff (Beigh et al., 2016).

Cattle

There are many pros and cons of feeding cattle in animal livestock. Most cattle in the world depend on fodder composed of grass, legumes, plant-based products, and essential nutrients (Archer et al., 1999). Potatoes are thought to be the main ingredient in the cattle diet as they are a rapidly digestible starch consisting of 70% dry matter. The potato should be half the cattle's dry feed matter. Cattle reared in feedlots are fed hay supplemented with grain, soy, and other essential nutritional components to enhance the feed's energy density. Cattle raised on a primarily foraged diet are termed grass-fed or pasture-raised; for example, meat or milk may be called grass-fed beef or pasture-raised dairy (Bigliardi et al., 2005). In comparison to pigs or chickens, both cattle and goats are better equipped to digest crude fibers easily. Adult animals need less protein than young animals while nonlactating animals need higher protein content in milk (Tona, 2018).

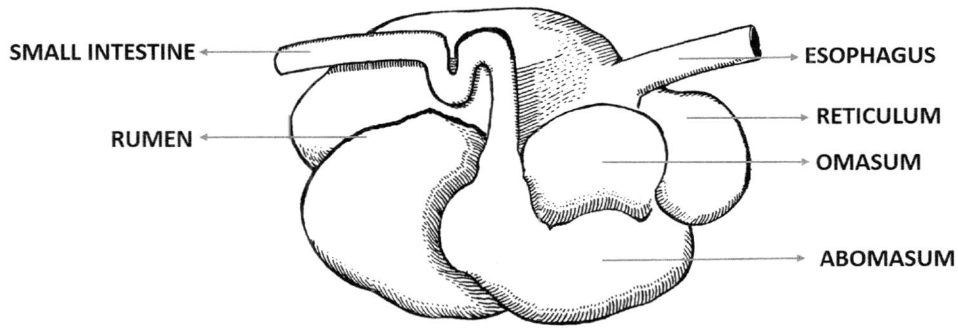

Fig. 1 Diagrammatic illustration of a ruminant digestive system.

Sheep

Two pounds of potato with grain and alfalfa hay per day is the requirement of any finishing lamb; 2–4 pounds of potatoes and alfalfa hay with grain are essential for any growing ewe lambing. An excess amount of potatoes does not add much nutritional value to feed as it may cause acidosis in the lambs. Eventually, it can result in the death of lambs. Therefore, potatoes should be introduced gradually to diets (Umberger, 2009).

Swine

Swine are monogastric animals. Their feed potatoes should be properly cooked and dried (Asmare, 2014; Wilson, 1950). The boiled potatoes of swine have one-half to two-thirds the nutritional value of the dried potatoes. Thirty pounds of grains can be replaced by 100 pounds of potatoes. Although grasses and sunshine are the best nutrient sources for cattle and sheep, they cannot provide sufficient nutritional value to other monogastrics and ruminants. Furthermore, animals can benefit from grassland feed if this is integrated with other feedstocks (legumes, shrubs) (Adesehinwa, 2008).

Feed availability and nutrition

Feeds can be chemically analyzed for their nutrition values; feeds also depend on the animal response when they are fed the feed. The response toward feed can be termed the biological response of the animal and their ability to derive nutrition from the chemical composition of feeds (Singh et al., 2019). To meet requirements, the livestock feed must contain a definite amount of nutritive value to keep the stomach healthy, along with nutrients. The feed should also contain essential micro- and macronutrients. The dry matter and trace elements in the feed should not be contaminated with harmful microbes or mixed with dirt/soil, and also should not contain pesticides, herbicides, or crop residues, which are harmful to the animal stomach (Asmare, 2014; Owen and Jayasuriya, 1989).

The feed composition depends on the animal breed being fed at the stage of breeding. All types of feed contain gross energy (GE). It works as a fuel for animals, as GE is available when the feed is burned. GE is calculated in dry matters only. Fresh sugarcane forage has a GE of 18.2 MJ per kg. The GE and digestible energy (DE) vary in different types of animals. For example, cattle can digest forage for 68%, and the DE of cattle is 11.3 MJ per kg DM. The pigs can only utilize 37%, so the DE of pigs remain very low at 6.7 MJ per kg DM (National Research Council, 2001). The animals cannot utilize DE as it is lost as methane gas or in the urine. The remaining energy in animals is metabolizable energy (ME). Animals cannot utilize heat from ME as it is lost; animals can use only net energy (NE). The use of NE differs from animal functions. For example, 1 kg of dry matter of sugarcane delivers 8.1 MJ; for growth, the same kg of sugarcane forage delivers 8.6 MJ, and lactation contains 6.7 MJ per kg DM. So, the net energy of the sugarcane forage for cattle varies from 6.7 to 8.6; for pigs, it varies from 4 to 5 (Tona, 2018). The fate of gross, digestible, metabolizable, maintenance, and production energy within the animal is shown in Fig. 2 (Egan, 2017).

Applications of biotechnology in livestock nutrition

Biotechnology is applied to enhance the feed nutrition value by preserving the essential proteins, amino acids, and fats (Adesehinwa, 2008). To improve the nutrient availability of feed and reduce waste generation from feed enzymes, it is being used in the form of prebiotics and probiotics. These enzymes enhance the good microbes and inhibit the growth of harmful enteric pathogens in the animal gut, thus improving the feed and fodder nutrition value. Genetic manipulation of these enzymes and bacteria is also achieved by biotechnological interventions to improve animal health and nutritional value. Biotechnology also has many roles to meet the increasing demand for livestock products and feed requirements for the rapidly increasing population as well as income growth in developing countries (Alemneh, 2019). In animal feed production, biotechnological techniques that impact animal health, breeding, reproduction, and nutrition can be divided into chemical, physical, and biological techniques (Asmare, 2014).

Several biotechnological techniques have been applied in most developed countries. To improve animal nutrition and reduce environmental waste generated by animals, biotechnological interventions in livestock have also changed the feed and feeding practices over time. Improvements have largely been done with maintenance in dilution while others were nutrient availability, improved digestibility, and genetic selection for feed efficiency (de Rufino et al., 2016; Onteru et al., 2010). Through biotechnology, livestock productivity improvements via increasing growth, reproduction, improved nutrition, changes in quality of feed, and reduction of waste generated in livestock can be achieved. Moreover, the use of biotechnology in livestock production is growing faster than any other sector; consequently, livestock is thought to be the most important value-added commodity (Thornton, 2010).

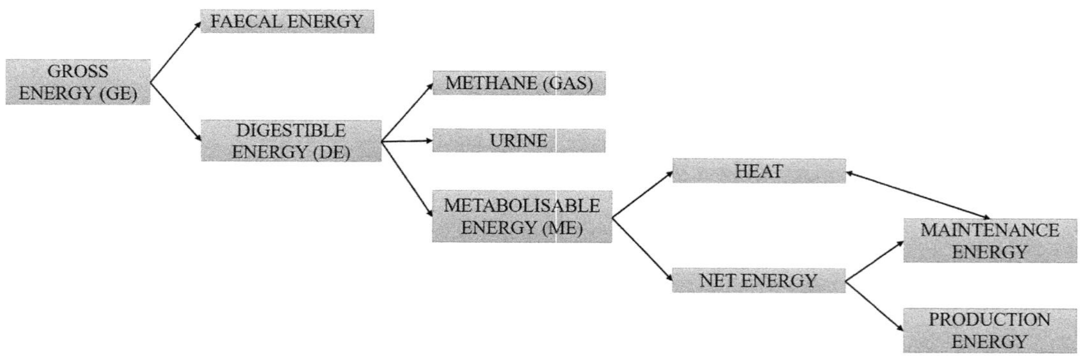

Fig. 2 Fate of feed energy within the animal.

Biotechnology in animal nutrition and feed improvement

Increasing livestock feeds, improvement in nutrient content, nutritional value, and digestibility of forage through the use of feed additives have major roles in biotechnology (Segal et al., 2012). Different enzymes are also exploited to increase the nutritional availability of feed, develop supplements to inhibit bacterial growth, use plant biotechnology to enhance the nutritional value of plant-based material, use recombinant DNA technology to increase nutritive values of product feed and fodder, develop mono and polyclonal antibodies to protect animals from disease, and genetically engineer gut microbes of the rumen to improve livestock health and growth. (Kier and Petrick, 2008).

In addition to essential nutrient requirements, many other aspects such as hygiene, palatability, digestibility, absorption, and the effect of feed on intestinal health play major roles in livestock health. Feed additives are used to enhance material use and achieve effectiveness of nutrients in the animal gut. Animal gut microbiota can be engineered using organic acids like lactic acid, acetic and sorbic acidas feed additive in diet. Prebiotics, probiotics, and synbiotics can be widely used to improve animal nutrition (Onteru et al., 2010). Probiotics help suppress the growth of harmful and unwanted pathogens in the animal gut, and they also enhance the efficiency of gut microbiota by secreting immune modulators and antibodies in the intestinal mucosa (Madsen et al., 2001; Garnsworthy and Wiseman, 2012).

Prebiotics are feed additives that stimulate the growth of beneficial microbes and inhibit the growth of unwanted or harmful microbes. Fructooligosaccharides (FOS), galactooligosaccharides (GOS), xylooligosaccharides (XOS), isomaltooligosaccharide (IMO), and lactulose alter the microflora, increasing the level of bifidobacteria and lactobacilli. They inhibit the entry of pathogens by competing them on receptor sites of the gut wall, reducing any potential risk of infection (Yirga, 2015). They may also compete effectively for nutrients with pathogens. Prebiotics manipulate the microbial gut environment, preventing infectious bowel disease incidences in animals (Gaggìa et al., 2010; Abdel-Aziz et al., 2015). Prebiotics can directly or indirectly affect the intestinal fermentation and improvement of growth of certain members of the gut microbiota. Synbiotics refer to nutritional supplements combining probiotics and prebiotics in a synergism, hence synbiotics (Liao and Nyachoti, 2017).

Pleurotus ostreatus is a white-rot fungus isolated from anaerobic bacteria; its enzymes decompose the plant-based material (crop residues and forage) such as lignin, cellulose, etc. *Aspergillus niger, A. terreus, Fusarium moniliforme,* and *Chaetomium celluloyticum* are the other fungal strains widely used for the hydrolysis of lignocellulosic products in animal feeds and fodder. Moreover, only white-rot fungi were reported for the decomposition of roughage due to their ability to open up the aromatic ring and release low molecular weight fragments by attacking lignin polymers, resulting in a complete breakdown of lignin (Sarnklong et al., 2010).

Other alternative approaches to improve fibrous plant-based material digestion in livestock are achieved by manipulating microbial inoculants in feed and fodder. The introduction of cellulose, hemicellulose, and amylase has been shown to improve lactic production by releasing sugars to grow certain beneficial bacteria (i.e., *Lactobacillus*). Additionally, the inoculation of the genetically engineered silage bacteria *Lactobacillus plantarum* as a starter culture and thus the production of such enzymes have been proposed to obtain better results in terms of pre- and postdigestion of plant materials to lead to better digestibility in the rumen (Hooper et al., 1989; Alagawany et al., 2018).

Biotechnology in forage breeding

Forages that are genetically modified, with many benefits for human health and the environment, have been developed by using biotechnology. Recombinant DNA technology has been used to modify forage crops to pursue desirable nutritional values and quantity of plants and seeds (Onteru et al., 2010). Transgenic forage crops are developed to provide many benefits to the producers, developers, and consumers. Moreover, products produced from transgenic crops and used by animals as feed and fodder are not themselves transgenic; on the other hand, they only provide health benefits to animals and humans (Kaasschieter et al., 1992).

Some bacterial attacks have developed secondary compounds that detract from these organisms colonizing the leaf tissues. Researchers at the Noble Foundation have successfully manipulated lignin composition and levels in alfalfa and other forages to improve their digestibility and the conversion of biomass to biofuels. Some shrubs and trees respond to leaf damage by grazing by producing greater secondary compounds, making them inedible. Antinutritive factors in plant tissues include protease inhibitors, tannins, phytohaemagglutinins and cyanogens in legumes, glucosinolates, tannins, etc. Sand pine in oilseed rape (*Brassica napus*) and other compounds in feeds belong to the Brassica group (Loosli, 1980).

Biotechnology products as feed additives

Enzymes, antibiotics, prebiotics, and probiotics are used as feed additives in animal feeds to increase the gut's feed utilization. Enzymes are commonly used feed additives; their preparation and large-scale production have become fast and inexpensive due to recent biotechnology advancements. Hence, this makes enzymes a common dietary supplement. However, enzymes as feed additives work differently in ruminants and nonruminants, although the purpose of enzymes remains the same: improving the nutritional value of feed. The most common example of an enzyme is phytase, a major ingredient of monogastric animal diet (Miller, 2002).

Role of biotechnology in feed improvement

Yeast cultures are very important in the feed; they enhance the microbial activities in the rumen environment. Yeast supplements in feed increase beneficial microbe growth while suppressing harmful microbes in the animal rumen. They also increase the numbers of anaerobic and cellulolytic bacteria in the ruminal gut. In this way, they influence fatty acid production along with increasing the digestibility of feed (Göncü and Güngör, 2018). Much progress is needed in forage improvement as low-quality forage remains a big concern in the livestock diet (den Hartog and Sijtsma, 2013).

The major disadvantage of feed is the low amount of essential amino acids and moderate protein content in livestock rations. To overcome this, several genetic engineering studies are under way in many developing countries to increase the level of lysine in barley. Some genetic modification in plants such as gene insertion in rice protoplast and transformed rice plant production have been achieved (Crews, 2005). The preservation of plant-based food as feedstock usually depends on the sugar fermentation of the feed materials influenced by naturally occurring lactic acid bacteria in the existing environment with limited nutrients (Beigh et al., 2016).

To increase weight, the carcass composition in the livestock recombinant bovine somatotropin is used extensively. Researchers have also developed porcine somatotropin, which reduces fat and increases muscle growth in swine, making swine a greater value product in the market. Porcine somatotropin also improves the milk yield in lactating animals and the lean-to-fat ratio while decreasing the animal waste of livestock (Miller, 2002).

Feed utilization and rationing

In finalizing feed rationing, the species and breed of the livestock are also considered, along with their nutrient requirements, feeding habits, the nutrient value of feed, the management practices of the farmer, the cost and availability of feeds, and the apt amount of feed being fed to animals with the ration. These considerations are important because of their effect on farmer profit and animal health. In balancing feed rations, digestible energy, metabolizable energy, and crude protein values should be used to measure the protein percentage in the feed. In the feed ratio of animal, all the ingredients should fall in any of the following categories; crop residue, forage, roughage, minerals, pasture, vitaminsrange plants, the energy feeds, vitamins mineral, silages, and feed additives (Belay and Negesse, 2018).

Feed utilization

For their maintenance and production, the animal needs a daily supply of nutrients. The livestock animal diet must have all the required components for their activities such as their growth, gestation, lactation, and milk and meat production. Sufficient feed is required to keep them physically active and alive, produce body heat, and maintain body tissues and respiration. The utilization of feed also depends on several factors, including:

A. Growth: Growth is determined by the increase in weight, height, and length that occurs in a young animal; to attain the right growth, an immense amount of energy is required, which will help in the development of new body tissue, bone, and muscle. The growth rate plays a crucial role as it determines the age when the animal gains the acceptable weight for sale and slaughter. Additionally, body weight is also important in determining when an animal can commence breeding. The larger the size, the higher the feed and maintenance demand. The size of an animal is considered an important factor in feed utilization.

B. Milk Production: Poor feed supplies to animals result in the reduction of milk production. Due to this, calves may die at a very young age as they will not get enough milk supplies from their dams. Older calves can be weaned if there is a severe feed shortage, and preference can be given to them for good quality and quantity of feed.

C. Reproduction: The nutritional value and right quantity and quality of the food ingested by the animals significantly influence the growing heifer's weight. The right quantity and quality of feed determines the right time of estrus of animals to commence breeding. Poor nutritional value can delay breeding by 12 months or more. Nutrition also controls the heating process and cycles of heat generation in calves. A lack of proper nutrition can preclude mating approximately 6 months after the parturition and adversely affect the fertilization process. On the other hand, the properly fed cow can mate with the bull after 2 months of parturition and have a greater chance of being fertile. Reproduction is the very first activity that gets affected by poor feed and nutritional values.

D. Weight gain: Enough feed and other essential nutrient requirements usually turns into weight gain in animals. An increase in muscle fibers achieves weight gain because of excess fat storage and energy within them.

E. Climate: Body heating of animals requires a high amount of energy during the cold season and rainy weather.

F. Type of food: High fiber-containing feed such as grasses and leaves requires a great amount of energy to break them down into basic feed components.

Biological treatments to improve feed utilization

Several biological methods have been used to better the feeding value of agricultural crop residues. For proper nutrient preservation and better fermentation of foliage of the crop residue, actions such as grinding, chopping, drying, reconstitution of moisture and nutrient contents, and mixing of useful additives such as cellulolytic and Lactobacillus bacteria have been widely used, and a large number of dry crop residues have also been profitably ensiled with the introduction of certain useful microorganisms (Huang et al., 2018).

Bacterial additives

The bacterial feed additive plays a major role in feed improvement and utilization by animals. The bacterial strains used can be categorized as direct-fed microbial (DFM). Such examples are bacteria that produce/utilize lactic acid or other microorganism species. Lactobacillus, Propionibacterium, Bifidobacterium, Enterococcus, Streptococcus, and Bacillus are the major microbial species used as bacterial additives in feed and fodder (Wanapat et al., 2013). A summary of microorganisms used as feed additives and growth products in animal feed and fodder is given in Table 1.

Exogenous enzymes

The introduction of enzymes is very important in feed to degrade some complex products such as cellulose, lignocelluloses, and hemicelluloses,

Table 1 List of microorganisms and unicellular fungi and their products along with their dosage that are used as growth promoters in animal feeding.

Microorganisms	Products	Doses
Bacilli		
B. subtilis	Sporigen	$9.6 \times 109\,\mathrm{CFU\,g}^{-1}$
B. licheniformis	Sporigen	$9.6 \times 109\,\mathrm{CFU\,g}^{-1}$
B. CIP 5832	Paciflor	$100\,\mathrm{g\,t}^{-1}$
Lactobacilli		
L. scidophillus	Lacto-Sacc	$1\,\mathrm{g\,kg}^{-1}$
L. plantarum	Biosavor	$25\,\mathrm{mg\,L}^{-1}$
L. sporogenes	Biosavor	$25\,\mathrm{mg\,L}^{-1}$
L. casei	TSC-66	$2 \times 109\,\mathrm{CFU\,g}^{-1}$
L. bulgaricus	Biosavor	$25\,\mathrm{mg\,L}^{-1}$
Lactic acid bacteria	Probics-R	$1\,\mathrm{g\,kg}^{-1}$
Saccharomyces		
S. carlsbergensis	Babybiol	$1\,\mathrm{g\,kg}^{-1}$
S. cerevisiae	Ergomyces	$3 \times 108\,\mathrm{g}^{-1}$

and these enzymes are also beneficial in crude protein value enchantment. Enzymatic treatment methods to enhance forage and roughage quality have not been implemented until now. However, this may be the most promising approach in the future. Animals use enzymes to digest feed; these enzymes are either present in gut microbiota or produced by the animal itself. Hence, to expand the animal's digestive system, enzymatic supplements should be introduced in animal feed to increase digestion efficiency. Biological methods include the treatment of fungi such as white/brown rot for the breakdown of lignin, cellulose, and hemicellulose, enabling breaking or loosening the bonds between lignin and carbohydrates. Hence, this means breaking it into simpler forms and decomposing the lignin. Furthermore, applying these methods in some developing countries can be difficult because of the difficulty in growing and producing fungi or their enzymatic products on a large scale.

Biotechnological interventions to improve feed utilization

Biotechnology plays a crucial role in improving animal nutrition (Table 2), and its effective feed utilization and different roles of

Table 2 Summary of biotechnological developments in developing countries.

Areas in biotechnology	Implementation in developing countries
Livestock health	This is a very important area in developing countries because of high disease challenge • Focus is mainly on production of vaccines against "orphan" diseases of the developing world such as theileriosis and babesiosis, and also on the early diagnosis of disease • Live attenuated vaccines, subunit recombinant vaccines, combined vaccines, and naked DNA vaccines have been used to control livestock diseases • Development of heat-stable vaccines not requiring refrigeration
Animal breeding	One issue that is of importance in developing countries is improvement of animal productivity through cross-breeding while at the same time preserving genotypes of native animals that are adapted to the conditions in developing countries • There are efforts to characterize local breeds • Problems associated with inadequate infrastructure hinder the utilization of marker-assisted selection and single nucleotide polymorphism technologies in developing countries • Use of genomics to predict best cross-bred performance
Animal nutrition	High fiber and lignocelluloses in ruminant feeds lead to poor nutrient availability and inefficient utilization of feed. Grain-based feeds for monogastric animals are also deficient in some key nutrients • Digestibility of certain fodder plants has been improved by conventional selection methods • Research on lignin modification is ongoing at several research institutes but the technology has largely not been adopted in the field • There is extensive use of feed additives to correct nutrient deficiencies in animal feeds • Research on the manipulation of rumen microorganisms to increase feed efficiency and reduce methane gas production is still limited to developed countries
Livestock reproduction	A major limitation of livestock production systems in developing countries is low fertility and less genetic potential of animals • Artificial insemination is the most widely adopted reproduction technology. It has been used to upgrade the genetic potential and increase the productivity of animals. Technical as well as extension constraints limit the complete success of the technology in some areas • In many developing countries, multiple ovulation and embryo transfer is still confined to research centers. There is potential to use the technology to combat disease. A few countries are importing embryos on a commercial basis • Some developing countries have successfully cloned animals

biotechnology are explained in Fig. 3. Further, recombinant DNA technology has been widely accepted in developing new strains of bacteria to improve fiber digestibility and feed utilization (Rege, 1995). The potential of recombinant strategies can be explored in the following ways:

(i) Increasing the competitiveness of *F. succinogenes* and *Ruminococcus* cellulolytic organisms by conferring their capability to utilize xylose and pectins. Furthermore, allowing their suitable colonization of particulate matter.

(ii) Insertion of the cellulose-degrading gene into *B. ruminicola* (numerically predominant species).

(iii) Increasing the competitiveness of *C. polysaccharolyticum*, a cellulolytic bacterial species present in low numbers in the animal gut.

(iv) Inserting a cellulase gene into acid-tolerant bacteria (Lactobacillus) allows fiber fermentation at a ruminal pH less than 6.

(v) Development of a cutinase activity and efficiency in predominant gut bacteria.

(vi) Overcoming the rate-limiting effect of lignin by manipulating predominant rumen species to degrade arabinose side chains (Abraham and Pal, 2014).

Until now, only some antibiotic resistance genes have been successfully transferred and expressed in the rumen microbiome. Several factors such as the integration site, episomal maintenance in the host genome, posttranscriptional modifications, and requirement of essential DNA sequences to be able to encode and transport vital proteins synthesized by the gene of interest (GOI) play crucial roles in the transfer and expression of GOI in ruminal bacteria (Abraham and Pal, 2014). Genetically altered organisms can be used to synthesize amino acids (tryptophan, threonine, etc.) in huge quantities. Further, using these amino acids as feed additives will require less utilization of crude proteins by 5%, thus reducing feeding costs (Xiong et al., 2019).

Bionanotechnology for animal health and feed utilization

Bionanotechnology plays a key role in technology advancement, enabling development and sustainable influence on the food, agriculture, and health sectors. In this technology, nanomaterials are utilized that can generate safer and highly nutritious foods. Bionanotechnologies in food processing are superior to traditional technologies, as they have improved food product shelf life, prevented food contamination, and produced

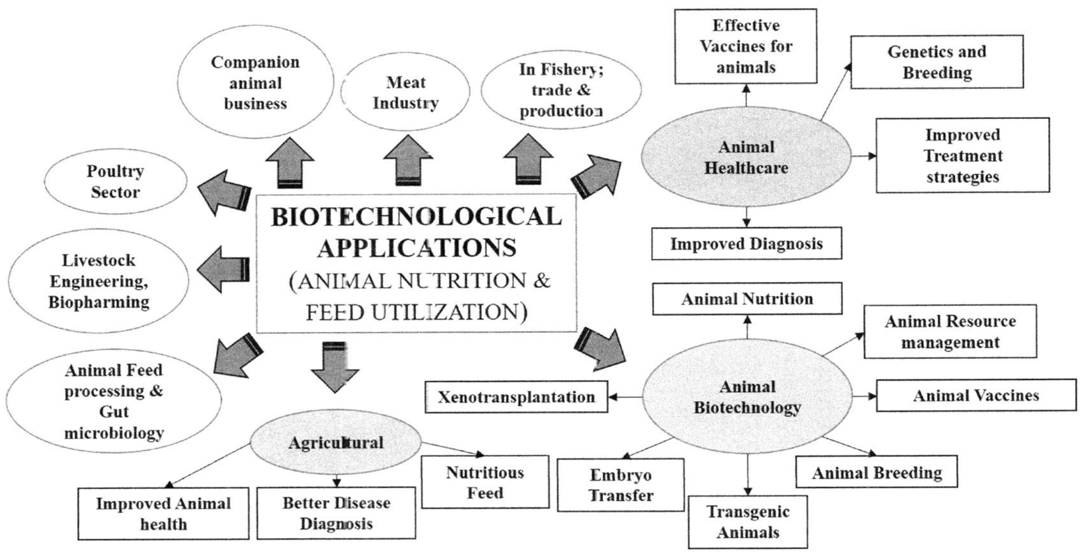

Fig. 3 Broad-spectrum application of biotechnology in animal nutrition and feed utilization.

nutritious food. Bionanotechnological applications improve feed utilization by enhancing feed bioavailability, which is gained by modifications in particle size, surface charge, and cluster formation (Nile et al., 2020). Nanobiotechnology offers potential application in livestock management by utilizing nanomaterial-based sensors to detect contaminants in the feed rapidly. There are many other benefits to these technologies, namely enrichment of nutrients in the feed, safety, reduction in agricultural inputs, and enhancement of output by absorbing miniature/nano-ranged nutrients from the soil, water, etc. The utility of nanomaterials in animal feed management and nutrition reduces the use of chemicals, minimizes nutrient losses, and increases yield. However, the basic objection in nanomaterial utilization is nanotoxicity and performance, but to overcome these issues, it is urgent to address them properly. After working on these factors, it is also very much needed to spread awareness among livestock industries to use novel nano-based technologies to manage animal nutrition, feed utilization, production, and health (Prasad et al., 2017; He et al., 2019).

In nanobiotechnological aspects, quantum dot nanoparticle (QD NP)-based products and processes are very beneficial for animal nutrition, feed utilization, and animal production and health. QD NPs can improve ova movement and sperm motility (Feugang et al., 2012). Furthermore, QD NPs can identify a male pig's gametes as they have unique biocompatibility and bioluminescence (Feugang et al., 2015). Weldon et al. (2018) suggested an urgent need to explore the health risk assessment by performing in vivo and in vitro experiments of cadmium-selenium-containing QD nanomaterials as commercial products, and they might pose potential toxicity. Tao et al. (2015) found that cytosine-phosphate-guanine QD nanocomposites hold promise to act as an optical probe that exhibits unique cellular uptake. Li et al. (2018) investigated aptamer-modified fluorescent-magnetic multifunctional nanoprobes for their application in the diagnosis and treatment of communicable diseases. Stanisavljevic et al. (2014) investigated CdTe QD interaction with the genomic DNA of a chicken. The genome sequence data of pathogens and humans can be used as biomarkers for the early detection of cancer. Nanoproteomics is also evolving, and it is a field in which proteomics is integrated with nanotechnology. It uses various nanomaterials, namely QDs, gold (Au) NPs, and carbon nanotubes (CNTs). Nanomaterials production, either commercially or industrially, is now a topic of concern because of their environmental risks on animal nutrition, feed utilization, production, and health (Singh et al., 2012).

Carbon-based nanoparticles (CNPs) exhibit low toxicity and are widely used for therapeutic applications; CNPs include different types, namely graphite, diamond, fullerenes, graphene, and carbon nanotubes. The biodegradability of CNPs is the major cause of concern for their broad-spectrum utility (Lamprecht et al., 2009). Mycotoxins cause the majority of diseases in domesticated animals. Gibson et al. (2011) studied various mycotoxins, namely aflatoxin-B1 and ochratoxin-A, by immobilizing them on nanodiamond substrates for detection and removal. Parvongnukul and Lumb (1978) investigated the utilization of polytetrafluoroethylene-graphite for coating hip prostheses in animals. Moreover, atomic absorption spectroscopy based on a graphite furnace was utilized to detect heavy metals (lead, cadmium, and selenium) in blood samples of cow, sheep, goat, and buffalo milk (Rahimi, 2013). Graphene is a carbon allotrope that is commonly used for drug delivery and cellular imaging. Further, Lee et al. (2015) stated that the use of reduced graphene oxide (rGO) coated polydimethylsiloxane film for the treatment and prevention of diseases is very effective, whereas Ye et al. (2015) reported the use of graphene oxide for the treatment and prevention of pseudorabies virus and porcine diarrhea virus. Fullerene (C60) is also an allotrope of carbon that is utilized to prevent disease in animals (Taylor et al., 1990). Shungite, a fullerene-based compound, is widely used as an antioxidant and antiinflammatory agent for animal healthcare (Sajo et al., 2017).

Human-made polymers such as dendrimers have enormous biomedical applications (Chahal et al., 2016). A poly-L-lysine dendrimer is biocompatible, biodegradable, water-soluble, and flexible as it has two primary amines, and they can also act as a carrier of the gene for therapeutic applications. On the other hand, a poly (propylene imine) dendrimer can be used for diagnostic applications as it has a 1; 4-diaminobutane/ethylenediamine core and a propylene imine monomer branching unit (Yandrapu et al., 2013; Shukla et al., 2010). Hence, dendrimers can become novel nanocarriers for drug delivery systems for the treatment of viral and parasitic infections in animals. They can also be used as a vaccine carrier to prevent schistosomiasis infection and increase interleukin-2 and interferon production. Moreover, dendrimer utilization as a nanocarrier is a promising alternative that can reduce drug toxicity, enhance drug solubility, and offer unique degradability, which will improve the strategy of leishmaniasis treatment in animals (Menezes et al., 2015).

Liposome nanoparticles are phospholipid bilayer colloidal structures and spherical vesicles. They have a unique encapsulation capability to load hydrophilic and lipophilic drugs in an aqueous phase. They are also permeable and have extraordinary charge density. Apart from these properties, they are biocompatible, less toxic, and biodegradable. Due to their unique properties, these nanoparticles can play a very important role in applications such as feed utilization, animal nutrition, livestock management, animal health, and production. Malam et al. (2009) investigated drug delivery based on liposomes to prevent and treat multidrug-resistant cancers. Sallovitz et al. (1998) formulated liposomal nanoparticles for bacterial prevention in animals. MacLeod and Prescott (1988) studied liposome-entrapped gentamicin for the treatment of mastitis in lactating cows. Leenders and de Marie (1996) investigated amphotericin-B-entrapped lipid formulation to treat fungal infections in animals. Later, Rhee et al. (2012) studied synthetic peptide hemagglutinin-derived DNA-liposomal nanoformulation to inhibit influenza virus. Miao et al. (2017) gave a rabies vaccine made of liposome and studied its effect in BALB/c mice, and the result suggested that these liposomal rabies vaccines are an immunopotentiator. Tada et al. (2018) investigated pneumococcal surface protein A, nanoformulated with 1, 2-dioleoyl-3-trimethylammonium-propane (DOTAP) and a cholesteryl 3-N-(dimethylaminoethyl)-carbamate (DC-chol) liposomal complex as a nasal vaccination, and the DOTAP/DC-chol liposome vaccine was very effective in the prevention of pneumococcal infection in animals.

Metallic nanoparticles are of various types, namely zero dimensional, nanowires, rods, nanosheets, and nanostructures. Physicobiochemical approaches can synthesize these metallic nanoparticles, and they can be engineered with functional groups of interest to bind with bioactive constituents (Shukla et al., 2010; Singh et al., 2012a; Singh, 2017, 2019). These nanoparticles exhibit enormous properties, making them an interesting agent to dodge drug resistance (a life-threatening issue). These nanoparticles also exhibit enormous applications toward animal nutrition, production, feed utilization, and health. Silver nanoparticles are widely used for diagnosis and treatment as antimicrobial, antiinflammatory, anticancer, and anti-angiogenesis elements. Metallic nanoparticles used in animal production are for fabricating sensors/nanobiosensors, bioimaging, drug delivery, and tissue engineering. Vaccines loaded with biologically synthesized AgNPs, when injected into mice and dogs, have shown promising results; thus, AgNP-loaded rabies vaccines are well established in veterinary vaccine (Stankic et al., 2016). For animal production and health, antibiotic-resistant bacteria are harmful. To overcome this issue, combination therapy is the best alternative, in which silver nanoparticles in combination with antibiotics are used to enhance antibacterial activity without showing any toxicity (Gurunathan et al., 2014, 2015). Smekalova et al. (2016) investigated a silver nanoparticle and gentamicin conjugate to inhibit Actinobacillus pleuropneumonia. Yuan et al. (2017) demonstrated silver nanoparticle utilization to treat mastitis in goats. Fondevila et al. (2009) demonstrated that silver nanoparticles reduce coliform bacteria in weaned piglets very efficiently. Gholami-Ahangaran and Zia-Jahromi (2013) studied silver nanoparticles to demonstrate the complete removal of aflatoxin in broiler chickens, and they showed excellent results. Pineda et al. (2012) investigated silver nanoparticles in broiler chickens as an antimicrobial growth-promoter supplement when given through drinking water using immunoglobulin (IgG). Bhanja et al. (2015) studied silver nanoparticles by administering them in chicken embryos, and from their findings, they suggested that silver nanoparticles alone or with amino acids will improve immunity in the chicken. However, Asgary et al. (2016) suggested using silver nanoparticles as an adjuvant in rabies vaccine, as they do not exhibit any toxicity. Gurunathan et al. (2018) suggested that silver nanoparticles can be successfully utilized to treat endometritis in cattle as these nanoparticles exhibit extraordinary antibacterial properties.

To deal with biomedical aspects in animal production, health, nutrition, and feed, polymeric nanoparticles can be successfully employed. As suggested by Yang and Alexandridis (2000), polymeric nanoparticles are a great tool as a drug carrier because of their unique targeting, biocompatibility, controlled release kinetics, and degradation. Cevher et al. (2015) investigated chitosan-pullulan nanocomposites as a vaccine, which can be delivered by the nasal tract for the treatment and prevention of diphtheria in livestock. Esposito et al. (2018) formulated a nanogel to heal wounded livestock by utilizing hyaluronic acid and retinyl palmitate. Margaroni et al. (2017) demonstrated PLGA (Poly (lactic-*co*-glycolic acid))-sLiAg-MPLA conjugate vaccine for visceral leishmaniasis by utilizing soluble Leishmania infantum antigens (sLiAg) and monophosphoryl lipid A (MPLA); the result obtained from their study was very much effective. Kang et al. (2017) reported silver nanoparticle-siRNA-glycol chitosan-taurocholic acid nanocomposite formulated from Akt2 siRNA to treat colorectal liver metastases. Gu et al. (2019) demonstrated Angelica Sinensis polysaccharide (ASP)-encapsulated PLGA nanoparticles for the treatment and prevention of infectious diseases in livestock animals.

Prospects

For a very long time, animal feed utilization and nutrition have proved to be one of the greatest challenges to the production and maintenance of animals, with limits arising from livestock feed's value and capacity. The majority of the livestock feeds are fibrous but have a varying level of crude protein, digestibility, and nutritional values. Animal nutritionists have utilized biotechnological aspects to improve feed's nutritional values by enhancing digestibility and acceptability by livestock. However, apart from targeting only livestock, manipulating the rumen microbial population through chemical treatment, microbial balancing, and genetic manipulation of feed is probably the most reliable means of enhancing the dietary value of feed fodder and also effectively raising the utilization of feed by livestock. The current methods used in treating animal feed to improve intake and digestibility are mainly chemical, physical, mechanical, and biological. For the effective production of animal products, manipulation of the livestock environment or their genotype remains a major component of the future means to provide sufficient food for the growing world sustainably. The use of biotechnological approaches plays a crucial part in reaching these objectives. As the biotechnological methods continue to evolve with time, biotechnology's contribution is presumed to increase in the future for suitable animal nutrition and effective feed utilization.

Conclusions

In conclusion, to sustain the livestock industry, it is very necessary to constantly maintain the quality and quantity of nutrition values in their feed. For this purpose, the utilization of newly evolved biotechnological applications is very much needed to promote the development of current and future livestock production globally. In livestock production, the expensive part is making concentrate feeds, and to overcome this problem, alternate feed resources such as grazing in open fields, forages, and easy availability of plant-based materials should be utilized properly. The quality of nutrition in feed can be improved using various biotechnological approaches for better utilization by livestock. Biotechnological approaches such as optimizing high-quality forages such as genetically modified forages, enhancement of feed nutritional content by genetic manipulation, and increasing the efficiency of digestible cell wall components by genetic manipulation in gut microbes can be used to increase feed utilization by livestock. Thus, the general focus should be on fulfilling the nutritional demands of livestock through genetic alteration in the feed.

Moreover, to get economically viable and biologically safe animal feeds/products, chemical additives should be minimal. For this, further improvement in biotechnology applications such as pre- and probiotics could improve animal performance. Further, for improvement and getting safe animal feed and products, nanotechnological applications can also be involved by using nanoparticles accompanied by genetic manipulations in feeds given to livestock, which will improve the nutritious value of feed. This will result in excellent growth and performance, and it will also result in a waste reduction of feed. However, nanotechnology's nanotoxicological aspects need to be studied before utilizing these applications for feed improvement and utilization.

Biotechnology and nanotechnology have revolutionized animal production and health by providing proper nutrition and feed utilization through modern tools. Nanomaterials and biosensors can also help solve problems pertaining to animal nutrition requirements for improving their health and production. It is very urgent to maintain and improve animal nutrition and utilization, which will enhance animal production and health, as this is the backbone of the economy. The current global status is that there is much demand in animal production due to the increasing human population, and to fulfill the demand, the use of antibiotics is increasing to improve yield. However, in the future this will result in microbial antibiotic resistance, and thus it will cause severe problem to both animals as well as humans (Singh, 2019). Hence, nanomaterial utility can help overcome these difficulties due to their enormous applications. However, the major problem with using these nanomaterials is that the mode of mechanism data related to its utility in animals is very scant and needs to be explored. Thus, we suggest that studies related to nanomaterial utilization for enhancing animal nutrition and feed utilization must be carried out.

References

Abdel-Aziz, N.A., Salem, A.Z.M., El-Adawy, M.M., Camacho, L.M., Kholif, A.E., Elghandour, M.M.Y., Borhami, B.E., 2015. Biological treatments as a mean to improve feed utilization in agriculture animals—an overview. J. Integr. Agric. 14, 534–543. https://doi.org/10.1016/S2095-3119(14)60829-7.

Abraham, H., Pal, S.K., 2014. Animal biotechnology options in improving livestock production in the horn of Africa. Int. J. Interdiscip. Multidiscip. Stud. 1, 1–8.

Adesehinwa, A.O.K., 2008. Energy and protein requirements of pigs and the utilization of fibrous feedstuffs in Nigeria: a review. Afr. J. Biotechnol. 7, 4798–4806.

Ajila, C.M., Brar, S.K., Verma, M., Tyagi, R.D., Godbout, S., Valéro, J.R., 2012. Bioprocessing of agro-byproducts to animal feed. Crit. Rev. Biotechnol. 32, 382–400.

Alagawany, M., Elnesr, S.S., Farag, M.R., 2018. The role of exogenous enzymes in promoting growth and improving nutrient digestibility in poultry. Iran. J. Vet. Res. 19, 157–164.

Alderman, C., 1989. Broadening horizons. Nurs. Stand. 4, 24–25.

Alemneh, T., 2019. The application of biotechnology on livestock feed improvement. Arch. Biomed. Eng. Biotechnol. 1, 1–7.

Archer, J.A., Richardson, E.C., Herd, R.M., Arthur, P.F., 1999. Potential for selection to improve efficiency of feed use in beef cattle: a review. Aust. J. Agric. Res. 50, 147.

Asgary, V., Shoari, A., Baghbani-Arani, F., Shandiz, A.S., Khosravy, M.S., Janani, A., Bigdeli, R., Bashar, R., Cohan, R., 2016. Green synthesis and evaluation of silver nanoparticles as adjuvant in rabies veterinary vaccine. Int. J. Nanomedicine 11, 3597–3605.

Asmare, B., 2014. Biotechnological advances for animal nutrition and feed improvement. World J. Agric. Res. 2, 115–118.

Asmare, B., Mekuriaw, Y., 2015. Potential of biotechnology for livestock feed improvement: a review. J. Agric. Environ. Sci. 1, 57–69.

Beigh, Y.A., Ganai, A.M., Ahmad, H.A., 2016. Prospects of complete feed system in ruminant feeding: a review. Vet. World 10, 424–437.

Belay, G., Negesse, T., 2018. Feed resource availability and their nutrient contribution for livestock evaluated using feed assessment tool (FEAST) in Burie Zuria District, North Western Ethiopia. Agric. Res. Technol. Open Access J. 17, 556022.

Bhanja, S., Hotowy, A., Mehra, M., Sawosz, E., Pineda, L., Vadalasetty, K., Kurantowicz, N., Chwalibog, A., 2015. In ovo administration of silver nanoparticles and/or amino acids influence metabolism and immune gene expression in chicken embryos. Int. J. Mol. Sci. 16, 9484–9503.

Bigliardi, B., Nosella, A., Verbano, C., 2005. Business models in Italian biotechnology industry: a quantitative analysis. Technovation 25, 1299–1306.

Bonneau, M., Laarveld, B., 1999. Biotechnology in animal nutrition, physiology and health. Livest. Prod. Sci. 59, 223–241.

Bryan, A.H., 1957. Trace elements in human and animal nutrition. Am. J. Public Health Nations Health 47, 496.

Cevher, E., Salomon, S.K., Somavarapu, S., Brocchini, S., Alpar, H.O., 2015. Development of chitosan–pullulan composite nanoparticles for nasal delivery of vaccines: in vivo studies. J. Microencapsul. 32, 769–783.

Chahal, J.S., Khan, O.F., Cooper, C.L., McPartlan, J.S., Tsosie, J.K., Tilley, L.D., Sidik, S. M., Lourido, S., Langer, R., Bavari, S., Ploegh, H.L., Anderson, D.G., 2016. Dendrimer-RNA nanoparticles generate protective immunity against lethal Ebola, H1N1 influenza, and Toxoplasma gondii challenges with a single dose. Proc. Natl. Acad. Sci. 113, E4133–E4142.

Cheah, P.L., Looi, L.M., 2008. Detection of the human papillomavirus in cervical carcinoma: a comparison between non-isotopic in-situ hybridisation and polymerase chain reaction as methods for detection in formalin-fixed, paraffin-embedded tissues. Malays. J. Pathol. 30, 37–42.

Crews, D.H.D., 2005. Genetics of efficient feed utilization and national cattle evaluation: a review. Genet. Mol. Res. 4, 152–165.

de Rufino, L.M.A., Detmann, E., Gomes, D.Í., dos Reis, W.L.S., Batista, E.D., de Valadares Filho, S.C., Paulino, M.F., 2016. Intake, digestibility and nitrogen utilization in cattle fed tropical forage and supplemented with protein in the rumen, abomasum, or both. J. Anim. Sci. Biotechnol. 7, 11.

den Hartog, L.A., Sijtsma, S.R., 2013. Challenges and Opportunities in Animal Feed and Nutrition. http://edepot.wur.nl/306279.

Egan, A.R., 2017. Animal nutrition and feed science. Engineering 3, 586–587.

Esposito, E., Pecorelli, A., Sguizzato, M., Drechsler, M., Mariani, P., Carducci, F., Cervellati, F., Nastruzzi, C., Cortesi, R., Valacchi, G., 2018. Production and characterization of nanoparticle based hyaluronate gel containing retinyl palmitate for wound healing. Curr. Drug Deliv. 15, 1172–1182.

FAO, 2012. Balanced feeding for improving livestock productivity—increase in milk production and nutrient use efficiency and decrease in methane emission. In: Garg, M. R. (Ed.), FAO Animal Production and Health Paper No. 173. Rome, Italy.

Feugang, J.M., Youngblood, R.C., Greene, J.M., Fahad, A.S., Monroe, W.A., Willard, S. T., Ryan, P.L., 2012. Application of quantum dot nanoparticles for potential non-invasive bio-imaging of mammalian spermatozoa. J. Nanobiotechnol. 10, 45.

Feugang, J.M., Youngblood, R.C., Greene, J.M., Willard, S.T., Ryan, P.L., 2015. Self-illuminating quantum dots for non-invasive bioluminescence imaging of mammalian gametes. J. Nanobiotechnol. 13, 38.

Fondevila, M., Herrer, R., Casallas, M.C., Abecia, L., Ducha, J.J., 2009. Silver nanoparticles as a potential antimicrobial additive for weaned pigs. Anim. Feed Sci. Technol. 150, 259–269.

Gaggìa, F., Mattarelli, P., Biavati, B., 2010. Probiotics and prebiotics in animal feeding for safe food production. Int. J. Food Microbiol. 141, S15–S28.

Garnsworthy, P.C., Wiseman, J., 2012. Recent Advances in Animal Nutrition. Nottingham University Press, Nottingham, UK.

Gholami-Ahangaran, M., Zia-Jahromi, N., 2013. Nanosilver effects on growth parameters in experimental aflatoxicosis in broiler chickens. Toxicol. Ind. Health 29, 121–125.

Gibson, N.M., Luo, T.J.M., Brenner, D.W., Shenderova, O., 2011. Immobilization of mycotoxins on modified nanodiamond substrates. Biointerphases 6, 210–217.

Göncü, S., Güngör, C., 2018. The innovative techniques in animal husbandry. In: Yucel, B. (Ed.), Animal Husbandry and Nutrition. IntechOpen, United Kingdom, pp. 3–22.

Gu, P., Wusiman, A., Wang, S., Zhang, Y., Liu, Z., Hu, Y., Liu, J., Wang, D., 2019. Polyethylenimine-coated PLGA nanoparticles-encapsulated Angelica sinensis polysaccharide as an adjuvant to enhance immune responses. Carbohydr. Polym. 223, 115128.

Gurunathan, S., Han, J.W., Kwon, D.-N., Kim, J.-H., 2014. Enhanced antibacterial and anti-biofilm activities of silver nanoparticles against gram-negative and gram-positive bacteria. Nanoscale Res. Lett. 9 (373), 1–17.

Gurunathan, S., Han, J.W., Park, J.-H., Kim, E.S., Choi, Y.-J., Kwon, D.-N., Kim, J.-H., 2015. Reduced graphene oxide-silver nanoparticle nanocomposite: a potential anticancer nanotherapy. Int. J. Nanomedicine 10, 6257.

Gurunathan, S., Choi, Y.-J., Kim, J.-H., 2018. Antibacterial efficacy of silver nanoparticles on endometritis caused by Prevotella melaninogenica and Arcanobacterum pyogenes in dairy cattle. Int. J. Mol. Sci. 19, 1210.

Guyer, P.Q., Owen, F.G., 1979. G79-476 Composition of Feeds in Relation to Cattle Nutrition. Historical Materials from University of Nebraska-Lincoln Extension, p. 290.

Halas, V., Gerrits, W.J., Milgen, J.V., 2018. Models of feed utilization and growth for monogastric animals. In: Moughan, P.J., Hendriks, W.H. (Eds.), Feed Evaluation Science. Wageningen Academic Publishers, The Netherlands, pp. 423–456.

Halmemies-Beauchet-Filleau, A., Rinne, M., Lamminen, M., Mapato, C., Ampapon, T., Wanapat, M., Vanhatalo, A., 2018. Review: alternative and novel feeds for ruminants: nutritive value, product quality and environmental aspects. Animal 12, s295–s309.

He, X., Deng, H., Hwang, H., 2019. The current application of nanotechnology in food and agriculture. J. Food Drug Anal. 27, 1–21.

Hooper, P.G., Sharp, R., Rowlinson, P., Armstrong, D.G., 1989. An assessment of the potential of a lactobacilli inoculant as a silage additive. Proc. Br. Soc. Anim. Prod. 1989, 63.

Huang, Q., Liu, X., Zhao, G., Hu, T., Wang, Y., 2018. Potential and challenges of tannins as an alternative to in-feed antibiotics for farm animal production. Anim. Nutr. 4, 137–150.

Kaasschieter, G.A., de Jong, R., Schiere, J.B., Zwart, D., 1992. Towards a sustainable livestock production in developing countries and the importance of animal health strategy therein. Vet. Q. 14, 66–75.

Kang, S.H., Revuri, V., Lee, S.-J., Cho, S., Park, I.-K., Cho, K.J., Bae, W.K., Lee, Y., 2017. Oral siRNA delivery to treat colorectal liver metastases. ACS Nano 11, 10417–10429.

Kier, L.D., Petrick, J.S., 2008. Safety assessment considerations for food and feed derived from plants with genetic modifications that modulate endogenous gene expression and pathways. Food Chem. Toxicol. 46, 2591–2605.

Lamprecht, C., Liashkovich, I., Neves, V., Danzberger, J., Heister, E., Rangl, M., Coley, H. M., McFadden, J., Flahaut, E., Gruber, H.J., Hinterdorfer, P., Kienberger, F., Ebner, A., 2009. AFM imaging of functionalized carbon nanotubes on biological membranes. Nanotechnology 20, 434001.

Lee, C., Kim, J., Zhang, Y., Jeon, M., Liu, C., Song, L., Lovell, J.F., Kim, C., 2015. Dual-color photoacoustic lymph node imaging using nanoformulated naphthalocyanines. Biomaterials 73, 142–148.

Leenders, A.C., de Marie, S., 1996. The use of lipid formulations of amphotericin B for systemic fungal infections. Leukemia 10, 1570–1575.

Li, L., Li, Q., Liao, Z., Sun, Y., Cheng, Q., Song, Y., Song, E., Tan, W., 2018. Magnetism-resolved separation and fluorescence quantification for near-simultaneous detection of multiple pathogens. Anal. Chem. 90, 9621–9628.

Liao, S.F., Nyachoti, M., 2017. Using probiotics to improve swine gut health and nutrient utilization. Anim. Nutr. 3, 331–343.

Loosli, J.K., 1980. Recent advances in animal nutrition—1979. Anim. Feed Sci. Technol. 5, 351–352.

MacLeod, D.L., Prescott, J.F., 1988. The use of liposomally-entrapped gentamicin in the treatment of bovine Staphylococcus aureus mastitis. Can. J. Vet. Res. 52, 445–450.

Madsen, K., Cornish, A., Soper, P., McKaigney, C., Jijon, H., Yachimec, C., Doyle, J., Jewell, L., De Simone, C., 2001. Probiotic bacteria enhance murine and human intestinal epithelial barrier function. Gastroenterology 121, 580–591.

Mahima, Verma, A.K., Kumar, A., Kumar, V., Roy, D., 2012. Scope of biotechnology in animal nutrition. Asian J. Anim. Sci. 6, 316–318.

Malam, Y., Loizidou, M., Seifalian, A.M., 2009. Liposomes and nanoparticles: nanosized vehicles for drug delivery in cancer. Trends Pharmacol. Sci. 30, 592–599. https://doi.org/10.1016/j.tips.2009.08.004.

Margaroni, M., Agallou, M., Athanasiou, E., Kammona, O., Kiparissides, C., Gaitanaki, C., Karagouni, E., 2017. Vaccination with poly(D,L-lactide-co-glycolide) nanoparticles loaded with soluble Leishmania antigens and modified with a TNFα-mimicking peptide or monophosphoryl lipid A confers protection against experimental visceral leishmaniasis. Int. J. Nanomedicine 12, 6169–6184.

McSweeney, C., Dalrymple, B., Gobius, K., Kennedy, P., Krause, D., Mackie, R., Xue, G., 1999. The application of rumen biotechnology to improve the nutritive value of fibrous feedstuffs: pre- and post-ingestion. Livest. Prod. Sci. 59, 265–283.

Menezes, J.P.B., Guedes, C.E.S., Petersen, A.L.O.A., Fraga, D.B.M., Veras, P.S.T., 2015. Advances in development of new treatment for Leishmaniasis. Biomed. Res. Int. 2015, 1–11.

Mertz, W., 1987. Trace Elements in Human and Animal Nutrition, fifth ed. Academic Press, Elsevier.

Miao, L., Yang, Y., Yan, M., Li, Y., Zhao, J., Guo, J., Zheng, D., 2017. Enhanced immune response to rabies viruses by the use of a liposome adjuvant in vaccines. Viral Immunol. 30, 727–733.

Miller, E.L., 2002. Protein nutrition requirements of farmed livestock and dietary supply. In: Protein Sources for the Animal Feed Industry, pp. 1–27. http://www.fao.org/docrep/007/y5019e/y5019e00.htm#Contents.

Morris, J.G., 2001. Vitamins in animal and human nutrition. Am. J. Clin. Nutr. 74, 413.

National Dairy Board, 2012. Nutritive Value of Commonly Available Feeds and Fodders in India. National Dairy Development Board, pp. 74–75.

National Research Council, 2001. Nutrient Requirements of Dairy Cattle, seventh ed. The National Academies Press, Washington, DC.

Nile, S.H., Baskar, V., Selvaraj, D., Nile, A., Xiao, J., Kai, G., 2020. Nanotechnologies in food science: applications, recent trends, and future perspectives. Nano-Micro Lett. 12, 45.

Onteru, S.K., Ampaire, A., Rothschild, M.F., 2010. Biotechnology developments in the livestock sector in developing countries. Biotechnol. Genet. Eng. Rev. 27, 217–228.

Owen, E., Jayasuriya, M.C.N., 1989. Use of crop residues as animal feeds in developing countries. Res. Dev. Agric. 6, 129–138.

Parvongnukul, K., Lumb, W.V., 1978. Evaluation of polytetrafluoroethylene-graphite-coated total hip prostheses in goats. Am. J. Vet. Res. 39, 221–228.

Perry, F.G., 1995. Biotechnology in animal feeds and animal feeding: an overview. In: Wallace, R.J., Chesson, A. (Eds.), Biotechnology in Animal Feeds and Animal Feeding. Wiley-VCH Verlag GmbH, Weinheim, Germany, pp. 1–15.

Pineda, L., Chwalibog, A., Sawosz, E., Lauridsen, C., Engberg, R., Elnif, J., Hotowy, A., Sawosz, F., Gao, Y., Ali, A., Moghaddam, H.S., 2012. Effect of silver nanoparticles on growth performance, metabolism and microbial profile of broiler chickens. Arch. Anim. Nutr. 66, 416–429.

Prasad, R., Bhattacharyya, A., Nguyen, Q.D., 2017. Nanotechnology in sustainable agriculture: recent developments, challenges, and perspectives. Front. Microbiol. 8 (1014), 1–13.

Rahimi, E., 2013. Lead and cadmium concentrations in goat, cow, sheep, and buffalo milks from different regions of Iran. Food Chem. 136, 389–391.

Rege, J.E.O., 1995. Application of Biotechnology in Genetic Improvement, Characterization and Conservation of Livestock. Cgiar https://cgspace.cgiar.org/handle/10568/16478.

Rhee, J.W., Kim, D., Park, B.K., Kwon, S., Cho, S., Lee, I., Park, M.-S., Seo, J.-N., Kim, Y.-S., Choi, H.S., Lee, Y., Kwon, H.-J., 2012. Immunization with a hemagglutinin-derived synthetic peptide formulated with a CpG-DNA-liposome complex induced protection against lethal influenza virus infection in mice. PLoS One 7, e48750.

Sajo, M.E.J., Kim, C.-S., Kim, S.-K., Shim, K.Y., Kang, T.-Y., Lee, K.-J., 2017. Antioxidant and anti-inflammatory effects of shungite against ultraviolet B irradiation induced skin damage in hairless mice. Oxidative Med. Cell. Longev. 2017, 1–11.

Sallam, S.M.A., Kholif, A.E., Amin, K.A., El-Din, A.N.M.N., Attia, M.F.A., Matloup, O. H., Anele, U.Y., 2020. Effects of microbial feed additives on feed utilization and growth performance in growing Barki lambs fed diet based on peanut hay. Anim. Biotechnol. 31 (5), 447–454.

Sallovitz, J.M., Zonco Menghini, M.I., Lanusse, C.E., 1998. Impact of liposomes as delivery systems in veterinary medicine. Vet. Res. 29 (5), 409–430.

Sarnklong, C., Cone, J.W., Pellikaan, W., Hendriks, W.H., 2010. Utilization of Rice straw and different treatments to improve its feed value for ruminants: a review. Asian Australas. J. Anim. Sci. 23, 680–692.

Segal, S.P., Knight, T.J., Beitz, D.C., 2012. Use of biotechnology to increase food production and nutritional value. In: Temple, N.J., Wilson, T., Jacobs, D.R. (Eds.), Nutritional Health. Humana Press, Totowa, NJ, pp. 505–522.

Shukla, V.K., Singh, R.P., Pandey, A.C., 2010. Black pepper assisted biomimetic synthesis of silver nanoparticles. J. Alloys Compd. 507, L13–L16.

Singh R.P., 2017. Application of nanomaterials towards development of nanobiosensors and their utility in agriculture, Springer Publisher, New York, USA, pp 293-303. In book

"Nanotechnology: An Agricultural Paradigm—Ch. 14" Eds.: Prasad, Ram, Kumar, Manoj, Kumar, Vivek (Eds.).

Singh, R.P., 2019. Utility of nanomaterials in food safety. In: Singh, R.L., Mondal, S. (Eds.), Food Safety and Human Health. Elsevier, pp. 285–318.

Singh, R.P., Choi, J.W., Tiwari, A., Pandey, A.C., Turner, A.P.F., 2012. Utility and potential application of nanomaterials in medicine. In: Tiwari, A., Ramalingam, M., Kobayashi, H. (Eds.), Biomedical Materials and Diagnostic Devices. John Wiley & Sons, Inc, Hoboken, NJ, USA, https://doi.org/10.1002/9781118523025.ch7.

Singh, R.P., Kumar, K., Rai, R., Tiwari, A., Choi, J.W., Pandey, A.C., 2012a. Synthesis, characterization of metal oxide-based nanomaterials and its application in biosensing. In: Rai, R. (Ed.), Synthesis, Characterization and Application of Smart Material—Ch. 11. International. Nova Science Publishers, Inc, USA, pp. 225–238.

Singh, M., Lathwal, S.S., Kotresh Prasad, C., Dey, D., Gupta, A., Saini, M., Lathwal, I., Sharma, B., Kumar, M., Sharma, V., 2019. Availability of feed sources and nutritional status of Hariana cattle in different seasons in the breeding tract. Biol. Rhythm Res. 52 (6), 1–7. https://doi.org/10.1080/09291016.2019.1607222.

Smekalova, M., Aragon, V., Panacek, A., Prucek, R., Zboril, R., Kvitek, L., 2016. Enhanced antibacterial effect of antibiotics in combination with silver nanoparticles against animal pathogens. Vet. J. 209, 174–179.

Spears, J.W., Weiss, W.P., 2014. Invited review: mineral and vitamin nutrition in ruminants 1. Prof. Anim. Sci. 30, 180–191.

Stanisavljevic, M., Chomoucka, J., Dostalova, S., Krizkova, S., Vaculovicova, M., Adam, V., Kizek, R., 2014. Interactions between CdTe quantum dots and DNA revealed by capillary electrophoresis with laser-induced fluorescence detection. Electrophoresis 35, 2587–2592.

Stankic, S., Suman, S., Haque, F., Vidic, J., 2016. Pure and multi metal oxide nanoparticles: synthesis, antibacterial and cytotoxic properties. J. Nanobiotechnol. 14, 73.

Tada, R., Suzuki, H., Takahashi, S., Negishi, Y., Kiyono, H., Kunisawa, J., Aramaki, Y., 2018. Nasal vaccination with pneumococcal surface protein A in combination with cationic liposomes consisting of DOTAP and DC-chol confers antigen-mediated protective immunity against Streptococcus pneumoniae infections in mice. Int. Immunopharmacol. 61, 385–393.

Tao, Y., Zhang, Y., Ju, E., Ren, J., Qu, X., 2015. One-step synthesized immunostimulatory oligonucleotides-functionalized quantum dots for simultaneous enhanced immunogenicity and cell imaging. Colloids Surf. B Biointerfaces 126, 585–589.

Taylor, R., Hare, J.P., Abdul-Sada, A.K., Kroto, H.W., 1990. Isolation, separation and characterisation of the fullerenes C60 and C70: the third form of carbon. J. Chem. Soc. Chem. Commun., 1423–1425. https://doi.org/10.1039/C39900001423.

Thornton, P.K., 2010. Livestock production: recent trends, future prospects. Philos. Trans. R. Soc. B Biol. Sci. 365, 2853–2867.

Tona, G.O., 2018. Current and future improvements in livestock nutrition and feed resources. In: Yucel, B. (Ed.), Animal Husbandry and Nutrition. InTechOpen, United Kingdom.

Umberger, S.H., 2009. Feeding Sheep. VCE Publishers, pp. 410–853.

van Deynze, A., Bradford, K.J., van Eenennaam, A., 2004. Crop Biotechnology: Feeds for Livestock. University of California, Agriculture and Natural Resources.

Wanapat, M., Chenost, M., Munoz, F., Kayouli, C., 1996. Methods for improving the nutritive value of fibrous feed: treatment and supplementation. Ann. Zootech. 45, 89–103.

Wanapat, M., Kang, S., Polyorach, S., 2013. Development of feeding systems and strategies of supplementation to enhance rumen fermentation and ruminant production in the tropics. J. Anim. Sci. Biotechnol. 4, 32.

Weldon, B.A., Griffith, W.C., Workman, T., Scoville, D.K., Kavanagh, T.J., Faustman, E. M., 2018. In vitro to in vivo benchmark dose comparisons to inform risk assessment of quantum dot nanomaterials. Wiley Interdiscip. Rev. Nanomed. Nanobiotechnol. 10, e1507.

Wilson, J., 1950. How to feed those low-grade spuds. Root Crops 5, 37.

Xiong, X., Tan, B., Song, M., Ji, P., Kim, K., Yin, Y., Liu, Y., 2019. Nutritional intervention for the intestinal development and health of weaned pigs. Front. Vet. Sci. 6, 46.

Yandrapu, S.K., Kanujia, P., Chalasani, K.B., Mangamoori, L., Kolapalli, R.V., Chauhan, A., 2013. Development and optimization of thiolated dendrimer as a viable mucoadhesive excipient for the controlled drug delivery: an acyclovir model formulation. Nanomed. Nanotechnol. Biol. Med. 9, 514–522.

Yang, L., Alexandridis, P., 2000. Physicochemical aspects of drug delivery and release from polymer-based colloids. Curr. Opin. Colloid Interface Sci. 5, 132–143.

Ye, S., Shao, K., Li, Z., Guo, N., Zuo, Y., Li, Q., Lu, Z., Chen, L., He, Q., Han, H., 2015. Antiviral activity of graphene oxide: how sharp edged structure and charge matter. ACS Appl. Mater. Interfaces 7, 21571–21579.

Yirga, H., 2015. The use of probiotics in animal nutrition. J. Probiotics Heal. 03 (2), 1–10.

Yuan, Y.-G., Peng, Q.L., Gurunathan, S., 2017. Effects of silver nanoparticles on multiple drug-resistant strains of Staphylococcus aureus and Pseudomonas aeruginosa from mastitis-infected goats: an alternative approach for antimicrobial therapy. Int. J. Mol. Sci. 18, 569.

Zaman, A., 2013. Livestock improvement using biotechnology: world and Bangladesh perspective. J. Livest. Sci. 4, 78–96.

CHAPTER FIFTEEN

Biotechnology in livestock reproduction

Bhoopendra Singh[a], Kaushlendra Pratap Singh[b], Mozammel Hoque[c], and Sukanta Mondal[d]

[a]Department of Veterinary Gynaecology & Obstetrics, College of Veterinary Science & Animal Husbandry, Acharya Narendra Deva University of Agriculture & Technology, Ayodhya, Uttar Pradesh, India
[b]Government Veterinary Hospital, Deornia, Bareilly, Uttar Pradesh, India
[c]Division of Surgery, ICAR-Indian Veterinary Research Institute, Izatnagar, Uttar Pradesh, India
[d]Animal Physiology Division, ICAR-National Institute of Animal Nutrition and Physiology, Adugodi, Bengaluru, Karnataka, India

Introduction

Farm animals and their products contribute significantly to human nutrition, health, research, and improved life expectancy. Reproductive inefficiency is one of the most common causes of economic losses in animal industries and it is realized throughout the world. Despite considerable development in the field of reproductive physiology in recent years, infertility due to low conception rate and high embryonic mortality rate remains a major problem. Biotechnology has given us unprecedented opportunities to improve the reproductive efficiency of livestock and penetrate their genetic ceiling to enhance milk production. To meet future needs and to sustain agricultural production and agricultural research, there is a need to use all emerging technologies, especially modern reproductive biotechnologies.

Various reproductive technologies such as artificial insemination (AI), sire evaluation, marker-assisted selection (MAS), semen sexing, estrous synchronization, multiple ovulation and embryo transfer (MOET), in vitro embryo production (IVEP), gamete intra fallopian tube transfer (GIFT), intra cytoplasmic sperm injection (ICSI), embryo sexing, embryo cloning, production and multiplication of transgenic animals, and stem cell technology singly or in combination can play prominent roles in increasing the productivity of dairy animals (Mishra and Prashad, 2005).

https://doi.org/10.1016/B978-0-12-822265-2.00013-2

Artificial insemination (AI)

AI is a technique in which semen is manually placed inside the reproductive tract of the female by a method other than natural mating. It is one of the most common assisted reproduction technologies (ART), in which offspring are produced by facilitating the meeting of gametes (spermatozoa and oocytes). It is a simple, successful, economical, and most important reproductive biotechnology that helps in the dissemination of desired genes in a population and plays a key role in breeding animals. Presently, this is an established technology in commercial dairy cattle programs in both developed and developing countries.

AI is the most commonly used ART in livestock, and it revolutionized the animal breeding industry during the 20th century. AI is the most common method of breeding in intensively kept domestic livestock, such as dairy cattle (approximately 80% in Europe and North America), pigs (more than 90% in Europe and North America), and turkeys (almost 100% in intensive production). AI has also been used in the conservation and breeding of rare or endangered species. In animals, the other ARTs are confined to specialist applications or for research purposes.

History of AI

In a bitch, the first successful insemination was done by Spallanzani (1784). Pioneering efforts to establish AI as a practical procedure were begun in Russia in 1899 by Ivanoff (1922). Ivanoff studied AI in domestic farm animals, dogs, foxes, rabbits, and poultry. He also studied AI in horses and published papers in the *Journal of Agriculture Science* in 1922. Ivanoff developed semen extenders and gave training to technicians for the selection and multiplication of superior stallions through AI. Most of the AI work in Russia was taken over later by Milovanov (1938). He established most of the projects related to sheep and cattle breeding. Milovanov designed and made practical artificial vaginas and other items in his own workshop.

The Japanese scientist Ishikawa studied with Ivanoff. After returning to Japan, he started a similar program in horses in 1912 (Nishikawa, 1962). Later on, AI was also performed in cattle, sheep, goats, swine, and poultry. Following the report from Ivanoff (1922), AI became widespread in the Western world. Edward Sorensen in collaboration with Gylling-Holm established the first cooperative dairy AI organization in Denmark in 1936. The method of rectovaginal fixation of the cervix, allowing semen to be deposited deeply into the cervix or into the body of the uterus, was developed by Danish veterinarians. This technique had an advantage because fewer sperm were required for the insemination of each cow. AI was started in the 1940s in the United States. The procedures developed in the United States became established worldwide (Salisbury and Lodge, 1978). The use of frozen semen (Polge et al., 1949) revolutionized AI through the worldwide transport of semen.

In India, AI was introduced in the early 1940s, but its wider application started in the late 1950s. At the initial stage, AI was used to improve indigenous breeds in the late 1960s. AI was used to introduce cross-breeding through 60 functional frozen semen stations and more than 54,000 AI centers; this is the latest breeding infrastructure in the world.

Merits and demerits

In animals, AI was developed to control the spread of disease by avoiding the transport of animals with potential pathogens to other animal units and by avoiding physical contact between individuals. The use of semen extenders containing antibiotics also helped to prevent the transmission of bacterial diseases. The merits and demerits of AI are as follows:

Merits

i. AI technology maximizes the use of outstanding males.
ii. It helps in the dissemination of superior genetic material.
iii. AI prevents the spread of infectious or contagious diseases that can be passed from one animal to other when animals are in close contact or share the same environment.
iv. It improves the rate and efficiency of genetic selection.
v. It helps in the introduction of new genetic material by the import of semen rather than live animals and thus reduces international transport costs.
vi. It enables breeding between animals in different geographic locations.
vii. It enables the use of frozen semen even after the death of the donor.
viii. Breeding can occur in the presence of physical, physiological, or behavioral abnormalities.
ix. AI reduces the risk of spreading sexually transmitted diseases.
x. AI is a powerful tool when linked to other reproductive biotechnologies such as sperm cryopreservation and sperm sexing.
xi. It can be used in the conservation of rare breeds or endangered species.

Demerits

i. Genetic abnormalities, if spread, do so at a very fast rate in AI programs.
ii. AI requires special facilities such as labs and equipment.
iii. If inseminators are not careful, they spread infections from animal to animal and from herd to herd.
iv. Some males shed virus in semen without clinical signs of disease ("shedders").
v. Some bacterial pathogens are resistant to the antibiotics in semen extenders or can prevent their effects by forming biofilms.
vi. Reduction in the conception rate has been reported in dairy cattle and horses associated with an increase in AI.

The conception rate of AI in developing countries is very low, and therefore the desired effect in terms of animal improvement has not been obtained. The main reasons for this are a lack of proper management and technical skill. AI will be more effective when farmers get better technical and organizational facility.

The selection of individuals to become parents in the next generation of a genetic improvement program on the basis of the results of DNA marker tests is known as marker-assisted selection. DNA tests are performed to identify those genes associated with a positive impact on the trait of interest and can result in an improvement in the observed phenotype for that trait. It is assumed that the use of MAS will increase exponentially as the industry evaluates and integrates the data from the bovine genome sequencing project. Several markers for milk production traits as well as resistance to mastitis and several diseases caused by recessive alleles such as a deficiency of uridine monophosphate synthatase (DUMPAS) and bovine leucocyte adhesion deficiency syndrome have been identified.

Embryo technologies

Embryo technologies (ET) can play important roles in improving the genetic potential of dairy animals and the fertility. Embryo transfer technologies (ETT) increase the efficiency of genetic selection, conserve the germplasm, allow transgenesis, and improve the fertility of infertile animals. Embryo transfer is considered the female counterpart of AI. It is an important tool to improve livestock at a faster rate. It is a technique of controlled breeding where embryos collected from donor females are transferred to recipients that serve as foster mothers for development throughout gestation. In this technique, the genetic contributions of both males and females are utilized simultaneously. The most important application of ETT is the production of AI bulls from the best proven bull and cow/buffalo available. Multiple ovulation embryo transfer (MOET) is a closed breeding system in which ET is used to increase the average rates of genetic gain.

Multiple ovulation and embryo transfer (MOET)

The mammalian ovary contains several thousand oocytes but the number of progeny a female produces is very small. Only one or two ova are usually shed per cycle in nonletter-bearing animals. The number of offspring that a female can produce in her lifetime can be greatly increased by repeatedly aligning her to become temporarily pregnant, recovering the embryo in early pregnancy, and transferring it to the reproductive tract of another animal to complete gestation. The term embryo transfer refers solely to the collection of an embryo from a donor animal and its placement into the oviduct or uterus of a recipient. This procedure depends entirely upon the availability of a source of good-quality embryos and the proper uterine environment in the recipient at the time of transfer.

History of embryo transfer

The first successful embryo transfer in a rabbit was carried out by Heape in 1890. Warwick and Berry performed the first successful embryo transfer in a sheep and a goat in 1949. The first successful embryo transfer in a pig was done by Kyansnickii in 1951. In cattle, the first successful embryo transfer was carried out by Willett et al. in 1951. Offspring were produced from long-term frozen mouse embryos by Whittingham et al. in 1972. Willmut and Rowson in 1973 produced offspring from the frozen embryos of cattle. A human girl was born after an embryo transfer by Steptoe and Edwards in 1978.

Steps involved in embryo transfer technology

(i) Selection of donor and recipient.
(ii) Superovulation.
(iii) Insemination.
(iv) Collection of embryo.
(v) Selection of embryo.
(vi) Transfer to embryo

Selection of donor and recipient

The main purpose of donor selection is to ensure genetic superiority and the success of the program. Certain criteria are used before selecting a donor:

(a) Superior individual performance.
(b) Good productive performance of offspring.
(c) Regular cyclicity.
(d) Ovary and cervix should be normal.
(e) Fallopian tubes and uterus also must not have any abnormalities or infections.
(f) No detectable genetic defects in the donor.
(g) Donor should be young (4–8 years).
(h) Donors must be healthy and of average body weight.
(i) Donor must have calved at least 60 days back.
(j) For repeated superovulation, donors should be selected at least 60 days after the first superadulatory treatments.

Selection of recipient

The following points should be kept in mind while selecting the recipients of ET work. The recipients should be healthy and free from infections and diseases.

(a) Recipient must be regular cyclic.
(b) Recipient must be without any reproductive abnormalities of the genital tract.
(c) Recipients must have good cyclic CL of the desired stage at the time of embryo transfer.

The estimation of P_4 should be done on the day of gonadotropin treatment, and it should be 3 ng/mL during pretreatment. If the P_4 level is more than 1 ng/mL at the time of estrus, it will adversely affect embryo quality.

Superovulation

This is one of the critical steps in the embryo transfer program. It is defined as the development and ovulation of more than the normal number of follicles. Superovulatory treatments are widely used in embryo transfer programs to increase the supply of embryos from animals of superior genetic merit and to exploit the use of the female germ pool of the ovary to the maximum. Superovulation can be divided into two phases. The first phase includes the development of follicles after FSH administration and the second phase includes the rupture of mature follicles with the subsequent release of ova.

Usually, S/C (subcutaneous) or I/M (intramuscular) injections of PMSG (pregnant mare serum gonadotropin), FSH (follicle stimulating hormone), or hMG (human menopausal gonadotrophin) are used to stimulate additional follicular growth. This treatment is often followed by the I/V administration of luteinizing hormone (LH) or hCG several days later to induce ovulation of the follicles. However, exogenous LH or hCG is not required in adult cattle, buffalo, sheep, and goats to induce ovulation. FSH has a shorter biological half-life than PMSG; therefore the total dose of FSH is divided and injected at intervals of 12h for 3–4 days to get the same response that would result from a single injection of PMSG. PMSG is used alone or in combination with antisera. Antisera bind with excess PMSG in the circulation after the onset of estrus. Superovulation increases the yield of a normal embryo about fivefold in cows, goats, sheep, and rabbits but only slightly in pigs and horses. In donor animals, PMSG is given in a single dose at 1500–3000IU on the 10th day of the cycle, whereas FSH is given at 20–50mg for 4days (between day 10 and day 13) in the morning and evening in eight divided dosages.

LH is given at 75–100mg as a single injection at the start of heat or hCG at 1500–2000IU at the start of heat. The single greatest advance in superovulation methodology in the last decade has been the use of prostaglandin $F_2\alpha$ or $PGF_2\alpha$ analogues such as cloprostenol. Superovulatory treatment can be initiated any time between day 6 of the cycle and natural CL regression. The optimal time for treatment is between day 8 and 12 of the cycle in cattle. $PGF_2\,\alpha$ increases the flexibility of the timing of superovulation and produces large numbers of normal embryos. Most donors are in estrus 2–3days after prostaglandin injection. The response to superovulation is the same in the first, second, and third treatments; however, the response to subsequent treatment is less in some individuals because of the development of antibodies against gonadotrophins.

Insemination

Donors are inseminated with a higher dose of sperm than usual AI upon detection of estrum after prostaglandin treatment. The fertilization rate of ova from a superovulated donor is considerably lower than ova from an untreated donor. This may be due to suboptimal sperm transport, ovulation over a period of time, defective oocytes, or other causes.

Collection of embryo

The embryo can be collected from the oviduct or uterus by surgical or nonsurgical methods. The surgical method is used to collect ova from lab animals and small farm animals, whereas the nonsurgical method is used in large farm animals. Recovery rate is 40%–80% of the CL formation. Low recovery rate is due to the loss of ova from the reproductive tract resulting from altered steroid levels, formation of corpora lutea without ovulation, and failure to recover all ova present. Some oocytes may not be picked up by the fimbria because of a greatly enlarged and superovulated ovary.

Ova reside in the oviduct for 3–4 days after estrus and then migrate to the uterus. Flushing the oviduct 1–3 days after estrus yields more ova than flushing the uterus 5 or more days after estrus. Uterine embryos are often collected because they result in higher pregnancy rates and can be frozen more successfully with some species than younger embryos.

Surgical technique

The surgical technique is most commonly used in sheep, goats, and pigs. In this technique, the reproductive tract is exposed by midventral incision under general anesthesia to collect embryos. Oviduct embryo can be flushed either from fimbria toward the uterus or from the uterotubal junction toward the fimbria. In species except pigs and horses, flushing toward the fimbria is the preferred method because a high percentage of embryos can be recovered with little damage to the reproductive tract. The uterotubal junction is value like in swine and mare. The flushing medium is introduced into the ampulla of the oviduct through the ostium of the fimbria with a small glass tube connected to a syringe. The medium is flushed toward the tip of the terrine horn and collected through a blunt needle or a fire-polished fine glass tube punctured through the uterine wall.

The embryo may be recovered from the uterine horn after it has left the oviducts, usually 5 days after estrus or later. The flushing medium (phosphate buffer saline) is introduced into the base of the uterine horn and flushed toward the tip, where the medium is collected through a blunt syringe needle or a small glass tube inserted into the lumen through a puncture wound in the uterine wall. A volume of 2–20 mL is used to flush oviducts and 10 mL is used to flush the uterus. The disadvantage of this method is the frequent formation of postoperative adhesions of the uterus or oviduct.

Nonsurgical method

The transcervical method of embryo collection is used in cattle, buffalo, and horses. In this technique, a three-way Foley catheter is used for nonsurgical embryo recovery. At first, the cervix of the animal is dilated with an expander or dilator to facilitate the insertion of a device. By rectal palpation, the Foley catheter with a stilette is guided through the cervix. A catheter is placed either in the uterine body or in one horn. Most scientists prefer to flush individual horns. After the catheter is positioned into one horn, the balloon is inflated. Each uterine horn is filled with 30–60 mL of medium, which is then allowed to flow into collection vessels while the uterus is gently massaged through the rectum. This process is repeated until 300–800 mL of the medium has been used.

The same technique can be used to collect embryos from mares, except that the balloon is inflated in the cervix and both horns are flushed simultaneously. The Foley catheter is then inserted into the other uterine horn and the process is repeated. Embryos are also collected in sheep (Mylne et al., 1992) and goats by this technique. PGF_2 alpha and oxytocin injection facilitate the insertion of the catheter in the goat (Pereira et al., 1998). Laparoscopy was used to collect embryos from the uterus of the sheep (McKelvey et al., 1986), goat, and pig (Besenfelder et al., 1997). In a laparoscopy, the instrument for flushing is inserted through a stab wound in the skin. After visualization of the uterus, a two-way Foley catheter is inserted through another stab wound and guided into one uterine horn, then the balloon is inflated. Then an intravenous catheter is passed into the uterine lumen close to the uterotubal junction. Through the intravenous catheter, about 50 mL of fluid are injected and fluid is flushed via the Foley catheter. In this way, the other horn is also flushed.

Selection of embryo

The flushings obtained from sheep, goats, and pigs are directly examined under a stereomicroscope. To prevent evaporation of the culture medium, embryos should be kept in a container. In order to prevent evaporation of the culture medium, paraffin oil is used to cover the medium. Only morphologically normal embryos are transferred to recipients; however, a few embryos that appear morphologically abnormal may develop into abnormal young.

Fertilized eggs of any stage (from one cell to the hatched blastocyst) can develop to term after transfer to a suitable recipient. Embryos between the eight-cell (four-cell in pigs) and blastocyst stage result in the maximum conception rate. Bovine embryos are collected from the uterus from the morulla to the expanded blastocyst stage. After collection, embryo grading is done on the basis of certain characteristics such as the state of the zona pellucid, the number of visible blastomere cells, the compactness of the blastomere cells, the form and size of blastomere cells, the state of perivitalline space, the color and appearance of the cytoplasm, the stage of development of the embryo, and the age of embryo in days. They are graded as excellent, good, fair, poor and unfertilized. Only excellent and good-quality embryos are selected for transfer into a recipient. Ova shows structural abnormalities, unfertilized, degenerating or fragmenting or having empty zona pellucida should be discarded.

Morphological description of embryos

Excellent

Embryo showing even granulation, a well-defined and distinct outline, no blastomere extrusion, and perfect symmetry.

Good

Embryo showing even granulation, well-defined and distinct outline, some blastomere extrusion, and somewhat asymmetric.

Fair

Intact embryo with a hazy outline, some blastomere extrusion and degeneration, and occasional asymmetry.

Poor

Embryo with uneven granulation with a hazy outline, some blastomere extrusion or degeneration, and an abnormal shape.

Unfertilized

Granular appearance of the cytoplasm, no indication of cleavage, and the formation of blastomere cells.

Transfer of embryo

Embryos fewer than eight cells should be transferred to the oviduct because younger embryos are especially susceptible to damage in the uterus, possibly because uterine secretion is toxic to them. Embryos with more than eight cells should be transferred to the uterine horn. There are two methods of embryo transfer:

Nonsurgical transfer Initially, embryos were transferred through laparoscopy under general or local anesthesia. Currently, the transcervical method is used to deposit the embryo in the uterus through the cervix with an AI gun 6–12 days after estrus. The method was developed by Sreenan (1978). In this method, the recipient is palpated through the rectum to determine the ovary-containing corpus luteum. Then, posterior epidural anesthesia is given to prevent straining during the passing the gun through the cervix. Under aseptic conditions, the embryo, loaded in a 0.25 mL straw, is fitted into the Cassou transfer gun and covered by a sheath. The gun is inserted into the vagina and passed through the cervix by rectal manipulations and guided into the uterine horn epsilateral to the corpus luteum. Then the semen is deposited into the uterine horn.

Surgical transfer The embryo is usually transferred surgically. A mid-ventral laparotomy under a local or general anesthetic is the most common method, although flank and lumbar approaches are often used with cattle and rabbits. For transfer into the oviduct, the tip of a capillary pipette containing the embryos is inserted into the infundibulum and ampulla of the oviduct, where embryos are then deposited in a drop or two of the medium. When the embryo is transferred into the uterus, the wall of the uterine horn is punctured with a blunt needle and embryos are expelled from the tip of the capillary pipette inserted into the uterine lumen. For bovine embryo transfer, a laparotomy may be performed under local anesthetic. The tip of the uterine horn is exposed through an incision in the flank. The ovum is then deposited in the uterine lumen.

In sheep and goats, the recovery and transfer of the embryo is done by a surgical procedure under general or local anesthetic from the midventral incision. Embryos are transferred with one or two drops of medium into the oviduct or the uterine horn extracted out of the incision.

Synchronization of Estrus between donor and recipient

For successful embryo transfer, synchronization between the stage of ovum and the reproductive tract of the recipient is necessary. This can be done by selecting recipients that are in estrus at the same time as the donor, either naturally or as a result of estrus synchronization. For optimum results, the recipient should be in estrus within 12 h of the donor. Pregnancy rates decrease drastically if the difference is greater than 24 h in cows and 48 h in sheep and goats.

Use of ETT

i. In vitro manipulation of ova/embryo, which will facilitate experimentation to study the physiology, pathology, and biochemistry.
ii. More offspring from superior females.
iii. Faster genetic improvement.

iv. Transport of a frozen embryo instead of an adult animal.
v. Rapid increase in rare blood lines.
vi. Twin production.

Limitations of ETT

i. Average number of cows per superovulation is 3–4 calves.
ii. A cow may be superovulated for 4–5 times only.
iii. The genetic program is lower in comparison to AI.
iv. Reprodŭction in the recipient is reduced.

In vitro maturation, fertilization, and embryo production

The extensive investigations of the mammalian embryo conducted during the last 3 decades have increased our depth of understanding of normal bio-physiological events that take place during fertilization and early embryonic development. Species differences have been marked in the characteristics of the estrous cycle, the ovulatory mechanism, ovulation times, egg viability, cleavage time, and the stage and location of the morula and blastocyst in the female reproductive tract. Although the mammalian ovary contains hundreds of thousands of oocytes, the number of progeny a female produces is small. In farm animals, the number of pregnancies is limited by the extended duration of gestation. In vitro techniques are powerful tools for studying the physiology of maturation, fertilization, development of per-iimplantation embryos, and increasing production, as it give access to the micromanipulation of embryos.

In vitro maturation, fertilization, and culture (IVMFC) involves the collection of oocytes from slaughterhouse ovaries or from living animals through transvaginal aspiration, in vitro maturation of selected oocytes in media containing serum and hormones, in vitro fertilization either with frozen thawed semen or freshly collected in vitro capacitated spermatozoa and in vitro culture of embryos in media containing oviduct or cumulus cell up to the morulae or blastocyst stage, at which stage they can be either transferred to synchronized recipients or frozen and stored (Palta and Chauhan, 1998).

In the past few decades, there has been an unprecedented evolution of technology for in vitro embryo production of farm animals. The rate of progress has intensified in the last decade with the development of defined and semidefined media for different species. The first successful IVF was achieved in rabbits in 1959 (Chang, 1959), and the next success was with mice in 1968. In humans, the first IVF production was done in 1978 and the first calf was produced with IVF in 1981 (Brackett et al., 1982). These technologies are helpful in the production of high genetic merit animals and provide an excellent source of embryos for emerging biotechnologies such as cloning, embryo sexing, transgenesis, nuclear transfer, etc. Furthermore, it allows analyzing the developmental potential of embryos, including the pattern of gene expression, epigenetic modifications, and cytogenetic disorders during the development (Galli and Lazzari, 2008). The development of a bovine embryo in the early stages shows many similarities with human embryos. That's why bovine embryos are used as a model organism (Niemann and Wrenzycki, 2000). Because only 30%–40% blastocyst development has been obtained from oocytes after in vitro maturation, fertilization, and embryo culture (Sirard et al., 2006), its efficiency is low.

Despite several advantages of IVP, the initial applications in both cattle and buffalo have been limited by the ability to recover oocytes. However, the recent developments of low invasive ultrasound-guided transvaginal oocyte retrieval (TVOR) and oocyte pick up (OPU) have removed these difficulties to a large extent. With the help of OPU, oocytes can be aspirated from live animals. In many countries, OPU-IVP is being used for embryo production at a commercial level. Through this technique, large numbers of embryos can be produced compared to standard embryo transfer (ET) practice (Galli and Lazzari, 2008). TVOR can be used for the repeated collection of oocytes from endangered species of livestock or livestock of high economic importance. However, due to high production costs and low overall efficiency under field conditions, the practical use of IVEP is limited.

Methods for in vitro maturation (IVM), fertilization (IVF), and culture (IVC)

Crucial steps of in vitro embryo production include:

1. Oocyte collection (oocyte pick up from living animal or postmortem from slaughterhouses).
2. Selecting and cleaning oocytes and placing oocytes in a maturation medium for 18–24 h.
3. Sperm purification using a percoll gradient.
4. Inseminate matured oocytes with purified sperm cells for 8–24 h.
5. Removal of cumulus cell complexes can be done by mechanical means (vortexing, pipeting) or by enzymatic digestion.
6. Placing putative zygotes in culture medium for 7–9 days.
7. Obtaining early bovine embryos that are ready to transfer to surrogate mothers.

Oocyte collection from ovum pick up technique

The ultrasound-guided ovum pick-up technique is a noninvasive technique used to collect oocytes from antral follicles in live animals. The ultrasonic-guided aspiration of bovine follicular oocytes was first discovered in Denmark in 1987. In 1988, a real OPU was first established in cattle by a Dutch scientist. It does not interfere with the normal reproduction and production cycles of the donor. The female from 6 months of age to the third month of pregnancy and even soon after calving (2–3 weeks) can be used as a suitable donor. This has been shown to be a feasible and practical alternative to the conventional multiple ovulation and embryo transfer (MOET) program. It is being used more and more for commercial applications worldwide. The OPU technique consists of the transvaginal removal of the oocytes by aspiration of the ovarian follicles with the aid of an ultrasound probe (Hoque, 2011). This procedure is absolutely harmless to the donor. In this technique, 15–20 min is required for oocyte collection from the donor animal. During collection, the donor is kept inside a crate. After oocyte collection, they are selected in the laboratory and transferred to an incubator in a suitable culture medium to complete the maturation phase. Just after retrieval, the oocytes are fertilized in vitro with the semen of the chosen bull and kept in an incubator for about a week until the embryo stage suitable for transfer to receiving cows or freezing is reached.

Advantage of OPU

- The retrieval of eggs can be repeated many times twice a week.
- Each group of oocytes taken can be fertilized with the seed of a different reproducer. In this way, we can get embryo groups with different paternities from the same donor in a very short time.
- It does not require any preventive hormonal treatment and does not interfere with the normal physiology of the donor.
- The embryos can be produced by donors of any age, even during lactation and in the first 3 months of pregnancy without the possible side effects of superovulation.
- In donors suffering from long-standing sterility, OPU can exert a therapeutic effect by removing the cystic formations that are often present on the ovaries of these animals.
- All donors return to estrus 7–10 days after the last withdrawal and can be immediately fertilized.

Collection and processing of abattoir ovaries

Ovaries collected from local abattoir in normal saline solution (NSS) with 100 IU/mL penicillin and 100 microgram/ml streptomycin sulfate at 37 °C are brought to the laboratory and washed with physiological normal saline solution and Dulbecco's phosphate buffer saline (DPBS, pH 7.3–7.4) before oocyte recovery.

Recovery/collection of oocytes

Different methods are used for the recovery/collection of oocytes with variable success rates.

- *Scoring method*: A hemostat is attached to the base of the ovary to hold the ovary in place. The excess tissue is cut from the ovarian stalk. The ovary is held above the beaker containing the oocyte collection medium and a 2–3 mm deep incision across all visible follicles (2–6 mm in diameter) is made with a sterile surgical blade, with instant rinsing and tapping of the ovary to release the maximum number of oocytes.
- *Slicing method*: Ovaries are placed in a plastic petri dish containing a saline solution and are chopped into small pieces with a surgical blade. The cumulus oocyte complexes are selected from the saline solution.
- *Puncture method*: A follicle visible on the surface ranging from 2 to 6 mm in diameter is punctured with an 18-gage sterile needle. The cumulus oocyte complexes are selected from the follicular fluid.
- *Aspiration method*: The follicular fluid from the surface follicle (2–6 mm) is aspirated through a sterile 18-gage needle attached with a 5 mL syringe containing a sterile solution. The aspirated content is expelled into a fresh petri dish containing the oocyte collection media, and cumulus oocyte complexes are selected from it.

Oocyte maturation

The oocytes are recovered by puncturing the visible surface of the atretic follicle (1–5 mm diameter) with an 18-gage needle into a 35-mm sterile glass petri dish containing the oocyte collection media (OCM). The oocytes are graded as excellent, good, fair, and poor depending upon the cumulus layers and cytoplasm. Only excellent and good-quality oocytes (grades A and B) are subjected to in vitro maturation.

The selected cumulus oocyte complexes (COCs) are washed 10–20 times in a 100 μL drop of oocyte handling media (OHM) followed by 8–10 washings in a 100 μL drop of maturation medium (MM), which was preequilibrated in a CO_2 incubator at 38.5 + 1 °C for 2 h (Singh et al., 2009). About 10–15 COCs are placed in a microdrop of 50 μL of the maturation medium (TCM-199, M-7528) containing a 20% concentration of EGS; a 20% NCS or 10% NCS plus hormone (FSH @ 1 μg/mL + LH @ 100 IU/mL + E_2 @ 1 μg/mL); or a 20% CFF that was equilibrated at 38.5 °C in a CO_2 incubator. The oocytes in the 50 μL droplets are covered with sterile mineral oil and cultured in a humidified atmosphere of 5% CO_2 in air at 38.5 + 1 °C for 27 h in a CO_2 incubator. Oocyte maturation is confirmed by cumulus expansion and the presence of the first polar body after 27 h of incubation with maturation media.

In vitro capacitation

50 μL of neat proven semen with a gross motility of +5 or +4 are transferred into a centrifuge tube containing 5 mL of fertilization media, fert-TALP. This is subjected to centrifugation at 1800 rpm for 10 min, after which a clear pellet of spermatozoa is formed. The supernatant is discarded by gentle pipetting without disturbing the semen pellet. The semen pellet is resuspended into 5 mL of fertilization media and centrifugation is repeated. The supernatant is again discarded and the pellet is resuspended in an equal volume of fertilization media. This is then kept for equilibration in a CO_2 incubator for 30–45 min. Following equilibration, the semen sample is again centrifuged (1800 rpm for 10 min) and the supernatant is discarded. The washing is essential to get rid of seminal plasma, impurities, contamination of cell debris, etc. After a final washing, 50 μL of the semen pellet is added into 400 μL of preequilibrated (in a CO_2 incubator for 2 h) fertilization media (Singh et al., 2010). Finally, the sperm concentration is assessed using Neubaur's hemocytometer.

In vitro fertilization of in vitro matured oocytes

After 27 h of culture, oocytes are evaluated for cumulus expansion and the presence of the first polar body. The mature oocytes are taken and denuded by 0.1% hyaluronidase and by repeated pipetting through a fine boar pipette. The denuded oocytes are washed 8–10 times in a fert-TALP medium that was preequilibrated in a CO_2 incubator at 38.5 + 1 °C for 2 h. They are then transferred into 25 μL drops of fert-TALP medium under sterile mineral oil and equilibrated for 2 h at 38.5 + 1 °C in a CO_2 incubator (Singh et al., 2009). The capacitated semen has a final concentration of 1×10^6 live sperm per ml; of this 50 μL are used for in vitro fertilization into each fertilization drop containing mature oocytes with a denuded cumulus cell mass. Oocytes are coincubated with spermatozoa for 18 h at 38.5 + 1 °C in an atmosphere of 5% CO_2 in humidified air.

In vitro culture of in vitro fertilized oocytes

Eighteen hours after insemination, the oocytes are washed in a culture media (embryo development medium, EDM) 8–10 times to separate adhering sperm cells. The oocytes are finally transferred either on monolayers (equilibrated with EDM under oil for 18 h) or a 50 μL drop of EDM coincubated with oviductal epithelial cells under sterile mineral oil (equilibrated for 2 h at 38.5 + 1 °C in a CO_2 incubator) for 48 h in a drop (Singh et al., 2010). They are cultured at 38.5 + 1 °C at 5% CO_2 in a CO_2 incubator for 7 days.

Assessment of cleavage rate and embryo development

The cleavage rate is evaluated 48 h post insemination by counting the number of cleaved $\geq$2 cell embryos/number of oocytes used for fertilization. After evaluation of the cleavage rate, embryo development is seen morphologically every 24 h under a stereo zoom microscope and classified as two-cell, four-cell, eight-cell and more than eight-cell embryos depending on a possible in vitro cellular block. After every 24 h, 50% media is replaced with a fresh preequilibrated EDM media.

Oocyte/embryo cryopreservation and vitrification

This is an essential component of the embryo transfer program. We can preserve surplus embryos with the help of this technique, which can be used for the conservation of endangered species and breed. By using this technique, the complete genotype can be conserved as an embryo.

The continuous availability of viable, developmentally competent oocytes and/or embryos has been critical to recent progress in IVP because of the relatively short fertile life span of mammalian oocytes and/or embryos. Hence, the storage of unfertilized oocytes would generate a readily available source that allows experiments to be carried out at a convenient time. It could therefore be of practical importance for the establishment of gamete bank from which particular genetic combinations could be derived. During the past few decades, significant progress in the cryopreservation of mammalian oocytes and embryos has been achieved. Live offspring of at least 25 species from the transfer of cryopreserved embryos or oocytes have been successfully produced (Gajda and Smorg, 2009).

Principles of embryo freezing

The principles of embryo freezing are similar to freezing of other living cells. It is a process of heat and water transport between the cell and its surrounding medium. During freezing and thawing, damage to the embryo occurs due to large intracellular ice crystal formation or increased intracellular concentration of solutes resulting from the dehydration of cells during freezing (solution effects). Fast freezing minimizes the damage from the solution effect, but it helps in the formation of large intracellular ice crystals that damage the embryo. On the other hand, slow freezing prevents the formation of large ice crystals but it leads to damaged spermatozoa through the solution effect. Therefore, the optimum freezing rate should be slow enough to prevent the formation of intracellular ice crystals and should be rapid enough to minimize exposure of the cells to the higher concentration of solutes.

The freezing point of the phosphate buffer saline media, which contain glycerol as a cryoprotectant, is 0.3 °C. If this solution is cooled at a cooling rate of 0.3 °C, supercooling occurs. To avoid supercooling, crystallization is induced in the extracellular medium at about 2–3 °C below the freezing point by seeding the medium with ice crystals. At seeding, most of the water outside the cells quickly turns to ice. The small amount of solution between the ice crystals becomes hypertonic and does not freeze because of the higher concentration of molecules. If cooling is continued at a slow rate, more water freezes and unfrozen water becomes more hypertonic. The cells of the embryo respond to this hypertonic solution by losing water through dehydration. If cooling is continued slowly, all the water is removed and none is left for freezing. This prevents the formation of intracellular ice crystals. But dehydration at low temperature is greater and can damage spermatozoa.

The major problems with cryopreservation of the germplasm are mechanical as well as osmotic damage during the processing steps. With the advent of glycerol, which is a cryoprotecting (CP) agent, cryopreservation became more feasible. Other CPs are also in widespread use, alone or in various combinations. These include permeating CPs such as ethylene glycol and propylene glycol (Chen et al., 2005; Luz et al., 2009) and non-permeating ones such as sucrose, glucose, or fructose (Barcelo-Fimbres and Seidel, 2007). With the development in traditional cryopreservation methods, vitrification of the germplasm was introduced (Rall and Fahy, 1985).

Vitrification is a simple, faster, less-expensive technology compared to slow freezing. Moreover, it was shown to be more effective than slow freezing for material more sensitive to chilling (Vajta et al., 1996). The cryopreservation of oocytes by vitrification was tested with variable success for oocytes from bovine (Hochi et al., 2000), swine (Huang and Holtz, 2002) and equine (Hurt et al., 2000). Freezing of the embryo is an established commercial practice and its major aim is to preserve the embryo in a viable condition from which it may be revived to continue normal development and avoid ice crystal formation that can be harmful for the embryo when thawing. Embryo freezing is an essential component in the commercial embryo transfer (ET) programs,as there is an obvious need to store the embryos on a temporary basis until they are transferred, as embryo viability starts to decline after 12h of storage in holding media. With regard to embryo sexing naturally, both male and female embryos should be stored to ensure the representation of both sexes and wide genetic diversity. The preservation of oocytes reduces the risk and expense involved in the transport of live animals as well as the hazards of disease transmission, and also provides insurance against catastrophes and natural disasters. On the other hand, the preservation of oocytes and embryos of endangered species safeguards from extinction. Cryopreservation can be considered a useful conservation strategy for endangered breeds. For these species, cryobanking of embryos could be helpful in establishing founder populations with the aim of eventual reintroduction into the wild (Ptak et al., 2002).

Zygote intrafallopian transfer (ZIFT) and gamete intrafallopian transfer (GIFT)

Zygote intrafallopian transfer (ZIFT), also referred to as tubal embryo transfer (TET), is an ART technique in which embryos are transferred into the fallopian tubes to achieve pregnancy. Meanwhile, gamete intrafallopian transfer (GIFT) allows the transfer of the gamete into the fallopian tubes. The obvious advantage of ZIFT over GIFT is that as in IVF, it is possible to document fertilization. On the other hand, both ZIFT and GIFT require the female to have at least one functioning fallopian tube, which is a disadvantage when compared with IVF. These different techniques are used to achieve pregnancy in high genetic merit animals with reproduction problems or to achieve implantation of produced embryos in surrogate mothers.

Sexing of embryo and sperm

The desire in man to control the sex/sex ratio seems as old as man himself. Sexing of the preimplantation embryo has a significant impact on the economics of livestock production. Embryo sexing has been carried out in cattle, buffalo, sheep, goats and pigs. The cattle and buffalo industries benefit from a higher proportion of male progeny (beef stud breeders). In the dairy industry, female progeny are preferred for milk production. This task can be completed in two ways:

(i) Separation of sperm.
(ii) Diagnosis of the sex of the preimplantation embryo.

An ideal and efficient method to control the sex ratio is:

(i) AI of females with a homogeneous population of either X or Y chromosome-enriched spermatozoa.
(ii) IVF with sexed spermatozoa and subsequent ET.

Effective methods for spermatozoa separation and the cost factor its use in AI for the production of young ones of the desired sex (Seidel Jr., 1988; Johnson et al., 1994). The genetic sex of the mammal is set at the time of fertilization. Reproductive biotechnologies such as artificial insemination, embryo transfer, cryopreservation, micromanipulation, in vitro fertilization, and gene transfer have been incorporated into livestock production systems. These technologies in association with embryo sexing can alter the livestock population.

Methods of sex determination

Sex can be determined in three stages:

(i) Predetermination using X or Y spermatozoa.
(ii) Prior to transfer.
(iii) During early gestation.

Sex can be predetermined by two methods:

(i) Sperm separation.
(ii) Nuclear transplantation.

Sexing of sperm

Sperm sexing was done for the first time by Erikson in 1972. Spermatozoa can be separated by the following methods:

(i) Nonequilibrium sedimentation.
(ii) Equilibrium sedimentation.
(iii) Electrophoretic separation.
(iv) Separation based on motility differences.
(v) Separation by flow cytometry.
(vi) Y-specific DNA probes.

Sexing of embryos

There are two major methods of embryo sexing:

(A) Noninvasive.

(B) Invasive.

Noninvasive methods

The noninvasive methods are optimal because they maintain the integrity of the embryo. This method is less likely to impair the potential for ET and implantation. Noninvasive methods include:

i. Enzymatic assay.

ii. Immunological method.

iii. Hormonal assay.

iv. Based on cleavage and development.

Embryonic sex selection by enzymatic assay

The ratio of X-linked enzyme activity to autosomal enzyme activity is higher in females than males. In a mouse embryo, G6PD, HPRT, and APRT activity were studied by Monk and Hadyside (1988). By the G6PD study, 57% of the male and 72% of the female embryos can be correctly identified. With both HPRT and HPRT, the APRT activities show 91% accuracy for females and 100% for males. Tiffin et al. (1991) reported that the metabolism of glucose and glutamine is higher in female embryos than male embryos.

Disadvantage

✓ Differences between male and female are slight.

✓ X-chromosome inactivation occurs during the embryonic stage.

✓ Initially, embryonic enzymes are of maternal origin.

Immunological method (Identification of HY-Ag)

HY-Ag is found on the surface of the male embryo. HY-Ag have a testicular organizing factor, which helps in the differentiation of indifferent gonads into testes. HY-Ag is highly conserved and present in heterogenous sex. This Ag is readily detectable at the eight-cell stage. Ag can be detected at the morula stage but not at the blastocyst stage. There are two methods of Ag detection in embryos:

Cytotoxicity assay Embryos are exposed to dilute HY-Ag and complement. Embryos with HY-Ag show lysis and are diagnosed as male embryos.

Immunological assay Embryos are mixed with HY-Ab (primary Ab). The mixture is kept for 30 min and then secondary Ab and FITC are added to the mixture. Then the mixture is seen in a phase contrast microscope. If there is fluorescence, then it is a male embryo; no fluorescence means it's female. 59 porcine embryos were sexed by IIFT (White et al., 1985). Out of 59, 48 (81%) embryos were correctly sexed.

Disadvantage

Poor quality embryos with lysed or dead cells also exhibit fluorescence.

Hormonal assay for sexing blastocyst

Fluid from the blastocyst is subjected to a hormonal assay (Hare and Betteridge, 1978). Hormones are measurable, present in blastocyst fluid, and differ between sexes. Estrogen in a pig embryo can be detected as early as day 12. Estrogen and androgen in mares can be detected as early as day 14.

Disadvantage

(a) It is not known whether steroid production varies with sex.
(b) It needs more exploration.

Sexing based on cleavage and development

DNA content is higher in male embryonic cells than females. If the DNA content is higher, it will take more time for duplication, resulting in a longer cell cycle that will affect the cleavage and development rate in males and females. Male embryos cleave early and develop faster to attain the morulae and blastocyst stage compared to female embryos.

Disadvantage

(a) In *in vivo* conditions, the cleavage time cannot be known.
(b) High skill is required in the separation of fast and slow embryos.

Invasive method The invasive method of embryo sexing was proposed by Monk and Hadyside (1988). It includes the following methods:

(i) Identification of the sex chromatin.
(ii) Sex chromatin analysis.
(iii) Y-specific DNA probes.
(iv) PCR method.

Identification of sex chromatin The sex chromatin, also known as the "Barr body," is formed during embryonic life. X-chromosome inactivation and appearance of the sex chromatin occur at the 4–8 cell stage. This method depends upon identification of the dark staining body adjacent to the nuclear membrane. The Barr body can be identified in rabbit embryos that are 5–6 days old at transfer.

Disadvantage

This method cannot be used in livestock species.

Sex chromosome analysis

The sex chromosome can be analyzed in metaphase cells. The sex chromosome can be analyzed by four methods:

(a) Biopsy from trophoblast (12–15 day embryo).

(b) Biopsy from morulae (6–7 day embryo).

(c) Half embryo on day 6–7.

(d) Few blastomeres from 8 to 16 cell stage.

Biopsy from trophoblast (12–15 day embryo)

This method was successfully used in 12–15 day bovine embryos at transfer (Hare and Betteridge, 1978). Using this method, 58.5% of embryos can be sexed with certainty. Betteridge (1981) reported that 68% bovine embryos can be sexed by this method.

Disadvantage

(i) Relative number of cells in metaphases.

(ii) Quality of metaphases.

(iii) Diagnosis time.

The relative number of cells in metaphases and the quality of metaphases can be increased by improving culture conditions, treating with a mitotic arresting agent, and taking large biopsies. Using this method, 83.9% of embryos were accurately sexed.

Biopsy from morulae (6–7 day embryo)

By using this method, a 63% sex ratio in 6 day embryos and a 53% sex ratio in 7 day embryos were obtained.

Half embryos on day 6–7

This method involves bisection of the embryo for sex chromosome analysis. Using this method, 26.6% of bovine embryos can be sexed.

Disadvantage
(i) Zona ruptured embryo cannot be frozen.
(ii) One of the twins is destroyed.

Sexing of blastomeres from morulae This method was given by Bondioli et al. (1990). A single blastomere is responsible for the development of the morulae and blastocyst. The nuclear transfer technique was used for multiplication of superior genotypes. Once the technique is perfect, a large number of embryos of the same genotype can be sexed. One such embryo can be used for sex chromosome analysis.

Embryo sexing using Y-specific DNA probes A complementary DNA sequence (probe) is produced that binds to the gene present on the chromosome. Labeled probes are used to identify the presence of the Y-chromosome in cell preparation. Three repeated Y-specific DNA probes were used to sex a 6–7 day embryo. In this method, cells from the embryo are mixed with the labeled DNA sequence specific to the Y-chromosome. Hybridization of the total cellular DNA to a labeled DNA sequence specific to the Y-chromosome occurs. Positive hybridization indicates the presence of the Y-chromosome (male sex).
Advantage
(a) Small number of cells required for DNA preparation.
(b) Accuracy of embryo sexing is 95%.
Disadvantage
(a) Invasive technique.
(b) Decreases viability and limits other potential manipulations.

Sexing of embryo by PCR method

Most of the methods of embryo sexing are inaccurate and time consuming. To overcome these disadvantages, the PCR method was developed in the 1980s. Male embryonic cells can be distinguished from female cells by amplification of the Y-chromosome specific repetitive sequences. This method was used to sex embryo in cattle, small ruminants, and buffalo. The isolation of different Y-chromosome specific DNA fragments paves the way for designing primers for embryo sexing. The accuracy of this method for sexing bovine embryos is 95% (Peura et al., 1991).

Protocol of PCR for embryo sexing

Sampling of embryo cell (4-5 cell) — Biopsy by microrazor / Aspiration by pipette

⇩

Rinsing of embryo by Kcl having BSA

⇩

Biopsy in microtube + buffer having proteinase K and primers

⇩

Heating at 55 °C for 1 hr (DNA released)

⇩

Heating at 95 °C for 15 min (Proteinase K destroyed)

⇩

Add reaction medium (Nucleotide and TAQ polymerase)

⇩

Microtube in PCR machine for amplification

⇩

Heating at three steps –

⇩

- At 96 °C for 20 sec (Denaturation)
- At 60 °C for 20 sec (Hybridization of primers)
- at 73 °C for 10 sec (Amplification)

⇩

Lot of copies of DNA in microtube

⇩

Agarose gel electrophoresis (For ½ hr)

⇩

See the band

We can also see the reaction tubes by passing UV light

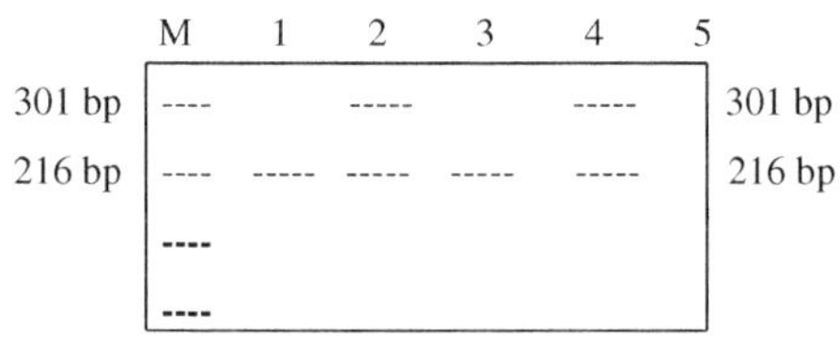

Interpretation

(a) In buffalo, two primers are used

- BRY-1 (301 bp) specific to the Y-chromosome.
- Satellite DNA (216 bp) specific to the male and female both.

(b) If the band coincides with the 301 and 216 bp marker band, the embryo is male.

(c) If it coincides only with the 216 bp band, the embryo is female

(d) When a tube is seen by passing UV light, then pink fluorescence indicates a male embryo.

Advantages of tube method

(a) It eliminates the cost, time, and labor of gel electrophoresis.

(b) There is no risk of contamination.

(c) It can be stored for reevaluation.

The cytological methods are very accurate and will remain applicable for testing the results of other effective methods available or emerging in the future. The Y-chromosome specific DNA probes are likely to emerge as the most preferred method. The noninvasive methods are not of much practical value. The sexing of embryos based on cleavage and development is possible in in vitro produced embryos. The use of Y-chromosome specific DNA probes along with the cytological method will emerge as the most successful embryo sexing technology in farm animals.

Sperm sexing

Semen sexing is one of the most desirable reproductive technologies. Males are required for the production of quality semen for the artificial insemination program and the meat industry. Dairy farmers are interested in females only. The X-chromosome bearing sperm (responsible for the production of heifers) contain 3.8% (3.6%–4.2%) more DNA than Y-chromosome bearing sperm, which produce bull calves. The X- and Y-chromosome bearing sperm can be separated with the help of a cell sorter/cell flow cytometer. Sexed semen is now available commercially from companies such as XYINC (Colorado, United States), Cogent (Chester, United Kingdom), and Goyaike (ESuobar, Argentina). Siedel (2003) reported that the accuracy of this technique is about 90% for the predetermination of sexes of calves. The process of cell sorting (15 million spermatozoa per hour) using this technique is slow. The laser light also lowers the visibility of the sexed sperm. So, due to the high cost of cell sorting and the low output, this technique has not reached wider use for artificial insemination purposes. However, sexed sperm are being used for IVF and ICSI, in which the requirement of sperm is much less. The first calf produced from sex-sorted spermatozoa was reported by Cran et al. (1993) in cattle.

Cloning

A clone can be defined as "a group of two or more individuals with identical genetic makeup, which are produced from a single common parent or ancestor by sexual reproduction." A clone can be produced either by embryo splitting (as occurs in nature) or by the transfer of embryonic blastomeres or somatic cells as donor nuclei. The production of offspring by nuclear transfer (NT), where the new individual is formed from the nuclear genome of other individuals, has tremendous potential for enhancing livestock productivity and the conservation of endangered animals. Animal cloning has several applications in research, medicine, and agriculture. The importance of animal cloning is to propagate valuable genotypes at a faster rate, induce genetic modification in animals, and use them for the production of transgenic animals for pharmaceutical proteins and tissues/organs for xenotransplantation.

The major breakthrough in cloning occurred when Willadsen (1986) produced a lamb from the transplanted nuclei of eight-cell stage blastomeres. The cloning efficiency of the mammalian embryo was reported to be 1%–4% in cows, 1% in pigs, and 4% in sheep. The birth of Dolly (the first mammal clone) by Wilmut et al. (1997) from adult cells generated interest in somatic cell nuclear transfer (SCNT) and clones of other species, including cattle and equine. Cloned buffalo embryos have been produced using embryonic (Singla et al., 1997) and fetal cells (Meena and Das, 2005).

Animal cloning provides new opportunities for the genetic engineering of eukaryotic cells, tissue regeneration, animal genome preservation, and the development of embryonic stem cells for therapeutic applications. Cloning is an important tool for studying gene functions, the regulation of embryonic development, genetic disease, and nuclear programming. Presently, this technique is being applied in a wide variety of domestic and wild animal species.

Methods of cloning

Currently, to produce cloned animals, two methods are being used: somatic cell nuclear transfer (SCNT) and handmade cloning (HMC). Somatic cells are derived from a single genome; however, the cells are extremely diverse in phenotypes and functions.

Somatic cell nuclear transfer (SCNT)

Animal cloning is also known as nuclear cloning. It is a highly complex process with the following steps:

i. Ooplast formation

In this process, the nuclear genetic materials (enucleation) of in vitro matured oocytes obtained from slaughtered female animals or through transvaginal OPU from live animals are removed. Removal of the chromosomal genome is essential for maintaining the diploid status of the cloned embryo.

ii. Transfer of donor nucleus

Somatic cells are cultured in vitro and then used as nuclear donors. Propagation in culture provides millions of cells that could be used for producing a number of initial offspring or for genetic modification for use in transgenic animal production.

On the basis of the transfer of the donor nucleus into the ooplast, the animal cloning method is either known as SCNT (somatic cell nuclear transfer), where the donor cell is mechanically placed underneath the zona pellucida, or HMC (handmade cloning), where nuclear somatic cell is attach to the zona free ooplast. The zona of in vitro matured oocytes is removed enzymatically (by pronase treatment) and the genetic material of oocytes is removed to obtain an ooplast. The process of removing the nuclear material is known as enucleation. Although this technique requires technical skills and sophisticated instruments (microtools and micromanipulation systems attached to an inverted microscope), most of the animal clones are produced through SCNT. Handmade cloning does not require microtools (holding pipettes, cutting needles, and transferring pipettes) and micromanipulators.

iii. Electrofusion of donor cell with ooplast

This process is similar to protoplast fusion, which involves the induced fusion of a diploid donor cell with the ooplast in the presence of direct current pulses, which should be optimized for different species. The resulting product is the oocyte cell complex, a diploid cell comparable to a zygote. Electrofusion of the somatic cell with the ooplast is an indispensable step of the adult animal cloning process. It is being used as a powerful tool for studying the molecular mechanism of epigenetic reprogramming.

iv. Activation of reconstructed zygote

This is performed by using electrofusion, through a specific chemical signal, or both to enable the reconstituted zygote to undergo cell division, differentiate, and form an embryo with all types of cells in the right place at the right time. This process is called reprogramming the genome, and is important for the survival of the cloned embryo and the development of the fetus to term. Extensive failure and/or error of genome reprogramming are the major causes of cloned embryo mortality.

v. In vitro culture and transplantation of NT embryos

The cloned embryos are cultured in vitro in a chemically defined cell culture or embryo culture media until the blastocysts stage of embryonic development, and then transferred into estrus-synchronized recipient females. About 7–10 days are needed to reach the blastocyst stage after initiating the in vitro culture (IVC). If the pregnancy is established, the offspring born will be the clone of the animal from which the somatic cell was derived and used as a donor nucleus.

Applications of animal cloning

1. Basic research on cell differentiation

To learn how reprogramming and reactivation of the genome occur during early embryonic development.

2. Conservation of threatened or endangered animals

Cloning can be used to conserve indigenous livestock breeds. Somatic cells from endangered animals can be propagated and maintained in vitro for use as a nucleus donor to produce animal clones.

3. Improvement in the productivity of dairy animals

Animal cloning is used for the transfer of superior genotypes from nuclear breeding flocks and herds to commercial farmers. Genotypes for high milk yield, disease resistance, or extreme environmental conditions are selected and disseminated. It is useful in the multiplication of F1 cross-bred animals or composite breeds to increase the benefits of both heterosis and uniformity within the clonal family.

4. Development of research animal models

Animal cloning can be used for the development of animal models to study the genetic basis of several human diseases. The transgenic mammalian model, particularly genetic knockout mice, have been used in medical science.

5. Formation of therapeutic embryonic stem cells

Therapeutic cloning entails the reprogramming of somatic cells (such as skin cells) into the blastocyst (cloned) stage. The next step is using the inner cell mass of the blastocyst as embryonic stem cells, which can produce multiple tissue types with perfect histocompatibility to the patient and thus bypass issues associates with the immunorejection of the transplanted tissue (Lanza et al., 2002). Presently, there is no therapy for curing mitochondrial gene-related disorders. Nuclear transfer cloning gives a suitable alternative to get rid of such genetic diseases.

6. Animal cloning and gene pharming

Transgenic livestock that are able to produce recombinant proteins can produce variable proteins such as pharmaceutical proteins for the treatment or prevention of human diseases or biomaterials for therapeutic applications.

Limitations

The loss of the majority of pregnancies/neonates is due to developmental abnormalities, which are commonly called "large calf syndrome." The major abnormalities in the clones are possibly epigenetic in nature. These abnormalities do not transmit to the offspring, even when male and female clones are mated together.

Transgenesis

Transgenic animals are those animals in which there is a modification of the genome, in contrast to spontaneous mutation. In transgenesis, either DNA is transferred into the animal or the DNA of the animal is altered. Transgenic animals are genetically modified and contain a gene from other species following gene transplantation or results from the molecular manipulations of endogenous genomic DNA. This technique is used to introduce exogenous genes into the animal genome by genetic engineering technology so that these genes can be expressed and inherited by offspring. By using recombinant DNA technology, foreign DNA is introduced into the animal, which is transmitted through the germ line so that every cell, including germ cells, of the animal contains the same modified genetic material.

The incorporation of foreign genes into the genome of an animal allows its genotypic manipulation in a desired way and leads to the development of transgenic technology. Therefore, the animal then produced is called transgenic. Gene transfer in animals has been used for modifying the fat or protein synthesis in mammary glands, transfer of the growth hormone gene into pigs (Pursei and Rexroad, 1993), transferring the cysteine synthesis gene into sheep to enhance wool production, impairing resistance to the influenza virus in pigs (Muller et al., 1992), human antitrypsin in sheep (Carver et al., 1993), a lactalbumin in cow (Colmon, 1996), tissue plasminogen activator and antithrombin in goat (Ziomek, 1998).

Methods for producing transgenic animals

Transgenic animals are produced by the introduction of a foreign gene or genes. The genes, which are inserted, are called transgenes. In order to get foreign genes in every cell, including germ cells, hey should be transmitted through the germ line.

The methods to create transgenic animals are gene knockdown using RNA interference, nuclear transfer, retroviral-mediated gene transfer, pronuclear microinjection, somatic cell cloning, sperm as a vector, embryonic stem cell-mediated gene transfer, etc.

Pronuclear microinjection

This method was first described by Gordon and colleagues. After the entry of the sperm into the oocyte, male and female pronuclei are microscopically visible for several hours. The transgene may be microinjected into male or female pronuclei. Big Blue animals and Muta Mouse have been produced by this technique. Only 0.5%–3% of the microinjected embryos produce transgenic offspring. This method includes the following steps:

i. Isolate the piece or pieces of DNA and clone it into a vector such as a plasmid.

ii. Harvest newly fertilized eggs before the pronuclei fuse. A pronucleus is the nucleus of the sperm cell (male) or egg cell (female) before fertilization.

iii. The piece of DNA that has been isolated is placed in a syringe and injected into the pronucleus of the sperm cell.

iv. When the pronuclei have fused to become the nucleus of the new zygote, the cells are allowed to divide to form two embryonic cells after transfer of the embryo into the uterus of a pseudopregnant mouse.

Embryonic stem cell-mediated gene transfer

The term embryonal carcinoma cells (EC) was derived from teratocarcinomas while the term embryonic stem cells (ES cells) is used to denote a cell line isolated directly from mouse embryos. From the inner cell mass (ICM) of mouse blastocysts, embryonic stem cells (ES cells) are harvested. These cells can be grown in culture and can produce all the cells of the mature animal, including its gametes. This method includes the following steps:

i. Recombinant DNA technology inserts DNA fragments from a variety of sources, having a desirable gene sequence by appropriate vector and promoter and enhancer sequences to help the gene to be expressed by host cells.

ii. Transform ES cells in culture to expose cultured cells to the DNA so that some will incorporate it.
iii. Select successfully transformed cells.
iv. Inject these cells into the inner cell mass (ICM) of mouse blastocysts.
v. Embryo transfer.
- Prepare a pseudopregnant mouse.
- The stimulus of mating elicits the hormonal changes needed to make her uterus receptive.
- Transfer the embryos into her uterus.

vi. Test her offspring.
- Remove a small piece of tissue from the tail and examine its DNA for the desired gene.

vii. Establish a transgenic strain
- Mate two heterozygous mice and screen their offspring for the 1:4 ratio that will be homozygous for the transgene.
- Mating these will give rise to a transgenic strain.

Retrovirus-mediated gene transfer

By the retroviral transduction of male germ line stem cells, transgenic mice can be produced. Male germ line stem cells are capable of self-renewal. To study the biology of self-renewal and the differentiation processes of male germ line stem cells and to produce a wide range of transgenic animal male germ lines, the stem cells should be genetically modified. A retrovirus is a virus that carries its genetic material in the form of RNA. These viruses are used as vectors to transfer genetic material into the host cell, resulting in the formation of chimera (an organism consisting of tissues or parts of diverse genetic constitutions).

Nuclear transfer method

Transgenic goats are formed by the nuclear transfer of fetal somatic cells. The karyoplasts of the donor were obtained from a primary fetal somatic cell line. The primary fetal somatic cell line is obtained from a 40-day transgenic female fetus that is produced by the artificial insemination of a nontransgenic adult female with semen from a transgenic male. Two types of nuclear transfer methods are used to produce transgenic animals:

i. Oocytes at the arrested metaphase II stage are enucleated, electrofused with donor somatic cells, and simultaneously activated.
ii. Activated in vivo oocytes are enucleated at the telophase II stage, electrofused with donor somatic cells, and simultaneously activated a second time to induce genome reactivation.

Transfection of gametes

The first transfection method was developed in the early 1960s. There are different types of transfection methods:

(a) In vitro procedure, in which foreign genes are introduced into cultured cells or tissues.
(b) In vivo method, in which genes are directly introduced into the tissue (by injection, aerosol, etc.).
(c) Ex vivo system, in which cells are transfected in vitro and then introduced into a living organism.

Artificial chromosome-mediated gene transfer

A group of nuclei is injected with transgene DNA. The eggs are transferred into a medium of incubation and visual evaluation is done within a few hours. The animal produced after receiving the transgene DNA is referred to as the founder of a new transgenic lineage. By using pronuclear microinjection, yeast artificial chromosome (YAC) transgenic mice were produced, representing the latest generation of vectors. This method was successful in mice and rabbits.

Testis cell transplantation method

This method includes the following steps:

(a) A single-cell suspension is produced from a fertile donor testis.
(b) The cells are cultured.
(c) The cells are microinjected into the lumen of seminiferous tubules of an infertile recipient mouse.
(d) Only a spermatogonial stem cell can produce a colony of spermatogenesis in the recipient testis. When testis cells carry a reporter transgene, which takes a blue color on staining, colonies of donor cell-derived spermatogenesis are identified easily in recipient testes as blue stretches of tubule.
(e) Mate the recipient male to a wild-type female.
(f) Produces progeny, which carry donor genes.

Lentiviral transfer of oocytes and zygotes

This method is employed to beat the limitations of viral-mediated gene transfer, containing the silencing of the transgenic locus and low expression levels. Transgenic cattle production by lentiviruses requires microinjection into the oocytes. These vectors (lentiviruses) have several merits compared to conventional retrovectors. They carry large amounts of transgene (10 kb), can infect nondiving cells, and can show stable expression within the tissue where they are introduced.

Chimera generation by injecting pluripotent cells

Embryonic stem cells have the ability to require part in organ and germ cell production after injection into the blastocysts. Embryonic stem cells are important for producing gene knock-ins and gene knockouts as large chromosomal rearrangements. Cattle and chimeric animals can be produced by primordial germ cells.

Applications of transgenic animals

1. Drug and industrial production

Transgenic animals are used for the production of proteins such as alpha-1-antitrypsin that may be employed in the treatment of emphysema or cystic fibrosis. Recombinant human proteins are produced within the mammary glands of transgenic animals. Goats are accustomed to produce silk, milk and they secrete tiny strands from their body. To obtain monoclonal antibodies, the mammary gland of the transgenic goat is used. By using transgenic cattle, a recombinant bispecific antibody is produced within the blood.

2. As a disease model

Transgenic animals are produced as disease models for conditions such as AIDS, Alzheimer's disease, cancer, etc. Transgenic animals help scientists know the role of genes in specific diseases. Transgenic animals such as mice are used for the analysis of different hereditary diseases.

3. As food

Cloned animals and their products are safe to eat for humans. Due to muscle hypertrophy, they are associated with dystocia requiring caesareans, poor viability of calves, and poor fertility.

4. Vaccine testing

Transgenic animals are used for vaccine testing.

5. Disease control

In Australia, by altering the genes of the mousepox virus, mice were developed that do not have the disease. Genetically modified mosquitoes were produced so they cannot produce malaria.

6. Xenotransplantation

Presently, because of the successful transplantation of an appropriate allotransplantation, approximately 250,000 people are alive. Sometimes there is a problem of appropriate organs or the rejection of live organ donation. So, to unravel this problem, porcine xenografts from domesticated pigs are considered the best choice. The genetically modified pigs are often used as a source animal for tissue and organ transplantation in people in general. In order to prevent hyperacute rejection and acute vascular rejection, human regulators of complement activity are produced in transgenic pigs.

7. Toxicity testing

Transgenic animals are utilized in toxicity testing.

8. Blood replacement

Transgenic swine are accustomed to producing human hemoglobin. Porcine blood, which is analogous to human hemoglobin, can be used for the purification of protein obtained from transgenesis.

9. Agriculture

Transgenic pigs are utilized to produce pork by using the spinach desaturase gene. The spinach desaturase gene produces a great deal of nonsaturated fatty acids. These nonsaturated fatty acids are used for diet purposes and are advantageous in cutting back the dangers of stroke and coronary disease. Transgenic animals are used for milk production. By altering the composition of lactose, milk production was increased by transgenic pigs. Scientists are trying to produce disease-resistant animals, such as influenza-resistant pigs, but a very limited number of genes are currently known to be responsible for resistance to diseases in farm animals.

Stem cell technology

Stem cells are the cellular structures found in multicellular organisms. Every single cell in the body originates from this type of cell. These cells are unspecialized cells, which can multiply themselves through mitotic cell division and can differentiate into special types of cells. These cells remain in an unspecialized state until they receive a specific signal to become specialized. There are two types of stem cells: adult stem cells, which are found in adult tissues, and embryonic stem cells, which are found in the inner cell mass of the blastocyst. In the embryo, the stem cell can differentiate into different types of specialized embryonic tissues. In adult organisms, the stem cell and progenitor cell act as the repair system of the body.

In the blastocyst, the stem cells produce multiple specialized cell types, which form heart, lung, skin, and other tissues. In bone marrow, brain, and muscle, the adult stem cells helps in replacements for the cells lost during normal wear and tear, injury, or disease. These cells are not just being used in cell-based therapies, but also for screening new drugs and understanding birth defects. Embryonic stem cells are being used in developmental biology, tissue regeneration, disease pathogenicity, regenerative medicine, and drug discovery (Wolf et al., 2004). Stem cells are obtained from an embryo or individual.

Characteristics of stem cells

Stem cells are different from other types of cells in the body. Irrespective of their source, all stem cells have three common characteristics:

i. Unspecialized

The stem cell does not have tissue-specific structures, which allows them to create specialized structures. A stem cell such as a red blood cell cannot carry oxygen molecules through the bloodstream. Like a nerve cell, it cannot fire electrochemical signals to other cells that allow the body to speak or move, and like a heart muscle cell, it cannot pump blood throughout the body. However, under certain experimental or pathologic conditions, their potency can be induced.

ii. Give rise to specialized cells

Through differentiation, the unspecialized stem cells give rise to specialized cells. Stem cell differentiation is induced by internal and external signals. The internal signals are controlled by genes, whereas the external signals include physical contact with adjacent cells, chemicals secreted by other cells, and certain molecules in the microenvironment.

iii. Self-Renewal

Unlike muscle cells, blood cells, and nerve cells, stem cells have the ability of self-renewal. They renew themselves for a longer period of time through cell division. When cells replicate themselves persistently it is called proliferation.

Types of stem cells

Normally, two types of stem cells are found in nature: adult or somatic stem cells and embryonic stem cells.

Adult stem cells

An adult or somatic stem cell is an undifferentiated cell found among differentiated cells in a tissue or organ. It has the ability of self-renewal and differentiation into major specialized cell types of the tissue or organ. In a living organism, the first role of adult stem cells is to take care of and repair the tissue within which they are found. They have been identified in several tissues and organs. A number of adult stem cells found in organs and tissues are very small. The adult stem cells are found in the brain, bone marrow, peripheral blood vessels, skeletal muscles, skin, and liver. The adult stem cells are found in a specific area of a tissue, where they are found in a quiescent state (nondividing stage) for many years until they are activated by disease or tissue injury.

Characterization of adult stem cells

To identify or characterize adult stem cells, there are three methods:

- Labeling the cells in a living tissue and then determining the specialized cell types they generate.
- Removing the cells from the living animal, labeling in a cell culture, and transplanting them back into another animal to identify whether the cells repopulate their tissue of origin.
- Isolating the cells, growing them in cell culture, and manipulating them by adding growth factors or introducing new genes to determine what differentiated cell types they can become.

Differentiation of adult stem cells

The adult stem cells enter into normal pathways to form the specialized cell type of the tissue in which they are found. Sometimes, adult stem cells may have the ability to form specialized cell types of other tissue, which is known as transdifferentiation or plasticity.

Normal differentiation

In a living animal, adult stem cells can divide for an extended period of time and may create mature cell types, which have the characteristic shapes and specialized structures and functions of a particular tissue. For example:

- Neural stem cells in the brain give rise to nerve cells (neurons), astrocytes, and oligodendrocytes.
- Bone marrow stromal cells (mesenchymal stem cells) give rise to a variety of cells such as osteocytes, chondrocytes, adipocytes, and other kinds of connective tissue cells such as those in tendons.
- Hematopoietic stem cells give rise to all types of blood cells such as red blood cells, B lymphocytes, T lymphocytes, NK cells, neutrophils, basophils, eosinophils, monocytes, macrophases, etc.

- Epithelial stem cells, which are found in the lining of the digestive tract in deep crypts, give rise to several cells such as absorptive cells, goblet cells, paneth cells, and entero endocrine cells.
- Skin stem cells are present in the basal layer of the epidermis and the base of hair follicles. The epidermal stem cells create keratinocytes, which migrate to the surface of the skin and form a protective layer. The follicular stem cells can create both hair follicles and the epidermis.

Plasticity or transdifferentiation of adult stem cells

The power of adult stem cells to differentiate into multiple cell types is termed plasticity or transdifferentiation. For example:

- Hematopoietic stem cells can differentiate into three major styles of brain cells (neurons, oligodendrocytes, and astrocytes), skeletal muscle cells, and liver cells.
- Brain stem cells may differentiate into cardiac muscle cells and skeletal muscle cells.
- Bone marrow stromal cells can differentiate into blood cells and skeletal muscle cells.

Embryonic stem cells

Embryonic stem cells have the ability to propagate indefinitely in nature, maintaining a normal karyotype and their undifferentiated state. They have the ability to differentiate into any specialized cell type in the body (Rao and Stice, 2004). They are derived from the inner cell mass of blastocysts and are pluripotent in nature, which suggests that they can differentiate into all derivatives of the three primary germ layers: the ectoderm, endoderm, and mesoderm. Embryonic stem cells maintain pluripotency through multiple cell divisions. Pluripotent stem cells are capable of producing gametes in vitro and the differentiation of the trophoectoderm. Embryonic stem cells are employed in many fields such as developmental biology, regenerative tissue, regenerative medicine, disease pathogenicity, and drug discovery.

Culture of embryonic stem cells

Embryonic stem cells are isolated by transferring the inner cell mass into a plastic laboratory culture dish containing a culture medium. The cells divide and spread over the surface of the dish. The inner surface of the culture dish is coated with mouse embryonic skin cells that are treated so that they will not divide. This layer of the cells is termed the feeder layer (Kuldip et al.,

2008). The presence of a mouse cell in the bottom of the culture dish provides a sticky surface to the inner cell mass cell for attachment. The feeder cell also reduces the nutrient into the culture medium. After several days, the inner cell mass proliferates and starts to form a confluence on the culture dish. Subsequently, they are gently removed and plated into several fresh culture dishes. The method of replating the cells is termed subculturing and is repeated many times for several months. Each subculturing cycle of cells is referred to as a passage. After 6 months or more, the original cells of the inner cell mass yield millions of embryonic stem cells. Embryonic stem cells that have proliferated in a cell culture for six 6 more months without differentiating, as pluripotent and appear genetically normal are spoken as embryonic stem cell line. Once cell lines are established, or even before that stage, batches of them can be frozen for future use.

Characterization of adult stem cells

During the generation of embryonic stem cell lines, their fundamental properties are checked. This process is named characterization. The cells are characterized by the following tests:

- Growing and culturing the stem cells for several months to confirm that cells have the capability of long-term self-renewal. The cells within the culture are checked with the assistance of a microscope to determine that they are healthy and remain undifferentiated.
- Identification of cell surface and intracellular markers.
- Karyotyping by examination of chromosomes under a microscope to see whether the chromosomes are damaged or changed.
- Freezing and thawing to understand whether the cells can be subcultured.
- Testing of the pluripotency of embryonic stem cells by allowing the cells to differentiate spontaneously in a cell culture, manipulating the cells, or injecting the cells into immunosuppressed mice to check for the formation of a benign tumor or for detecting the presence of pluripotent gene markers (nanog and alkaline phosphatase markers).

Application of stem cells

- Stem cells may be utilized in basic research, clinical research, and for production improvement.
- Pluripotent stem cells would allow drug testing in a wider range.
- Generation of cells and tissues that might be used for cell-based therapies.

- Adult stem cells are used extensively to analyze gene function in development via gene targeting.
- Differentiation of embryonic stem cells or adult stem cells in cell culture to make insulin-producing cells that would be employed in transplantation therapy for diabetics.
- Production of cloned and transgenic animals.
- Production of in vitro gametes from pluripotent stem cells.

Challenges and way forward

Technological advancements have made tremendous strides in animal reproduction. However, these techniques face various challenges such as paucity of databases on indigenous livestock and its biodiversity, including production, reproduction, disease-resistant traits within species and breeds, infrastructure, high cost, lack of expertise, and interface between a public and private partnership (Choudhury et al., 2016). Ultrasonography and other monitoring modalities have revolutionized animal reproduction technologies. Various innovative micro- and nanosensing devices such as chip/bio microelectromechanical system (bioMEMS) microprocessors will come into play to monitor the health of the fetus to augment fertility in farm animals.

References

Barcelo-Fimbres, M., Seidel, G.E., 2007. Effects of fetal calf serum, phenazineethosulfateand either glucose or fructose during in vitro culture of bovine embryos on embryonic development after cryopreservation. Mol. Reprod. Dev. 74, 1395–1405.

Besenfelder, U., Modl, J., Muller, M., Brem, G., 1997. Endoscopic embryo collection and embryo transfer into the oviduct and the uterus of pigs. Theriogenology 47, 1051.

Betteridge, K.J., 1981. An historical look at embryo transfer. J. Reprod. Fertil. 62 (1), 1–13.

Bondioli, K.R., Westhusin, M.E., Looney, C.R., 1990. Production of identical bovine offspring by nuclear transfer. Theriogenology 33 (1), 165–174.

Brackett, B.G., Bousquet, D., Boice, M.L., Donawick, W.J., Evans, J.F., Dressel, M.A., 1982. Normal development following in vitro fertilization in the cow. Biol. Reprod. 27, 147–158.

Carver, A.S., Dalrymple, M.A., Wright, G., Cotton, D.S., Reeves, D.B.D., Gibson, Y.H., Keenan, J.L., Barrass, J.D., Scot, A.R., Colman, A., Garner, I., 1993. Transgenic livestock as bioreactor: stable expression of human alpha-l -antitrypsin by a flock of sheep. Bio/Technology 1 (I), 1263–1270.

Chang, M.C., 1959. Fertilization of rabbit ova in vitro. Nature 184, 466–467.

Chen, S.U., Lien, Y.R., Chen, H.F., Chang, L.J., Tsai, Y.Y., Yang, Y.S., 2005. Observational clinical follow-up of oocyte cryopreservation using a slow freezing method with 1, 2-propanediolplus sucrose followed by ICSI. Hum. Reprod. 20, 1975–1980.

Choudhury, K.K., Kavya, K.M., Jerome, A., Sharma, R.K., 2016. Advances in reproductive biotechnologies. Vet. World 9 (4), 388–395.

Colmon, A., 1996. Production of proteins in the milk of transgenic livestock: problems, solutions and success. Am. J. Clin. Nutr. 63, 639–645.

Cran, D.G., Johnson, L., Miller, N.G.A., Cochrane, D., Polge, C.E., 1993. Production of bovine calves following separation of X-and Y-bearing sperm and in vitro fertilization. Vet. Rec. 132, 40–41.

Gajda, B., Smorg, Z., 2009. Oocytes and embryos cryopreservation-state of art and recent development in domestic animals. J. Anim. Feed Sci. 18, 371–387.

Galli, C., Lazzari, G., 2008. The manipulation of gametes and embryos in farm animals. Reprod. Domest. Anim. 43, 1–7.

Hare, W.C.D., Betteridge, K.J., 1978. Relationship of embryo sexing to other methods of prenatal determination in farm animals: a review. Theriogenology 9, 27–43.

Hochi, S., Ito, K., Hirabayashi, M., Ueda, M., Kimura, K., Hanada, A., 2000. Effect of nuclear stages during IVM on the survival of vitrified warmed bovine oocytes. Theriogenology 49, 787–796.

Hoque, M., 2011. Ultrasonography—a diagnostic tool for examination of ovarian and uterine structures in animals. In: Singh, S.K., Ghosh, S.K., Kumar, H., Agarwal, S.K. (Eds.), Advances in Reproductive Technologies to Augment Fertility in Farm Animals. Indian Veterinary Research Institute, Izatnagar, India, pp. 188–191.

Huang, W.T., Holtz, W., 2002. Effect of meiotic stages, cryoprotetants, cooling and vitrification on the cryopreservation of porcine oocytes. Asian Australas. J. Anim. Sci. 15 (4), 485–493.

Hurt, A.E., Landim-Alvarenga, G.E., Siedel, J.R., Squires, E.L., 2000. Vitrification of immature and mature equine and bovine oocytes in ethylene glycol, ficoll and sucrose solution using open-pulled straws. Theriogenology 54, 119–128.

Ivanoff, E.I., 1922. On the use of artificial insemination for zootechnical purposes in Russia. J. Agric. Sci. 12, 244–256.

Johnson, L.A., Cran, D.G., Polge, C., 1994. Recent advances in sex preselection in cattle: flow cytometric sorting of X- and Y-chromosome bearing sperm based on DNA to produce progny. Theriogenology 41, 51–56.

Kuldip, S.S., John, P.R., Bernie, E.T., 2008. Derivation of a new human embryonic stem cell line, Endeavour-1, and its clonal propagation. Stem Cells Dev. 17 (1), 41–52.

Lanza, R.P., et al., 2002. Generation of histocompatible tissues using nuclear transplantation. Nat. Biochnol. 20, 689–696.

Luz, M.R., Holanda, C.C., Pereira, J.J., Teixeira, N.S., Vantini, R., Freitas, P.M.C., Salgado, A.E.P., Oliveira, S.B., Guaitolini, C.R.F., Santos, M.C., 2009. Survival rate and in vitro development of in vivo produced and cryopreserved dog embryos. Reprod. Fertil. Dev. 22, 208–219.

McKelvey, W.A.C., Robinson, J.J., Aitken, R.P., Robertson, I.S., 1986. Repeated recoveries of embryos from ewes by laparoscopy. Theriogenology 25, 855.

Meena, C.R., Das, S.K., 2005. *In vitro* development of reconstructed water buffalo *(Bubalus bubalis)* oocytes after fetal skin fibroblast cell nuclear transfer. Reprod. Fertil. Dev. 17 (2), 248.

Milovanov, V.K., 1938. The Artificial Insemination of Farm Animals. Seljhozgiz, Mowcow.

Mishra, A.K., Prashad, S., 2005. Application of reproductive biotechnology to augment productivity of dairy animals: a review. Indian J. Anim. Sci. 75 (7), 893–904.

Monk, M., Hadyside, A.H., 1988. Sexing of pre-implantation mouse embryos by measurement of X-linked gene dosage in a single blastomere. J. Reprod. Fertil. 82, 365–368.

Muller, M., Brenig, B., Winnacker, E.L., Brem, G., 1992. Transgenic pigs carrying c DNA copies encoding the murine MxL protein which confers resistance to influenza virus infection. Gene 121, 263–270.

Mylne, M.J.A., McKelvey, W.A.C., Fermie, K., Mathews, K., 1992. Use of a transcervical technique for embryo recovery in sheep. Vet. Rec. 130, 450.

Niemann, H., Wrenzycki, C., 2000. Alterations of expression of developmentally important genes in preimplantation bovine embryos by in vitro culture conditions: implications for subsequent development. Theriogenology 53, 21–34.

Nishikawa, Y., 1962. Fifty Years of Artificial Insemination of Farm Animals in Japan. English Bull. 2. Kyoto University, Japan.

Palta, P., Chauhan, M.S., 1998. Laboratory production of buffalo (*Bubalus bubalis*) embryo. Reprod. Fertil. Dev. 10, 379–391.

Pereira, R.J., Sohnrey, B., Holtz, W., 1998. Nonsurgical embryo collection in goats treated with prostaglandin F_2 alpha and oxytocin. J. Anim. Sci. 76, 360.

Peura, T., Hyttinen, J.M., Turunen, M., Janne, J., 1991. A reliable sex determination assay for bovine preimplantation embryos using the polymerase chain reaction. Theriogenology 35, 547–555.

Polge, C., Smith, A.U., Parkes, A.S., 1949. Revival of spermatozoa after vitrification and dehydration at low temperatures. Nature 164, 666.

Ptak, G., Clinton, M., Barboni, B., 2002. Preservation of the wild European mouflon: the first example of genetic management using a complete program of reproductive biotechnologies. Biol. Reprod. 66, 796–801.

Pursei, V.G., Rexroad, C.E., 1993. Status of research with transgenic farm animals. J. Anim. Sci. 71 (Suppl. 3), 10–19.

Rall, W.F., Fahy, G.M., 1985. Ice-free cryopreservation of mouse embryos at −196 °C by vitrification. Nature 313, 573–575.

Rao, R.R., Stice, S.L., 2004. Gene expression profiling of embryonic stem cells leads to greater understanding of pluripotency and early developmental events. Biol. Reprod. 71, 1772–1778.

Salisbury, G.W.N.L.V.D., Lodge, J.R., 1978. Physiology of Reproduction and Artificial Insemination of Cattle, second ed. W.H. Freeman Co, San Francisco.

Seidel, G.E. Jr., 1988. Sexing spermatozoa and embryos—state of art. In: Proceeding of XI International Congress of Animal Reproduction and Artificial Insemination (Dublin), vol. 5 University College, Dublin. Pg. 136–144.

Siedel, G.E., 2003. Sexing mammalian sperm—interwining of commerce, technology and biology. Anim. Reprod. Sci. 79, 145–156.

Singh, K.P., Saxena, A., Kharche, S.D., Singh, P., 2009. Studies of *in vitro* maturation, fertilization and cleavage rate of prepubertal and pubertal goat oocytes. Indian J. Anim. Sci. 79 (6), 550–553.

Singh, K.P., Saxena, A., Kharche, S.D., Singh, P., 2010. Effect of granulosa cell monolayer and oviductal epithelial cell co-culture on cleavage rate and embryo development of *in vitro* fertilized goat oocytes. Indian J. Anim. Sci. 80 (3), 209–212.

Singla, S.K., Manik, R.S., Madan, M., 1997. Micromanipulatjon and cloning studies on buffalo oocytes and embryo using nucleus transfer. Indian J. Exp. Biol. 35, 1273–1283.

Sirard, M.A., Richard, F., Blondin, P., Robert, C., 2006. Contribution of the oocyte to embryo quality. Theriogenology 65, 126–136.

Spallanzani, L., 1784. Dissertations Relative to the Natural History of Animals and Vegetables. Trans. by T. Beddoes in Dissertations Relative to the Natural History of Animals and Vegetables, vol. 2 J. Murray, London, pp. 195–199.

Sreenan, J.M., 1978. Non-surgical embryo transfer in the cow. Theriogenology 9, 69.

Tiffin, G.J., Rieger, D., Betteridge, K.J., Yadav, B.R., King, W.A., 1991. Glucose and glutamine metabolism in pre-attachment cattle embryos in relation to sex and stage of development. J. Reprod. Fertil. 93, 125–132.

Vajta, G., Holm, P., Greve, T., Callesen, H., 1996. Overall efficiency of in vitro embryo production and vitrification in cattle. Theriogenology 45, 683–689.

White, K.L., Anderson, G.B., Berger, P.J., Bon Durant, R.H., Pashen, R.L., 1985. Expression of a male-specific factor (H-Y antigen) on preimplantation porcine embryos. J. Anim. Sci. 81 (1), 408–416.
Willadsen, S.M., 1986. Nuclear transplantation in sheep embryos. Nature 320, 63–65.
Wilmut, I., Schnieke, A.E., McWhir, J., Kind, A.J., Campbell, K.H.S., 1997. Viable offspring derived from fetal and adult mammalian cells. Nature 385 (6619), 810–813.
Wolf, D.P., Kuo, H.C., Francis, P.K.Y., Lester, L., 2004. Progress with non human primate embryonic stem cells. Biol. Reprod. 71, 1766–1771.
Ziomek, C.A., 1998. Commercialization of proteins produced in the mammary gland. Theriogenology 49, 139–144.

CHAPTER SIXTEEN

Targeted genome editing by CRISPR/Cas9 for livestock improvement

Darshika Nigam and Kumari Bhoomika
School of Health Sciences, Quantum University, Roorkee, India

Introduction

In view of the ever-increasing worldwide population, we are facing the biggest challenge in fulfilling the increasing demand for food. Apart from the agriculture sector, the livestock sector is a prime component of the world's agrarian economy, providing food, nutritional security, and livelihood (FAO, 2009). Hence, genetic improvement programs for livestock have been an incredible impetus for improving efficiencies and to retreat the ecological footprint and outflow power of milk, meat and egg creation. To increase the productivity of livestock by improving nutrition, health, immunity, genomic information, and reproduction, classical breeding approaches have been used for many decades. But, these are not enough to meet the current needs of the world, so scientists have been continuously working to develop novel targeted genome editing tools and applications for the last decade (Clark and Whitelaw, 2003; Gaj et al., 2016; Ruan et al., 2017).

Advanced genome editing and transgenic technologies provide the most promising tools to hasten the rate of genetic gain in livestock by creating novel, more advantageous alleles or by introducing desirable alleles from other breeds or species more precisely than the traditional breeding technologies by removing the consequences of linkage drag. Genome editors contain a DNA-binding domain and a cleavage domain, which can be designed to bind to almost any known sequence of DNA. Gene knockout, the targeted insertion of a specific sequence, or single nucleotide polymorphism (SNP) can be achieved by selecting for different outcomes of DNA modification. More than nine genome editors are available and four families of engineered nucleases are primarily being used: meganucleases (MNs)

Emerging Issues in Climate Smart Livestock Production
https://doi.org/10.1016/B978-0-12-822265-2.00002-8

(Silva et al., 2006); zinc finger nucleases (ZFNs) (Smith et al., 2000); transcription activator-like effector-based nucleases (TALENs); (Wu et al., 2015) and the clustered regularly interspaced short palindromic repeats (CRISPR) and associated protein 9 (CRISPR/Cas9) system (Doudna and Charpentier, 2014). The CRISPR/Cas9 system is the most recent one.

The CRISPR/Cas9 system is faster, cheaper, more precise, and more efficient than other existing genome editing methods and is the most widely used genome editing tool in different organisms for various biological applications. The CRISPR/Cas9 system has been adapted from a naturally occurring genome editing system in bacteria and archaeal species, where it is a system of adaptive immunity and functions to protect against invading viruses (Jinek et al., 2012). This system targets a Cas nuclease to specific DNA sequences using small noncoding RNAs. This novel and most promising CRISPR/Cas9 system is based on the system of Streptococcus bacteria and uses a short guide RNA (sgRNA) complexed with Cas9 nuclease. Based on different mechanisms in generating CRISPR RNA (crRNA) and Cas proteins, there are several types of CRISPR/Cas systems. They are primarily categorized as Class 1 and Class 2 systems. The Class 2 system was adapted by researchers for genome editing due to its simplicity in requiring only the CRISPR-associated protein 9 (Cas9) protein from *Streptococcus pyogenes* and two other RNA components, crRNA and trans-activating crRNA (tracrRNA), that fused to become guide RNA (gRNA) (Wei et al., 2015). Some details of both classes will be discussed later in the chapter.

The discovery of CRISPR/Cas9 has revolutionized the field for targeted genomic engineering in a short span since its appearance. The number of reports has radically increased since the first description of the successful use of CRISPR/Cas for targeting a specific genomic locus. Most researchers have switched to the CRISPR/Cas9 system as a genome editor. Due to the CRISPR-Cas9 system's high significance and contribution in targeted genomic engineering, the Nobel Prize in Chemistry in 2020 was awarded to Emmanuelle Charpentier and Jennifer Doudna, who were the first ones to propose that the system could be used for the programmable editing of genomes (Nobel Foundation, 2020). So far, the CRISPR/Cas technology has been extensively adapted to modify the genomes of livestock, such as pigs, cattle, goat, sheep, etc. This chapter will provide an understanding of the CRISPR/Cas9 system as a new tool in molecular biology, its mechanism of action, a comparative study with other available genome editors, and its applications for livestock improvement and human welfare.

Genome editors

In the genome or gene editing process, DNA is inserted, deleted, or replaced in the genome of a living organism. Unlike early genetic engineering techniques that arbitrarily incorporated genetic material into a host genome, novel genome editing tools precisely insert or delete a gene at desired locations with a targeting rate of 10,000–100,000 folds in compared to the old techniques being used earlier. Genome editors (GEs) can manipulate almost any genomic sequence to create animal models for scientists to study the causes of human diseases and formulate new possibilities for human gene therapy. Further, GEs can be applied in livestock for their welfare and improvement in the productive traits and in plants as well to improve their shelf life and nutritional qualities and to control pests in agriculture (Nicholl, 2008). These GEs create site-specific double-strand breaks (DSBs) at desired locations in the targeted genome. Induced DSBs are then repaired through two key pathways: nonhomologous end joining (NHEJ) and homology directed repair (HDR). This results in targeted mutations. NHEJ utilizes a variety of enzymes to directly join the DNA ends, though HDR utilizes a homologous DNA template for the regeneration of missing DNA sequences at the DSB point, which is more accurate (Jasin, 1996; Lieber, 2010). However, restriction enzymes used to create DSBs are effective, but usually recognize and cut target DNA at multiple sites. To overwhelm this problem, four classes of novel programmable DNA nucleases have been employed for genome editing purposes: meganucleases (MNs), zinc finger nucleases (ZFNs), transcription activator-like effector (TALE)-based nucleases (TALENs), and clustered regularly interspaced short palindromic repeats (CRISPR) with the RNA-guided Cas9 endonuclease. Table 1 shows a comparison of the various properties of these four GEs.

Meganucleases (MNs)

The most specific naturally occurring restriction enzymes are meganucleases (endodeoxyribonucleases). MNs are characterized by a large recognition site of 12–40 base pairs and thus have the potential to excise large pieces of the DNA sequence (Silva et al., 2006). MNs are used to modify targeted DNA sequences of all living organisms in a highly specific manner. MNs represent two types of homing endonucleases: intron endonucleases and intein endonucleases, which collectively produce breaks in complementary intron/intein free alleles. On the basis of sequence and structure motifs, MNs

Table 1 Comparative properties of genome editors.

Properties	ZFNs	TALENs	Meganucleases	CRISPR/Cas9 system
Nuclease	FokI	FokI	Homing endonucleases	Cas9
Architecture	Modular/dimer	Modular/dimer	Integrated	Modular/RNA-protein
Type of recognition	Protein–DNA	Protein–DNA	Protein–DNA	RNA–DNA
Recognition site	18–36 bp per ZFN pair	28–40 bp per TALEN pair	18–22 bp	22 bp
Targeting constraints or limitations	Prefer GC-rich targets	Prefer GC-rich targets	Central four bases intolerant to change	PAM required
Repair	NHEJ	HDR	HDR	NHEJ
Off-target effect/ specificity	More potential off-target effects than TALENs/nonspecific	Less observed off-target effects/specific	Very specific	More potential off-target effects than TALENs and Z FNs/ nonspecific
Coding sequence size	Short (1000 bp per ZFN monomer)	Long and repetitive (3000 bp per TALEN monomer)	Short (1000 bp)	Long (4200 bp + gRNA)
Multiplex potential	Low	Low	Medium	High
Efficiency	~12%	~76%	~80%	76%–100%
Ease of design	Difficult	Moderate	Easy but problematic	Easy
Programmable	Moderate	Moderate/high	Moderate/high	Moderate/high
Dimerization required	Yes	Yes	No	No
Vector packaging	Difficult	Difficult	Easy	Moderate
Therapeutic delivery	Moderate	Moderate	Moderate	Easy
Cost-effective	No	Moderate	No	Yes

can be divided into five families: LAGLIDADG, GIY-YIG, HNH, His-Cys box, and PD-(D/E)XK (Stoddard, 2006; Zhao et al., 2007). Among these families, the LAGLIDADG family of endonucleases has become a valuable tool for genome editing. They either function as RNA maturases involved in facilitating the splicing of their own intron or they function as highly specific endonucleases capable of recognizing and cleaving the exon-exon junction sequence where their intron resides (Chatterjee et al., 2003). MNs are more stringent at recognizing DNA sequences than ZFNs and thus cause less toxicity in cells. However, unlike ZFNs and TALENs, MNs are not modular, which limits their utility for genome editing. Further, they are costlier and more time consuming in genome editing. A fusion form of GEs known as megaTALs (fusion of homing endonuclease of MN and TALE binding domain) is now being used that has great specific gene modification ability (Boissel et al., 2014).

Zinc finger nucleases (ZFNs)

ZFNs are artificial restriction enzymes generated by the fusion between a zinc finger DNA-binding domain (containing the Cys2-His2 zinc-finger protein) and the nonspecific DNA-cleavage domain (*Fok*I restriction endonuclease mainly) (Doyon et al., 2011; Kim et al., 1996; Porteus and Carroll, 2005). A functional ZFN has dimerized the FoxI cleavage domain. Two ZFNs are required to target nonpalindromic DNA sites to increase the number of targeted base pairs twice. They fuse with the targeted DNA by their C-terminal parts in a tail-to-tail orientation. These two ZFNs are 5–7 base pairs apart from each other on the targeted DNA. The DNA-cleavage domain makes a cut in targeted DNA within a 5–7 bps spacer sequence (Smith et al., 2000). To increase the efficiency and specificity of the zinc finger, multiple ZFNs can be used to make a bigger DNA domain to be recognized (Doyon et al., 2011; Miller et al., 2007).

One major concern associated with ZFNs is that they are not specific enough for their target sites and hence, off-target mutations may occur. As a result, enough DSBs are produced and consequently yield chromosomal rearrangement or cell death. Off-target mutations also promote the random integration of donor DNA. To resolve this problem, *Fok*I with heterodimers can be used to minimize the potential of ZFNs to dimerize at off-target sites (Miller et al., 2007; Pattanayak et al., 2011).

Transcription activator-like effector endonucleases (TALENs)

TALENs are naturally produced by plant pathogens such as Xanthomonas (Gram-negative bacteria) when they infect a wide variety of plant species (Boch and Bonas, 2010; Voytas and Joung, 2009). TALEs bind to their host DNA and activate the expression of plant genes that help bacterial infection. TALENs are restriction enzymes that can be engineered to cut specifically recognized DNA sequences. Like ZFNs, TALENs are also designed by fusing the N-terminal of the TALE DNA-binding domain (effector domain) to the C-terminal of the DNA cleavage domain that is the nonspecific FokI endonuclease (Christian et al., 2010; Miller et al., 2011). Like ZFNs, in TALENs, the functional FokI cleavage domain is dimerized, which makes a cut within a 12–19 bp spacer sequence that separates each TALE binding site (Miller et al., 2011). The TALE DNA-binding domain contains repeats of a highly conserved 33–34 amino acid sequence with divergence (highly variable) at the 12 and 13 amino acid positions. This variable is known as the repeat variable di-residues (RVDs). The RVD recognizes the specific nucleotide sequence of the target DNA to which TALE will bind. By creating a slight change in RVD, its target specificity can be improved. The most commonly used RVDs are NI (Asn/Ile) for adenine, HD (His/Asp) for cytosine, NG (Asn/Gly) for thymine, and NN (Asn/Asn) or HN (His/Asn) for guanine or adenine (Deng et al., 2012; Moscou and Bogdanove, 2009). The nuclease of TALENs induces DSBs at the target site that can be repaired by NHEJ or HDR. However, NHEJ ligation is error-prone and results in gene disruptions via the introduction of small insertions or deletions (indels) or chromosomal rearrangement (Christian et al., 2010; Miller et al., 2011).

Clustered regularly interspaced short palindromic repeats (CRISPRs) and associated proteins (Cas)

The latest gene editing technology is the CRISPR/Cas system. CRISPRs are a family of DNA sequences found in the genomes of prokaryotic organisms such as archaea and bacteria that confer acquired immunity to protect against viruses. They are short sequences of DNA that originate from viral genomes and have been incorporated into the bacterial genome. A CRISPR-associated protein such as Cas9 is an endonuclease enzyme that uses CRISPR sequences as a guide to recognize and cleave the target (pathogenic) DNA or RNA. This is carried out by introducing plasmids containing Cas genes and precisely constructed CRISPRs into eukaryotic cells, so the eukaryotic genome can be cut at any desired position (Wei et al., 2015). The structure and mechanism of action of the CRISPR/Cas9 system is further detailed in the next section.

CRISPR/Cas9 system

Journey of CRISPR/Cas9 system from its discovery to development as a genome editing tool

The presence of CRISPR short repeated DNA sequences was first noticed in 1987 by Yoshizumi Ishino and his colleagues, when they accidentally cloned part of such a sequence together with their target gene of alkaline phosphatase, the "iap" gene (isozyme conversion of alkaline phosphatase), but they did not report their function (Ishino et al., 1987). In 2002, the most remarkable discovery of the association of CRISPRs with a set of homologous genes, known as CRISPR-associated systems or *Cas* genes, was reported by Jansen et al. (2002). They also were the first ones who used the acronym CRISPR as the universal name for these DNA repeats in a whole publication and reported four associated *Cas* genes (*Cas1–4*). Though they could not interpret the role of CRISPR loci, they suggested a possible association of the Cas proteins with the function of CRISPRs, as the Cas proteins showed helicase and nuclease motifs in studies.

Later on, many researches have attempted to find their function. Among such attempts, Mojica et al. (2005) took much interest in CRISPRs. Finally, in 2005, they proposed a role of spacer transcripted RNAs in recognition of target sequences in a mechanism, which could be somehow similar to the RNA interference (RNAi) mechanism found in eukaryotic cells. However, in 2007, the first experimental evidence of the involvement of the CRISPR-Cas system in the adaptive immune system of prokaryotes against viruses was published by Barrangou et al. (2007).

The detailed mechanism of the CRISPR/Cas system was published by Brouns et al. (2008). They could identify the processing of CRISPR RNA (crRNA) in *E. coli*. They observed that the CRISPR RNA precursor was cut within the repeats into mature spacer-containing RNA molecules known as CRISPR RNA (crRNA) by a complex of Cas proteins (called Cascade) and remained bound to the protein complex. They also showed that the crRNA guides were targeting dsDNA as both the orientations of crRNA (sense/antisense) provided immunity to the host bacteria. It was interestingly noted that the Cas proteins complex, crRNA, and a helicase/nuclease (Cas3) altogether were required by the host bacteria to provide immunity against DNA viruses.

In 2011, a Cas9 protein-based simpler CRISPR/Cas system was identified by researchers from *S. pyogenes* (Deltcheva et al., 2011). In 2012, one of the most promising discoveries in biological studies was published by Jennifer Doudna and Emmanuelle Charpentier. They determined that the CRISPR/Cas9 system could be programmed with RNA for genome DNA editing. In 2020, Doudna and Charpentier were awarded the Nobel Prize in Chemistry for the development of the CRISPR/Cas9 system-based genome editing technique (Press release, October 2020).

Subtypes of CRISPR/Cas system

CRISPR/Cas systems have been primarily categorized into two classes, Class 1 and Class 2, based on the involvement of Cas proteins in the degradation of foreign DNA. In Class 1 systems, a cascade of multiple Cas proteins is involved in the foreign DNA cleavage while a single large Cas protein is involved for the same in Class 2 systems. Based on the presence of a specific *Cas* gene and its protein, both classes have been divided into three types each, where types I, III, and IV come under Class 1 and types II, V, and VI fall under Class 2 (Wright et al., 2016). These six types have been further divided into 19 subtypes, which would be named as I-A, I—B, I—C, II-A, II-B, II-C and so on (Westra et al., 2016). It has been noted that the *Cas1* and *Cas2* genes are universally found in almost all types and subtypes of CRISPR/Cas systems, but *Cas3*, *Cas9*, and *Cas10* are exclusively found in type I, type II, and type III CRISPR/Cas systems, respectively (Makarova et al., 2015).

Of all the CRISPR/Cas systems, the type II of Class 2 is the most studied one due to its simplicity of mechanism, as these systems rely on a single Cas protein, the endonuclease Cas9, to recognize and degrade the target DNA sequence. Moreover, the Cas9 endonuclease system could be restructured in a much simpler form for genome editing, which has now become the most popular and widely used genome editing tool.

CRISPR/Cas9 system components

CRISPR array

A typical CRISPR array is composed of an AT-rich leader sequence trailed by a 28–37 base pairs (bps) long CRISPR and intermediary nonrepetitive 32–38 bps long spacer sequences. The length of CRISPR and the spacer sequences varies greatly in different species and even in strains. The range of CRISPR sequences may vary from 23 bps to 55 bps and that of spacers may range between 21 and 72 bps (Barrangou and Marraffini, 2014). In general, less than 50 sets of CRISPR-spacer sequences are found in a CRISPR array. The spacer sequences are derived from the DNA of bacteriophages that attacked the prokaryotes earlier, and the rapid appearance of new spacers has been noted on new viral infections (Tyson and Banfield, 2008).

Cas9

The gene *Cas9* encodes a large 160 kDa, 1368-amino-acid long protein, known as CRISPR associated protein 9 (Cas9), which has also been referred to earlier as Cas5, Csn1, or Csx12 (Jinek et al. 2014). Cas9, the exclusive protein of the CRISPR/Cas9 system, is naturally found in *S. pyogenes* so it is also referred to as SpCas9 by some researchers. It plays a prime role in the adaptive immune system of bacteria against DNA viruses. Basically, it is a dual RNA-mediated DNA nuclease enzyme that may create site-directed double-strand breaks in genomic DNA or plasmid DNA, so it may alter the cellular genome. In general, the Cas9 protein remains inactive in the prokaryotic cell until it binds with guide RNA (gRNA; For details, refer to next heading) and is generally referred to as "Apoenzyme" (Jinek et al. 2014). After the binding of gRNA, it changes its conformation to an active form, which then detects and binds to the target DNA to create a site-directed double stranded break (DSB). Hence, the Cas9 protein has been observed and studied in its three conformations,which are apoenzyme, the apoenzyme:gRNA hybrid, and the apoenzyme:gRNA:target dsDNA hybrid (Jinek et al. 2014).

The Cas9 apoenzyme is composed of two lobes, the α-helical recognition (REC) lobe and the nuclease (NUC) lobe, which remains connected through two linking segments. One linking segment is made up of the arginine-rich bridge helix and the other is of a disordered linker. The REC as well as NUC lobes further contain three domains each. In the α-helical REC lobe, Hel-I (REC I), Hel-II (REC II), and Hel-III (REC III) are present while in the NUC lobe, the conserved HNH domain, the split RuvC domain, and the more variable C-terminal domain (CTD) are present (Jinek et al. 2014). A Cas9-specific fold is also present as an elongated CTD sequence that contains PAM-interacting sites required for PAM detection (the details of PAM sequences will be discussed later in the chapter). Though both the lobes and linking segments are involved in the gRNA:target DNA hybrid binding to the protein, the REC lobe is majorly involved in the gRNA:target DNA binding to the Cas9 protein while the NUC lobe has a prime role in the cleavage of target DNA. During gRNA interaction with the protein, major contacts are made between Hel—I, the arginine-rich bridge helix, and the CTD domain. And, in the nuclease action of protein, the HNH domain takes part in the cleavage of the target DNA strand while the RuvC domain is responsible for the cleavage of the nontarget strand. Here, target and nontarget DNA strands refer to the complementary DNA strand of the gRNA sequence and the opposite complementary strand, respectively. Research studies have revealed that the PAM-recognition site remains highly disordered in the apoenzyme and after binding with gRNA, a major conformational change takes place in the Hel-III domain of the REC lobe. However, upon binding to target DNA, much less conformational change takes place.

Guide RNA

Guide RNAs (gRNA, sgRNA) are noncoding short RNA sequences responsible for the activity of the Cas9 protein and its target-specific binding, as they convert the inactive apoenzyme into its active form upon binding and contain the complementary target DNA sequences. In natural systems, gRNAs are composed of two separately encoded RNAs, the CRISPR RNA (crRNA) and the transactivating crRNA (tracrRNA), where the tracrRNA is accountable for the Cas9 endonuclease activity while the crRNA primarily binds to the target-specific DNA region (Deltcheva et al., 2011) (the details of crRNA and tracrRNA are given in the mechanism section). However, a more manageable single-guide RNA (sgRNA) by fusing the crRNA and tracrRNA was developed by Doudna and Charpentier, which indeed works in a similar manner and could find and cut the DNA target when combined with the Cas9 endonuclease. This development of sgRNA made the artificial CRISPR/Cas9 system so programmable that any DNA sequence could be targeted by manipulating the specific nucleotide sequence of this sgRNA (Jinek et al., 2012) which ultimately led this discovery to win the Nobel Prize in Chemistry in 2020. In a single sgRNA, more than one crRNA and the tracrRNAs can be packed together (Ly, 2013). This sgRNA can be inserted into a plasmid in the vicinity of the cas9 gene to transfect the target cells. Effective sgRNA sequences may be designed easily with the help of several available online tools (Brazelton et al., 2015; Mohr et al., 2016).

The artificially engineered sgRNA is a single strand of RNA that is similar to a T-shape. Starting from its 5′ end, it is comprised of a 20 nucleotides (nts) long crRNA targeting sequence including a 10 nts seed region toward the 3′ end, around 12 nts long crRNA repeat sequence, 4 nts (GAAA) long tetra loop, approx. 64 nts long tracrRNA sequence including 12 nts long crRNA antirepeat (forms repeat-antirepeat duplex) and stem loops 1, 2 and 3. There is a four nucleotide long linker between stem loop 1 and 2 (Jinek et al. 2012; Nishimasu et al. 2014). The 10 nts long seed region lying in the 20-nts long spacer sequence of crRNA at the 5′ end of sg-RNA plays a prime role in its target specificity (Semenova et al., 2011; Wiedenheft et al., 2011).

PAM sequences

The protospacer adjacent motif (PAM) is a short conserved 2–6 nt long DNA sequence present in the target (viral) DNA and is generally present 3–4 nts downstream from the targeted cleavage site. These sequences are exclusively recognized by the Cas9 protein to identify the site of cleavage in the target DNA (Bolotin et al., 2005). In all CRISPR/Cas systems, PAMs have been identified. PAMs with a specific nucleotide sequence are specific to the particular type of CRISPR system (Mojica et al., 2009). For example, the PAM in *S. pyogenes* is 5′-NGG-3′ (Anders et al., 2014; Jinek et al., 2014).

Mechanism of action

The CRISPR/Cas9 system is a naturally occurring system in prokaryotes, where it is used as an adaptive immunity system and provides natural defense against bacteriophage attacks. We also know that the CRISPR/Cas9 system falls into Class 2, Type II CRISPR/Cas systems where a single large Cas9 protein is used for the cleavage of invading bacteriophage DNA. This cleavage is strictly reliant on the presence of a PAM in the target (viral) DNA. The DNA cleavage at specific sites by Cas9 is carried out with the help of guide RNA, and this has made CRISPR/Cas9 a versatile tool for genome engineering. So, we will know how this system works in a natural system of prokaryotes for their protection, which will help to understand how this system can be useful as a genome modification tool (Fig. 1).

In all prokaryotes, CRISPR/Cas systems function in three major steps: (1) Acquisition of phage sequences, (2) CRISPR RNA (crRNA) processing, and (3) target interference and interaction. Cas9 helps in adaptation, participates in crRNA processing, and cleaves the target DNA supported by crRNA and an additional transencoded small RNA, called tracrRNA (collectively known as sgRNA).

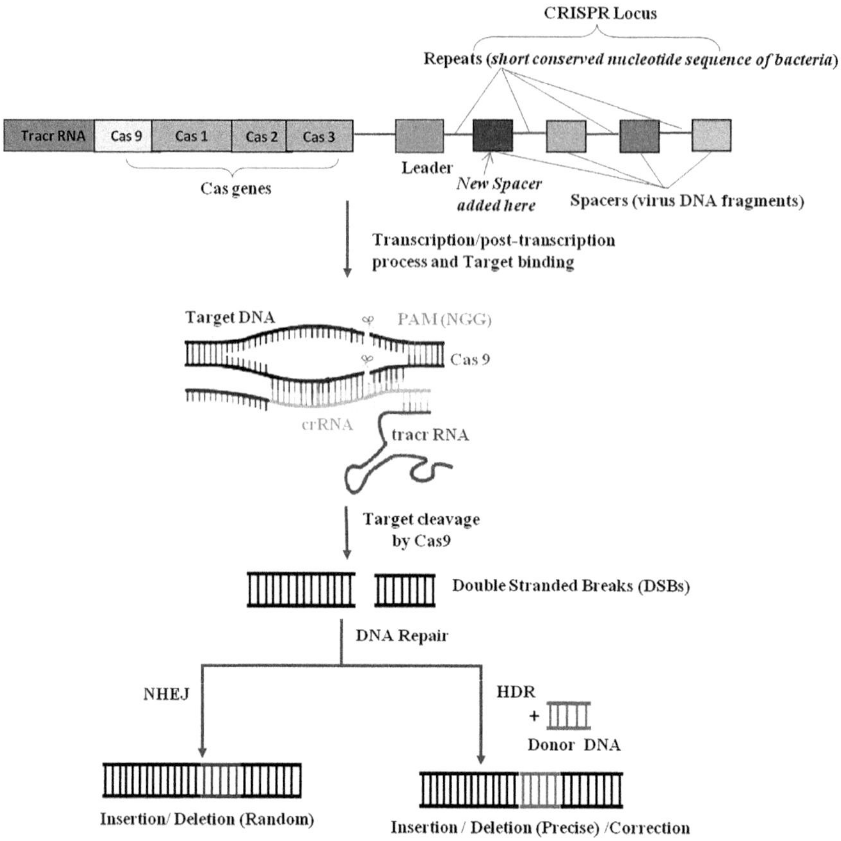

Fig. 1 Schematic representation of the CRISPR/Cas 9 mechanism.

Acquisition or adaption phase

The acquisition or adaptation phase is the very first step of CRISPR/Cas functioning once the prokaryotic cell is attacked by any bacteriophage. In this step, the DNA from an invading phage virus is processed into small segments that are integrated at the CRISPR locus as new spacers. This provides immunological memory for the ensuing processing and interference phases that counteract the secondary exposure of the same virus. It is the process in which DNA from an invading phage virus is processed into small segments that are integrated into the CRISPR sequence as new spacers. Spacers are transcribed and converted into individual crRNA, which helps effector Cas nucleases (Cas 9) to recognize and cleave specific strands of DNA that are a complementary CRISPR sequence during the subsequent infection by similar phages. The key factors of this step of spacer integration are Cas1 and

Cas2 proteins (nucleases), which are highly conserved for all CRISPR/Cas systems (as was mentioned earlier in the "subtypes of CRISPR/Cas system" section of this chapter). On the invasion of phage DNA, Cas1 and Cas2 proteins identify the PAM sequences in the phage DNA to capture spacer sequences. The spacer sequences in the phage DNA near PAM sequences are referred to as "protospacers." These protospacers are cleaved by Cas1 and Cas2 proteins to get them ligated in the CRISPR array as new spacer sequences. A new spacer gets integrated at the leader-repeat boundary of the CRISPR array while the first repeat of the array is duplicated (as new repeat–new spacer unit forms). Cas1 binds with CRISPR DNA in a Cas2-dependent manner (Anders et al., 2014; Jinek et al., 2012, 2014).

CRISPR RNA processing

After integration of the spacer in the CRISPR array of the host prokaryotic cell, spacers get transcribed and converted into individual crRNA, which helps effector Cas nucleases (Cas9 in CRISPR/Cas9 systems) to recognize and cleave specific strands of invading phage DNA that are complementary to CRISPR spacer sequences during the subsequent infection by similar phages.

To enable immunity, CRISPR repeats and spacers in the bacterial DNA undergo transcription into a long precursor CRISPR RNA (pre-crRNA) containing repeats and different spacers at the leader region of the CRISPR arrays. This pre-crRNA chain is further processed into mature CRISPR RNA (crRNA) containing a single spacer (memorized sequences of invaders) and a partial CRISPR repeat. In CRISPR/Cas9 systems, crRNA biogenesis is dependent on host RNase III and tracrRNA that base pairs with the pre-crRNA. The tracrRNA is a noncoding RNA that transcribes separately from a specific upstream gene of the CRISPR/Cas domain. The tracrRNA has a prime role in the processing of pre-crRNA repeats for crRNA maturation, as its annealing to pre-crRNA triggers the activity of RNase III for the cleavage of the pre-crRNA-tracrRNA hybrid to form crRNA. The antirepeat sequence of tracrRNA facilitates the formation of an RNA duplex with each of the repeats of the pre-crRNA, which is stabilized by Cas9. This duplex is then recognized and processed by the host RNase III. It is processed into an intermediate form of crRNA that undergoes trimming at the 5′ end by unknown nucleases to reduce its length to 20 nt. It forms the small guide RNA (gRNA), which then directs the Cas9 protein for invading phage DNA degradation. The mechanism of crRNA processing into gRNA is still unknown (Jinek et al., 2012; Sternberg et al., 2014).

Interference or targeting

In the last phase of the CRISPR/Cas9 immune mechanism, small guide RNAs (gRNAs) guide the bacterial molecular machinery to trigger the degradation of the invading viral nucleic acids. The degradation is majorly performed by the Cas9 nucleases in CRISPR-Cas9 systems. As the small guide RNAs (gRNAs) contain the part of viral genomic DNA that was captured during the earlier phase of spacer acquisition or adaptation process, gRNAs serve as excellent guides for the degradation of invading viral DNA.

During the interference stage, CRISPR/Cas9 systems rely on a single multifunctional protein, Cas9. Inactive Cas9 becomes activated on binding with gRNA and tracrRNA to initiate cleavage of invading phage DNA by using its HNH and RuvC (RNaseH-like endonuclease) domains (Jinek et al., 2014). The recognition lobe (REC lobe) of Cas9 is important for binding gRNA with target DNA and the nuclease lobe containing the HNH and RuvC nuclease domain is the key domain to cleave the target DNA to form DSBs (Sternberg et al., 2014).

Here, one major question may arise in anyone's mind: how does the host prokaryotic cell prevent its own genomic DNA from degradation? So, the answer is, the self-defense mechanism of the host cell in which it prevents its own genomic DNA from degradation is performed by crRNA/gRNA. As was also mentioned earlier in the text, the crRNA and hence gRNA contain a partial CRISPR repeat sequence that prevents its binding with the host cell genomic DNA and prevents its degradation. Further, the identification of PAM sequences (which are the highly conserved motifs present in the phage DNA) by Cas9 proteins for identifying target DNA for degradation also prevents host prokaryotic cells from degradation (Jinek et al., 2012; Nishimasu et al., 2014).

Application of CRISPR/Cas9 in livestock improvement

Traditionally, producing improved livestock varieties through selective breeding is a very slow and less-effective process, as the transfer of improved traits may be achieved with slight progress through longer breeding cycles. Moreover, many traits such as disease resistance and fertility remain difficult to improve (Clark and Whitelaw, 2003).

In recent years, genome editing has offered a quick response approach for realizing genetic improvements in livestock as multiple traits can be attained within a single breeding cycle. In addition, the genome editing technologies are capable of conferring desired genetic traits that are not available in natural genetic sources through traditional selective breeding (Ruan et al., 2017).

Among the genome editors, CRISPR/Cas9 technology is an effective tool because of its precision and ease in handling to modify the genome of livestock species such as cattle, pigs, goats, and sheep for improving welfare and productive traits. These include disease-resistant animals, improvements in adaptation and resilience in animals, and generating large animal models for biomedical and agricultural research (Table 2).

Disease-resistant livestock

Diseases in the form of endemics, epidemics, or pandemics not only cause suffering and death of livestock, but they also affect the markets for meat, derived products, and crops across the globe. One such example is African swine fever (ASF),which is a contagious lethal hemorrhagic disease that causes severe economic losses in the pig industry globally (Costard et al., 2013). Increased demand for livestock production worldwide also has led to zoonotic infections due to the recurrence of animal diseases. Therefore, reducing the disease susceptibility in livestock has emerged as a prime focus of research.

With the CRISPR/Cas9 genome editor, the generation of disease-resistant animals can be achieved. The development of pigs resistant against reproductive and respiratory syndrome virus (PRRSV) infection via genetic knockout of the CD163 receptor (CD163-KO) is the best fitted example for this (Burkard et al., 2017; Whitworth et al., 2016). In the absence of effective treatment and vaccines along with the high rate of severe infection, PRRSV causes huge economic losses in the pig industry, as depopulation of the herd is the only effective option to control the outbreak (Wells and Prather, 2017). The macrophage surface protein CD163, a member of the scavenger receptor cysteine-rich superfamily (SRCR), mediates the entry of PRRSV into the host cell and expresses at high levels on the onset of macrophages in the respiratory system. Hence, the disruption in the *CD163* gene (CD163-KO) by CRISPR/Cas9 genetic scissors resulted in the completely resistant pigs to both Type 1 and Type 2 PRRSV isolates which also restores the reproductive losses associated with this infection (Wells et al., 2017; Yang et al., 2018). Another study reported that CRISPR/Cas9 gene editing in

Table 2 Applications of CRISPR/CAS9.

Livestock	Gene	Mutation	Function	References
Cattle	NRAMP1 gene inserted into intergenic region between FSCN1 and ACTB	HDR-KI	Controls mycobacterial infection	Gao et al. (2017)
	Repaired mutation in isoleucyl-tRNA synthetase (IARS) gene that causes IARS syndrome	HDR–single base substitution	Correction in IARS syndrome	Ikeda et al. (2017)
	Celtic or Friesian mutation in POLLED gene locus on BTA1 chromosome	NHEJ-KO a nd KI	Dehorning in cattle	Deykin et al. (2020)
	BLG (b-lactoglobulin gene)	NHEJ-KO	Allergen-free milk	Silaeva et al. (2020)
	BLG and mFAT-2 (from *C. elegans*)	NHEJ-KO and KI	Allergen-free milk with w-3 and w-6 fatty acids	Alessio et al. (2017)
Sheep	MSTN (Myostatin gene)	NHEJ–KO	Induce muscle mass	Crispo et al. (2015), Zhang et al. (2019), Ding et al. (2020)
	MSTN, agouti signaling protein (ASIP) and β-carotene oxygenase2 (BCO2)	NHEJ–KO	Desired muscle hypertrophy, coat color, and lessening of yellow fat	Wang et al. (2016)
	Fibroblast growth factor 5 (FGF5): negative regulator of wool length	NHEJ–KO	Longer wool length phenotype	Li et al. (2017)
	AANAT and ASMT genes (involved in synthesis of the antioxidant melatonin) with β-casein promoter into MSTN locus	Homology-independent DNA reparation–KI and KO	Antioxidant melatonin-enriched milk	Ma et al. (2018)

Goats	MSTN	NHEJ– KO	Induces muscle mass	Guo et al. (2016), Dowidar et al. (2018)
	MSTN and FGF5	NHEJ–KO	Muscle hypertrophy and enriched cashmere fiber	Wang et al. (2015b)
	Introduced point mutation in growth differentiation factor 9 (GDF9) gene	HDR–ssODN point mutation	Increasing litter size	Niu et al. (2018)
	C. elegans fat-1 gene (convert n-6 PUFA into n-3 PUFA) into MSTN locus	HDR–KI and KO	Safe animal breeding	Zhang et al. (2018)
	BLG and HLF genes (human lactoferrin)	HR- KO and KI	Milk free from allergen β-lactoglobulin and containing human lactoferrin protein	Zhou et al. (2017b)
Pigs	MSTN	NHEJ–KO	Production of myostatin mutant to increase muscle mass	Bi et al. (2016), Wang et al. (2015a)
	FBXO40	NHEJ–KO	Muscle hypertrophy	Zou et al. (2018)
	Insulin-like growth factor 2 (IGF2): promotes growth site on intron 3	NHEJ – KO repressor ZBED6 binding	Improved meat production	Xiang et al. (2018)
	NANOS2 (required for sexual differentiation with male fate	NHEJ–KO	Generation of offspring with mono-allelic and bi-allelic mutations	Wells and Prather (2017), Park et al. (2017)
	CD1 63	NHEJ–KO	Resistance to highly pathogenic porcine reproductive and respiratory viruses in pigs	Whitworth et al. (2016) Burkard et al. (2017), Yang et al. (2018)

Continued

Table 2 .Applications of CRISPR/CAS9—cont'd

Livestock	Gene	Mutation	Function	References
	Amino peptidase N (ANPEP): receptor for transmissible gastroenteritis virus and porcine epidemic diarrhea virus	NHEJ–KO	Resistance for transmissible gastroenteritis virus and porcine epidemic diarrhea virus	Whitworth et al. (2019), Xu et al. (2020)
	von Willebrand Factor)	Cytoplasmic microinjection	Model the human von Willebrand disease	Hai et al. (2014)
	GGTA1 along with CD55, CD46, CD47	NHEJ-KO	Minimizing the rejection response in xenotransplantation	Petersen et al. (2016), Butler et al. (2016), Fischer et al. (2016), Terpiloswska et al. (2020)
Chickens	Ovalbumin (OVA) and ovomucoid (OVM) genes	NHEJ–KO	Remove egg white proteins–allergens	Oishi et al. (2016)

the SRCR 5 domain (scavenger receptor cysteine rich domain 5) of the CD163 gene receptorin zygotes made pigs completely resistant to PRRS infection, as this domain plays a key role in interactions with PSSRV (Burkard et al., 2017).

Another life-threatening porcine disease associated with alphacoronavirus infection is transmissible gastroenteritis virus (TGEV), which leads to significant morbidity and mortality in preweaned piglets. CRISPR/Cas9 has also been used to produce TGEV-resistant pigs by editing exon 2 of a putative receptor of TGEV, the porcine amino peptidase N (pAPN) or amino peptidase N (ANPEP) (Whitworth et al., 2019). Xu et al. (2020) reported complete resistance to genotype 2 PRRSV and TGEV and also to PDCoV (deltacoronavirus) in pigs by the double gene knockout (DKO) of the *CD163* and *pAPN* receptor genes.

Similarly, approximately 50 million cattle worldwide suffer from bovine tuberculosis, which is caused by a mycobacterial pathogen, *Mycobacterium bovis*. This leads to significant economic losses for livestock farmers as these cattle are slaughtered to avoid expansion of the disease and to avoid the zoonotic transmission of *M. bovis* through the consumption of nonpasteurized dairy products or close contact with infected cattle. Recently, a tuberculosis-resistant *NRAMP1* (natural resistance-associated macrophage protein1) gene has been incorporated between *FSCN1* and *ACTB* genes on chromosome 25 into the bovine genome through CRISPR/Cas9 nickase (Cas 9n), resulting in increased resistance to *M. bovis* infection in cattle (Gao et al., 2017). Similarly, Japanese black cattle which are known for the high-quality meat, suffer from a genetic disorder *viz.* isoleucyl-t RNA synthetase (IARS) syndrome in which impaired protein synthesis occurs which in turn exhibit muscular weakness and thus retards the growth of neonatal caves. IARS syndrome is corrected by disrupting the *IARS* gene (responsible for single nucleotide substitution in IARS syndrome) using CRISPR/Cas9 (Ikeda et al., 2017).

Improved performance of livestock

Classical genetics such as selective breeding improve the meat quality and quantity in livestock, but there are significant losses/damage in wool yield and quality. The myostatin gene (*MSTN*) or growth differentiation factor 8 (GDF8), which belongs to the TGF-B superfamily, is a negative regulator of the growth hormone family and thus limits muscle mass. Therefore, muscle mass can be increased by suppressing the function of MSTN (Thomas et al.,

2000). Cattle breeds such as the Italian Piedmontese, Texel sheep in the Netherlands, and the Belgian Blue bull are well known for the naturally occurring MSTN-KO phenotype (Clop et al., 2006; Fiems, 2012). The high growth rate and muscle production in these breeds are linked with an inactive *MSTN* gene encoding for myostatin (or GDF8). With the help of CRISPR/Cas9, the MSTN-KO phenotype was successfully induced in other livestock such as pigs (Bi et al., 2016; Wang et al., 2015a), sheep (Ding et al., 2020; Wang et al., 2016; Zhang et al., 2019), and goats (Dowidar et al., 2018; Guo et al., 2016; Wang et al., 2015b). This led to the double-muscled (DM) phenotype for enhancing the formation of skeletal muscles and for decreasing the accumulation of fat, which is beneficial for the meat industry. Another example is the superfine Merino breed of sheep, which are famous for yielding the finest quality of wool in the world. However, their smaller body size and slower growth rate in comparison to other breeds affect farmer income. The superfine Merino breed was also genetically improved through CRISPR technology by introducing an MSTN in Merino embryos to achieve greater growth rate and muscle mass with the same quality wool in wild Merino counterparts (Crispo et al., 2015).

Desired muscle hypertrophy along with alterations in coat color patterns and yellow fat color have been achieved by multiplex gene editing through CRISPR/Cas9 in sheep. Triple knockout (TKO) of genes by targeting exon 2 or 3 of the *MSTN* gene, exon 5 of the *ASIP* gene (encodes agouti signaling protein responsible for coat color pattern), and exon 2 of the *BCO2* gene (encodes β-carotene oxygenase 2 responsible for yellow fat color, which can cause yellow fat disease or panniculitis in sheep) in both somatic cells and gonads in sheep embryos without detectable off-target mutations improved the traits of sheep (Wang et al., 2016).

Other researchers have demonstrated the improved productive traits (both meat and hair production) in cashmere goats (Wang et al., 2015b) and sheep (Li et al., 2017), which are well recognized for high-quality hair, by editing both *MSTN* and *FGF5* (fibroblast growth factor 5) genes via CRISPR. The *FGF5* gene is responsible for cell proliferation and differentiation for the hair growth cycle.

Although the MSTN-KO phenotype is helpful in increasing the muscle mass in livestock effectively, but some homozygous MSTN-KO piglets of Landrace breeds showed severe side effects such as abnormal limbs and early mortality (Zou et al., 2018). Further, DM animals are more prone to respiratory disease, stress, dystocia, and lameness,which require additional care in animal husbandry (Fiems, 2012). To overcome this problem, a novel gene, the muscle-specific F-box protein (FBXO40), has been found to influence the muscle production in pigs. F-box proteins are the main components of SCF (Skp1–Cullin1–F-box protein) complexes, which exhibit E3 ubiquitin ligase activity and participate in cell cycle as well as signal transduction. FBXO40 knockout pigs mediated by CRISPR/Cas 9 genetic scissors were observed with muscle hypertrophy and no pathological changes in major organs, leading to a better survival rate (Zou et al., 2018). Similarly, another study showed gene mutation in insulin-like growth factor 2 (IGF2). With the help of CRISPR/Cas9, meat quality has been improved in Bama pigs without showing deleterious impacts on growth and development of piglets. IGF2 promotes the growth and proliferation of cells in many different tissues (Xiang et al., 2018).

The *Fat-1* gene (encodes n-3 fatty acid desaturase) from the nematode *Caenorhabditis elegans* shows a significant reduction in the ratio of n-6 to n-3 fatty acids, which have health benefits for human beings. Knock-in (HDR-KI) of the *C. elegans fat-1* gene into porcine Rosa 26 locus (pRosa 26) via CRISPR/Cas9 exhibits a significant increase in n-3 polyunsaturated fatty acid in pigs. The insertion of the *fat-1* gene in pigs holds great promise in improving the nutritional value of pork and serving as an animal model to investigate the therapeutic effects of n-3PUFAs on various diseases (Li et al., 2018). Similarly, double- edited Arab cashmere goats were generated via CRSPR/Cas9 by knockout of the endogenous gene *MSTN* and knock-in of the exogenous gene *fat 1* (Zhang et al., 2018).

Genome editing has noteworthy effects on the desired traits of economic importance on livestock genetics. With the help of CRISPR/Cas9, the male-to-female ratio in livestock can be changed at the desired level by only one sex, for example, milk from females and meat from males. Because growth in male beef cattle is faster and they are bigger than females, a male-only offspring strategy through CRISPR/Cas9 has been applied in beef cattle (Van Eenennaam, 2019). In another example, the female population in pigs was increased by suppressing testis development in the fetal gonadal ridges to avoid male piglets (Kurtz & Petersen, 2019).

Reproduction traits can also be improved through CRISPR/Cas9. A single base substitution (ssODN) in growth differentiation factor 9 (*GDF9*) gene goat zygotes via CRISPR/Cas9 increased the litter size by improving fertility (Niu et al., 2018).

Another interesting example of gene editing is to create livestock that can produce gametes from germ cells derived from genetically superior breeds to expand the genetic potential. An example of such an application considered the pig as an animal model that carries a knockout of the *NANOS2* (Nanos C2HC type zinc finger 2) gene. *NANOS2* is a protein coding gene associated with sexual differentiation of the germ cell by promoting the male fate. NANOS-KO male pigs serve as recipients for the transplantation of spermatogonial stem cells (Park et al., 2017; Wells and Prather, 2017). These and many other improvements have been proposed by various researchers as CRISPR applications in enhancing livestock efficiency.

Animal welfare

There were several attempts made in previous years by researchers for animal welfare to avoid unnecessary suffering and pain in livestock during routine procedures such as male castration for controlling population, abortion, offspring culling of unwanted genders, calf dehorning to avoid injuries to the herd and farmers, tail-docking in dairy cattle and sheep, and mulesing in sheep to prevent the parasitic infection of flystrikes. Gene editing can be a novel strategy in this regard to prevent animal suffering while increasing genetic adaptation and resilience against infection and the environment (Shriver and Connachie, 2018).

Horn removal is a traditional practice in calves to prevent injury to farm caretakers and animals, avoid economic losses, and remove headlock feeding barriers. This is quite painful and stressful event for the animals (Knierim et al., 2015). It is found that many cattle breeds such as Angus are naturally horn-free. Polled phenotype is an autosomal dominant trait because of the POLLED genetic locus located on the bovine chromosome 1(BTA1). In cattle, four polled allelic variants–Celtic POLLED (P_C), Friesian POLLED (P_F), Mongolian POLLED (P_M), and Guarani POLLED (P_G)–have been identified (Aldersey et al., 2020; Medugorac et al., 2012; Rothammer et al., 2014). Earlier, selective breeding techniques were not effective at introducing this polled trait into Holstein cattle. Using the latest CRISPR/Cas 9 technology, polled or hornless phenotypes by Celtic mutation (Pc) responsible for the polled trait in most of the breed and Friesian mutation (P_F) in the Holstein-Friesian breed are being produced to provide a noninvasive solution to this problem (Deykin et al., 2020).

Pests directly or indirectly affect the health of livestock and crops, meaning they affect humans too. In farm husbandry, these pests normally compete for the same food resources with livestock (Bernhardt et al., 2017). Therefore, pests affect the host directly or act as disease carriers. This causes economic loss, which has become a worldwide problem. Each year, tons of pesticides are sprayed to combat this problem. These pesticides have ill effects on the health of livestock and humans. CRISPR technology has the potential to control the number of insects by genetic modification to prevent them from acting as carriers for many diseases in livestock. Examples include disrupting the *Chtra* gene in the screwworm fly (Cochliomyia hominivorax) and the *Lctra* gene in the Australian sheep blowfly (*Lucilia cuprina*) with CRISPR/Cas9 to control the pest population (Doudna and Charpentier, 2014; Paulo et al., 2019).

CRISPR/Cas9 technology is now being used to reduce the cold susceptibility in neonatal pigs, as this can lead to death due to cold stress, especially in cold climatic regions. Pigs lack brown adipose tissues due to the absence of the functional *UCP1* (uncoupling protein 1) gene, which causes the absence of adaptive nonshivering thermogenesis (Berg et al., 2006; Jastroch and Andersson, 2015). This problem could be overcome by inserting mouse adiponectin-driven UCP1 into the porcine endogenous UCP1 locus by using a CRISPR/Cas9-mediated knock-in (KI) method combined with SCNT. Genetically modified UCP1-KI pigs not only showed thermal stability against acute cold conditions, but also demonstrated reduced fat deposition because of UCP1 promoted lipolysis (Zheng et al., 2017).

Production of allergen-reduced/free and enriched animal-derived products

Food allergies unfortunately affect a large number of people worldwide and can be life threatening in some cases. With the help of CRISPR/Cas, many allergy-causing foods such as wheat, peanut, eggs, and milk can be modified to be safe for everyone. The bovine whey protein β-lactoglobulin is encoded by the *BLG* gene. β-lactoglobulin is a major milk protein present in cow and goat milk but not in human milk. Therefore, it act as a dominant allergenic component. This allergen may lead to intestinal allergic reactions in human infants in the form of diarrhea. This β-lactoglobulin protein cannot be eliminated from milk by fermentation or by heat processing. This *BLG* gene was knocked out in cattle using CRISPR/Cas 9 to produce animals whose milk protein composition has shown elevated casein levels with a lack of β-lactoglobulin. This increases the number of consumers of milk (Silaeva et al., 2020; Zhou et al., 2017a). CRISPR/Cas9 is used to produce milk having high nutritional value with reduced allergenicity by the knock-in of the *mFAT-2* gene (encodes desaturase enzyme that catalyzes the synthesis of w-3 and w-6 fatty acids) derived from *C. elegans* and the knockout of exon1 of the *BLG* gene (Alessio et al., 2017).

The knock-in of the *AANAT* and *ASMT* genes (encode enzymes for melatonin production) by CRISPR/Cas9, which were cloned from the pineal gland of Dorper sheep, caused a significant increase in antioxidant, melatonin-rich milk in sheep enriched with high nutritional and medicinal values (Ma et al., 2018).

In another study, Zhou et al. (2017b) obtained biallelic mutants of goats through CRISPR/Cas 9 whose milk was completely free from the β-lactoglobulin protein while containing human lactoferrin (a glycoprotein involved in iron adsorption and in nonspecific immune reactions in the human intestinal tract) by knocking out the *BLG* gene and by knocking in the *HLF* (human lactoferrin) gene, which encode human lactoferrin.

Similarly, CRISPR was used to eliminate allergy-causing proteins in egg whites. For this, genes encoding for ovalbumin (*OVA*) and ovomucoid (*OVM*) were knocked out by CRISPR/Cas 9 to remove the two major allergenic components from egg whites. This increases the egg customer base by adding those who were previously allergic (Oishi et al., 2016).

Biomedical research

The regulatory system of CRISPR/Cas9 technology is continuously advancing in biomedical research to explore the causes of diseases and the successful xenotransplantation of organs for human welfare. Nowadays, livestock are being used as models as they are physiologically and genetically closer to humans than small animal models. Among farm animals, pigs are the most used livestock as a preclinical model for biomedicine and xenotransplantation. Other livestock species such as sheep, goats, and cattle are also becoming interesting models for these studies.

In biomedicine, many human diseases, including neurological disorders, cancer, cardiovascular diseases, hearing disorders, and diabetes, are caused by many gene variants. CRISPR has been applied by many researchers to precisely mimic the clinical phenotype of human diseases in livestock models as an interesting strategy for understanding disease progression and effective therapeutics.

Recently, CRISPR/Cas 9 was used to generate a knockout model of sheep (assisted by ssODN) by editing the otoferlin (*OTOF*) gene for human deafness (Menchaca et al., 2020). The absence of the *OTOF* gene is involved in deafness in humans, and a livestock model that replicated the hearing disorder is useful to assess the impact of the various therapies in order to reverse the hearing disorder. Similarly, CRISPR/Cas9 with SCNT was used to obtain a *CFTR* gene knockout sheep model for cystic fibrosis to develop new therapies for this progressive fatal genetic disease in humans (Fan et al., 2018).

Another application of CRISPR/Cas9 is to generate an ovine model of hypophosphatasia, a rare bone disease in humans. In this research, a single point mutation in the tissue nonspecific alkaline phosphatase (*TNSALP*) gene was carried out to induce the same disease as in humans (Williams et al., 2018).

Similarly, a pig model was generated to develop progressive common neurodegenerative Parkinson's disease (Zhu et al., 2018) and a sheep model was developed to generate an incurable neurodegenerative pediatric disease, infantile neuronal ceroid lipofuscinosis (CLN1 disease), by CRISPR/Cas9 genome scissors to imitate the human mutation (Eaton et al., 2018).

Several studies have been carried out to imitate human disease models in livestock by CRISPR/Cas9 to identify the root causes of various types of cancers and genetic disorders at gene levels so that better therapeutic approaches can be developed. Examples include the tumor suppressor gene *RUNX3*-KO pig model to induce gastric cancer (Kang et al., 2016), the *TP53* gene mutated animal models to induce certain types of cancer (Tanihara et al., 2018) and the *DMD* gene mutated livestock models for Duchenne type muscular dystrophy which is an inherited muscle disease (Yu et al., 2016). Also include the development of livestock models for cardiovascular diseases by the double knock out of *ApoE* (apolipoprotein E) and *LDLR* (low density lipoprotein receptor) genes (Huang et al., 2017) and for coagulation disorders by knocking out the gene coding for von Willebrand factor (vWF) (Hai et al., 2014).

Organ transplantation in humans is considered a prime life-saving strategy in organ failure patients. However, in many cases, the host destroys the xenograft within hours or days through acute xenograft rejection, which becomes fatal for the host. In recent years, the demand for vital organs for transplantation has increased,which in turn increased the shortage of donors. Therefore, livestock organs and tissues, especially those from pigs, are considered a solution for such shortages. CRISPR/Cas may be applied to produce genome edited pigs to avoid hyperacute xenograft rejection and to increase especially, pigs' tolerance of xenotransplantation in pigs to humans.

The prime reason for the hyperacute rejection of the xenograft is the presence of antibodies in human plasma that identify the Galactose α(1–3) Galactose (Galα(1–3)Gal) antigen on the surface of porcine endothelial cells, which is encoded by the α1,3-galactosyltransferase (*GGTA1*) gene. The generation of GGTA-1knockout (GTKO) pigs by using CRISPR/Cas9 can help partially overcome rejection during xenotransplantation (Butler et al., 2016; Petersen et al., 2016). In addition to GTKO pigs, various other gene edits are made to further overcome immune reactive barriers. This includes the genetic modification of membrane bound complement regulatory proteins such as CD46, CD55, CD59, and CD39, and inflammatory proteins such as thrombomodulin, heme oxygenase 1, A20, HLA-E, and CD47 (Fischer et al., 2016; Terpiłowska et al., 2020). However, the porcine endogenous retrovirus (PERV) in pigs may cause a risk of infection and rejection in the host during xenotransplantation. To combat this, the PERV genome in pigs is completely knocked out by the CRISPR/Cas vector (Yang et al., 2015).

Conclusions and prospects

The advent of CRISPR/Cas9 for genome editing has led to a range of novel ideas addressing challenges associated with modern livestock, including productive and fitness traits, animal health and welfare, environmental preservation, and impacts on human health. The simplicity and high efficiency of the CRISPR/Cas9 system allows affordable genome editing of livestock and holds great promise for livestock improvements in the near future.

Although great advances have been made in improving the specificity and expanding the application of CRISPR/Cas9 technology, there is still plenty of room for improvement and extension. A major concern with the application of this system is toxicity or off-target mutation of RNA-guided DNA endonucleases. Many entities, such as paired nickases, truncated sgRNAs, and conjugations to form dimeric *Fok*I–dCas9 nucleases, have been exploited to reduce off-target effects. Sequence requirements within the PAM sequence may limit some applications. Therefore, the evolution of the Cas9 protein should pave the way toward PAM independence, and may also provide a means to generate an even more efficient Cas9 endonuclease. Although these modifications have reduced off-target events significantly and enhanced flexibility, further improvement is needed, especially for more precise modifications or therapeutic applications.

Recently, other members of the CRISPR/Cas family such as CRISPR/Cas12a (Cfp1), which is an RNA-guided system (classified as type V of class II CRISPR system); CRISPR/Cas13a, which belongs to class II type VI;

and CRISPR/Cas14a, a highly compact protein that can be utilized as a guided GE tool for the cleavage of single-stranded DNA (ssDNA) have been considered as entirely new CRISPR types. Base editors are the most recent and quite different GE system, which change a single nucleotide within the genome in a precise and highly predictable manner without any requirement to introduce a double-stranded DNA template and rely on the inner cell repair mechanisms (no dependency on HDR and NHEJ). This could also be another step ahead. Base editors use CRISPR components, guide RNAs, and a mutant Cas9, but do not cut the DNA. Instead, they use deaminase enzymes such as TadA (tRNA specific adenosine deaminase) and ADAR (adenosine deaminase acting on RNA) enzymes to chemically convert single bases without altering the surrounding sequence.

References

Aldersey, J.E., Sonstegard, T.S., Williams, J.L., Bottema, C.D.K., 2020. Understanding the effects of the bovine POLLED variants. Anim. Genet. 51, 166–176.

Alessio, A., Pericuesta, E., Llamas-Toranzo, I., Forcato, D., Fili, A., Liaudat, C., et al., 2017. 203 genome modifications by sleeping beauty transposition and CRISPR/Cas9 to improve cow milk composition for human consumption. Reprod. Fertil. Dev. 30 (1), 242. https://doi.org/10.1071/RDv30n1Ab203.

Anders, C., Niewoehner, O., Duerst, A., Jinek, M., 2014. Structural basis of PAM-dependent target DNA recognition by the Cas9 endonuclease. Nature 513, 569–573.

Barrangou, R., Marraffini, L.A., 2014. CRISPR-Cas systems: prokaryotes upgrade to adaptive immunity. Mol. Cell 54 (2), 234–244. https://doi.org/10.1016/j.molcel.2014.03.011.

Barrangou, R., Fremaux, C., Deveau, H., Richards, M., Boyaval, P., Moineau, S., et al., 2007. CRISPR provides acquired resistance against viruses in prokaryotes. Science 315 (5819), 1709–1712. https://doi.org/10.1126/science.1138140.

Berg, F., Gustafson, U., Andersson, L., 2006. The uncoupling protein 1gene (UCP1) is disrupted in the pig lineage: a genetic explanation for poor thermoregulation in piglets. PLoS Genet. 2. https://doi.org/10.1371/journal.pgen.0020129, e129.

Bernhardt, E.S., Rosi, E.J., Gessner, M.O., 2017. Synthetic chemicals as agents of global change. Front. Ecol. Environ. 15, 84–90. https://doi.org/10.1002/fee.1450.

Bi, Y., Hua, Z., Liu, X., Hua, W., Ren, H., Xiao, H., et al., 2016. Isozygous and selectable marker-free MSTN knockout cloned pigs generated by the combined use of CRISPR/Cas9 and Cre/LoxP. Sci. Rep. 6, 31729. http://www.nature.com/articles/srep31729.

Boch, J., Bonas, U., 2010. Xanthomonas AvrBs3 family-type III effectors: discovery and function. Annu. Rev. Phytopathol. 48, 419–436.

Boissel, S., Jarjour, J., Astrakhan, A., Adey, A., Gouble, A., Duchateau, P., et al., 2014. megaTALs: a rare-cleaving nuclease architecture for therapeutic genome engineering. Nucleic Acids Res. 42 (4), 2591–2601. https://doi.org/10.1093/nar/gkt1224.

Bolotin, A., Quinquis, B., Sorokin, A., Ehrlich, S.D., 2005. Clustered regularly interspaced short palindrome repeats (CRISPRs) have spacers of extrachromosomal origin. Microbiology 151 (8), 2551–2561.

Brazelton, V.A., Zarecor, S., Wright, D.A., Wang, Y., Liu, J., Chen, K., et al., 2015. A quick guide to CRISPR sgRNA design tools. GM Crops Food 6 (4), 266–276. https://doi.org/10.1080/21645698.2015.1137690.

Brouns, S.J., Jore, M.M., Lundgren, M., Westra, E.R., Slijkhuis, R.J., Snijders, A.P., 2008. Small CRISPR RNAs guide antiviral defense in prokaryotes. Science 321 (5891), 960–964. https://doi.org/10.1126/science.1159689.

Burkard, C., Lillico, S.G., Reid, E., Jackson, B., Mileham, A.J., Ait-Ali, T., et al., 2017. Precision engineering for PRRSV resistance in pigs: macrophages from genome edited pigs lacking CD163 SRCR5 domain are fully resistant to both PRRSV genotypes while maintaining biological function. PLoS Pathog. https://doi.org/10.1371/journal.ppat.1006206.

Butler, J.R., Martens, G.R., Estrada, J.L., Reyes, L.M., Ladowski, J.M., Galli, C., et al., 2016. Silencing porcine genes significantly reduces human-anti-pig cytotoxicity profiles: an alternative to direct complement regulation. Transgenic Res. 25 (5), 751–759. https://doi.org/10.1007/s11248-016-9958-0.

Chatterjee, P., Brady, K.L., Solem, A., Ho, Y., Caprara, M.G., 2003. Functionally distinct nucleic acid binding sites for a group I intron encoded RNA maturase/DNA homing endonuclease. J. Mol. Biol. 329, 239–251.

Christian, M., Cermak, T., Doyle, E., Schmidt, C., Zhang, F., Hummel, A., Voytas, D., 2010. Targeting DNA double-strand breaks with TAL effector nucleases. Genetics 186, 757–761.

Clark, J., Whitelaw, B., 2003. A future for transgenic livestock. Nat. Rev. Genet. 4 (10), 825–833. https://doi.org/10.1038/nrg1183.

Clop, A., Marcq, F., Takeda, H., Pirottin, D., Tordoir, X., Bibé, B., et al., 2006. A mutation creating a potential illegitimate microRNA target site in the myostatin gene affects muscularity in sheep. Nat. Genet. 38 (7), 813–818. https://doi.org/10.1038/ng1810.

Costard, S., Mur, L., Lubroth, J., Sanchez-Vizcaino, J.M., Pfeiffer, D.U., 2013. Epidemiology of African swine fever virus. Virus Res. 173, 191–197. https://doi.org/10.1016/j.virusres.2012.10.030.

Crispo, M., Mulet, A.P., Tesson, L., Barrera, N., Cuadro, F., dos Santos-Neto, P.C., et al., 2015. Efficient generation of myostatin knock-out sheep using CRISPR/Cas9 technology and microinjection into zygotes. PLoS One. https://doi.org/10.1371/journal.pone.0136690.

Deltcheva, E., Chylinski, K., Sharma, C.M., Gonzales, K., Chao, Y., Pirzada, Z.A., et al., 2011. CRISPR RNA maturation by trans-encoded small RNA and host factor RNase III. Nature 471 (7340), 602–607. https://doi.org/10.1038/nature09886.

Deng, D., Yan, C., Pan, X., Mahfouz, M., Wang, J., Zhu, J.K., et al., 2012. Structural basis for sequence-specific recognition of DNA by TAL effectors. Science 335, 720–723.

Deykin, A.V., Bruter, A.V., Krivonogova, A.S., Shepelev, M.V., Koshchaev, A.G., 2020. Production of hornless dairy cattle from genome-edited blastocysts. E3S Web Conf. 176, 01008. https://doi.org/10.1051/e3sconf/202017601008.

Ding, Y., Zhou, S., Ding, Q., et al., 2020. The CRISPR/Cas9 induces large genomic fragment deletions of MSTN and phenotypic changes in sheep. J. Integr. Agric. 19 (4), 1065–1073.

Doudna, J.A., Charpentier, E., 2014. Genome editing. The new frontier of genome engineering with CRISPR-Cas9. Science 346, 1258096. https://doi.org/10.1126/science.1258096.

Dowidar, Y.A., El-Sayed, M.A., Elrefy, A.M., Shoura, H.E., 2018. Detection of myostatin gene MSTN in some goat breeds (Capra hircus). J. Genet. Eng. Biotechnol. 16 (2), 507–512.

Doyon, Y., Vo, T.D., Mendel, M.C., Greenberg, S.G., Wang, J., Xia, D.F., et al., 2011. Enhancing zinc-finger-nuclease activity with improved obligate heterodimeric architectures. Nat. Methods 8, 74–79.

Eaton, S.L., Proudfoot, C., Lillico, S.G., Skehel, P., Kline, R.A., Hamer, K., et al., 2018. CRISPR/Cas9 mediated generation of an ovine model for infantile neuronal ceroid lipofuscinosis (CLN1 disease). Sci. Rep. 9, 9891.

Fan, Z., Perisse, I.V., Cotton, C.U., Regouski, M., Meng, Q., Domb, C., et al., 2018. A sheep model of cystic fibrosis generated by CRISPR/Cas9 disruption of the CFTR gene. JCI Insight 3. https://doi.org/10.1172/jci.insight.123529.

FAO, 2009. How to Feed the World in 2050. Food and Agriculture Organization.

Fiems, L.O., 2012. Double muscling in cattle: genes, husbandry, carcasses and meat. Animals (Basel) 2, 472–506. https://doi.org/10.3390/ani2030472.

Fischer, K., Kraner-Scheiber, S., Petersen, B., Rieblinger, B., Buermann, A., Flisikowska, T., et al., 2016. Efficient production of multi-modified pigs for xenotransplantation by 'combineering', gene stacking and gene editing. Sci. Rep. 6, 29081. https://doi.org/10.1038/srep29081.

Gaj, T., Sirk, S.J., Shui, S., Liu, J., 2016. Genome-editing technologies: principles and applications. Cold Spring Harb. Perspect. Biol. 8 (12). https://doi.org/10.1101/cshperspect.a023754, a023754.

Gao, Y., Wu, H., Wang, Y., Liu, X., Chen, L., Li, Q., Cui, C., et al., 2017. Single Cas9 nickase induced generation of NRAMP1 knockin cattle with reduced off-target effects. Genome Biol. 18, 13.

Guo, R., Wan, Y., Xu, D., Cui, L., Deng, M., Zhang, G., Jia, R., et al., 2016. Generation and evaluation of Myostatin knock-out rabbits and goats using CRISPR/Cas9 system. Sci. Rep. 6, 29855.

Hai, T., Teng, F., Guo, R., Li, W., Zhou, Q., 2014. One-step generation of knockout pigs by zygote injection of CRISPR/Cas system. Cell Res. 24, 372–375.

Huang, L., Hua, Z., Xiao, H., Cheng, Y., Xu, K., Gao, Q., Xia, Y., et al., 2017. CRISPR/Cas9-mediated ApoE −/− and LDLR −/− double gene knockout in pigs elevates serum LDL-C and TC levels. Oncotarget 8 (23), 37751–37760. https://doi.org/10.18632/oncotarget.17154.

Ikeda, M., Matsuyama, S., Akagi, S., Ohkoshi, K., Nakamura, S., Minabe, S., Koji, K., Hosoe, M., 2017. Correction of a disease mutation using CRISPR/Cas9-assisted genome editing in Japanese black cattle. Sci. Rep. 19 (7), 17827. https://doi.org/10.1038/s41598-017-17968-w.

Ishino, Y., Shinagawa, H., Makino, K., Amemura, M., Nakata, A., 1987. Nucleotide sequence of the iap gene, responsible for alkaline phosphatase isozyme conversion in Escherichia coli, and identification of the gene product. J. Bacteriol. 169 (12), 5429–5433. https://doi.org/10.1128/jb.16.9.12.5429-5433.1987.

Jansen, R., Embden, J.D., Gaastra, W., Schouls, L.M., 2002. Identification of genes that are associated with DNA repeats in prokaryotes. Mol. Microbiol. 43 (6), 1565–1575. https://doi.org/10.1046/j.1365-2958.2002.02839.

Jasin, M., 1996. Genetic manipulation of genomes with rare-cutting endonucleases. Trends Genet. 12 (6), 224–228. https://doi.org/10.1016/0168-9525(96)10019-6.

Jastroch, M., Andersson, L., 2015. When pigs fly, UCP1 makes heat. Mol. Metab. 4, 359–362. https://doi.org/10.1016/j.molmet.2015.02.005.

Jinek, M., Chylinski, K., Fonfara, I., Hauer, M., Doudna, J.A., Charpentier, E., 2012. A programmable dual-RNA-guided DNA endonuclease in adaptive bacterial immunity. Science 337 (6096), 816–821. https://doi.org/10.1126/science.1225829.

Jinek, M., Jiang, F., Taylor, D.W., Sternberg, S.H., Kaya, E., Ma, E., et al., 2014. Structures of Cas9 endonucleases reveal RNA-mediated conformational activation. Science 343 (6176), 1247997. https://doi.org/10.1126/science.1247997.

Kang, J.T., Ryu, J., Cho, B., Lee, E.J., Yun, Y.J., Ahn, S., et al., 2016. Generation of RUNX3 knockout pigs using CRISPR/Cas9-mediated gene targeting. Reprod. Domest. Anim. 51 (6), 970–978. https://doi.org/10.1111/rda.12775.

Kim, Y.G., Cha, J., Chandrasegaran, S., 1996. Hybrid restriction enzymes: Zinc finger fusions to Fok I cleavage domain. Proc. Natl. Acad. Sci. U. S. A. 93, 1156–1160.

Knierim, U., Irrgang, N., Roth, B.A., 2015. To be or not to be horned consequences in cattle. Livest. Sci. 179, 29–37.

Kurtz, S., Petersen, B., 2019. Pre-determination of sex in pigs by application of CRISPR/Cas system for genome editing. J. Theriogenol. 137, 67–74. https://doi.org/10.1016/j.theriogenology.2019.05.039.

Li, W., Liu, C., Zhang, X., Chen, L., Peng, X., He, S., Lin, J., et al., 2017. CRISPR/Cas9-mediated loss of FGF5 function increases wool staple length in sheep. FEBS J. 284 (17), 2764–2773. https://doi.org/10.1111/febs.14144.

Li, M., Ouyang, H., Yuan, H., Li, J., Xie, Z., Wang, K., et al., 2018. Site-specific Fat-1 knock-in enables significant decrease of n-6PUFAs/n-3PUFAs ratio in pigs. Genetics 8, 1747–1754.

Lieber, M.R., 2010. The mechanism of double-strand DNA break repair by the non-homologous DNA end-joining pathway. Annu. Rev. Biochem. 79, 181–211.

Ly, J., 2013. Discovering Genes Responsible for Kidney Diseases (Ph.D.). University of Toronto. Retrieved 26 December 2016.

Ma, T., Tao, J., Yang, M., He, C., Tian, X., Zhang, X., et al., 2018. An AANAT/ASMT transgenic animal model constructed with CRISPR/Cas9 system serving as the mammary gland bioreactor to produce melatonin-enriched milk in sheep. J. Pineal Res. 63 (1). https://doi.org/10.1111/jpi.12406.

Makarova, K.S., Wolf, Y.I., Alkhnbashi, O.S., Costa, F., Shah, S.A., Saunders, S.J., et al., 2015. An updated evolutionary classification of CRISPR-Cas systems. Nat. Rev. Microbiol. 13 (11), 722–736. https://doi.org/10.1038/nrmicro3569.

Medugorac, I., Seichter, D., Graf, A., Russ, I., Blum, H., Göpel, K.H., et al., 2012. Bovine polledness – an autosomal dominant trait with allelic heterogeneity. PLoS One 7. https://doi.org/10.1371/journal.pone.0039477, e39477.

Menchaca, A., Santos-Neto, P.C.D., Souza-Neves, M., Cuadro, F., Mulet, A.P.L., et al., 2020. Otoferlin gene editing in sheep via CRISPR-assisted ssODN-mediated homology directed repair. Sci. Rep. 10, 5995.

Miller, J.C., Holmes, M.C., Wang, J., Guschin, D.Y., Lee, Y., Rupniewski, I., Beausejour, C.M., et al., 2007. An improved zinc-finger nuclease architecture for highly specific genome editing. Nat. Biotechnol. 25, 778–785.

Miller, J.C., Tan, S., Qiao, G., Barlow, K.A., Wang, J., Xia, D.F., et al., 2011. A TALE nuclease architecture for efficient genome editing. Nat. Biotechnol. 29, 143–148.

Mohr, S.E., Hu, Y., Ewen-Campen, B., Housden, B.E., Viswanatha, R., Perrimon, N., 2016. CRISPR guide RNA design for research applications. FEBS J. 283 (17), 3232–3238. https://doi.org/10.1111/febs.13777.

Mojica, F.J., Díez-Villaseñor, C., García-Martínez, J., Soria, E., 2005. Intervening sequences of regularly spaced prokaryotic repeats derive from foreign genetic elements. J. Mol. Evol. 60 (2), 174–182.

Mojica, F.J.M., D'ıez-Villasenor, C., Garc'ıa-Mart'ınez, J., Almendros, C., 2009. Short motif sequences determine the targets of the prokaryotic CRISPR defence system. Microbiology 155 (3), 733–740.

Moscou, M.J., Bogdanove, A.J., 2009. A simple cipher governs DNA recognition by TAL effectors. Science 326, 1501.

Nicholl, S.T., 2008. An Introduction to Genetic Engineering. Cambridge University Press, ISBN: 978-1-139-47178-7.

Nishimasu, H., Ran, F.A., Hsu, P.D., Konermann, S., Shehata, S.I., Dohmae, N., et al., 2014. Crystal structure of Cas9 in complex with guide RNA and target DNA. Cell 156 (5), 935–949. https://doi.org/10.1016/j.cell.2014.02.001.

Niu, Y., Zha, X., Zhou, J., Li, Y., Huang, Y., Cai, B., Liu, Y., et al., 2018. Efficient generation of goats with defined point mutation (I397V) in GDF9 through CRISPR/Cas9. Reprod. Fertil. Dev. 30 (2), 307–312. https://doi.org/10.1071/RD17068. 28692815.

Nobel Foundation, 2020. Press release: "The Nobel Prize in Chemistry 2020". Annual Review 2020. Nobel Foundation, p. 7. Retrieved 7 October 2020 https://www.nobelprize.org/uploads/2019/04/annual-review-fin-2020.pdf.

Oishi, I., Yoshii, K., Miyahara, D., Kagami, H., Tagami, T., 2016. Targeted mutagenesis in chicken using CRISPR/Cas9 system. Sci. Rep. 6, 23980.

Park, K.E., Kaucher, A.V., Powell, A., Waqas, M.S., Sandmaier, S.E.S., Oatley, M.J., et al., 2017. Generation of germline ablated male pigs by CRISPR/Cas9 editing of the NANOS2 gene. Sci. Rep. 7, 40176. http://www.nature.com/articles/srep40176.

Pattanayak, V., Ramirez, C.L., Joung, J.K., Liu, D.R., 2011. Revealing off-target cleavage specificities of zinc-finger nucleases by in vitro selection. Nat. Methods 8, 765–770.

Paulo, D.F., Williamson, M.E., Arp, A.P., Li, F., Sagel, A., Skoda, S.R., et al., 2019. Specific gene disruption in the major livestock pests Cochliomyia hominivorax and Lucilia cuprina using CRISPR/Cas9. G3 (Bethesda) 9 (9), 3045–3055. https://doi.org/10.1534/g3.119.400544.

Petersen, B., Frenzel, A., Lucas-Hahn, A., Herrmann, D., Hassel, P., Klein, S., et al., 2016. Efficient production of biallelic GGTA1 knockout pigs by cytoplasmic microinjection of CRISPR/Cas9 into zygotes. Xenotransplantation 23 (5), 338–346. https://doi.org/10.1111/xen.12258.

Porteus, M.H., Carroll, D., 2005. Gene targeting using zinc finger nucleases. Nat. Biotechnol. 23, 967–973.

Rothammer, S., Capitan, A., Mullaart, E., Seichter, D., Russ, I., Medugorac, I., 2014. The 80-kb DNA duplication on BTA1 is the only remaining candidate mutation for the polled phenotype of Friesian origin. Genet. Sel. Evol. 46, 1–5.

Ruan, J., Xu, J., Yanru, R.C., Li, K., 2017. Genome editing in livestock: are we ready for a revolution in animal breeding industry? Transgenic Res. 26 (6), 715–726. https://doi.org/10.1007/s11248-017-0049-7.

Semenova, E., Jore, M.M., Datsenko, K.A., Semenova, A., Westra, E.R., et al., 2011. Interference by clustered regularly interspaced short palindromic repeat (CRISPR) RNA is governed by a seed sequence. PNAS 108 (25), 10098–100103.

Shriver, A., Connachie, E.M., 2018. Genetically modifying livestock for improved welfare: a path forward. J. Agric. Environ. Ethics 31, 161–180.

Silaeva, Y.Y., Kubekina, M.V., Bruter, A.V., Isaeva, A.G., Koshchaev, A.G., 2020. Gene editing CRISPR/Cas9 system for producing cows with hypoallergenic milk on the background of a beta-lactoglobulin gene knockout. E3S Web Conf. 176. https://doi.org/10.1051/e3sconf /202017601006. 01006 (2020); IDSISA 2020.

Silva, G.H., Belfort, M., Wende, W., Piigoud, A., 2006. From monomeric to homodimeric endonucleases and back: engineering novel specificity of LAGLIDADG enzymes. J. Mol. Biol. 361, 744–754.

Smith, J., Bibikova, M., Whitby, F.G., Reddy, A.R., Chandrasegaran, S., Carroll, D., 2000. Requirements for double-strand cleavage by chimeric restriction enzymes with zinc finger DNA-recognition domains. Nucleic Acids Res. 28, 3361–3369.

Sternberg, S.H., Redding, S., Jinek, M., Greene, E.C., Doudna, J.A., 2014. DNA interrogation by the CRISPR RNA-guided endonuclease Cas9. Nature 507 (7490), 62–67. https://doi.org/10.1038/nature13011.

Stoddard, B.L., 2006. Homing endonuclease structure and function. Q. Rev. Biol. 38 (1), 49–95. https://doi.org/10.1017/S0033583505004063.

Tanihara, F., Hirata, M., Nguyen, N.T., Le, Q.A., Hirano, T., Takemoto, T., et al., 2018. Generation of a TP53-modified porcine cancer model by CRISPR/Cas9-mediated gene modification in porcine zygotes via electroporation. PLoS One. https://doi.org/10.1371/journal.pone.0206360.

Terpiłowska, N.A., Lipiński, D., Hryhorowicz, M., Juzwa, W., Jura, J., Słomski, R., 2020. Production of ULBP1-KO pigs with human CD55 expression using CRISPR technology. J. Appl. Anim. Res. 48 (1), 93–101.

Thomas, M., Langley, B., Berry, C., Sharma, M., Kirk, S., Bass, J., et al., 2000. Myostatin, a negative regulator of muscle growth, functions by inhibiting myoblast proliferation. J. Biol. Chem. 275, 40235–40243.

Tyson, G.W., Banfield, J.F., 2008. Rapidly evolving CRISPRs implicated in acquired resistance of microorganisms to viruses. Environ. Microbiol. 10 (1), 200–207. https://doi.org/10.1111/j.1462-2920.2007.01444.x.

Van Eenennaam, A.L., 2019. Application of genome editing in farm animals: cattle. Transgenic Res. 28, 93–100.

Voytas, D.F., Joung, J.K., 2009. Plant Science. DNA binding made easy. Science 326, 1491–1492.

Wang, K., Ouyang, H., Xie, Z., Yao, C., Guo, N., Li, M., Jiao, H., Pang, D., 2015a. Efficient generation of myostatin mutations in pigs using the CRISPR/Cas9 system. Sci. Rep. 5, 16623. https://doi.org/10.1038/srep16623.

Wang, X., Yu, H., Lei, A., Zhou, J., Zeng, W., Zhu, H., et al., 2015b. Generation of gene-modified goats targeting MSTN and FGF5 via zygote injection of CRISPR/Cas9 system. Sci. Rep. 5, 13878.

Wang, X., Niu, Y., Zohu, J., Yu, H., Kou, Q., Lei, A., Zhao, X., Yan, H., et al., 2016. Multiplex gene editing via CRISPR/Cas9 exhibits desirable muscle hypertrophy without detectable off-target effects in sheep. Sci. Rep. 6, 32271.

Wei, Y., Terns, R.M., Terns, M.P., 2015. Cas9 function and host genome sampling in type II-A CRISPR-Cas adaptation. Genes Dev. 29, 356–361. https://doi.org/10.1101/gad.257550.114.

Wells, K.D., Prather, R.S., 2017. Genome-editing technologies to improve research, reproduction, and production in pigs. Mol. Reprod. Dev. 84 (9), 1012–1017. https://doi.org/10.1002/mrd.22812.

Wells, K.D., Bardot, R., Whitworth, K.M., Trible, B.R., Fang, Y., Mileham, A., Kerrigan, M.A., Samuel, M.S., et al., 2017. Replacement of porcine CD163 scavenger receptor cysteine-rich domain 5 with a CD163-like homolog confers resistance of pigs to genotype 1 but not genotype 2 porcine reproductive and respiratory syndrome virus. J. Virol. 91 (2). https://doi.org/10.1128/JVI.01521-16.

Westra, E.R., Dowling, A.J., Broniewski, J.M., van Houte, S., 2016. Evolution and ecology of CRISPR. Annu. Rev. Ecol. Evol. Syst. 47 (1), 307–331. https://doi.org/10.1146/annurev-ecolsys-121415-032428.

Whitworth, K.M., Rowland, R.R., Ewen, C.L., Trible, B.R., Kerrigan, M.A., Cino-Ozuna, A.G., et al., 2016. Gene-edited pigs are protected from porcine reproductive and respiratory syndrome virus. Nat. Biotechnol. 34, 20–22.

Whitworth, K.M., Rowland, R.R.R., Petrovan, V., Sheahan, M., Cino-Ozuna, A.G., Fang, Y., et al., 2019. Resistance to coronavirus infection in amino peptidase N-deficient pigs. Transgenic Res. 28, 21–32.

Wiedenheft, B., van Duijn, E., Bultema, J.B., Bultema, J., Waghmare, S.P., et al., 2011. RNA-guided complex from a bacterial immune system enhances target recognition through seed sequence interactions. PNAS 108 (25), 10092–10097.

Williams, D.K., Pinzon, C., Huggins, S., Pryor, J.H., Falck, A., Herman, F., et al., 2018. Genetic engineering a large animal model of human hypophosphatasia in sheep. Sci. Rep. 8, 16945.

Wright, A.V., Nuñez, J.K., Doudna, J.A., 2016. Biology and applications of CRISPR systems: harnessing nature's toolbox for genome engineering. Cell 164 (1–2), 29–44. https://doi.org/10.1016/j.cell.2015.12.035.

Wu, H.B., Wang, Y.S., Zhang, Y., Yang, M.Q., Lv, J.X., Liu, J., et al., 2015. TALE nickase-mediated SP110 knockin endows cattle with increased resistance to tuberculosis. Proc. Natl. Acad. Sci. U. S. A. 112. https://doi.org/10.1073/pnas.1421587112. E15 Clark Wuan 30–E9.

Xiang, G., Ren, J., Hai, T., Fu, R., Yu, D., 2018. Editing porcine IGF2 regulatory element improved meat production in Chinese Bama pigs. Cell. Mol. Life Sci. https://doi.org/10.1007/s00018-018-2917-6.

Xu, K., Zhou, Y., Mu, Y., Liu, Z., Hou, S., Xiong, Y., et al., 2020 Sep. CD163 and pAPN double-knockout pigs are resistant to PRRSV and TGEV and exhibit decreased susceptibility to PDCoV while maintaining normal production performance. eLife 9. https://doi.org/10.7554/eLife.57132, e57132.

Yang, L., Güell, M., Niu, D., George, H., Lesha, E., Grishin, D., et al., 2015. Genome-wide inactivation of porcine endogenous retroviruses (PERVs). Science 350 (6264), 1101–1104. https://doi.org/10.1126/science.aad1191.

Yang, H., Zhang, J., Zhang, X., Shi, J., Pan, Y., Zhou, R., Li, G., et al., 2018. CD163 knockout pigs are fully resistant to highly pathogenic porcine reproductive and respiratory syndrome virus. Antivir. Res. 151, 63–70. https://doi.org/10.1016/j.antiviral.2018.01.004.

Yu, H., Zhao, H., Qing, Y., Pan, W., Jia, B., Zhao, H., et al., 2016. Porcine zygote injection with Cas9/sgRNA results in DMD-modified pig with muscle dystrophy. Int. J. Mol. Sci. 17 (10), 1668. https://doi.org/10.3390/ijms17101668.

Zhang, J., Cui, M.L., Nie, Y.W., Dai, B., Li, F.R., Liu, D.J., Liang, H., Cang, M., 2018. CRISPR/Cas9-mediated specific integration of fat-1 at the goat MSTN locus. FEBS J. 285 (15), 2828–2839. https://doi.org/10.1111/febs.14520.

Zhang, Y., Wang, Y., Yulin, B., Tang, B., Wang, M., Zhang, C., et al., 2019. CRISPR/Cas9 mediated sheep MSTN gene knockout and promote sSMSCs differentiation. J. Cell. Biochem. 120 (2), 1794–1806. https://doi.org/10.1002/jcb.27474.

Zhao, L., Bonocora, R.P., Shub, D.A., Stoddard, B.L., 2007. The restriction fold turns to the dark side: a bacterial homing endonuclease with a PD-(D/E)-XK motif. EMBO J. 26, 2432–2442.

Zheng, Q., Lin, J., Huang, J., Zhang, H., Zhang, R., Zhang, X., et al., 2017. Reconstitution of UCP1 using CRISPR/Cas9 in the white adipose tissue of pigs decreases fat deposition and improves thermogenic capacity. PNAS 114, E9474–E9482. https://doi.org/10.1073/pnas.170785311.

Zhou, W., Wan, Y., Guo, R., Deng, M., Deng, K., Wang, Z., et al., 2017a. Generation of beta-lactoglobulin knock-out goats using CRISPR/Cas9. PLoS One. https://doi.org/10.1371/journal.pone.0186056.

Zhou, W., Guo, R., Deng, M., Wang, F., Zhang, Y., 2017b. RS-1 enhanced the efficiency of CRISPR-Cas9 mediated knock-in of human lactoferrin. Chin. J. Biotechnol. 33 (8), 1224–1234. https://doi.org/10.13345/j.cjb.170105.

Zhu, X.X., Zhong, Y.Z., Ge, Y.W., Lu, K.H., Lu, S.S., 2018. CRISPR/Cas9-mediated generation of Guangxi Bama Minipigs harboring three mutations in α-synuclein causing Parkinson's disease. Sci. Rep. 8, 12420. https://doi.org/10.1038/s41598-018-30436-3.

Zou, Y., Li, Z., Zou, Y., Hao, H., Li, N., Li, Q., 2018. An FBXO40 knockout generated by CRISPR/Cas9 causes muscle hypertrophy in pigs without detectable pathological effects. Biochem. Biophys. Res. Commun. 498, 940–945.

CHAPTER SEVENTEEN

Stem cell technology: Application in animal health and livestock production

Ritambhara[a], Neha Dobriyal[b], S.I. Rizvi[a], and Munish Kumar[a]
[a]Department of Biochemistry, University of Allahabad, Prayagraj, UP, India
[b]Department of Biological Sciences, Indian Institute of Science Education and Research Bhopal, Bhopal, MP, India

Introduction

Stem cell biology

Stem cells have the capacity to proliferate into different lineages. These cells are classified into two categories: embryonic stem cells (ESC) and somatic stem cells (SSC). The ESCs originate from the inner cell mass of blastocysts and are pluripotent in nature, having the capacity to differentiate into various cell lineages (Jiang et al., 2002).

Compared to embryonic stem cells, somatic stem cells have limited capacity to differentiate into any cell lineage. They have self-renewal capacity and can be cultured for a longer time (Kuroda et al., 2010). SSCs can be isolated from mice, rats, dogs, monkeys, pigs, sheep, goats, rabbits, cats, chickens (Khatri et al., 2009), and human beings (Liechty et al., 2000).

There are other sources of stem cells such as cells derived from bone marrow and adipose tissue, which are widely used for disease treatment of animals (Marx et al., 2014; Torricelli et al., 2011). Mesenchymal stem cells (MSC) are isolated from the mesoderm and neuroectoderm (Ito and Suda, 2014) and are present uniformly in all vascularized adult tissues (adipose tissue, skin, heart, brain, vessels, bones, and cartilage) (Mimeault and Batra, 2008). Significantly, these cells display regenerative capacity.

MSCs can generate new cells. Various cellular growth factors and hormones provide these cells a transformative property in their location within the body (Ferraro et al., 2010). These new stem cells have less proliferative

Emerging Issues in Climate Smart Livestock Production
https://doi.org/10.1016/B978-0-12-822265-2.00018-1

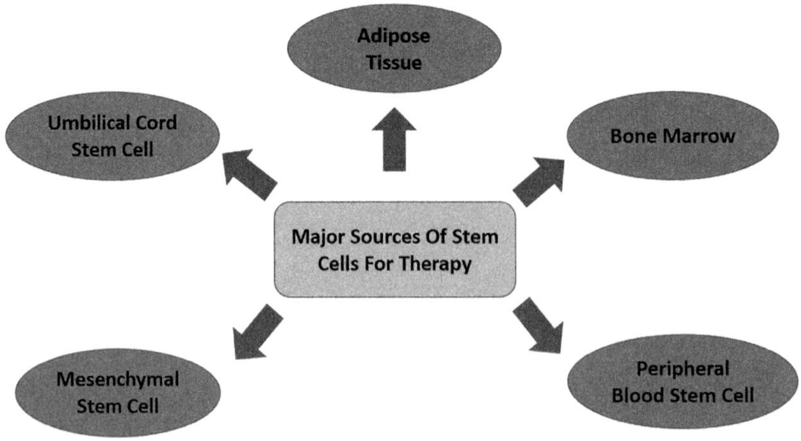

Fig. 1 Major resources of stem cells for stem cell therapy.

capacity than MSCs and are called progenitor cells. These progenitor cells differentiate into limited cells compared to MSCs. The MSC and progenitor cells are prominently responsible for tissue repair and the generation of new cells in case of injury and other cellular transduction pathways in response to various stimuli (Ito and Suda, 2014; Parekkadan and Milwid, 2010; Shin and Peterson, 2013). The widespread use of MSCs and progenitor cells in cell therapy in animals is shown in Fig. 1.

Several signaling factors affect MSCs such as exposure to various signals. They are involved in proliferation and cellular differentiation, including injury and malfunctioning. MSCs function by modifying molecular signals, repairing tissue, and the formation of new blood vessels.

MSCs have self-renewal capacity and can differentiate into other cell types. Cell signaling pathways related to autocrine, paracrine, and endocrine intercellular processes signals from MSCs to activate their mechanisms of cell proliferation and differentiation (Ito and Suda, 2014). Overall, these processes influence the interactions between stem cells and other neighboring differentiated cells. Furthermore, these interactions persist between stem cells and cellular adhesive networks such as extracellular matrix components. They are also affected by several other growth factors and various cytokines (Ferraro et al., 2010; Li and Xie, 2005) Fig. 2.

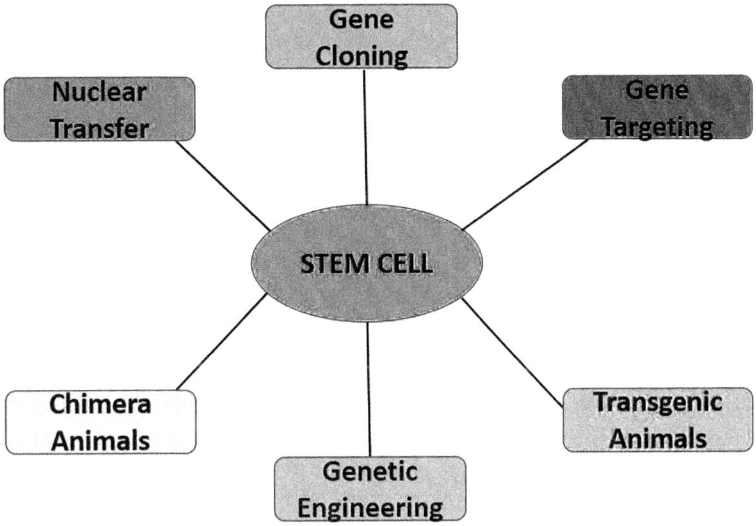

Fig. 2 Characteristics of stem cells.

Characteristics of mesenchymal stem cells

MSCs can induce the production of bioactive mediators and various adhesion molecules that help to prevent scar formation at an injury site. They also facilitate the formation of new blood cells, prevent apoptosis, and enhance the plasticity of progenitor cells (Lane et al., 2014; Porada and Almeida-Porada, 2010). The classification of stem cells includes embryonic stem cells (ESC) and adult (somatic) stem cells (ASC) (Tang et al., 2015), which includes mesenchymal stem cells, hematopoietic stem cells, and tissue-specific progenitor cells (Marx et al., 2014).

Mesenchymal stem cell culture

In studies, it is well established that MSCs adhere to plastic substrate culture plates (Dominici et al., 2006). The characteristic condition of MSC arises after tissue disaggregation. The enzymatic method involves disaggregation, in which proteolytic enzyme digestion is done to isolate cells from tissue.

Several protocols have been proposed to isolate, expand, and characterize MSCs. The generalized protocol uses the MSC property of cell adherence to the solid substrate during the first 48–72 h of culture, producing a heterogeneous population of cells (Hill et al., 2018; Kaiser, 2004; Schellenberg et al., 2012). To obtain a homogenous population of MSCs, different conditions for isolation can be used, including the use of different cell culture media (Wagner et al., 2006), cell sorting (Anderson et al., 2013; Yasumasa et al., 2010), and cell adherence to various substrates in the first 3 h of culture (Hill et al., 2018; Anson, 2004).While culturing MSCs, optimum conditions should be maintained such as an incubator with 5% CO_2 to maintain the physiological pH in the culture medium (Freshney, 1994) and other supplementations in the culture medium. The growth factors are supplemented in culture media for nutrition, adherence to the culture, plate and expansion of cells. Fetal bovine serum (FBS) is used in the cell culture medium as a growth factor (Schwarz et al., 2013). It is necessary to use antibiotics to prevent contaminations. Commonly used antibiotics are penicillin and streptomycin, and they are generally used at concentrations of 100 U/mL and 100 mg/mL, respectively (Martin, 1994).

Mesenchymal stem cell characterization

Most MSCs express CD105, CD73, and CD90 and lack the expression of hematopoietic markers such as CD45, CD34, CD14, CD11b, CD79α, CD19, and HLA-DR surface molecules (Dominici et al., 2006). These MSC markers vary in different tissues and in different species with varying self-renewal capacities (Trivanović et al., 2015).

In bovine species, MSCs from different tissues possess adhesion markers such as CD29, CD166, and CD105; surface enzymes such as CD73; receptors such as CD44; and glycoproteins such as CD90. Bovine MSC also expresses pluripotent stem cell markers such as OCT4, SOX2, and NANOG (Lara et al., 2017).

In equine species, MSC markers from bone marrow (BM-MSC), adipose tissue (AT-MSC), and the umbilical cord (UC-MSC) were characterized. Bone marrow, adipose tissue, and umbilical cord MSCs expressed CD105, CD90, and CD44. A lower expression of CD90 was expressed in UC-MSC compared to other sources. The differentiation capacity of BM-MSC and AT-MSCs was increased over UC-MSC (Barberini et al., 2014).

Therapeutic delivery of mesenchymal stem cells

MSC cryopreservation can be used for further cell culture and expansion as autologous stem cells (Martinello et al., 2001). The characteristics of bovine MSC do not change after cryopreservation for fresh as well as thawed cells (de Moraes et al., 2016). For mesenchymal stem cells, the delivery preparation should be rapid to prevent contamination and maintain cell viability (Webster et al., 2012). It would be advantageous to deliver stem cells before the formation of fibrous scar tissue (Schnabel et al., 2013). Delivery of these MSCs depends upon the condition to treat these cells. The differentiation property of MSCs elicits with intravenous delivery and can migrate across the endothelium as well as to injured tissues (Chamberlain et al., 2007; Gazit et al., 2011). The direct delivery of MSCs to injured tissues enhances the MSC concentration (Choi et al., 2014), and also cell migration can be prevented to other sites in the body (Gazit et al., 2011).

A substrate is needed as a carrier for delivery in defective structures such as segmental bone and articular cartilage to regulate cell adhesion in the particular location of the cell and to form a network for new tissue formation (Choi et al., 2014).

Decellularized tissues are the best tools for scaffold construction (Rana et al., 2017). In the bovine model, a large quantity of tissue can be decellularized in comparison to smaller animal models such as mice. The bovine models provide better translational applications in organ construction and tissue engineering and are also correlated with human conditions (Barreto et al., 2018).

Translational medicine and bovine stem cells

The bovine model could be an advantageous technique to study human disease. Cattle can serve as animal models and are used in cartilage repair, mammary gland regeneration, and hereditary genetic diseases of the blood and immune system.

Cartilage

Cartilage has poor regenerative capacity. Various approaches are used for regenerative therapy in cartilage. First, a three-dimensional (3D) cell culture employs adult chondrocytes that have reduced collagen formation and altered biochemical properties (Erickson et al., 2011). A study reported that the differentiation of MSCs into chondrocytes was elevated by transduction factors such as BMP-2, IGF1, and TGFb1 in the adenoviral system (Steinert et al., 2009). These chondrogenic proteins in MSCs produce cell aggregates and a higher level of cartilage-specific proteoglycans and collagen. This ultimately suggests that they can be explored for stem cell-based therapy for cartilage repair.

Another approach to cell-based therapy on cartilage can be achieved by cloning the chondroprogenitor, which is referred to as immature cartilage. These cells were isolated from bovine immature metacarpalphalangeal joints. Cloning resulted in the production of MSCs with similar differentiation factors for cartilage formation (Khan et al., 2009).

Mammary gland

Common mechanisms were found between mammary stem cells from humans and cattle. In a study it was reported that in vitro colony formation assay of bovine mammary stem cells lacking aldehyde dehydrogenase activity and also showed similarity with the luminal epithelial stem cell through the differentiation. The regenerated structures restate the organization of bovine mammary tissue. Various other cell types such as epithelial and myoepithelial progenitors showed importance in regeneration of the mammary gland.8

Embryonic stem cells (ESCs)

Embryonic stem cells (ESCs) are the pluripotent cells derived from the inner cell mass (ICM). These stem cells have characteristic expansion and a huge differentiation property, and they can be cultured. They have self-renewal and differentiation ability. ESCs were first established as cell lines from the ICM of mouse blastocysts. When combined with a host embryo (chimera production), ESCs can produce any tissue or organ type, including the germ cells of the resulting chimeric mouse (Talbot and Blomberg, 2008). These cells are considered totipotent or pluripotent due to their ability to form all cell types, including extraembryonic placental tissues. They can also be clonally propagated and maintained in a culture system indefinitely. These characteristics of ESCs have made them a vital tool in genetic engineering to study functional mammalian genetics and mammalian developmental biology as well as to produce animal models of human diseases.

Induced pluripotent stem cells (iPSCs)

Pluripotent stem cells (iPSCs) are reprogrammed from somatic cells by introducing pluripotency factors. In a 2006 meeting of the International Society for Stem Cell Research in Canada, Shinya Yamanaka announced the discovery that adult skin cells could be directly reprogrammed to become pluripotent by using four genes; Oct4, Sox2, c-Myc, and Klf4. These are also known as "Yamanaka factors." Various combinations of these genes were introduced to adult mouse skin fibroblasts via engineered retroviruses to screen for gene combinations that could reprogram skin cells to become pluripotent (Takahashi and Yamanaka, 2006). This method was termed direct reprogramming. The resulting stem cells that were reprogrammed were called induced pluripotent stem cells, or iPS cells. Further studies on iPSC technology will lead to important discoveries in disease modeling, drug discovery, and stem cell-based therapy (Tada, 2008). iPSC technology has also showed prominent potential applicability for regenerative medicine.

Adult mammary stem cells

The mammary stem cell is a quiescent, multipotent, and self-renewal population located at the top of the epithelium hierarchy. These mammary stem cell lineages to progenitor cells then terminally differentiated into ducts and lobules which finally constitute the functional mammary glands which is a unique exocrine glandular organ (Zhou et al., 2019). Adult mammary stem cells play a crucial role in improving reproductive performance (Baratta, 2013). The regenerative potential of one individual mammary stem cell into the complete mammary gland was first demonstrated in the late 1950s and early 1960s (Inman et al., 2015). In addition to human mammary glands, studies are ongoing on animal mammary stem cells. The recent discovery of the presence of stem cells and their precursors in the mammary gland of cows provides an opportunity to fulfill the need for milk in humans. A single mammary stem cell has strong regenerative potential to regenerate alveoli that secrete milk, thus increasing or maintaining the longer lactation of cows during the life span. This in turn benefits milk yield and persistence. The adult mammary stem cells are also very well studied in the female mouse, which has five pairs of mammary epithelial placodes and rudimentary placodes derived from the ectoderm during embryogenesis. It is shaped to a mammary bud that sprouts with a lumen. During puberty, the glands expand due to increased estrogen levels and undergo morphogenesis, which gives rise to a bilayer ductal structure by the myoepithelial cells. It further contracts and helps in milk secretion. There is a precisely choreographed process that involves proliferation, differentiation, and involution, hence constituting a mammary reproductive cycle (Chen et al., 2017). Further in-depth study will help in fully characterizing the mammary gland stem cells as well as their regulation, turnover, differentiation, and role in milk production. These studies showed an interesting perspective for livestock enhancement, specifically in dairy science as well as provides insights for enhanced milk production (Baratta, 2013). In humans, MaSCs are differentiated into multipotent and unipotent MaScs in the mammary gland (Zhou et al., 2019). The multipotent MaSCs further differentiate into either a myoepithelial or luminal lineage, whereas unipotent MaSCs differentiate into the alveolar cell, ductal cells, etc. (Chen et al., 2017; Zhou et al., 2019). Through the genetic lineage tracing technique, it is possible to directly observe all the progeny of a single stem cell in the mouse model (Zhou et al., 2019).

Germ cell transfer technique to livestock production

The germ cell transfer technique is also referred to as spermatogonial stem cell transplantation. It implies the harvesting of testis stem cells (spermatogonial stem cells) from donor animals and injecting them into the seminiferous tubules of the recipient. Thus, the recipient has the potential of producing donor sperm cells. It is considered a major development in transgenic livestock production with higher efficiency and less time. Postpubertal mammals produce millions of sperm each day through a complex process known as spermatogenesis, which is dependent on spermatogonial stem cells. The complete process of the development of male sperm from male germ cells is referred to as spermatogenesis. After a long period of prespermatogensis in the fetus and infant stages, spermatogenesis starts at puberty. It is divided into the following stages: spermatogoniogenesis, meiosis, spermiogenesis, and spermiation. Seminiferous tubules give rise to spermatogonia, spermatocytes, and spermatids. These cells are located within the invaginations of Sertoli cells. Sertoli cells are considered nursing cells for the testicles, helping in the production of sperm. Spermatogonia continuously undergo mitosis up to a certain point, then undergo meiosis. It comprises two divisions: the first division is known as primary spermatocytes and

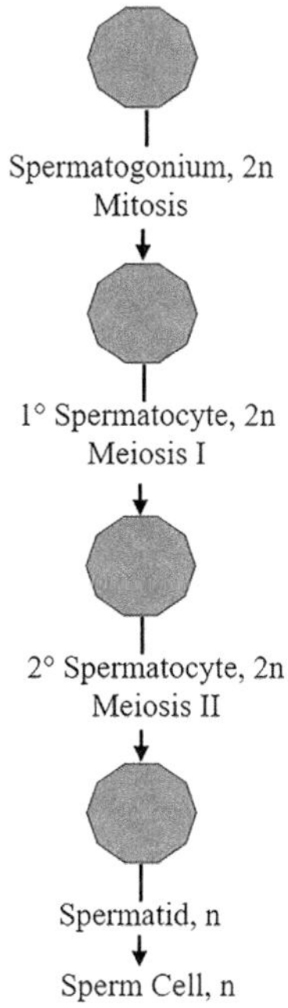

Fig. 3 Stages of spermatogenesis.

the second division is known as secondary spermatocytes. Secondary spermatocytes do not undergo DNA replication and divide quickly to give rise to spermatid. Spermiogenesis ends after the formation of spermatozoa from the cells releasing from the germinal epithelium. Spermiation occurs when the mature spermatids from the Sertoli cells enter the seminiferous tubules. Spermatozoa, or the male gamete/sperm, with its unique structure can penetrate the female gamete and finally, spermatogenesis is accomplished (Holstein et al., 2003). The spermatogonial stem cells are the foundation of spermatogenesis and provide a unique opportunity for animal transgenesis (see Fig. 3). It was first introduced in 1994 by Brinster and colleagues using a mouse model. The transfer of spermatogonial stem cells between different species is known as cross-species (xenogeneic) spermatogonial transplantation (McLean et al., 2001; Marx et al., 2014).

Applications of germ cell transplantation in livestock

Germ cell transplantation is successful between rats and mice where testes cells from donor rats are injected into the recipient mouse. This initiates the idea of using a laboratory mouse as a universal recipient for testes cells from donors of different species. The recipient can be made compatible with the donor by either compromising the immune system or generating immune-deficient animals such as pigs, goats, sheep, and bulls. The desired gene is inserted into the recipient using several techniques such as chemical, lipofection, electroporation, and viral vectors. This technique is further improved when the recipient endogenous germ cells are either depleted or suppressed. Recipient compatibility is achieved by using several techniques such as chemotherapy agents (busulfan), irradiation, etc. Like the recipient, the efficiency of transgenesis also depends on donor germ cells. In testis cells, the spermatogonial stem cells populate only 0.02%–0.2%, which is very low. Of this, only 1% can colonize in the recipient. These techniques are well pursued in farm animals. In vitro fertilization is now gaining importance in the animal breeding/cattle industry via creating embryos from oocytes by fertilizing with sperm cells in petri dishes (Honaramooz and Yang, 2011). This has a significant advantage compared to the conventional breeding system, especially in shortening generation intervals as well as increasing the number of female monotonous offspring and the selection intensity. The fundamental challenges for IVF are the generation of subsequent embryos and offspring in farm animals. Although this technology is still at the experimental level, it has had a huge impact on farm animals. Transgenic pigs are gaining importance because of their potential as donors for humans and to be able to act as a model for translational biomedical research. The goal of germ cell transplantation technology is to enhance male fertility in livestock. In comparison to conventional methods, 30–40 times more genetic gain has been possible by shortening the generation interval and enhancing the selection intensity by using this technique (Hou et al., 2018). It is also applied to azoospermic or tetrazoospermic animals, preserving the genetic material of the valuable species and cryopreserving the sperm; it is the main source for species continuation. It is also an alternative approach for artificial insemination and transgenesis (Savvulidi et al., 2019). Successful germ cell transplantation largely relies on the efficient production of donor-derived offspring, which further needs to be investigated and improved. This technique is mostly used in mice, rats, and goats. However, this mechanism is very slow and is just 0.7% in swine while being even lower in cattle and sheep (Hou et al., 2018). This study was started during 2002–2003, and has shown a significant outcome. This technique is still in the emerging phase with a remote chance of success but with a strong potential for providing new opportunities in livestock enhancement

(Honaramooz and Yang, 2011). The major obstacles among cross-species is incompetency and the in vitro maintenance of cells. This helps revolutionize the system for generating transgenic animals.

Transgenesis

Transgenesis is the transfer of specific genes from one organism into the genome of other organisms. It is further followed by germline transmission primarily in swine, sheep, goats, and cattle (Clark and Whitelaw, 2003). Transgenic technology provides an excellent approach to rapidly introduce new genes into livestock. This technology was initiated around 20 years ago. Recently, several transgenic lines have been developed to increase the potential or efficiency of the livestock (Maksimenko et al., 2013). There are several approaches to produce transgenic animals such as pronuclear injection, somatic cell nuclear transfer, etc. (Savvulidi et al., 2019); see Fig. 4. The outstanding application of the production of the transgenic animals includes enhanced prolificacy and reproductive performance; improved milk production, composition, and disease resistance; and enhanced growth of offspring, developmental biology, biomedicine, and agriculture (Wheeler, 2007).

Common techniques used for transgenic animal generation

Recombinant retroviruses

Retroviruses have an RNA genome that is further transcribed by their reverse transcriptase into the DNA in the recipient cells. It subsequently integrates into the genome of the infected cells. Here, the transgene is inserted in the place of a part of the retroviral genome and is integrated into

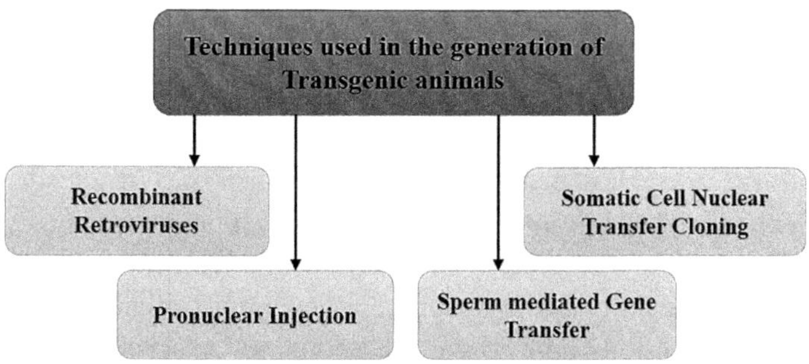

Fig. 4 Techniques used in the generation of transgenic animals.

the recipient genome. This integration occurs accidentally, and successful utilization has not been found in animals yet (Anson, 2004).

Pronuclear injection

Pronuclear injection is the process of microinjecting foreign DNA into the fertilized oocyte. This transfection includes cloning, followed by the preparation of the donor cell. After the recovery of the zygotes or fertilized oocytes, a DNA solution is microinjected into the pronuclei of the zygotes (Savvulidi et al., 2019). Finally, the zygotes are transferred into the oviducts of synchronized recipients. The insertion of the foreign DNA into the genome is detected from the new-born animals using the southern blot technique. Genetically modified sheep, rabbits, and pigs were generated using this technique. However, the major drawback of this technique is the random integration into the genome that results in variable gene expression, low efficiency, etc. Rodent and pig models are widely used for this technology and translational biomedical research because they resemble humans more closely than rodents (Tang et al., 2015).

Sperm-mediated gene transfer (SMGT)

The product of spermatogenesis is the production of sperm. SMGT is the process that involves the internalization of exogenous DNA into the sperm. Exogenous DNA specifically binds to the head of the sperm through binding proteins and integrates into the oocyte during fertilization. Different approaches have been used in combination with SMGT to increase its efficiency such as electroporation, liposomes, and restriction enzyme-mediated integration. The major drawbacks of this technology are the relatively low frequency, the random integration of DNA, mosaicism, and variable outcomes. To overcome some of these hurdles, intracytoplasmic sperm injection-mediated gene transfer (ICSI-GMT) is considered an extension of SMGT (Tang et al., 2015).

Somatic cell nuclear transfer cloning (SCNT)

SCNT is the technique of transferring a nucleus from the donor cell (somatic cell) into the oocytes, the early embryo from which the chromosomes have been depleted, or enucleated oocytes. It opened a new avenue to generate genetically modified livestock. This process involves the following steps. First, the enucleated matured oocytes are generated, which is followed by the injection or fusion of the donor cell, leading to the creation of transgenic animals. This technique is successful in many domestic, wild, and laboratory species (Tang et al., 2015). Currently, it is used to generate large transgenic animals. The transgenic sheep Dolly was generated in 1996 by Ian Wilmut and coworkers, and is a well-established SCNT example. To clone an identical animal, it is essential to begin with an egg without the nucleus and then inject the desired nucleus into it (Liu et al., 2020).

In addition to Dolly, transgenic swine were also generated to study the basic biology of milk production, secretion, and lactation (Wheeler, 2003). The genetically modified animals are now used as a model to study diabetes, cystic fibrosis, cardiovascular diseases, and neurodegenerative diseases (Tang et al., 2015). The application of transgenic animals includes improved milk production and composition, feed utilization, carcass composition, disease resistance, reproductive performance, and increased prolificacy; see Fig. 5. Genetically altered goats, sheep, and cows are of great importance in the production of biopharmaceutical proteins in their milk that benefit human/animal health, safety, production, etc. In addition provides benefits of milk safety by expression of human lysozyme and human alpha 1 antitrypsin in the mammary gland of transgenic goats and sheep respectively (Wheeler, 2007). This improvement in milk will have a profound impact on the survival and growth of newborn humans and animals. Human organ transplantation is a relatively emerging field that faces several significant challenges. Its clinical success will be a benchmark from the perspective of human donor organs.

Routes of transplantation of stem cells

Autologous route of transplantation (AUTO transplant)

In autologous route of transplantation a person can cryopreserve his or her cord blood cells for any biomedical situations. The cryopreservation technique allows the storage of bone marrow and placenta stem cells for further use by donors during a medical emergency such as cancer or organ transplant. It is considered the safest route of transplantation (Brown et al., 2019).

Allogeneic route of transplantation (ALLO transplant)

Allogeneic transplantation is a procedure involving the removal of an organ from one individual. The donor (allogeneic) can be the parent, child, sister, or brother, and the organ is transferred into the respective position in the second individual. The major drawback in allogeneic transplantation is the shortage

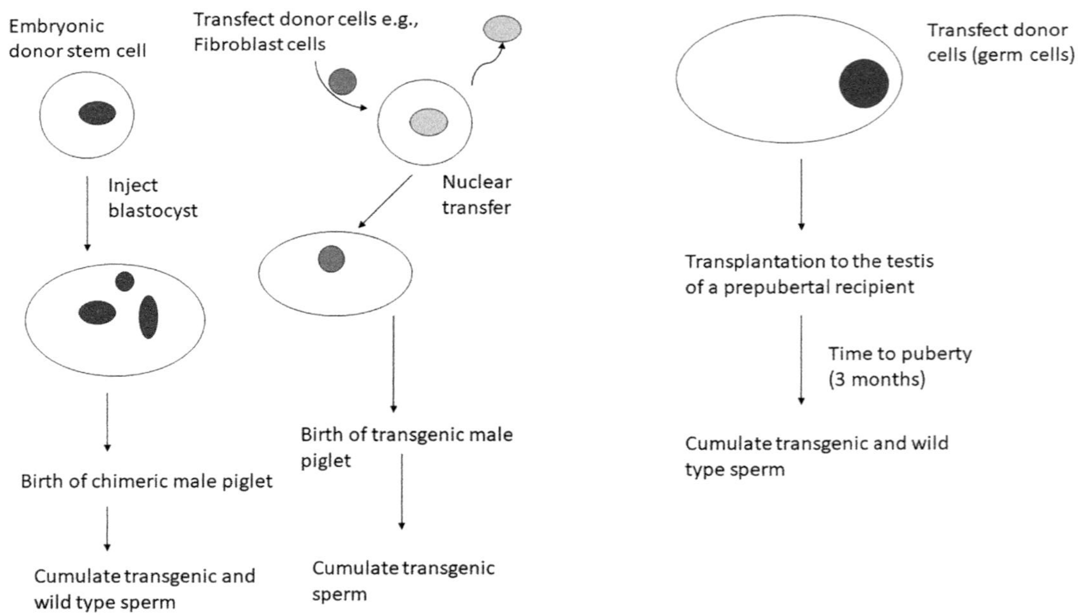

Fig. 5 Approaches to generate transgenic animals.

of organ donors and the ever-increasing need, so to overcome this drawback xenotransplantation is gaining importance (Yamada and Scalea, 2012).

Xenogeneic route of transplantation (XENO transplant)

The procedure of implanting cells, tissues, and organs from nonhuman animal sources into a human is known as xenotransplantation. Approximately 10 people die every day due to the shortage of human organ donations. Globally many people are unable to afford organ transplants due to age, health, and cost issues. Several biotechnological and pharmaceutical industries are continuing research to overcome the organ shortage problems (Yamada and Scalea, 2012). Despite significant awareness of the need, the donor frequency remains stagnant. During the 1960s, chimpanzee kidneys were transplanted into patients with kidney failure for the first time (Cooper, 2012). Xenotransplantation also involves the development of xenogeneic cells such as porcine islets cells for diabetes, porcine hepatocytes for liver failure, and porcine dopaminergic neurons for Parkinson's disease as well as injuries to the spinal cord, organs, and limbs; all have a cross-species physiological activity (Huang et al., 2018). However, the greatest hurdle in xenotransplantation is that the recipient body treats the xenograft as an infection and activates its immune system to get rid of it. Among cows, pigs, and monkeys, pigs are normally preferred because they easily mature and produce many litters. Also, their organ size is comparable to humans in both infancy and adulthood. Studies suggest the use of pancreatic islet cells and neural cells from pigs for treating diabetes. Huntington's patients also receive modified tissues from pigs as an experimental treatment. Still, they are at the experimental or preliminary stages (Tang et al., 2015). However, xenotransplantation should be done only when it meets ethical issues, which are further categorized into animal welfare, clinical experimentation, and public health implications (Kaiser, 2004; Samstein and Platt, 2001). There are various modes of rejection of xenotransplants, see Fig. 6.

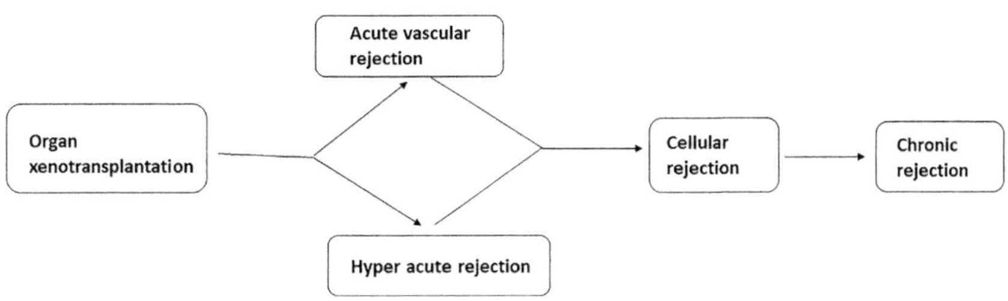

Fig. 6 Types of xenotransplant rejections.

Types of xenotransplant rejections

Hyperacute rejection

Hyperacute rejection is defined as the immune response that occurs within 24h after cross-species xenotransplantation. The antibodies attacks the endothelial lining, which results in the activation of the complement, coagulation system, thrombosis, interstitial hemorrhage and edema thus disrupt graft function (Cooper et al., 2015). It occurs in nearly every organ. The species in which this rejection rarely occurs are known as concordant, otherwise they are considered discordant (Bogaerde and White, 1997). The symptoms of hyperacute rejection include that the graft rapidly develops a beefy red or blue appearance. Swelling occurs due to a decrease in the blood flow in the graft and injury to epithelial cells.

Moreover, the graft rapidly develops a beefy red or blue appearance and swells due to a decrease in the blood flow in the graft, injury to epithelial cells, the formation of platelet thrombin in small blood vessels and interstitial hemorrhage. This is the first barrier to organ transplantation. Progress has been continuing in overcoming this problem. The first strategy is to deplete the xenoreactive antibodies from the circulation of the recipient. The second strategy is to deplete or inhibit the complement system (Samstein and Platt, 2001).

Acute humoral xenograft rejection

If hyperacute rejection is prevented, similar rejections can occur normally within a few days or weeks; these are known as acute vascular rejection, acute humoral xenograft rejection, or delayed xenograft rejection (Cooper et al., 2015). After averting the hyperacute rejection, the xenograft is subjected to acute vascular rejection, which is characterized by ischemia, fibrinoid necrosis, endothelial swelling, and different intravascular thrombosis (Porada and Almeida-Porada, 2010). In addition, it also involves the activation of antibodies, an innate immune response such as macrophages, NK cells, polymorphonuclear leukocytes to destruct the graft. However, their exact importance remains unclear (Cooper et al., 2015). The prevention of acute vascular rejection uses several approaches such as the depletion of xenoreactive antibodies from the recipient, which involves treatment with cyclophosphamide that inhibits antibody synthesis (Samstein and Platt, 2001). Another approach is tolerance induction, leading to the decreased production of antibodies against the graft.

Adaptive immune response

If both these rejections are prevented, then the adaptive immune responses are activated. This is comprised of two classes–antibody responses and cell-mediated immune responses. They are further carried out by the different classes of lymphocytes, namely B cells and T lymphocytes (Cooper et al., 2015). A nonvascularized xenograft rejection is where B cells secrete antibodies that bind to the graft and T cells activate macrophages and react with the graft for destruction (Yamada and Scalea, 2012).

Chronic rejection

If the graft normally survives for more than a few weeks, then features of chronic vasculopathy develop that result in chronic rejection (Cooper et al., 2015). Generally, the graft is lost within the fifth year of transplantation. Chronic rejection includes medication toxicity, jaundice, dark urine, itching, vomiting, abdominal swelling, coughing, breathing problems, frequent infections, arterial hypertension, proteinuria, fatigue, weight gain, premature death, anorexia, anemia, nausea, neoplasia, edema, and atherosclerosis. Several diseases, namely pulmonary disease, pancreatic failure, and liver and kidney failure, are also associated with chronic rejection. Immune cells such as macrophages, monocytes, etc., are the major contributors to this rejection. However, immunosuppressants do not prevent chronic rejection, suggesting that other mechanisms are also involved in chronic rejections. Human leukocyte antigen (HLA) matching, electrocardiogram, x ray and CT scanning, and urine and blood sampling between the donor and recipient make a transplant more successful (Justiz Vaillant et al., 2018). The major obstacle in xenogeneic cell therapy is the induction of the host immune system to reject xenografts. Now, scientists are developing novel immunosuppressive drugs that facilitate xenotransplantation or make it clinically feasible. Research is ongoing to understand how the immune system acts on the transplant organs derived from animal sources. Drugs like rapamycin, deoxyspergualin, and leflunomide are approved for clinical trials that enable the tolerance of grafts and prevent cellular rejection (Samstein and Platt, 2001). Xenotransplants are gaining importance in treating several diseases, including cancer, that bring collaborative damage to human health. Here, cells derived from animals such as pigs undergo xenotransplantation in healing the wound or restoring the host tissue functions in cancer patients (Cooper et al., 2015; Samstein and Platt, 2001).

To prevent chronic rejection, there is an interprofessional team that educates and manages such patients. The team includes nephrology, cardiology, and transplant nurses as well as pharmacists to monitor the patient status and direct feedback to the team, which ultimately improves outcomes. Chronic transplant is associated with several complications, namely kidney and liver failure, pulmonary disease, pancreatic problems, arteriosclerosis, blood dyscrasias, CMV infection, and infectious diseases caused by bacteria, viruses, fungi, protozoans, or combinations. A chronic rejection diagnosis generally includes acute rejection, infection, medication toxicity, posttransplant lymphoproliferative disease, or other problems with vascular or structural anastomotic complications. In the end, retransplantation of the organ or tissue is considered. The etiology of chronic rejection is not yet completely understood, but it is considered the leading cause of graft rejection. Phagocytic cells such as macrophages, monocytes, and immature dendritic cell subsets are contributors to chronic rejection. Chronic rejection is generally caused by antibodies and lymphocytes. Conventional cancer treatment such as surgery, radiotherapy, chemotherapy, and targeted therapy cure tumors but at the same time destroy normal tissues, thus having limited efficacy. Although current treatments are effective, they bring collaborative damage to human health, causing several complications and adverse effects. Consistent with this, the outcomes are mostly discouraging for most cancer patients. To get rid of these problems, xenogeneic cell therapy is gaining importance.

Stem cell therapy application in veterinary medicine

Regenerative medicine involves the utilization of stem cells in regenerating human and animal cells, tissues, or organs to restore normal function (Voga et al., 2020). Dental tissue-derived stem cells, intestinal epithelial cells, etc., are considered potential sources of regenerative medicine (Nakamura and Toshiro Sato, 2018; Feng and Lengner, 2013). It also helps in the development of novel therapeutics, including organ transplantation, surgery, treatment of genetic and autoimmune disorders, etc. Livestock play a myriad of functions in the livelihood of humankind, including food, work, and economic and social levels. Therefore, it is essential to enhance production, specifically in developing countries. As with human disease, stem cell technology has also been widely used in veterinary diseases and regenerative medicine; it is gaining equal importance in livestock enhancement. Several biomedical studies are under way in veterinary medicine, such as using water buffalo as an emerging large animal model. Even, the contribution of domesticated animals such as mice, nonhuman primates (monkeys), etc., in the livelihood and development of humankind and regenerative medicine in humans cannot be denied. Water buffalos are domesticated and economically important livestock. The following reproductive techniques have been successfully employed in water buffalo: in vitro fertilization, embryo transfer, somatic cell nuclear transfer, genetic engineering/editing, and the cryopreservation of embryos and sperm (Singh et al., 2020). Thus, water buffalos are also used for stem cell therapy in regenerative medicine. Besides this, epithelial cells, fibroblasts, fetal stem cells, adipose, bone marrow, and amnion fluid cells obtained from mice, goats, cattle, and sheep also have high therapeutic potential of reprogramming (Yadav et al., 2012). They are used in veterinary regenerative medicine, including the generation of cloned animals such as sheep, cattle, buffalo, mice, pigs, etc. (Yamada and Scalea, 2012; Cibelli et al., 2013).

Future aspects of stem cells and regeneration in veterinary science and translation to human health

Transgenic animals such as buffalos and cows are now considered a good source for the production of therapeutic proteins (Monzani et al., 2016; Houdebine, 2000). Consistent with this, genetic modification in bubaline perinatal stem cells has also opened a new avenue for clinical and pharmaceuticals trials. Also, genetic modifications help in adapting to climate change, resistance to infectious disease, germplasm conservation, conservation of endangered species, etc. This further suggests that the role of transgenic animals such as buffalo is not limited to milk and meat production, but is also very useful in other functions. Also, future research is focusing on a detailed understanding of the development of novel cell-based therapies and veterinary regenerative medicine via cellular reprogramming, cell differentiation, and in vitro cell proliferation (Singh et al., 2020). Further studies are also focusing on the development of in vitro gamete differentiation, gametogenesis or reconstitution, and in vitro oocyte production to increase the frequency of fertilization in animal reproduction. Research is ongoing that involves the transplantation of germinal stem cells into recipient testes to overcome the poor semen cryopreservation (Savvulidi et al., 2019). The major obstacles of this technology are the percentage of success is very low, early deaths, incompatibility between mates, and an unsustainable population, which means functional gametes with true reproductive potential are much less (Clark and Whitelaw, 2003). Other obstacles are that the techniques are insufficient and expensive. This is coupled with the fact that the most relevant traits are highly complex and usually controlled by more than one gene, which also restricts the use of stem cell technology. Insufficient knowledge of human stem cells and the absence of an animal model with a comparable organ size and physiology are the current limitations of these techniques. More research is required to increase the reproduction efficiency and genetic factors in animal production that present new, promising avenues for reproductive medicine and endangered species reproduction (Cibelli et al., 2013).

Conclusion

Stem cell technology provides an attractive platform to study developmental processes and disease pathophysiology because MSCs have self-renewal capacity and can differentiate into other cell types. Several studies are ongoing to increase the reproduction efficiency and genetic factors in animal production, which present new, promising avenues for reproductive medicine and endangered species reproduction. Cells isolated from livestock species could serve as potential tools to explore developmental biology and to optimize culture conditions for stem cells. Stem cell transplantation can provide an effective selective breeding tool to achieve excellent genetic makeup to enhance the efficiency of livestock production systems.

References

Anderson, P., Carrillo-Gálvez, A.B., García-Pérez, A., Cobo, M., Martín, F., 2013. CD105 (endoglin)-negative murine mesenchymal stromal cells define a new multipotent subpopulation with distinct differentiation and immunomodulatory capacities. PLoS One 8 (10), e76979.

Anson, D.S., 2004. The use of retroviral vectors for gene therapy-what are the risks? A review of retroviral pathogenesis and its relevance to retroviral vector-mediated gene delivery. Genet. Vaccines Ther. 2, 9.

Baratta, M., 2013. Adults mammary stem cell in cow's milk: new perspectives and future challenge. J. Vet. Sci. Anim. Husb. 1 (2), e202.

Barberini, D.J., Freitas, N.P.P., Magnoni, M.S., Maia, L., Listoni, A.J., Heckler, M.C., et al., 2014. Equine mesenchymal stem cells from bone marrow, adipose tissue and umbilical cord: immunophenotypic characterization and differentiation potential. Stem Cell Res Ther 5 (1), 25.

Barreto, R.D.S.N., Romagnolli, P., Mess, A.M., Miglino, M.A., 2018. Decellularized bovine cotyledons may serve as biological scaffolds with preserved vascular arrangement. J. Tissue Eng. Regen. Med. 124, e1880–e1888.

Bogaerde, J.V., White, D.J.G., 1997. Xenogeneic transplantation. Br. Med. Bull. 53, 904–920.

Brown, K.S., Rao, M.S., Brown, H., 2019. The future state of newborn stem cell banking. J. Clin. Med. 8 (1), 117.
Chamberlain, G., Fox, J., Ashton, B., Middleton, J., 2007. Concise review: mesenchymal stem cells: their phenotype, differentiation capacity, immunological features, and potential for homing. Stem Cells 25, 2739–2749.
Chen, X., Liu, Q., Song, E., 2017. Mammary stem cells: angels or demons in mammary gland? Signal Transduct. Target. Ther. 2, 16038.
Choi, Y.C., Choi, J.S., Woo, C.H., Cho, Y.W., 2014. Stem cell delivery systems inspired by tissue-specific niches. J. Control. Release 193, 42–50.
Cibelli, J., Emborg, M.E., Prockop, D.J., Roberts, M., Schatten, G., Rao, M., Harding, J., Mirochnitchenko, O., 2013. Strategies for improving animal models for regenerative medicine. Cell Stem Cell 12 (3), 271–274.
Clark, J., Whitelaw, B., 2003. 2003. A future for transgenic livestock. Nat. Rev. Genet. 4 (10), 825–833.
Cooper, D.K.C., 2012. A brief history of cross-species organ transplantation. Proc. (Baylor Univ. Med. Cent.) 25 (1), 49–57.
Cooper, D.K.C., Ekser, B., Tector, A.J., 2015. Immunobiological barriers to xenotransplantation. Int. J. Surg. 23, 211–216.
de Moraes, C.N., Maia, L., Dias, M.C., Dell'Aqua, C.P.F., da Mota, L.S.L.S., Chapwanya, A., Landim-Alvarenga, F.D.C., Oba, E., 2016. Bovine endometrial cells: a source of mesenchymal stem/progenitor cells. Cell Biol. Int. 40 (12), 1332–1339.
Dominici, M., LeBlanc, K., Mueller, I., Slaper-Cortenbach, I., Marini, F., Krause, D., Deans, R., Keating, A., Prockop, D.J., Horwitz, E., 2006. Minimal criteria for defining multipotent mesenchymal stromal cells. The international society for cellular therapy position statement. Cytotherapy 8 (4), 315–317.
Erickson, I.E., van Veen, S.C., Sengupta, S., Kestle, S.R., Mauck, R.L., 2011. Cartilage matrix formation by bovine mesenchymal stem cells in three-dimensional culture is age-dependent. Clin. Orthop. Relat. Res. 469, 2744–2753.
Feng, R., Lengner, C., 2013. Application of stem cell technology in dental regenerative medicine. Adv. Wound Care 2 (6), 296–305.
Ferraro, F., Celso, C.L., Scadden, D., 2010. Adult stem cells and their niches. In: The Cell Biology of Stem Cells. Advances in Experimental Medicine and Biology, vol. 695. Springer, Boston, pp. 155–168.
Freshney, R.I., 1994. Culture of Animals Cells: A Manual of Basic Technique, fourth ed. Wiley-Liss Press, New York, p. 672.
Gazit, Z., Pelled, G., Sheyn, D., Kimelman, N., Gazit, D., 2011. Mesenchymal stem cells. In: Principles of Regenerative Medicine, second ed. vol. 17. Academic Press, pp. 285–304.
Hill, A.B., Hill, J.E., Bressan, F.F., Miglino, M.A., Garcia, J.M., 2018. Derivation and differentiation of canine ovarian mesenchymal stem cells. J. Vis. Exp. 142, e58163.
Holstein, A.F., Schulze, W., Davidoff, M., 2003. Understanding spermatogenesis is a prerequisite for treatment. Reprod. Biol. Endocrinol. 1, 107.
Honaramooz, A., Yang, Y., 2011. Recent advances in application of male germ cell transplantation in farm animals. Vet. Med. Int., 657860.
Hou, Z., An, L., Han, J., Yuan, Y., Chen, D., Tian, J., 2018. Revolutionize livestock breeding in the future: an animal embryo-stem cell breeding system in a dish. J. Anim. Sci. Biotechnol. 9, 90.
Houdebine, L.M., 2000. Transgenic animal bioreactors. Transgenic Res. 9, 305–320.
Huang, C.P., Chen, C.C., Shyr, C.R., 2018. Xenogeneic cell therapy provides a novel potential therapeutic option for cancers by restoring tissue function, repairing cancer wound and reviving anti-tumor immune responses. Cancer Cell Int. 18, 9.

Inman, J.L., Robertson, C., Mott, J.D., Bissell, M.J., 2015. Mammary gland development: cell fate specification, stem cells and the microenvironment. Development 142, 1028–1042.

Ito, K., Suda, T., 2014. Metabolic requirements for the maintenance of self-renewing stem cells. Nat. Rev. Mol. Cell Biol. 15 (4), 243–256.

Jiang, Y.B.N., Jahagirdar, R.L., Reinhardt, R.E., Schwartz, C.D., Keene, X.R., Ortiz-Gonzalez, et al., 2002. Pluripotency of mesenchymal stem cells derived from adult marrow. Nature 418, 41–49.

Justiz Vaillant, A.A., Waheed, A., Mohseni, M., 2018. Chronic transplantation rejection. In: Stat Pearls. Stat Pearls Publishing, Treasure Island (FL), p. 14.

Kaiser, M., 2004. Xenotransplantation-ethical considerations based on human and societal perspectives. Acta Vet. Scand. 99 (99), 65–73.

Khan, I.M., Bishop, J.C., Gilbert, S., Archer, C.W., 2009. Clonal chondroprogenitors maintain telomerase activity and Sox9 expression during extended monolayer culture and retain chondrogenic potential. Osteoarthr. Cartil. 17, 518–528.

Khatri, M.T.D., O'Brien, J., Sharma, M., 2009. Isolation and differentiation of chicken mesenchymal stem cells from bone marrow. Stem Cells Dev. 18, 1485–1494.

Kuroda, Y.M., Kitada, S., Wakao, K., Nishikawa, Y., Tanimura, H., Makinoshima, M., Goda, et al., 2010. Unique multipotent cells in adult human mesenchymal cell populations. Proc. Natl. Acad. Sci. U. S. A. 107, 8639–8643.

Lane, S.W., Williams, D.A., Watt, F.M., 2014. Modulating the stem cell niche for tissue regeneration. Nat. Biotechnol. 32 (8), 795–803.

Lara, E., Rivera, N., Rojas, D., Rodríguez-Alvarez, L.L., Castro, F.O., 2017. Characterization of mesenchymal stem cells in bovine endometrium during follicular phase of oestrous cycle. Reprod. Domest. Anim. 52 (5), 707–714.

Li, L., Xie, T., 2005. Stem cell niche: structure and function. Annu. Rev. Cell Dev. Biol. 21, 605–631.

Liechty, K., Mackenzie, T., Shaaban, A., Radu, A., Moseley, A.M., Deans, R., Marshak, D.R., Flake, A.W., 2000. Human mesenchymal stem cells engraft and demonstrate site-specific differentiation after in utero transplantation in sheep. Nat. Med. 6, 1282–1287.

Liu, G., David, B.T., Trawczynski Mand Fessler, R.G., 2020. Advances in pluripotent stem cells: history, mechanisms, technologies, and applications. Stem Cell Rev. Rep. 16, 3–32.

Maksimenko, O.G., Deykin, A.V., Khodarovich, Y.M., Georgiev, P.G., 2013. Use of transgenic animals in biotechnology: prospects and problems. Acta Nat. 5 (1), 33–46.

Martin, B.M., 1994. Routine cell culture. In: Martin, B.M. (Ed.), Tissue Culture Techniques: An Introduction. Birkhäuser, Berlin, pp. 29–89.

Martinello, T., Bronzini, I., Maccatrozzo, L., Mollo, A., Sampaolesi, M., Mascarello, F., Decaminada, M., Patruno, M., 2001. Canine adipose-derived-mesenchymal stem cells do not lose stem features after a long-term cryopreservation. Res. Vet. Sci. 91, 18–24.

Marx, C., Silveira, M.D., Selbach, I., et al., 2014. Acupoint injection of autologous stromal vascular fraction and allogeneic adipose derived stem cells to treat hip dysplasia in dogs. Stem Cells Int., 391274. 6 pages.

McLean, D.J., Johnston, D.S., Russell, L.D., Griswold, M.D., 2001. Germ cell transplantation and the study of testicular function. Trends Endocrinol. Metab. 12 (1), 16–21.

Mimeault, M., Batra, S.K., 2008. Recent progress on tissue resident adult stem cell biology and their therapeutic implications. Stem Cell Rev. 4 (1), 27–49.

Monzani, P.S., Adona, P.R., Ohashi, O.M., Meirelles, F.V., Wheeler, M.B., 2016. Transgenic bovine as bioreactors: challenges and perspectives. Bioengineered 7 (3), 123–131.

Nakamura, T., Toshiro Sato, T., 2018. Advancing intestinal organoid technology toward regenerative medicine. Cell. Mol. Gastroenterol. Hepatol. 5, 51–60.

Parekkadan, B., Milwid, J.M., 2010. Mesenchymal stem cells as therapeutics. Annu. Rev. Biomed. Eng. 12, 87–117.

Porada, C.D., Almeida-Porada, G., 2010. Mesenchymal stem cells as therapeutics and vehicles for gene and drug delivery. Adv. Drug Deliv. Rev. 62 (12), 1156–1166.

Rana, D., Zreiqat, H., Benkirane-Jessel, N., Ramakrishna, S., Ramalingam, M., 2017. Development of decellularized scaffolds for stem cell-driven tissue engineering. J. Tissue Eng. Regen. Med. 11 (4), 942–965.

Samstein, B., Platt, J.L., 2001. Physiologic and immunologic hurdles to xenotransplantation. J. Am. Soc. Nephrol. 12 (1), 182–193.

Savvulidi, F., Ptacek, M., Vargova, K.S., Stadnik, L., 2019. Manipulation of spermatogonial stem cells in livestock species. J. Anim. Sci. Biotechnol. 10, 46.

Schellenberg, A., Stiehl, T., Horn, P., Joussen, S., Pallua, N., Ho, A.D., Wagner, W., 2012. Population dynamics of mesenchymal stromal cells during culture expansion. Cytotherapy 14 (4), 401–411.

Schnabel, L.V., Fortier, L.A., McIlwraith, C.W., Nobert, K.M., 2013. Therapeutic use of stem cells in horses: which type, how, and when? Vet. J. 197, 570–577.

Schwarz, C., Leicht, U., Rothe, C., Drosse, I., Luibl, V., Röcken, M., Schieker, M., 2013. Effects of different media on proliferation and differentiation capacity of canine, equine and porcine adipose derived stem cells. Res. Vet. Sci. 93, 457–462.

Shin, L., Peterson, D.A., 2013. Human mesenchymal stem cell grafts enhance normal and impaired wound healing by recruiting existing endogenous tissue stem/progenitor cells. Stem Cells Transl. Med. 2 (1), 33–42.

Singh, B., Mal, G., Kues, W.A., Yadav, P.S., 2020. The domesticated buffalo—an emerging model for experimental and therapeutic use of extraembryonic tissues. Theriogenology 15 (151), 95–102.

Steinert, A.F., Palmer, G.D., Pilapil, C., Noth, U., Evans, C.H., Ghivizzani, S.C., 2009. Enhanced in vitro chondrogenesis of primary mesenchymal stem cells by combined gene transfer. Tissue Eng. 15, 1127–1139.

Tada, T., 2008. Genetic modification-free reprogramming to induced pluripotent cells: fantasy or reality? Cell Stem Cell 3, 121–122.

Takahashi, K., Yamanaka, S., 2006. Induction of pluripotent stem cells from mouse embryonic and adult fibroblast cultures by defined factors. Cell 126, 663–676.

Talbot, N.C., Blomberg, A., 2008. The pursuit of ES cell lines of domestic ungulates. Stem Cell Rev. 4, 235–254.

Tang, L., González, R., Dobrinski, I., 2015. Germline modification of domestic animals. Anim. Reprod. 12 (1), 93–104.

Torricelli, P., Fini, M., Filardo, G., et al., 2011. Regenerative medicine for the treatment of musculoskeletal overuse injuries in competition horses. Int. Orthop. 35 (10), 1569–1576.

Trivanović, D., Jauković, A., Popović, B., Krstić, J., Mojsilović, S., Okić-Djordjević, I., et al., 2015. Mesenchymal stem cells of different origin: comparative evaluation of proliferative capacity, telomere length and pluripotency marker expression. Life Sci. 141, 61–73.

Voga, M., Adamic, N., Vengust, M., Majdic, G., 2020. Stem cells in veterinary medicine—current state and treatment options. Front. Vet. Sci. 7, 278.

Wagner, W., Feldmann Jr., R.E., Seckinger, A., Maurer, M.H., Wein, F., Blake, J., et al., 2006. The heterogeneity of human mesenchymal stem cell preparations evidence from simultaneous analysis of proteomes and transcriptomes. Exp. Hematol. 34 (4), 536–548.

Webster, R.A., Blaber, S.P., Herbert, B.R., Wilkins, M.R., Vesey, G., 2012. The role of mesenchymal stem cells in veterinary therapeutics–a review. N. Z. Vet. J. 60 (5), 265–272.

Wheeler, M.B., 2003. Production of transgenic livestock: promise fulfilled. J. Anim. Sci. 81, 32–37.

Wheeler, M.B., 2007. Agricultural applications for transgenic livestock. Trends Biotechnol. 25 (5), 204–210.

Yadav, P.S., Singh, R.K., Singh, B., 2012. Fetal stem cells in farm animals: applications in health and production. Agric. Res. 1, 67–77.
Yamada, K., Scalea, J., 2012. Current progress in xenogeneic tolerance. Curr. Opin. Organ Transplant. 17 (2), 168–173.
Yasumasa, K., Kitada, M., Wakao, S., Nishikawa, K., Tanimura, Y., Makinoshima, H., Goda, M., et al., 2010. Unique multipotent cells in adult human mesenchymal cell populations. Proc. Natl. Acad. Sci. U. S. A. 107 (19), 8639–8643.
Zhou, J., Chen, Q., Zou, Y., Zheng, S., Chen, Y., 2019. Stem cells and cellular origins of mammary gland: updates in rationale, controversies, and cancer relevance. Stem Cells Int. 4247168, 12.

CHAPTER EIGHTEEN

Way forward for sustainable livestock sector

Mozammel Hoque[a], Akash[a], Sukanta Mondal[b], and Satish Adusumilli[c]
[a]Division of Surgery, ICAR-Indian Veterinary Research Institute, Izatnagar, Uttar Pradesh, India
[b]Animal Physiology Division, ICAR-National Institute of Animal Nutrition and Physiology, Adugodi, Bengaluru, Karnataka, India
[c]Freimann Life Science Center, University of Notre Dame, Notre Dame, IN, United States

Introduction

Livestock is an important global sector contributing significantly to the economies of countries and the food security and livelihoods of millions of people around the globe. Demands for livestock product are ever increasing and is expected to double by 2050 due to the growing population and improved standard of living. Therefore, there is an urgent need to expand the livestock sector to meet the demand of a growing population. The livestock sector is facing a huge challenge due to climate change. Change in climate brings about changes in environmental factors such as temperature, solar radiation, rainfall, relative humidity, and wind speed, resulting in extreme weather conditions. These extreme weather conditions adversely affect livestock productivity directly and indirectly. Among the climatic variables, the most intriguing factor that adversely affects the production of livestock seems to be heat stress. Due to high temperatures, livestock production is jeopardized by reduced growth, reproductive efficiency, milk and meat production, and animal health. The quality and quantity of fodder and water supplies are affected by climate extremes and seasonal variations, and they are also vulnerable to new diseases. Livestock systems also have a large environmental footprint. They are a huge contributor to emissions that drive climate change. Livestock production has been estimated to contribute about 14.5% of the total GHG emissions. Global food security is threatened by climate change and its adverse impact on the production of livestock. Therefore, to ensure food security, our approaches to sustaining livestock productivity need a fundamental change to fulfill the demands of the increasing human population. Efforts are therefore required to improve the resilience capacity of livestock to give them the ability to manage climate change-related

https://doi.org/10.1016/B978-0-12-822265-2.00016-8

adversity and to sustain their productivity. The solution to this global challenge is climate-smart livestock production, which aims to sustainably increase efficiency and resilience (adaptation), minimize or eliminate GHGs (mitigation), and enhance food security and development goals (Bezuidenhout, 2019; GaitaÂn et al., 2016; UNEP, 2011).

Importance of livestock production

According to an estimate, the livestock sector accounts for 40% of global agricultural gross domestic product (GDP), supports more than 1 billion people, and provides more than 33% of the world's protein intake (FAO, 2013, 2017). The sector has been growing rapidly. Annual meat consumption per capita in the growing world more than tripled between 1960 and 2005; global meat demand is projected to more than double between 1990 and 2050 (to 465 million tonnes). Annual per capita milk consumption in the developing world nearly doubled between 1960 and 2005; global milk production is expected to increase by nearly 80% (to 1043 million tonnes). The size and rapid growth of the sector mean that livestock production requires significant resources. It occupies 30% of the world's land surface and 70% of the agricultural land surface (primarily in the cultivation of feed crops such as cereals and soybean used to feed cattle, particularly in the developed world). It accounts for more than 8% of global water use (mainly irrigation for feed crops). Livestock feed accounts for more than 35% of overall cereal use with cattle consuming more than 1 billion tons of grain each year.

Adverse effect of climatic change on livestock production

Heat stress induced by climate change-related drought directly affects livestock production. The decline in milk quality, meat production, reproductive performance, and animal health due to climatic change creates a major financial burden for livestock producers. The indirect influence of drought on livestock is primarily due to decreases in feed and water resource or nonavailability. In such a scenario, animals are subject to combined thermal and nutritional stress, which has a significant impact on the production, reproductive health, and immune status of animals. The morbidity and frequency of death of livestock animals is increased by heat stress due to increased temperature. The indirect impact of higher temperatures on microbial populations (pathogens or parasites), the spread of vector-borne and foodborne diseases, host resistance, and the lack of feed and water (Nordone et al., 2010; Thornton, 2010; Tubiello et al., 2008; PKSF, 2019). Increased temperature could increase the growth of pathogens that live outside their host, which affects livestock negatively (Harvell et al., 2002; Karl et al., 2009; Patz et al., 2000). Climatic changes can lead to changes in disease transmission, extreme disease outbreaks, or even the introduction of new diseases that can harm livestock that are not typically exposed to these types of diseases (Thornton, 2010).The effect of temperature and nutrition on the productive and reproductive success of ewes in India was studied by and explained by experts. The substantial decrease in live weight gain and reproductive efficiency, including the lambing rate of ewes, caused nutritional, thermal, or combined stresses.

Climate change affects livestock production and consequently food security. Livestock production is adversely affected (due to disease, water availability, etc.) specifically in arid and semiarid regions. Moreover, the nutritional content of livestock products, which are a source of global calories, proteins, and essential micronutrients, is affected by climate change. In comparison, livestock production also affects climate change. Deforestation contributes around 9.2% of overall livestock GHG emissions due to the expansion of crop lands and pasture lands for livestock production (Gerber et al., 2013). The feed production level, however, accounts for the largest fraction (nearly half) of GHG emissions in the entire livestock production process. It is predicted that, due to the intensification of livestock production, this stage would further increase its contribution. Meanwhile, in the livestock production process, enteric fermentation is found to be the largest contributor of GHG. Therefore, if livestock levels continue to grow and feeding practices do not change, global emissions will increase due to livestock production.

Role of livestock sector in climate change

Significant quantities of specific gases (5% of emissions of anthropogenic CO_2; 44% of emissions of anthropogenic methane; and 53% of emissions of anthropogenic nitrous oxide). Emission sources include direct sources such as ruminant enteric fermentation (39% of emissions) and manure fermentation (26%) as well as indirect sources that include the production, processing, and transportation of animal feed (45% of sector emissions).

Wider environmental issues include grazing land loss due to issues such as overgrazing (more than 50% of deforestation has been connected to livestock in the United States) as well as pollution from animal waste, veterinary waste, and pesticides/fertilizers used to grow feed crops (in the United States, 37% of pesticides and 50% of antibiotics are projected to be related to livestock production).

The quality and supply of fodder and feed are potentially affected by climate change and may intensify the degradation of pasture land (e.g., due to increased drought or flood risk) as well as the risk of disease (e.g., because of warmer temperatures). Arid and semiarid grazing systems in vulnerable regions such as sub-Saharan Africa are at serious risk (Hoffman and Vogel, 2008; Kadzere, 2018).

About 70% of all land used for agriculture, or 30% of the world's land area, is covered by livestock production. As calculated in CO_2 equivalents, this industry is a significant source of GHG, accounting for 18% of the world's GHG emissions. The United Nations Environmental Programme (UNEP) reports that under current practices and consumption trends, methane emissions from the global livestock sector are expected to increase by 60% by 2030.

Way forward

The two main approaches to make livestock production "climate-smart" are **adaptation** to climatic changes and **mitigation** of GHG (Rojas-Downing et al., 2017). At the beginning, these two approaches were initiated independently. Recently, there has been a growing realization from decision makers and policy makers to address them in a more integrated manner, as their agenda has moved toward the development of solutions. Integration of these two aspects can exploit synergies and minimize trade-offs between mitigation and adaptation or avoid maladaptation that could potentially increase the GHG footprint.

Adaptation approaches include the promotion of resilient livestock production, modifications in production and management systems, science and technological advances, institutional and policy changes, and changing farmers' perception and adaptive capacity (IFAD, 2010; USDA, 2013).

Promotion of resilient livestock production

Climate-smart (resilient) livestock production is guided by the animal's genotype within its environment; this exploits the animal's inherent competitive production advantage in the environment. More emphasis is also needed on climate-smart breeding programs that use heat stress-resistant genotypes with higher-than-average production performance and low carbon and water footprints per product unit.

Alterations in breeding strategies may help animals increase their resistance to heat stress and various diseases while enhancing their growth and reproduction (Henry et al., 2012). Advanced biotechnological tools are making significant progress to recognize important biological markers related to diversified production, adaptation, and low methane emission control characteristics. Therefore, in planning our future breeding programs to grow climate-resilient livestock breeds, it is time to rethink our strategies. Future research activities must be directed toward this objective by means of marker-assisted selection as well as integrating production, adaptation, and low methane emission characteristics into the breeding programs. Apart from imparting thermotolerance, these efforts will channel the energy resources available into the production pathway. These initiatives will help sustain livestock production in the climate change scenario, thereby ensuring both the protection of poor and marginal farmer livelihoods and the food security of the increasing human population (Rashamol and Sejian, 2018).

Due to their greater ability to withstand a stressful environment, indigenous breeds show more resilience potential than exotic breeds. By providing a secure, nonthreatening climate, the resilience ability of livestock species may be increased. It may be possible to enhance resilience by providing the animal with cognitive and emotional enrichment. In comparison to imported breeds, indigenous breeds excel with minimum production costs. They are crucial in the production of climate-smart livestock, especially in community-based and small-scale farming systems. Breeding more productive animals and in turn reducing the number of unproductive or less-productive animals contribute toward climate-smart livestock production.

According to the USDA, the number of dairy cows decreased worldwide from 133 million to 125 million between 1997 and 2007 while the volume of milk increased from 377 to 418 million tonnes. This shows an 18% improvement over 10 years in milk production efficiency.

Improved herd management to improve output, including improved herd health management with less reliance on antibiotics, also contribute tow climate-smart livestock production.

Climate-smart livestock production includes the following activities: (1) Reducing heat/thermal stress by ensuring animal comfort; (2) removing nutritional stress by increasing feed and fodder availability and their utilization; (3) reducing water stress by increasing availability; (4) mitigation, and (5) capacity building and knowledge management.

Mitigation approaches in animal husbandry practices

Capacity building and knowledge management

Provide training to selected farmers on climate-smart livestock rearing technologies and practices, the documentation of lessons learned, and workshops and seminars to share the lessons.

Increase access to market: Activities include linking with local traders and transportation and linking relevant industries and markets.

Reducing heat/thermal stress by ensuring animal comfort

Promote stress-tolerant genotype: The native varieties of livestock are proven heat-tolerant species. At the same time, some high-yielding climate adaptive varieties are practiced in different parts of the country. These varieties should be promoted in the drought vulnerable areas to increase productivity.

Promote heat stress resilient housing: Heat-protecting houses with adequate ventilation, cooling systems, free access to feed and water, and proper management of solid and liquid wastes ensure animal comfort, leading to expected increased productivity of livestock.

Ensure veterinary care: Increased temperature and drought enhance the activities of pathogens or parasites, spreading vector-borne and food-borne diseases and reducing host resistance. Preventive and curative veterinary services are needed to combat this.

Extending financial and risk mitigation services: The financial services include appropriate loans/credit for livestock production and management of livestock products; and risk management by introducing "animal protective services" that covers financial compensation together with preventive and curative veterinary care services.

Removing nutritional stress by increasing feed and fodder availability and their utilization

Improve feeding strategy

Promote drought-resilient fodder cultivation: Cultivation of animal/ perennial grass, legumes, tree fodder etc. Promote the production and utilization of hydroponic fodder. Introduce ensilage and hay production techniques.

Reducing water stress by increasing availability

Installation of submergible tube wells: Water scarcity causes disruptions in animal physiology leading to the low productivity of milk, meat, and calves/kids. It also causes a reduction in the immune system leading to disease outbreaks and lower productivity. Install submergible tube wells at the community level to ensure the availability of safe water for livestock.

Excavate ditch for rainwater harvesting

Strategic feed management: Strategic supplementation of rumen with fermentable energy and a fermentable N-diet can reduce enteric CH4 emission by 15%–20%. Introduce urea-molasses-straw (UMS) to large and small ruminants to reduce CH4 production and increase animal production.

Modifications in livestock production and management systems

Diversification of varieties of livestock and crops may increase tolerance to drought and heat waves, and may increase the production of livestock when animals are exposed to stress. In addition, these varieties of crops and livestock are successful in the fight against diseases associated with climate change and pest outbreaks (Batima et al., 2005; IFAD, 2010; Kurukulasuriya and Rosenthal, 2003).

Agroforestry: The planting of trees alongside crops and pastures as mixed cultivation as a land management approach can help maintain the balance between agricultural production, environmental protection, and carbon sequestration to offset emissions from the sector. Agroforestry can increase productivity and improve the air, soil, and water quality as well as biodiversity, pests, and diseases while enhancing nutrient cycling (Jose, 2009; Smith et al., 2012). Modifications in mixed crop-livestock systems are a measure of adaptation that could enhance food security (Herrero et al., 2010; Wani et al., 2009).

Better grassland management (e.g., sowing improved varieties of pasture and rotational grazing):

Enhancing diets to minimize enteric fermentation, increase carbon sequestration, improve manure management and use fertilizers more effectively are mitigation approaches to climate change (Steinfeld et al., 2006; UNFCCC, 2008). Mitigation measures need public policy support to be effective (Dickie et al., 2014).

Enteric strategies to reduce methane can be divided into three categories: feeding and management of nutrients, rumen modifiers, and genetic management. Research produces innovations that increase the efficiency of production and result in less consumption of feed and water per unit product.

Nutrigenomics and precision nutrition

Feed is attributed to 65%–75% of variable costs in livestock enterprises, and feed production accounts for 47% of livestock farming's GHG. That's why animal nutrition is a crucial area for climate change mitigation. Improving forage quality increases digestion and decreases methane enteric excretions (CH_4), which improves feed efficiency. Precision feeding is about getting the right nutrient to the right animal at the right time. Customized balanced feeding programs in grazing dairy cattle systems have increased productivity and reduced enteric methane emission intensity (15%–20%) as well as N excretion (20%–30%), which results in reduced emissions from manure (Belay, 2019). Precision nutrition reduces GHG pollution and decreases the quantity of excreta, which benefits both the environment and the climate.

If the production of enteric CH_4 can be reduced without harm to productivity, more feed energy is divided for productive purposes into the intermediate metabolism and improves efficiency.

Enteric methanogenesis affects the efficient use of feed and should be reduced without sacrificing the role of the rumen to alleviate global warming. Indeed, several technologies have evolved, such as feed additives to boost DMI and suppress the production of acetate and the quantity of hydrogen released; ensiling forages to enhance digestive efficiency and reduce GHG emissions at earlier regrowth stages; and grinding and pelleting the forage before feeding to reduce GHG. Strong associations between the production of enteric CH_4 and dry matter intake in cattle fed temperate forage were reported as far back as 1930. In cattle fed tropical forage diets, CH_4 production is higher, particularly on higher intakes, than in those fed temperate forages due to the low digestibility of the latter. Therefore, due to improved digestibility, improved forage quality would lower the lifetime GHG emissions per unit product. In addition, the rise in human activity-related GHG raises average temperatures that can lead to changes in vegetation, where nutritious plants may gradually be replaced by desert-adapted, woody plants whose stomata remain closed during the day to minimize evapotranspiration, but open to carbon dioxide collection at night. These modifications in the composition of grass species erode the feed base. They can increase GHG emissions by substituting less-digestible forage with a higher methane conversion rate. The ability to transform shrubs, grass, and foliage to food and fiber is the major reason why ruminants occupy 45% of the ice-free land area of the planet. About 7 to 22 % (average 14.5 %) of human-induced GHG come from livestock, specifically from enteric fermentation, manure, feed production and transport, and land-use shift. By altering gene expression, the new field of nutrigenomics investigates how feed-bioactive compounds influence the metabolism of the animal. Nutrigenomics offers technologies that allow feeding according to genetic potential, ensuring precise nutrition and covering several disciplines. At any point on the path from gene packaging and unpacking of DNA to protein processing and degradation, gene expression can be controlled by dietary means through nutrigenomics. In line with their genetic capacity and physiological and development states, this allows diet optimization for livestock.

In addition, substituting higher-quality and digestible varieties for naturalized grass decreases the amount of methane released per unit of milk or meat produced (Herrero et al., 2013). By managing soil erosion and restoring land through increased soil organic matter and improved microclimate, the vegetative cover of well-managed pastures improves ecosystem services (Peters et al., 2013). Brachiaria grasses are by far the most commonly used forage in the tropics for livestock. *B. humidicola* releases from its roots biological nitrification inhibitors (BNIs) that minimize emissions of nitrous oxides by inhibiting nitrification in the soil. Silvopastoral systems combine trees and shrubs with forage grasses, boost animal nutrition, and produce cobenefits such as improved soil productivity and increased accumulation of C (Murgueitio et al., 2011). Improved forage that is well maintained will accumulate carbon (C) in the soil at levels second only to a forest. Improved forage-based systems lead to adaptations to climate change. During drought or water logging, forage grasses and legumes that are immune to stress provide livestock feed. In crop-livestock systems, drought-adapted forage legumes may provide high-quality feed in the dry season.

Climate-smart farming plans provide incentives to mitigate and adjust to climate change, such as carbon credits. These proposals include policies that promote carbon storage in forests and on agricultural land. Although the livestock sector is the major contributor to GHG emissions in agriculture, there are no related initiatives. The first step will be to devise livestock production systems that minimize GHG emissions and increase C accumulation capacity. A second move would be to formulate a smallholder certification scheme that would allow them to benefit from the development of climate-smart livestock.

Improving diets so that animals produce more protein with less feed and lower emissions:

Enteric fermentation

Enteric fermentation is a source of methane emissions that can be minimized by activities such as animal nutrition enhancement and genetics (Waghorn and Hegarty, 2011). Mitigating enteric fermentation practices include increasing the fat content in diet (Beauchemin et al., 2008; Martin et al., 2010), providing higher-quality forage (Hristov et al., 2013), enhancing protein content (ICF International, 2013), providing essential supplements (e.g., bovine somatotropin, feed antibiotics) (Boadi et al., 2004), and the utilization of antimethanogens (vaccines to suppress methane emissions) (EPA, 2013).

However, the effectiveness of these procedures is highly questionable because multiple studies have shown the initial reductions in enteric fermentation achieved are only temporary (ICF International, 2013). A 1% increase in dietary fat content can decrease enteric methane emissions up to 4%–5% (Beauchemin et al., 2008; Martin et al., 2010). However, rumens need to reduce their fat content to 8% of dry matter to prevent a decrease in livestock efficiency (Hales et al., 2012). Providing higher-quality forage often decreases methane emissions as it improves digestibility (Hristov et al., 2013). Increasing the protein content of feed can also increase digestibility and decrease total emissions of methane per product unit (ICF International, 2013). Supplements such as feed antibiotics that appear to increase weight gain and decrease feed intake per metric tonne of meat produced can decrease enteric fermentation (Boadi et al., 2004). In milk, bovine somatotropin (bovine growth hormone) enhances production. The increase in milk production means that fewer animals are required to produce the same quantity of milk and less pollution is produced (EPA, 2013). Antimethanogen vaccinations are another practice that directly decreases rumen methane emissions. However, this is a modern technology with minimal data on emission reduction efficiency and animal health (EPA, 2013).

The feed and nutrition-related interventions (supplementation with leguminous shrubs, use of urea molasses multinutrient blocks (UMMB), use of urea-treated crop residues, supplementation with high protein or energy concentrate) result in a reduction in emission intensities between 16% and 44% (FAO, 2017). Treatment of crop residues with urea results in a 44% reduction in emission rate compared to the baseline. The addition of UMMB to the feed of lactating cows results in a decrease in emission intensities to 20%–27% (FAO, 2017). These reductions are the result of improved feed digestibility, increased feed intake, and subsequent increases in milk output. All measures returned a positive productivity result with a milk output increase of between 8% and 70% (FAO, 2017).

The research aims to contain the production of methane in rumen by modifying the microbial flora of rumen as well as the addition of antimethanogenic feed ingredients. Some plant extracts can alter the rumen microbial population to make nitrogen and energy use more efficient. This micromanipulation of the rumen environment contributes not only to more effective processing of meat and milk but also to a correspondingly lower emission of GHG and ammonia byproducts. Research has shown that the enzyme in red clover, *Trifolium pratense,* widely grown in temperate countries, improves the ability of ruminants to use dietary proteins more effectively. The field trial shows that dairy cows fed more clover in their diets increased their feed intake and had greater milk production. In Australia, during the dry autumn most of the pasture plants possess poor feed value sheep nibble on deep rooted perennial tar bush, *Eremophila glabra* have inherent natural benefits. The tar bush has been reported to fight gastrointestinal nematodes and acidosis to reduce rumen methane emissions. High production can be accomplished by using a minimum quantity of high-yielding elite germplasms. The amount of unproductive or low yields should be reduced to reduce methane production and feed and fodders (Hoque, 2018).

Carbon sequestration

Carbon sequestration can be achieved by reducing deforestation rates, reversing deforestation by replanting (Carvalho et al., 2004), targeting higher-yield crops with better suited varieties for climate change, and improving land and water management (Steinfeld et al., 2006). Organic soil carbon can be preserved in cultivated soils by the conservation of tillage, erosion control, soil acidity management, double-cropping, crop rotation, higher crop residues, and mulching (Paustian et al., 1997; Steinfeld et al., 2006). Improving pasture management will also lead to carbon sequestration by incorporating plants, improving plant varieties, interseeding legumes, adding earthworms, and fertilization (Conant et al., 2001). In addition, grass production and soil carbon sequestration could also be increased by increasing grazing pressure in grasslands with a lower number of grazing animals than the livestock holding capacity (Holland et al., 1992). Improving grazing land management could sequester around 0.15 gigatonnes CO_2-eq yr_1 globally (Henderson et al., 2015).

Better manure management (e.g., composting)

Manure management: Most methane emissions from manure management are connected to storage and anaerobic treatment. Anaerobic digestion, while producing biogas, will minimize methane emissions (Gerber et al., 2008). Anaerobic digesters are lagoons or tanks that sustain manure under anaerobic conditions for the capture and combustion of biogas to produce energy or flaring. This method decreases the potential for GHG emissions by transforming methane to CO_2 (ICF International, 2013). Anaerobic digesters are expensive for producers, so the best way to introduce digesters is through policies that provide ample incentives for adaptation (Dickie et al., 2014).

Adjusting livestock diets may also be used as a mitigation measure by adjusting the amount and composition of manure. GHG emissions can be minimized by adjusting dietary proteins and feed supplements. If protein intake is decreased, the nitrogen excreted by animals may also be reduced. Supplements such as tannins are also considered to have potential to minimize emissions. Tannins are able to transfer the excretion of nitrogen from urine to feces in order to minimize total emissions (Dickie et al., 2014; Hess et al., 2006).

Promote vermin compost: A traditional manure management system has high potential for GHG emissions, particularly ammonia, methane, and nitrous oxide. The production of vermin compost significantly reduces GHG emissions. Normal compost can reduce 22% of GHG gases whereas vermin compost can reduce 52%. Thus, the activity will contribute to reducing GHG emissions. The compost contains high nutrients for soil and crop and is suitable for use in agricultural land.

Promote a biogas plant: A biogas plant will reduce GHG emissions by producing methane gas. This gas may be used for cooking at the household level.

Fertilizer management: Fertilizer application on animal feed crops increases nitrous oxide emissions (Bouwman, 1996). Therefore, mitigation steps such as increasing the effectiveness of nitrogen usage, plant breeding, and genetic modifications (Dickie et al., 2014); the use of organic fertilizers (Denef et al., 2011); periodic soil monitoring; the use of technologically advanced fertilizers; and the replacement of legumes with grasses in pasture areas can also minimize GHG emissions in feed production (Dickie et al., 2014).

Shifting human dietary trends: Beef accounts for a significant proportion of GHG emissions from the livestock sector and is the least resource-efficient producer of animal protein (Stehfest et al., 2009), although the mitigation capacity for the livestock aspect of beef is strong. So, a reduction in meat consumption may significantly reduce GHG emissions.

Conclusion

Livestock production remains one of the world's most important sectors concerning economy, food security, and the livelihoods of people. On the one hand, there is a growing appetite for animal-based products but on the other, the effect of livestock production on climate change and the negative contribution of livestock to climatic changes. The livestock sector must become more sustainable while adjusting to climatic changes while at the same time meeting the rising population's demands for animal-based products. The two main approaches to make livestock production "climate-smart" have been adaptation to climatic changes and mitigation of GHG. Adaptation steps include improvements to the production and management systems, breeding methods, institutional and policy adjustments, advances in science and technology, and changing the understanding and adaptive ability of farmers. Carbon sequestration, enhancing diets in order to avoid enteric fermentation, enhancing the management of manure, and the more effective use of fertilizers are mitigation steps to reduce GHG emissions. With integrated adaptation and mitigation approaches, livestock can ascertain its role in achieving food security and the livelihood of millions of people around the globe under climate change.

References

Batima, P., Bat, B., Tserendash, L., Bayarbaatar, S., Shiirev-Adya, S., Tuvaansuren, G., Natsagdorj, L., Chuluun, T., 2005. Adaptation to Climate Change. vol. 90 ADMON Publishing, Ulaanbaatar.

Beauchemin, K.A., Eriksen, L., Norgaard, P., Rode, L.M., 2008. Nutritional management for enteric methane abatement: a review. Aust. J. Exp. Agric. 48, 21–27.

Belay, G., 2019. Climate smart livestock feed improvement promising and prioritized technologies and intervention options: a review. Agric. Res. Technol. 20 (2). ARTOAJ.MS. ID.556125.

Bezuidenhout, R., 2019. Towards climate-smart livestock farming. In: Farmer's Weekly. Craighall, Johannesburg, South Africa. September 17, 2019.

Boadi, D., Benchaar, C., Chiquette, J., Massé, D., 2004. Mitigation strategies to reduce enteric methane emissions from dairy cows: update review. Can. J. Anim. Sci. 84, 319–335.

Bouwman, A.F., 1996. Direct emission of nitrous oxide from agricultural soils. Nutr. Cycl. Agroecosyst. 46 (1), 53–70.

Carvalho, G., Moutinho, P., Nepstad, D., Mattos, L., Santilli, M., 2004. An Amazon perspective on the forest–climate connection: opportunity for climate mitigation, conservation and development? Environ. Dev. Sustain. 6 (1–2), 163–174.

Conant, R.T., Paustian, K., Elliott, E.T., 2001. Grassland management and conversion into grassland: effects on soil carbon. Ecol. Appl. 11, 343–355.

Denef, K., Archibeque, S., Paustian, K., 2011. Greenhouse Gas Emissions From U.S. Agriculture and Forestry: A Review of Emission Sources, Controlling Factors, and Mitigation Potential: Interim Report to USDA Under Contract #GS-23F-8182H. http://www.usda.gov/oce/climate_change/techguide/Denef_et_al_2011_Review_of_reviews_v1.0.pdf.

Dickie, A., Streck, C., Roe, S., Zurek, M., Haupt, F., Dolginow, A., 2014. Strategies for Mitigating Climate Change in Agriculture: Abridged Report. Climate Focus and California Environmental Associates, Prifadred With the Support of the Climate and Land Use Alliance. Report and Supplementary Materials. Available at: www.agriculturalmitigation.org.

Anon., 2013. Global Mitigation of Non-CO_2 Greenhouse Gases. EPA(U.S. Environmental Protection Agency). http://www.epa.gov/climatechange/Downloads/EPAactivities/MAC_Report_2013.pdf>.

FAO, 2013. Module 8: Climate-smart livestock. In: Climate-Smart Agriculture Sourcebook. FAO, Rome, Italy.

FAO, 2017. Feeding Strategies to Reduce Methane and Improve Livestock Productivity. Animal Production and Health Division, FAO, Rome, Italy, p. 5.

GaitaÂn, L., LaÈderach, P., Graefe, S., Rao, I., van der Hoek, R., 2016. Climate-smart LivestockSystems: an assessment of carbon stocks and GHG emissions in Nicaragua. PLoS One 11 (12). https://doi.org/10.1371/journal.pone.0167949, e0167949.

Gerber, P.J., Carsjens, G.J., Pak-uthai, T., Robinson, T.P., 2008. Decision support for spatially targeted livestock policies: diverse examples from Uganda and Thailand. Agric. Syst. 96, 37–51.

Gerber, P.J., Steinfeld, H., Henderson, B., Mottet, A., Opio, C., Dijkman, J., Falcucci, A., Tempio, G., 2013. Tackling, Climate Change Through Livestock: A Global Assessment of Emissions and Mitigation Opportunities. FAO, Rome.

Hales, K.E., Cole, N.A., MacDonald, J.C., 2012. Effects of corn processing method and dietary inclusion of wet distillers grain with solubles on energy metabolism and enteric methane emissions of finishing cattle. J. Anim. Sci. 90 (9), 3174–3185.

Harvell, C.D., Mitchell, C.E., Ward, J.R., Altizer, S., Dobson, A.P., Ostfeld, R.S., Samuel, M.D., 2002. Climate warming and disease risks for terrestrial and marine biota. Science 296, 2158–2162.

Henderson, B.B., Gerber, P.J., Hilinski, T.E., Falcucci, A., Ojima, D.S., Salvatore, M., Conant, R.T., 2015. Greenhouse gas mitigation potential of the world's grazing lands: modeling soil carbon and nitrogen fluxes of mitigation practices. Agric. Ecosyst. Environ. 207, 91–100.

Henry, B., Charmley, E., Eckard, R., Gaughan, J.B., Hegarty, R., 2012. Livestock production in a changing climate: adaptation and mitigation research in Australia. Crop Pasture Sci. 63, 191–202.

Herrero, M., Thornton, P.K., Notenbaert, A.M., Wood, S., Msangi, S., Freeman, H.A., Bossio, D., Dixon, J., Peters, M., van de Steeg, J., Lynam, J., ParthasarathyRao, P., Macmillan, S., Gerard, B., McDermott, J., Sere, C., Rosegrant, M., 2010. Smart investments in sustainable food production: revisiting mixed croplivestock systems. Science 327, 822–825.

Herrero, M., Havlik, P., Valin, H., Notenbaert, A., Rufino, M.C., Thornton, P.K., et al., 2013. Biomass use, production, feed efficiencies, and greenhouse gas emissions from global livestock systems. Proc. Natl. Acad. Sci. U. S. A. 110 (52), 20888–20893.

Hess, H.D., Tiemann, T.T., Noto, F., Carulla, J.E., Kreuzer, M., 2006. Strategic use of tannins as means to limit methane emission from ruminant livestock. In: International Conference on Greenhouse Gases and Animal Agriculture, Elsevier International Congress Series, Zurich, Switzerland. 1293, pp. 164–167.

Hoffman, M., Vogel, C., 2008. Climate change impacts on African rangelands. Rangelands 30, 12–17.

Holland, E.A., Parton, W.J., Detling, J.K., Coppock, D.L., 1992. Physiological responses of plant populations to herbivory and their consequences for ecosystem nutrient flow. Am. Nat. 140 (4), 85–706.

Hoque, M., 2018. The way ahead. (chapter 13). In: Singh, R.L., Mondal, S. (Eds.), Biotechnology for Sustainable Agriculture-Emerging Approaches and Strategies. Woodhead Publishing, Elsevier Inc., USA, pp. 375–397. ISBN:978-0-12-812160-3 (print), ISBN:978-0-12-812238-9.

Hristov, A.N., Oh, J., Firkins, J.L., Dijksta, J., Kebreab, E., Waghorn, G., Makkar, H.P.S., Adesogan, A.T., Yang, W., Lee, C., Gerber, P.J., Henderson, B., Tricarico, J.M., 2013. Mitigation of methane and nitrous oxide emissions from animal operations: I a review of enteric methane mitigation options. Am. Soc. Anim. Sci. 91.

ICF International, 2013. Greenhouse Gas Mitigation Options and Costs for Agricultural Land and Animal Production Within the United States. http://www.usda.gov/oce/climate_change/mitigation_technologies/GHG_Mitigation_Options.pdf.

IFAD (International Fund for Agricultural Development), 2010. Livestock and Climate Change. http://www.ifad.org/lrkm/events/cops/papers/climate.pdf.

Jose, S., 2009. Agroforestry for ecosystem services and environmental benefits: an overview. Agrofor. Syst. 76, 1–10.

Kadzere, C.T., 2018. Environmentally smart animal agriculture and integrated advisory services ameliorate the negative effects of climate change on production. S. Afr. J. Anim. Sci. 48 (5), 842–857. https://doi.org/10.4314/sajas.v48i5.5.

Karl, T.R., Melillo, J.M., Peterson, T.C., 2009. Global Climate Change Impacts in the United States. U.S. Global Change Research Programme. Cambridge University Press.

Kurukulasuriya, P., Rosenthal, S., 2003. Climate Change and Agriculture: A Review of Impacts and Adaptations. Climate Change Series Paper No. 91, World Bank, Washington DC.

Martin, C., Morgavi, D.P., Doreau, M., 2010. Methane mitigation in ruminants: from microbe to the farm scale. Animal 4 (3), 351–365.

Murgueitio, E., Van der Hoek, R., LaÈderach, P., Hyman, G., Tapasco, J., Strassburg, B., et al., 2011. Challenges andopportunities for improving eco-efficiency of tropical forage-based systems to mitigate greenhouse gasemissions. Trop. Grassl. 1, 137–148.

Nordone, A., Ronchi, B., Lacetera, N., Ranieri, M.S., Bernabucci, U., 2010. Effect of climate change on animal production and sustainability of livestock system. Livest. Sci. 130 (1), 57–69. https://doi.org/10.1016/j.livsci.2010.02.011.

Patz, J.A., Graczyk, T.K., Geller, N., Vittor, A.Y., 2000. Effects of environmental change on emerging parasitic diseases. Int. J. Parasitol. 30, 1395–1405.

Paustian, K., Andren, O., Janzen, H.H., Lal, R., Smith, P., Tian, G., Tiessen, H., Van Noordwijk, M., Woomer, P.L., 1997. Agricultural soils as a sink to mitigate carbon dioxide emissions. Soil Use Manag. 13 (4), 230–244.
Peters, M., Herrero, M., Fisher, M., Erb, K.-H., Rao, I., Subbarao, G.V., et al., 2013. Challenges and opportunities for improving eco-efficiency of tropical forage-based systems to mitigate greenhouse gas emissions. Trop. Grassl. 1, 137–148.
PKSF (Palli Karma-Sahayak Foundation), 2019. Concept Note- Green Climate Fund. Climate Resilient Livestock Production for the Climate-Vulnerable Regions of Bangladesh, Bangladesh. 22 January 2019.
Rashamol, V.P., Sejian, V., 2018. Climate resilient livestock production: way forward. Dairy Vet. Sci. J. 5 (5), 5556673. https://doi.org/10.19080/JDVS.2018.05.555673.
Rojas-Downing, M.M., Nejadhashemi, A.P., Harrigan, T., Woznicki, S.A., 2017. Climate change and livestock: impacts, adaptation, and mitigation. Clim. Risk Manag. 16, 145–163.
Smith, J.P., Pearce, B.D., Wolfe, M.S., 2012. Reconciling productivity with protection of the environment: is temperate agroforestry the answer? Renewable Agric. Food Syst. 28 (1), 80–92.
Stehfest, E., Bouwman, L., van Vuuren, D.P., den Elzen, M., Eickhout, B., Kabat, P., 2009. Climate benefits of changing diet. Clim. Chang. 95 (1–2), 83–102.
Steinfeld, H., Gerber, P., Wassenaar, T., Castel, V., Rosales, M., Haan, C., 2006. Livestock's Long Shadow: Environmental Issues and Options. FAO, Rome.
Thornton, P., 2010. Livestock production: recent trends, future prospects. Philos. Trans. R. Soc. B 365, 2853–2867.
Tubiello, F., Schmidhuber, J., Howden, M., Neofotis, P.G., Park, S., Fernandes, E., Thapa, D., 2008. Climate Change Response Strategies for Agriculture: Challenges and Opportunities for the 21st Century. The World Bank, Washington, DC.
UNEP, 2011. Towards a Green Economy: Pathways to Sustainable Development and Poverty Eradication (2011). UNEP. www.unep.org/greeneconomy.
UNFCCC (United Nations Framework Convention on Climate Change), 2008. Challenges and Opportunities for Mitigation in the Agricultural Sector: Technical Paper. http://unfccc.int/resource/docs/2008/tp/08.pdf.
USDA (United States Department of Agriculture), 2013. Climate Change and Agriculture in the United States: Effects and Adaptation. USDA Technical Bulletin, Washington, DC. http://www.usda.gov/oce/climate_change/effects_2012/CC%20and%20Agriculture%20Report%20%2802-04-2013%29b.
Waghorn, G., Hegarty, R., 2011. Lowering ruminant methane emissions through improved feed conversion efficiency. Anim. Feed Sci. Technol. 166–167, 291–301.
Wani, S.P., Rockstrom, J., Oweis, T., 2009. Rainfed Agriculture: Unlocking the Potential. CAB International, Wallingford, UK.

Further reading

Beauchemin, K.A., McAllister, T.A., McGinn, S.M., 2009. Dietary mitigation of enteric methane from cattle. CAB reviews: perspectives in agriculture, veterinary science. Nutr. Nat. Resour. 4 (35), 1–8.

Index

Note: Page numbers followed by *f* indicate figures and *t* indicate tables.

C

H

M

N

O

P

R

S

T

U

V

9780128222652